Systematics and the Properties
of the Lanthanides

NATO ASI Series

Advanced Science Institutes Series

A series presenting the results of activities sponsored by the NATO Science Committee, which aims at the dissemination of advanced scientific and technological knowledge, with a view to strengthening links between scientific communities.

The series is published by an international board of publishers in conjunction with the NATO Scientific Affairs Division

A	Life Sciences	Plenum Publishing Corporation
B	Physics	London and New York
C	Mathematical and Physical Sciences	D. Reidel Publishing Company Dordrecht, Boston and Lancaster
D	Behavioural and Social Sciences	Martinus Nijhoff Publishers
E	Engineering and Materials Sciences	The Hague, Boston and Lancaster
F	Computer and Systems Sciences	Springer Verlag
G	Ecological Sciences	Heidelberg

Series C: Mathematical and Physical Sciences No. 109

Systematics and the Properties of the Lanthanides

edited by

Shyama P. Sinha

Hahn-Meitner-Institute, Berlin, F.R.G.

D. Reidel Publishing Company

Dordrecht / Boston / Lancaster

Published in cooperation with NATO Scientific Affairs Division

CHEMISTRY

7302-9063

Proceedings of the NATO Advanced Study Institute on
Systematics and the Properties of the Lanthanides
Braunlage, Germany
July 11-25, 1982

Library of Congress Cataloging in Publication Data

NATO Advanced Study Institute on Systematics and the Properties of the
    Lanthanides (1982: Braunlage, Germany)
    Systematics and the properties of the lanthanides.

    (NATO ASI series. Series C, Mathematical and physical sciences; no. 109)
    "Published in cooperation with NATO Scientific Affairs Division."
    Includes index.
    1. Rare earth metals—Congresses. I. Sinha, Shyama, P. II. North
Atlantic Treaty Organization. Scientific Affairs Division. III. Title.
IV. Series.
QD172.R2N36   1982   546'.41   83-8716
ISBN 90-277-1613-7

Published by D. Reidel Publishing Company
P.O. Box 17, 3300 AA Dordrecht, Holland

Sold and distributed in the U.S.A. and Canada
by Kluwer Academic Publishers,
190 Old Derby Street, Hingham, MA 02043, U.S.A.

In all other countries, sold and distributed
by Kluwer Academic Publishers Group,
P.O. Box 322, 3300 AH Dordrecht, Holland.

D. Reidel Publishing Company is a member of the Kluwer Academic Publishers Group

Printed in The Netherlands

CONTENTS

PREFACE

Science is not a mere collection of facts. It is the correlation
of facts, the interpretative synthesis of the available knowledge
and its application that excite the imagination of a scientist.
Even in these days of modern technology, the need for quick and
accurate dissemination of new information and current concepts
still exists. Conferences and Symposia offer one direct method of
communication. The Summer Schools are another approach. The success
of a Summer School is mainly due to that human factor and understanding that goes with it and allows for extensive and often
time-unrestricted discussions.

During the course of the past 20 years, one of the most in-
tensively studied groups of elements in the Periodic Table is the
Lanthanides. In this period, we have increased our knowledge on
these once exotic elements, which were once considered to be a
part of a lean and hungry industry, many-fold due to the involve-
ment of scientists from various disciplines.

The purpose of our Summer School was to bring a group of ex-
perts and participants together for the exchange of ideas and information in an informal setting and to promote interdisciplinary
interactions. Out of many conceivable topics, we selected the
following five as the main basis to broaden our knowledge and
understanding

1) Systematics 2) Structure 3) Electronic and Magnetic Properties 4) Spectroscopic Properties and 5) Lanthanide Geochemistry.

The first thirteen chapters of this Proceedings are the result of
the lectures delivered by the Plenary Lecturers. The Lecturers

S. P. Sinha (ed.), Systematics and the Properties of the Lanthanides, vii–viii.
Copyright © 1983 by D. Reidel Publishing Company.

were asked not only to review their topics but to add newer materials, current concepts and treat their topics from a pedagogical point of view. Extensive discussion followed the Plenary Lectures and during the Group Sessions and Panel Discussion Sessions. It is almost impossible to include all these discussions verbatim in this Proceedings and still keep the volume to its normal size. Hence, questions and answers are pooled together to give a uniform character and are added at the end of the chapters concerned.

The Lecturers were asked to speculate on the future developments. Four reports summarizing the future trends were presented during the School and these are collected in the penultimate chapter of this Proceedings.

A small collection of the Posters brought by the participants were displayed through the entire duration of the School and provoked interesting discussions. This medium provided another channel for effective personal communication and dissemination of scientific information. Abstracts of these Posters are also included in the last chapter.

The highlight of our School was the delivery of the opening lecture by the Guest of Honour, the 86 year-old Professor W. Klemm, who in the early thirties developed the very first systematics of the lanthanides.

I am deeply indebted to my colleagues, the Lecturers, Hahn-Meitner-Institut, and all individuals who gave their support and active help during the planning and execution of this Summer School. Last but not least, without the generous sponsorship of the NATO Advanced Study Institute our Summer School would have never materialized. It was my great pleasure to work with such a distinguished and knowledgeable group of scientists. I hope that our School has in its own way contributed towards international understanding.

                                        Shyama P. Sinha

Systematics

EARLY ATTEMPTS ON THE SYSTEMATICS OF THE LANTHANOIDS

Wilhelm Klemm

Institut für Anorganische Chemie, Westfälische
Wilhelms-Universität Münster, Corrensstr. 36,
D-4400 Münster, Fed. Rep. Germany

ABSTRACT

Historical report on the attempts of the author
and his coworkers to elaborate a systematics of the
lanthanoids; the discovery of $YbCl_2$; the preparation
of the metals, their crystal structures and magnetic
qualities.

Natural science is developing very fast. Many
scientists have no great tendency to look back to the
early attempts in their field of research. Therefore
I am very happy that you have invited me to give a
review of our work of the Systematics of the Lanthanoids. These investigations have been performed about
50 years ago, that means at a time at which many of
you were children and many of you were not yet born.

Born 1896 in Silesia I have made my studies in
chemistry in Breslau, and received 1923 the degree
of Dr. phil. with a thesis elaborated in the field
of organic chemistry under the guidance of Heinrich
Biltz. One day Biltz said to me:"I believe your
interest belongs more to inorganic chemistry - don't
you want to go to my brother Wilhelm, head of the
Institute of Inorganic Chemistry at the Technical
University of Hannover?" I accepted, and by this way
I was transmitted within the family Biltz. The change
to Hannover was of great importance for me. Wilhelm
Biltz was an outstanding inorganic chemist and I have
a great debt of gratitude to him. With Biltz I investi-

gated the electric conductivity and other properties
of molten salts. For my independent investigations I
chose the chemistry of rare elements: Ga, In and rare
earths.

For the rare earths a certain stage of development was attained. The separation of these more than
a dozen elements which are more similar than in any
other groups is an admirable work of chemists which
has required about 100 years. But at the beginning of
this century chemists were not able with absolute
certainty to say how many "lanthanoids" exist. This
problem was finally solved by the atomic theory and
by the discovery of hafnium.

Now the number being certain - La + 14 lanthanoids! - the question arose whether there are periodic
properties inside this group. Older attempts to
establish "a small periodic system of the lanthanoids"
had not given clear results; therefore it was necessary to discuss the problem from the very beginning. Are
there periodic properties? Indeed, there are some
physical properties (colour, magnetic moments of the
ions), but periodical chemical properties? At first
glance there seemed only one, that is the existence
of compounds of anomalous valency.

In 1927 there appeared a small book of the
discoverer of Hf, the Nobel-prize laureate Georg von
Hevesy, concerning rare earths [1]. In this book
Abb. 5, a copy of which is given in Fig. 1, demonstrated the compounds with valencies other than 3 known up

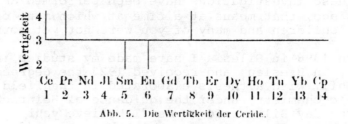

Abb. 5. Die Wertigkeit der Ceride.

Fig. 1   Anomalous valencies

to this time. Hevesy noted only the experimental facts,
but he gave no comment. This figure fascinated me!
There was no doubt that the distribution of these
anomalous compounds was not arbitrary. At this time

it was already known that CeO_2 had a quadrivalent cation and was not a peroxide. This means that Ce has a certain tendency to accept the electronic configuration of La^{3+}. This is not surprising as La^{3+} has, like Ba^{2+}, the configuration of a noble gas (s^2p^6) and such configurations are especially stable. Therefore the existence of CeO_2 was not surprising.

But astonishing is that a situation similar to the neighbourhood of La^{3+} exists in the surroundings of Gd^{3+}: the existence of a higher oxide of Tb and a rather stable $EuCl_2$ indicate the tendency of Tb and Eu to accept the configuration of Gd^{3+}. By this consideration we come to the conclusion that Gd^{3+} has also a special stability, similar to La^{3+}. This has to be discussed (Fig. 2).

Tabelle 2.
Elektronenanordnungen der Ionen.

Ion-quantenzahl	Elektronenanordnung								Magnetonen nach Weiss	Bemerkungen
	1	2	3	4_1	4_2	4_3	4_4	$5_1,5_2$		
Ba++							0		0	
La+++							0		0	
Ce++++							0		0	CeO_2: beständig, nahezu farblos, Fluoritgitter
Ce+++							1		11,4	
Pr++++							1		11,5	PrO_2: wenig beständig, schwarz, Fluoritgitter
Pr+++							2		17,3—17,8	
Nd+++							3		17,5—18,0	
Il+++							4		—	
Sm+++							5		7,0—8,0	
Sm++	2	8	18	2	6	10	6	2 6	20,2	$SmCl_2$: wenig beständig, tief rotbraun
Eu+++							6		15,5—17,9	
Eu++							7		—	$EuCl_2$: beständig, farblos
Gd+++							7		40,0—40,2	
Tb++++							7		< 44.8	Tb_4O_7: wenig beständig, schwarz, Fluoritgitter
Tb+++							8		47,1	
Dy+++							9		52,2—53,0	
Ho+++							10		51,9—52,0	
Er+++							11		46,7—47,0	
Tu+++							12		35,6—37,5	
Yb+++							13		21,9—23,0	
Cp++								14	0	
Hf++++								14	0	

Fig. 2 Electronic configurations

Incidentally we may note, that Ce^{3+} and Eu^{3+} must have also a similar but not equally high stability as La^{3+} and Gd^{3+}; but this must be let aside.

All these considerations are correct only if $EuCl_2$ and $SmCl_2$ are really dihalogenides with Me^{2+}-ions. This was by no means sure for $SmCl_2$, for this compound has a deep colour and some authors [3] assumed that

it might be a colloid solution of metal in the trichloride.

To investigate this problem we needed samarium.
In our institute we had only a very small sample the
purity of which was doubtful. The separation of rare
earths was at that time very hard work. Therefore I
wrote to Baron Auer von Welsbach, the well-known
pioneer, and asked him to send me some grams of samarium oxide. For many months nothing happened, but one
day I received a telegram: "Samarium is on the way".
Of course I was happy and, together with Dr. Rockstroh,
we prepared $SmCl_2$ and $SmBr_2$ by reduction of the tri-
halogenides with hydrogen at elevated temperatures
[2-4]. Like the other authors we had difficulties; our
best specimens had contents of ca. 95% $SmCl_2$. The
reason for the incomplete reduction was not the impurity of the samarium, because an investigation of
Dr. Beuthe, Charlottenburg demonstrated that our sample
contained only circa 1% of impurities (Eu and Gd).
Therefore we made our measurements on specimens of a
content of ca. 90% $SmCl_2$. We obtained the following
results:

mol. volume: something lower than $SrCl_2$ and $CaCl_2$,
                    near to $PbCl_2$
behaviour to NH_3: similar to $SrCl_2$
magnetic susceptibility
at room temperature: like Eu(III)-compounds.

The elucidation of the crystal structure and of the
optical properties were not possible because we did
not have the necessary experimental equipment.

Although our results were incomplete there was
no doubt that $SmCl_2$ and $SmBr_2$ were real dichlorides
with Sm^{2+}-ions. It was of great importance that soon
later we were able to elucidate the crystal structures;
we found that $SmCl_2$ and $EuCl_2$ had structures similar
to $PbCl_2$. In this way the conclusions discussed earlier
were confirmed [5].

Considering once more the existence of electronic
configurations of special stability known till ca.
1929 (Fig. 1 u. 2), the question was whether these
were the only compounds with anomalous valencies.
Especially, it was to assume that the tervalent ion
of the element 71 with the totally occupied shell
(14f-electrons) has also a special stability. This
element has today the official name lutetium. In

former time the name cassiopeium was in use; therefore
one finds in our tables the symbol Cp in the place of
Lu. The configuration with 14 f-electrons is realized
in the quadrivalent Hf-compounds. Should it not be
possible to prepare compounds with Yb^{2+} ions? Again
we had no Yb, and again I wrote to Baron Auer v. Welsbach who was helping once more and sent us Dy_2O_3,
Er_2O_3 and Yb_2O_3, the latter unfortunately being rather
impure ($\sim 9\%$ impurities). As we expected, $DyCl_3$ and
$ErCl_3$ were not reducable by hydrogen, $YbCl_3$, on the
contrary could easily be reduced to a nearly colourless
$YbCl_2$ which gave a yellow solution in water and a red-
yellow addition compound with dry NH_3. $YbCl_2$ was practically nonmagnetic [6].

Including $YbCl_2$, Fig. 3 gives a complete survey
of the valencies of the compounds under the experimental conditions mentioned above (oxidation with gaseous
oxygen, reduction with hydrogen).

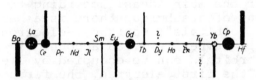

Schematische Darstellung des Auftretens 2- und 4wertiger Verbindungen
bei den seltenen Erden: Ein Strich oberhalb der Wagerechten bedeutet das
Auftreten von 4-Wertigkeit, ein Strich nach unten 2-Wertigkeit. Die Dicke
der Striche soll die Beständigkeit der Verbindungen veranschaulichen. Die
Größe der Punkte gibt ein Maß für die Vorzugsstellung der einzelnen
Elektronenkonfigurationen.

Fig. 3 Compounds with anomalous valency states

Of course with stronger reducing agents (e.g.
with the metals of the lanthanoids!) all elements of
the lanthanoids give a lot of compounds with still
lower valency as 2+ and with metallic nature, but in
1929 this way was not yet known.

The facts comprehended in Fig. 3 are the basis of
a Systematics of the Rare Earths proposed in 1929 [7]
(Fig. 4). In this "small periodic system" La^{3+}, Gd^{3+}
and Cp^{3+} play a similar role as the noble gases in the
normal periodic system: the tervalent ions of the
neighbouring elements of these ions have a certain
tendency to supply or to pick up one electron, that
means to form quadrivalent resp. divalent compounds
with the electronic configuration of La^{3+}, Gd^{3+} and
Cp^{3+}. The reason for this behaviour is easy to

1.   . . .   (Ba^{+++}) **La**$^{+++}$
2. Ce^{+++} Pr^{+++} Nd^{+++} 61^{+++} Sm^{+++} Eu^{+++} **Gd**$^{+++}$
3. Tb^{+++} Dy^{+++} Ho^{+++} Er^{+++} Tu^{+++} Yb^{+++} **Cp**$^{+++}$
4. (Hf^{+++}) . . .

Fig. 4   Systematics of the Rare Earths

understand for the configurations of La^{3+} and Cp^{3+}.
Both have fully occupied electron shells (S = O, L = O,
J = O), and it corresponds to generally accepted rules
that fully occupied shells have a high energy of ionisation.

But surprising was the fact that the configuration
of Gd^{3+} which is only <u>half</u>-occupied has a special
stability too! It happened very fortunately that
physicists interested in spectroscopy had just in these
years elucidated the general situation. Shortly we can
mention some of the main ideas: according to the number of electrons of a subgroup there is a certain
possibility of "orbitals"; for the f-electrons of the
lanthanoids this number is 7. On account of the spin
each of these orbitals can be occupied by 2 electrons.
Beginning with the first electron, the first orbital
is occupied with one electron (l = 3); the second
electron chooses the orbital with l = 2, the next one
with l = 1, the last (7.) orbital has l = -3. In this
way by the presence of 7 electrons each orbital is
occupied with one electron. The system is electrostatically of spherical symmetry (L = O) and has a low
energy.

But this system with 7 electrons cannot have also
magnetic symmetry (S = O) because the magnetic moment
S cannot be compensated in systems with an odd number
of electrons; in the contrary S has for Gd^{3+} its
maximal value, and S becomes smaller, when the double
occupancy is beginning (with Tb^{3+}). The filling up of
the shell is complete if all orbitals are occupied
with 2 electrons (Cp^{3+}). Fig. 5 may interprete the
situation.

The state of a "half-occupied" configuration
corresponds therefore not totally to a "closed" shell,
but it is similar to it. Important is that the energy
of ionisation has a maximum not only for fully closed,
but also for half-occupied shells. Fig. 6 demonstrates

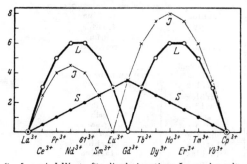

S-, *L*- und *J*-Werte für die dreiwertigen Ionen der seltenen Erden.

Fig. 5   S, L and J for Ln^{3+}-ions

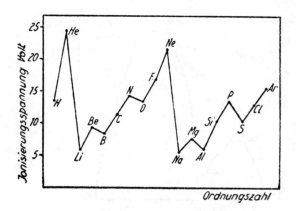

Fig. 6   Energy of ionisation

this for N and P (p-shells) [8].

The "periodic system" of the lanthanoids derived
from the existence of compounds with anomalous valencies, that means from chemical facts, is in harmony
with some physical properties. Fig. 7 demonstrates
this for the colours of the salt-like compounds; the
Fig. needs no comment. - The magnetic behaviour is
a little more complicated (Fig. 8). On the first view
one comes to the conclusion that there is a caesura
between Sm^{3+} and Eu^{3+}. But F. Hund [9] has demonstrated that the situation is more complicated. Let us
refer to Fig. 5. The curves for L and S are totally
symmetrical. But in the first half one has to subtract
S from L in order to come to the total momentum J,

	Oxyd	Ion
La	farblos	farblos
Ce	?[1]	farblos
Pr	hellgrün	grün
Nd	lichtblau	rotviolett
Il	?	?
Sm	gelblich weiß	gelb
Eu	farblos?[2]	nahezu farblos (rosa)
Gd	farblos	farblos
Tb	farblos	farblos
Dy	farblos	gelb
Ho	gelb	gelb
Er	rosa	rosa
Tu	grünlich weiß	grün
Yb	farblos	farblos
Cp	farblos	farblos

Fig. 7   Colour of the tervalent compounds

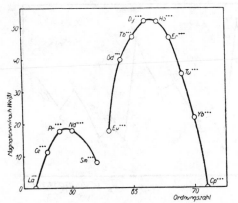

Paramagnetismus der Ionen der seltenen Erden nach CABRERA.

Fig. 8   Magnetic moments (Cabrera)

whereas in the second half J is L + S. The result is in excellent agreement with the experimental values.

Whereas the compounds with anomalous valency discussed in the former chapter exist only with certain elements and therefore indicate periods, the majority of the qualities of the lanthanoid ions has no periodic character at all or only in a very weak manner. For instance the special character of Gd^{3+} is very weakly pronounced concerning the distances in crystals. Only in some compounds (Fig. 9) one finds a small bending of the volume curve at Gd^{3+}, e.g. for

the C-forms of the oxides [10] and for the nitrides
[11]. Normally there is no peculiarity at Gd^{3+}.

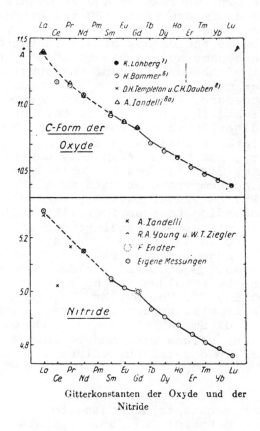

Fig. 9   Volume curves for oxides and nitrides

THE METALS

At the end of my talk allow me to give a short
report on the metals. Our "Periodic System of the
Lanthanoids" refers to the <u>ions</u> in salt-like compounds. The periodic system of all elements is valid
for the totality of the qualities of a certain element. Especially the elements themselves are included (compare the curve of the atomic volumes of
Lothar Meyer!). The question was therefore whether
our small periodic system of the lanthanoids has
relations on the metals. In order to answer this
question, it was necessary to know the properties of

the metals of all lanthanoids. This knowledge was
fifty years ago imperfect: The metals of La, Ce, Pr,
Nd and Gd were known, whereas Sm and Er metals had been
prepared only in an impure state. By the kindness of
Baron von Welsbach and some other colleagues we came
in the possession of the oxides of all lanthanoids
(except Ho which became accessible to us only some
time later!). But of the majority of the substances we
had only a few grammes and this was not enough for the
electrolytic procedure. Therefore a new method had to
be elaborated. I had the privilege to have at this
time an extraordinarily able coworker - Dr. Heinrich
Bommer; he had later a leading position in the German
chemical industry, but he unfortunately passed away
very early! Bommer [12] reduced the chlorides or bromides at increased temperatures with metallic alkaline
metals (Na, K, Rb, Cs) under extremly careful conditions
and obtained, according to reactions like $LaCl_3$ + 3K
$\longrightarrow$ La + 3KCl, a mixture of metal powder and alkali
chloride. Now we could determine by X-rays the crystal
structure and the magnetic susceptibility. After test
experiments with La, Ce etc. had yielded good consistency with the best values in the literature we
investigated Sm, Eu, Gd, Tb, Dy, Er, Tm, Yb and Cp
and obtained the crystal structures of all elements
with the exception of Sm, which gave no good diagrams.
A comparison of our values with modern values (Tab. 1)
demonstrates that in general our values were satisfactory.

Table 1   Lattice constants according to
           Klemm and Bommer (Kl./Bo.) compared
           with modern values (Lit.) (in Å)

	a		c	
	Kl./Bo.	Lit.	Kl./Bo.	Lit.
Eu	4,573	4,578	–	–
Gd	3,622	3,632	5,748	5,777
Tb	3,585	3,599	5,664	5,696
Dy	3,578	3,592	5,648	5,655
Ho	3,557	3,576	5,620	5,617
Er	3,532	3,559	5,589	5,592
Tm	3,523	3,537	5,564	5,562
Yb	5,468	5,481	–	–
Lu	3,509	3,505	5,559	5,549

The measurements of the crystal structures allowed to draw the curve of the atomic volumes (Fig. 10).

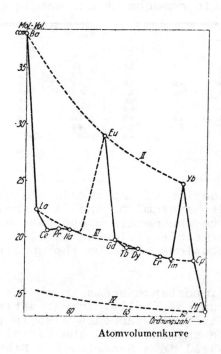

Atomvolumenkurve

Fig. 10   Atomic volumes

One comes to the conclusion that the majority of the
elements contain Me^{3+}-ions and 3 electrons in the
electronic gas. The elements with the tendency to
form quadrivalent compounds (Ce and to a weaker extent
Pr and Tb) seem to have in the metal a definite, but
low content of Me^{4+}-ions. Quite different is the behaviour of Eu and Yb which have the tendency to form
divalent ions. Here it is obvious that the metals
contain Me^{2+}-ions.

The magnetic measurements yielded values which
were in general in agreement with the conclusions drawn
from the atomic volumes. The higher elements, beginning
with Eu, did not follow the Curie-law; but the Curie-
Weiss-law is in the majority of cases applicable,
although the behaviour at low temperatures is different.
The moments evaluated at higher temperatures are easy
to understand: in the majority they correspond to
Me^{3+}-ions, Eu and Yb with the strongly deviating

volumes have Me^{2+}-ions (Tab. 2).

Table 2   Magnetic moments of the metals

	Θ-Wert	$\mu_{gef.}$	$\mu_{ber.}$ für		
			Me^{4-}	Me^{3+}	Me^{2+}
Eu. . . .	+ 15	8,3	1,5	3,4	7,9
Gd . . .	+302	7,8	3,4	7,9	9,7
Tb. . . .	+205	(9.0)	7,9	9,7	10,6
Dy . . .	+150	10.9.	9,7	10,6	10,6
Ho . . .	—	—	10,6	10,6	9,6
Er. . . .	+ 40	9,5	10,6	9,6	7,6
Tm . . .	+ 10	7,6	9,6	7,6	4,5
Yb . . .	—	0	7,6	4,5	0
Cp. . . .	—	0	4,5	0	$\geqslant$0

The investigation of the metals has made clear that the "periodic system" of the lanthanoid-ions is also applicable to the metals - one can say that the behaviour of the metals yields an excellent confirmation of the conclusions based on the qualities of the ions.

Of special importance seems to me the influence of the "half-occupied" shells. It was satisfactory to learn that this influence appears also in the d-electron shells. Here the Mn^{2+}- and the Fe^{3+}-ion have a half-filled 3d shell in a S-state. The Figures 11, 12 and Table 3 shall demonstrate the influence according to examples given by W. Biltz and myself in 1934 [13].

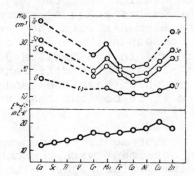

Raumbeanspruchung von Chalkogeniden (d. h. O-, S-, Se-, Te-Verbindungen) zweiwertiger Manganiden. Zum Vergleiche: Ionisierungsspannungen bei Abgabe des zweiten Elektrons.

Fig. 11 Volume of chalcogenides

	Ca²⁺	Cr²⁺	Mn²⁺	Fe²⁺	Co²⁺	Ni²⁺	Cu²⁺	Zn²⁺
O	NaCl-Typ	NaCl-Typ	NaCl-Typ	NaCl-Typ	NaCl-Typ	NaCl-Typ	eig. Typ	ZnS-Typ
S	NaCl-Typ	NiAs-Typ	NaCl-Typ	NiAs-Typ	NiAs-Typ	NiAs-Typ	eig. Typ	ZnS-Typ
Se	NaCl-Typ	NiAs-Typ	NaCl-Typ	NiAs-Typ	NiAs-Typ	NiAs-Typ	--	ZnS-Typ
Te	NaCl-Typ	NiAs-Typ	NiAs-Typ	NiAs-Typ³⁾	NiAs-Typ	NiAs-Typ	--	ZnS-Typ

Fig. 12   Structure of chalcogenides

Table 3   Colours of solutions of Me(II)-compounds

Ca²⁺   V²⁺ Cr²⁺   Mn²⁺   Fe²⁺   Co²⁺   Ni²⁺ Cu²⁺ Zn²⁺
farblos viol. blau fastfarblos grünlich rot-viol. grün blau farblos

At the end I wish to express my thanks to Baron Auer v. Welsbach and other colleagues who made this work possible by sending rare earths and to those who assisted us by their advice.

REFERENCES

1. Georg v. Hevesy, Die Seltenen Erden vom Standpunkt des Atombaues. Verlag Springer, Berlin 1927

2. cp. G. Jantsch, H. Rüping and W. Kunze, Z. Anorg. Allg. Chem. 161, 210 (1927)

3. cp. Wilhelm Prandtl and Hans Köhl, Z. Anorg. Allg. Chem. 172, 265 (1928)

4. Wilhelm Klemm and Joachim Rockstroh, Z. Anorg. Allg. Chem. 176, 181 (1928)

5. Walter Döll and Wilhelm Klemm, Z. Anorg. Allg. Chem. 241, 239 (1939)

6. Wilhelm Klemm and Wilhelm Schüth, Z. Anorg. Allg. Chem. 184, 352 (1929)

7. Wilhelm Klemm, Z. Anorg. Allg. Chem. 184, 345 (1929); 187, 29 (1930)

8. According to Rodebush, Chem. Rev. 5, 514 (1928);
   Peierls, Z. Phys. 55, 738 (1929)

9. F. Hund, Z. Phys. 33, 855 (1925)

10. Heinrich Bommer, Z. Anorg. Allg. Chem. 241, 273
    (1930); cp. l.c. 11

11. W. Klemm and G. Winkelmann, Z. Anorg. Allg. Chem.
    288, 87 (1956)

12. W. Klemm and H. Bommer, Z. Anorg. Allg. Chem. 231,
    138 (1937); 241, 264 (1939); 242, 277 (1939)

13. cp. Wilhelm Biltz, Raumchemie der festen Stoffe,
    Verlag L. Voss, Leipzig 1934, pp. 140-142

2

SYSTEMATICS OF THE PROPERTIES OF THE LANTHANIDES

Leo Brewer

Lawrence Berkeley Laboratory & Dept. of Chemistry,
University of California, Berkeley

ABSTRACT

The influence of electronic configurations upon the stability of
various crystal structures of the lanthanides and actinides and
upon their melting and boiling points is examined. A unified
model that is applicable to lanthanides, actinides, and all of
the transistion metals, is presented that provides predictions
of the thermodynamic properties, magnetism, and a variety of
properties. Methods of calculation of phase diagrams are discussed. Spectroscopic and thermodynamic data for the elemental
lanthanides are tabulated to provide the base for binary system
predictions. Thermodynamic data are tabulated for the solid and
gaseous lanthanide oxides.

1. INTRODUCTION

An examination of the systematics of the properties of the
lanthanide elements and their compounds is of particular value
in improving our understanding of all the elements. For elements other than the lanthanides or the second half of the actinides, the change from one element to the adjoining element in
the Periodic Table, or the change from one element to another in
the same group, involves not only an increase in nuclear charge
but also a change in the type of valence electrons that are
available. Very substantial and irregular variation of properties result from the change in the type of valence electrons and
change in size of the atom. As one illustration, consider the
enthalpies of formation of the halides and oxides (1). For the
alkali metals, the enthalpies of formation of the iodides become

17

S. P. Sinha (ed.), Systematics and the Properties of the Lanthanides, 17–69.
Copyright © 1983 by D. Reidel Publishing Company.

more negative from lithium to cesium, while the fluorides and oxides show the opposite trend. For the alkaline earths, the iodides increase in stability from Be to Ra as for the alkali metals, but the fluorides and oxides initially increase in stability going down from beryllium, reach a maximum, and then decrease in stability. For the alkali and alkaline earth compounds, these irregular trends can be understood through use of the Born-Haber cycle. There are two contributions to the enthalpy of formation that vary in opposite directions. One is the enthalpy of formation of the gaseous ions which decreases as one goes from lithium to cesium, for example. The other is the effect on the lattice energy of increasing size from lithium to cesium ion. If one examines the data for the transition metal compounds, the behavior becomes more complex due to crystal field splitting of the d orbitals and one observes two maxima and a minimum in the lattice energies of the halides between divalent calcium and divalent zinc, for example.

In contrast, the enthalpies of formation of many lanthanide compounds vary slowly and smoothly from lanthanum to lutetium with minor perturbations for europium and ytterbium. This is illustrated by Table Al in the Appendix, which presents the enthalpies of formation of trivalent ions and gaseous electrons from the solid elements, the enthalpies of formation of the solid oxides, and their lattice energies. The trivalent lanthanides all have filled 5s and 5p orbitals. The only difference is the addition of 4f electrons beyond lanthanum until fourteen have been added for lutetium. As the 4f orbitals are very localized within the closed 5s-5p shell, they have little influence upon the bonding. As the electrons largely neutralize the increase in nuclear charge, there is only a very slow change in ionic radius.

The enthalpies of formation of the oxides from the solid metals show some irregularity. Except for Eu and Yb, the irregularity is small compared to variations across the Periodic Table for other elements. Even the small variations as well as the deviations of the Eu and Yb oxides are gone when one considers the lattice energies which correspond to forming the trivalent oxides from trivalent gaseous ions with no change in electronic configuration. As will be demonstrated shortly, this is a general principle that is useful in considering all types of bonding. One should examine the formation of a solid phase or a gaseous species in a process that involves little or no change in electronic configuration. Thus the lattice energies of the oxides in Table Al correspond to the condensation of trivalent ions with the electronic configuration f^{n-3} to a solid containing ions of the same electronic configuration. In the formation of the oxides from the solid lanthanide metals, the number of f electrons per atom is the same for the metal as for the oxide, except for Eu and Yb. The simple behavior of lanthanide compounds is not restricted to the solid phase. Table A2 of the Appendix presents the enthalpies of formation of the gaseous oxides, LnO, Ln_2O, Ln_2O_2, and LnO_2. Here again, a slow and smooth variation is observed with minor perturbations for europium and ytterbium which will be discussed shortly. When abnormal behavior is observed for the lanthanides or the actinides, there must be some change in the localization of the valence electrons which allows the f electrons to participate in bonding, or there must be promotion of 4f electrons to the 5d or 6s,p orbitals to allow their participation in bonding. Changing the character of bonding due to change of localization of existing valence electrons or due to promotion to less localized orbitals is a general phenomenon for all of the elements, but the lanthanides provide a clear demonstration.

2. ABNORMAL BEHAVIOR OF LANTHANIDE GASES

As an example of abnormal behavior of the lanthanides, we will examine the bonding in the elemental metallic solids and liquids and in the gaseous Ln_2 species. The understanding of lanthanide behavior is valuable in understanding the irregular variation of properties of other elements. As one illustration, consider the boiling points of the elements (2). For the alkali metals, the boiling points decrease with increasing nuclear charge from lithium to cesium. For the sixth group transition metals, the opposite trend is observed with increase of boiling point from chromium to tungsten. For the 5d transition metals, there is a smooth variation of boiling points from lutetium to a maximum at tungsten, and then a decrease to gold. For the 3d transition metals, there is a maximum at vanadium, then a drop to a minimum at manganese, and then an increase to a maximum at nickel. Any reasonably simple model capable of reconciling these discordant trends would have to describe the atomic interactions in terms of at least two factors that would vary in opposing directions. The behavior of the lanthanide metals provides an insight into the important factors that must be considered.

We will first consider the behavior of the lanthanide elements as gaseous diatomic species (3) and then as crystalline solids (2,4). In contrast to the behavior of the lanthanides in the trivalent state, the vaporization processes of the neutral elements display a very discordant behavior. This is illustrated by values of $\Delta H_0^0/R$ in kilokelvin for $Ln_2(g) = 2Ln(g)$.

La	Ce	Pr	Nd	Pm	Sm	Eu	Gd	Tb	Dy	Ho	Er	Tm	Yb	Lu
29	29	18	10	(8.5)	7	4	20.5	15	8	8	8	6	2	(20)

The bonding of the crystalline solids can be represented by
values of the enthalpy of sublimation for the stable crystal
structures at 0 K: the double hexagonal close-packed (dhcp)
structure for La, Pr, Nd, Pm, and the cubic close-packed (ccp)
structure for cerium, the rhombohedral Sm-type structure for
samarium, the body-centered cubic (bcc) structure for europium,
and the hexagonal close-packed (hcp) structure for all others.
The values of $\Delta H_0^0/R$ in kilokelvin for Ln(s) = Ln(g) from Table 2
are given here.

La	Ce	Pr	Nd	Pm	Sm	Eu	Gd	Tb	Dy	Ho	Er	Tm	Yb	Lu
52	50.2	43	39.5	(37)	24.8	21.5	48	47	35.3	36.4	38	28	18.7	51.4

If the enthalpies of sublimation of the lanthanide metals
are taken as a measure of the strength of cohesion, they indicate a rapid reduction of cohesion from Ce to Eu with a sudden
increase at Gd, followed by a drop to Yp, and a jump again at
Lu. However, if the melting points (2) are taken as a measure of
cohesion, one finds a steady increase in melting point from Ce
to Lu, with the exception of low melting Eu and Yb.

3. ROLE OF ELECTRONIC CONFIGURATION

It has been recognized by many in the past that the anomalous behavior of the lanthanide metals is due to the rapid stabilization of electrons in the f orbitals as the nuclear charge
is increased. However, the spectra of the lanthanides are so
complicated that it has been difficult to establish which electronic states are important to the metal compared to the gas.
It has been possible to help establish the relative stabilities
of different electron configurations of the lanthanides through
use of the Engel theory of metals (5).

It is well established in chemistry that many compounds
which are isoelectronic have similar coordination numbers or
structures. If we move horizontally in the Periodic Table from
sodium to argon, there is a change in crystal structure for each
element which clearly indicates a strong correlation between
electrons per atom and structure. Fifty years ago, Hume-Rothery
(6) noted that many intermetallic compounds with varying compositions had the same crystal structure if they had the same
average number of valence electrons per atom. From examination
of alloy systems, where the bonding is due to only s or p electrons, the tetrahedral diamond or ZnS structure can be correlated with four s,p electrons per atom, the cubic close-packed
(ccp) structure can be correlated with a range of approximately
2.5 to 3 s,p electrons per atom (e/a). The hexagonal close-
packed (hcp) structure can be correlated with 1.7 to 2.1 s,p

electrons per atom and the body-centered cubic (bcc) structure
is limited to less than 1.5 s,p electrons per atom. He encountered difficulties in using the correlation for transition
metals in regard to the number of electrons. Using the total
number of valence electrons gave too high an electron count.
Engel (7) resolved this dilemma by recognizing the difference
between the outer-shell s,p electrons and the inner-shell d and
f electrons. In contrast to the change from bcc to hcp to ccp
to diamond structure for the adjoining nontransition elements,
Na, Mg, Al and Si, the transition metals from Rb to Mo all have
the bcc structure as one of their structures. There is a change
to hcp for Tc and Ru, and finally the fcc structure is found for
Rh, Pd, and Ag. Engel, in agreement with Pauling (8), recognized that the unfilled d shells of the first half of the transition elements act as sinks for electrons and, thus, tend to
keep the s,p electron concentration below 1.5 e/a. After the
d^5s configuration is reached for Mo, the addition of another
electron for Tc required promotion to d^5sp to allow all seven
electrons to be used in bonding and thus a change from a configuration with less than 0.5 p electrons per atom (bcc) to one
with one p electron per atom (hcp). Engel pointed out that the
Hume-Rothery correlation could be extended to the transition
metals if only the s,p electrons were counted. The Engel correlation is based on the concept that the inner-shell d or f
orbitals are somewhat localized and interact primarily with
nearest neighbors and do not influence long-range order. The
outer-shell s,p electrons range far out and are decisive in
controlling long-range order which determines the crystal structure.

Appendix 3 presents tables of energies of the lowest level
of each electronic configuration of the neutral gaseous atoms
and of the univalent to trivalent ions of the lanthanides. One
would expect that one would want to use some average energy of
the various levels corresponding to a given electronic configuration to represent the energy of the configuration. It was
demonstrated (9) for the transition metals that there was a
simple and smooth relationship between the energy of the lowest
level and the average energy of a configuration. For a number
of elements and particularly for the lanthanides and actinides,
the positions of the various levels corresponding to a given
electronic configuration are poorly known and it is convenient
to use the lowest level energies as has been done in Appendix 3.

An examination of the electronic configurations of the
trivalent ions and the neutral atoms provides a clear explanation of the difference in behavior. For the trivalent ions, the
only valence electrons beyond the filled $5s^2 5p^6$ core are electrons in 4f orbitals varying from $4f^1$ for Ce^{3+} to $4f^{14}$ for
Lu^{3+}. These 4f electrons are largely localized inside the 5s,5p

core and do not participate significantly in bonding. As the radii of the trivalent ions decrease slowly and reasonably smoothly with increasing nuclear charge, the properties of the trivalent compounds vary slowly and smoothly from La to Lu. For the neutral atoms, there are several low-lying electronic configurations involving 5d, 6s, and 6p orbitals that play the decisive role in determining the strength of bonding. The ground electronic states are $f^{n-3}ds^2$ or $f^{n-2}s^2$, where n is the total number of valence electrons, varying from three for La to seventeen for Lu. However, the $6s^2$ electrons are nonbonding, and two types of promotion are necessary to achieve optimum bonding. For lanthanides other than Eu and Yb, the promotion from a divalent state with only two non-f electrons to a trivalent state such as $f^{n-3}d^2s$ or $f^{n-3}ds$ takes less energy than the additional bonding obtained through use of three bonding electrons. For La, Ce, and Gd, which have the trivalent $f^{n-3}ds^2$ ground state, only the single 5d electron can provide bonding, and promotion to the $n^{n-3}d^2s$ or $f^{n-3}dsp$ states requires less energy than the additional bonding energy provided by two additional bonding electrons.

The Engel correlation allows an assignment of electronic configuration on the basis of crystal structure. The hcp structure of a trivalent lanthanide metal with n valence electrons would correspond to an electronic configuration between $f^{n-3}d^{1.3}sp^{0.7}$ and $f^{n-3}d^{0.9}sp^{1.1}$. If the metal had only two valence electrons not in an f orbital, the range would be $f^{n-2}d^{0.3}sp^{0.7}$ to $f^{n-2}sp$. When the bcc structure is not the sole crystal structure, it would have a configuration close to $f^{n-3}d^{1.5}sp^{0.5}$. If it used only two non-f electrons as for Eu, there would be one more f and one less d electron. The 5d orbital is somewhat localized compared to the 6s and 6p orbitals and can not contribute as strongly to bonding. Bonding capabilities of electrons in the various orbitals have been evaluated (9) and one can calculate the reduction in energy upon bringing together gaseous atoms in the various valence states corresponding to the electronic configurations of the bcc or hcp structures. The energy required to vaporize the metal to a gaseous atom with the same electronic configuration, or the same valence state, is the proper measure of the strength of cohesion.

We can now show that the melting point variation does indicate reasonably well the correct variation of cohesion or bonding strength for the lathanide metals, because the electronic configurations are reasonably close for the solid and liquid phases. The abnormal trend of the dissociation energies of the diatomic gases and boiling points of the metals is not to be attributed to the diatomic gas or the metallic phase, but is due to abnormality of the gaseous atoms. Kant and Lin (10) noted that the abnormal trends of the dissociation energies of the diatomic lanthanides were parallel to those for the enthalpies of sublimation of the metals, and that the trends were due to the increasing difficulty of promotion of 4f electrons to bonding orbitals with increasing nuclear charge. The quantitative comparison based on the presently available spectroscopic data tabulated in Appendix 3 confirms the role of promotion of 4f electrons, but there are significant differences between the bonding in the M_2 gas and in the metal, particularly from Sm and beyond. Table 1 lists the ground electronic configurations for the gaseous atom, the electronic configuration corresponding to the valence state responsible for bonding in the Ln_2 gas (3), and the electronic configuration corresponding to the metallic phase in equilibrium with the liquid (5).

TABLE 1

Element	Ground Electronic State of Atom	Valence State for Ln_2	Valence State for Ln Metal at m.p.
La	ds^2	d^2s	$d^{1.5}sp^{0.5}$
Ce	fds^2	fd^2s	$fd^{1.5}sp^{0.5}$
Pr	f^3s^2	f^2d^2s	$f^2s^{1.5}sp^{0.5}$
Nd	f^4s^2	f^3d^2s	$f^3d^{1.5}sp^{0.5}$
Pm	f^5s^2	f^4d^2s	$f^4d^{1.5}sp^{0.5}$
Sm	f^6s^2	$f^6d^{0.5}sp^{0.5}$	$f^5d^{1.5}sp^{0.5}$
Eu	f^7s^2	$f^7d^{0.5}sp^{0.5}$	$f^7d^{0.5}sp^{0.5}$
Gd	f^7ds^2	f^7d^2s	$f^7d^{1.5}sp^{0.5}$
Tb	f^9s^2	f^8d^2s	$f^8d^{1.5}sp^{0.5}$
Dy	$f^{10}s^2$	$f^{10}sp$	$f^9d^{1.5}sp^{0.5}$
Ho	$f^{11}s^2$	$f^{11}sp$	$f^{10}d^{1.5}sp^{0.5}$
Er	$f^{12}s^2$	$f^{12}sp$	$f^{11}dsp$
Tm	$f^{13}s^2$	$f^{13}sp$	$f^{12}dsp$
Yb	$f^{14}s^2$	$f^{14}sp$	$f^{14}d^{0.5}sp^{0.5}$
Lu	$f^{14}ds^2$	$f^{14}dsp$	$f^{14}dsp$

The major differences to be noted are the number of f electrons. For all of the metals except Eu and Yb, there are three bonding electrons in d, s, and p orbitals, while for all of the gaseous atoms except for La, Ce, and Gd, all of the valence electrons are in f orbitals except for an s^2 pair. As the nuclear charge is increased, it requires more energy to promote from an f^{n-2} to an f^{n-3} configuration. For Eu and Yb, even in the metal, the bonding due to an extra bonding electron does not offset the high promotion energy. For the diatomic gas, where the bonding is restricted to a pair of atoms and is less effective than in the metal, promotion from f^{n-2} to f^{n-3} can not take place for Sm, Eu, Dy, Ho, Er, Tm, and Yb (3).

Table 2 lists the enthalpy of sublimation to the ground electronic state of gaseous atoms for each of the known structures at normal pressures of the lanthanide metals and the enthalpies of sublimation to the indicated valence state. In addition, the enthalpy of sublimation has been calculated for the hcp and bcc structures even when they are not stable, through use of the spectroscopic data of Appendix 3 and the tabulated bonding energies for 5d, 6s, and 6p electrons (9).

TABLE 2

	Atomization Enthalpy to Gaseous Ground State			Atomization Enthalpy to Gaseous Valence State	
	$\Delta H^{\circ}/R$ in kilokelvin				
	at 298.15 K		at 0 K	at 0 K	
Sc bcc	45.0	±0.5	44.9	64.5	$d^{1.5}sp^{0.5}$
hcp	45.4	±0.5	45.2	67.8	dsp
Y bcc	50.22	±0.4	50.17	68.8	$d^{1.5}sp^{0.5}$
hcp	50.83	±0.4	50.72	72.2	dsp
La bcc	51.4	±0.5	51.5	63.0	$d^{1.5}sp^{0.5}$
cpp	51.8	±0.5	51.8	(72.5)	$d^{0.5}sp^{1.5}$
dhcp	51.9	±0.5	51.9	71.0	dsp
Ce bcc	49.62	±0.3	49.72	61.1	f $d^{1.5}sp^{0.5}$
γccp	49.92	±0.3	49.97	(72.5)	f $d^{0.5}sp^{1.5}$
dhcp	49.92	±0.3	49.97	69.2	fdsp
αccp	--		50.17	(73.0)	f $d^{0.5}sp^{1.5}$

TABLE 2 (Cont'd)

	Atomization Enthalpy to Gaseous Ground State		Atomization Enthalpy to Gaseous Valence State	
	$\Delta H^{o}/R$ in kilokelvin			
	at 298.15 K	at 0 K	at 0 K	
Pr bcc	42.4 ±0.3	42.6	(61.)	$f^2 d^{1.5} sp^{0.5}$
dhcp	42.77 ±0.3	42.92	(69.)	$f^2 dsp$
Nd bcc	39.1 ±0.3	39.3	60.2	$f^3 d^{1.5} sp^{0.5}$
dhcp	39.4 ±0.3	39.5	68.7	$f^3 dsp$
Pm bcc	(36.9) ±1.5	(37.0)	(60.)	$f^4 d^{1.5} sp^{0.5}$
dhcp	(37.2) ±1.5	(37.3)	(68.)	$f^4 dsp$
Sm bcc	24.5 ±0.3	24.5	(59.)	$f^5 d^{1.5} sp^{0.5}$
α	24.86 ±0.3	24.81	--	
hcp	(24.7) ±0.5	(24.9)	(67.)	$f^5 dsp$
Eu bcc	21.33 ±0.1	21.54	41.0	$f^5 d^{1.5} sp^{0.5}$
Gd bcc	47.4 ±0.3	47.7	62.4	$f^7 d^{1.5} sp^{0.5}$
hcp	47.9 ±0.3	48.1	68.2	$f^7 dsp$
Tb bcc	46.1 ±0.3	46.5	(63.4)	$f^8 d^{1.5} sp^{0.5}$
hcp	46.7 ±0.3	47.0	(68.4)	$f^8 dsp$
Dy bcc	34.5 ±0.3	35.0	(64.)	$f^9 d^{1.5} sp^{0.5}$
hcp	35.0 ±0.1	35.3	68.4	$f^9 dsp$
Ho bcc	35.6 ±0.3	36.1	(66.4)	$f^{10} d^{1.5} sp^{0.5}$
hcp	36.2 ±0.1	36.4	(70.4)	$f^{10} dsp$
Er bcc	(37.3) ±0.3	(37.7)	(68.)	$f^{11} d^{1.5} sp^{0.5}$
hcp	38.0 ±0.1	38.1	71.2	$f^{11} dsp$
Tm bcc	(27.3) ±0.5	(27.7)	(68.)	$f^{12} d^{1.5} sp^{0.5}$
hcp	27.9 ±0.5	28.1	70.3	$f^{12} dsp$
Yb bcc	18.37 ±0.2	18.47	48.5	$f^{14} d^{0.5} sp^{0.5}$
ccp	18.57 ±0.2	18.62	--	
hcp	18.6 ±0.4	18.7	43.5	$f^{14} sp$
Lu bcc	(49.) ±1.	(49.)	(76.)	$f^{14} d^{1.5} sp^{0.5}$
hcp	51.43 ±0.1	51.43	76.5	$f^{14} dsp$

Examination of the energies of various electronic configurations tabulated for the neutral atom in Appendix 3 indicates that the three lowest trivalent configurations are $f^{n-3}ds^2$, $f^{n-3}d^2s$, and $f^{n-3}dsp$. The $f^{n-3}ds^2$ configuration is always the lowest in energy for the gaseous atom, but it is unimportant in the metal because the filled s^2 orbital is nonbonding. The $f^{n-3}d^2s$ configuration corresponding to the bcc structure is lower than the $f^{n-3}dsp$ configuration corresponding to the hcp structure, but the configuration with the most outer shell electrons is the most strongly bonding and the hcp metal is more stable than the bcc metal. However, the bcc structure with a coordination number of eight has lower vibrational frequencies and a higher heat capacity and entropy. Thus upon increasing the temperature, the bcc structure becomes lower in Gibbs energy than the hcp structure (9,11,12) and is the high temperature form for most of the lanthanide metals. For the gaseous atom, the $f^{n-3}d^2s$ configuration rises rapidly compared to the $f^{n-3}dsp$ through the second half of the lanthanides, and the bcc structure becomes so much higher in energy than the hcp structure that even its higher entropy does not cause stabilization before the melting point. The bcc structure is the high temperature structure for Sc, Y, and La to Dy and possibly for Ho, but not for Er, Tm, and Lu.

4.   EFFECT OF LOCALIZATION OF INNER SHELL ORBITALS

As one goes to the right in the Periodic Table, the increasing nuclear charge increases the localization of inner shell d or f orbitals so that overlap of orbitals from adjacent atoms is rather poor. This causes the poorer bonding for d electrons compared to outer shell s and p electrons and the very poor bonding contribution from f electrons. The order (9,13,14) of decreasing bonding ability is: 1s > 2s,p > 3s,p > 4s,p > 5s,p > 6s,p > 7s,p > 6d > 5d > 4d > 3d and the first half of 5f > 4f and the second half of 5f. Magnetism arises when the orbitals are so localized that overlap with orbitals of neighbors is very small and there is little bonding. The crystal field effect of neighbors will cause splitting of the inner shell orbitals with some becoming more localized and others becoming more extended and better bonding. When there are not enough electrons to fill the very localized orbitals with non-bonding electron pairs, the occurrence of single electrons in localized orbitals causes ferromagnetism as with Cr to Ni, Ce to Tm, and for the heavier actinides (13,14,15). Another consequence of somewhat localized inner shell orbitals is the effect of pressure upon improving the overlap of the orbitals and improving the bonding. For the transition metals of Group IV through Group XI, there is a 100% correlation of the prediction that among the bcc, hcp, and fcc structures, pressure stabilizes the structure with the most bonding d electrons. As noted above, the localization of the inner shell d orbitals decreases from 6d to 3d and decreases for any one group with increasing nuclear charge as one goes from left to right in the Periodic Table. Thus the effects due to localization of the d orbitals should diminish as one moves to the left in the Periodic Table and particularly if one moves toward the lower left hand corner which brings us to the lanthanides and actinides.

One would normally expect that increasing the pressure would stabilize the close-packed structures over the bcc structure. However for transition metal groups IV to VI, pressure stabilizes the bcc structure because the $d^{n-1}s$ configuration has more bonding d electrons than the $d^{n-2}sp$ configuration for $n < 6$. Because of the relatively poor overlap of the inner shell d orbitals, compression considerably improves the bonding and in effect offsets the steep repulsive potential energy to yield much softer net repulsive potential at small internuclear distances. With the smaller degree of delocalization in the lower left hand corner, this effect should be greatly reduced for the lanthanides and actinides. For most of these elements for which data are available, increasing the pressure destabilizes the bcc phase relative to the close-packed structure or has close to zero effect. (16)

The effect of pressure upon improving the overlap and bonding of f orbitals must now be considered. The effect of pressure upon improving the bonding of localized orbitals is most dramatically illustrated by the effect of compression upon fcc cerium. Cerium metal has only one electron in an f orbital which is sufficiently localized that it does not overlap significantly with neighboring orbitals. Thus there is no significant contribution from bonding and the single electron is responsible for the magnetism of cerium. However, if the internuclear distance is sufficiently reduced by either cooling or by application of pressure, the overlap of f orbitals will become sufficient to provide for significant f electron bonding. This enhanced bonding results in a substantial contraction and elimination of magnetism. As the pressure is increased the maximum temperature at which the dense non-magnetic ccp phase of Ce is stable increases, but the less dense magnetic ccp phase is more compressible and approaches the density of the non-magnetic phases. A critical point is reached at which the two phases merge and at higher pressures, one has only a dense non-magnetic phase. (16,17) At higher pressures, the same phenomenon is observed for Pr. (17) At the beginning of the actinides series, the 5f orbitals are considerably more extended than the 4f orbitals of the corresponding lanthanides. The contribution of the 5f electrons to bonding as well as the ability to promote to configurations with more than three non-f electrons results in early actinides which are more strongly bonded than the

corresponding lanthanides and which are non-magnetic. As the
nuclear charge is increased, the 5f orbitals are rapidly stabilized and become more localized. As a result, the heavier
actinides are magnetic and many more of them are divalent like
Eu and Yb because the additional bonding upon promotion of an
electron from the 5f orbital to a 6d or 7p orbital does not
offset the large promotion energy.

The variety of electronic configurations of comparable
stability in the solid state for the lanthanide and actinide
metals results in some unusual properties. Experimental observations of many of the electronic states have been made in
recent years and Appendix 3 presents the available data for the
lanthanides. Similar tabulations are available for the actinides (5). From the tabulated values for the gaseous atoms and
the tabulated values of bonding energies per electron for the
various types of electrons (9), one finds that for many elements
such as Mo or W, for example, a single electronic configuration
contributes significantly in the metal. However, for elements
such as U, Np, and Pu, one calculates that four or more electronic configurations are of comparable stability in the
metal. This results in two types of unusual behavior due to the
fact that atoms in different electronic configurations will have
different effective sizes. When one mixes atoms of equal size,
a close-packed structure provides the best space utilization.
However, a mixture of atoms of different sizes can obtain better
space utilization by achieving coordination higher than twelve.
The Laves phases (hP12, hP24, and cF24) and the σ phase (tP30),
are examples. Normally these phases do not form unless elements
of different sizes are mixed. However, β-uranium achieves the σ
structure with only uranium atoms which are of different electronic configurations and therefore different sizes in different
lattice sizes. Pa, U, Np, and Pu all have structures of this
type. For Mn, the configurations d^6s, d^5sp, and d^4sp^2 are of
comparable stability in the metal, and the formation of α- and
β-Mn is another example. When there are size differences,
structures with equivalent lattice sites such as the bcc, hcp,
and ccp structures are destabilized by the resulting strain
energy. On the other hand, the liquid is stabilized by a mixture of sizes because it can then achieve better space utilization. A most striking example is that of plutonium which has an
abnormally low melting point because of the abnormally dense
liquid resulting from the mixture of atoms with different electronic configurations and different sizes. Uranium also melts
abnormally low compared with its cohesion energy. Cerium metal
is an example of one of the trivalent lanthanide metals which
has an abnormally low melting point because of the variety of
electronic configurations. As pressure is applied to the
metals, the relative stability of the different electronic configurations will change as the bonding of the more localized

orbitals is enhanced. Thus application of pressure will bring different electronic configurations to comparable energies resulting in a mixture of configurations and thus sizes in the liquid. The resulting densification of the liquid reverses the usual increase of melting point with pressure, resulting in a maximum melting point, with reduction in melting point with further increase in pressure. This reduction in melting point with increasing pressure will continue until a solid structure with denser packing than close-packing such as the σ or β-U structure or the α- or β-Mn structures. These solids will be denser than the liquid and the melting point will then increase upon further increase of pressure. These are just a few of the examples that can be given of the important role of electronic configurations in determining the behavior of actinides, lanthanides, and transition metals, generally.

5. THERMODYNAMIC STABILITY OF LANTHANIDE INTERMETALLIC BINARY SYSTEMS

If one applies regular solution theory (1,18) using energies of vaporization in the usual manner to calculate internal pressures, the results are quite meaningless. If the rapid decrease in energy of vaporization from La to Sm were taken as a measure of the internal pressure, one would predict immiscibility between the trivalent lanthanide metals. To obtain more meaningful values, one must use the energy of vaporization to the valence state, the gaseous atomic state with the same electronic configuration as in the condensed metal. Then, as can be seen in Table 2, the energies of vaporization and internal pressures vary only slowly among the trivalent metals. In the usual regular solution calculation, the interaction between unlike species is taken as the geometric mean of the interactions between like species. There is no good theoretical justification for this, and a simple modification yields much better results for the lanthanides. The partial molal Gibbs energy of component A in a solution of B of composition x_B is given by

$$\Delta \overline{G}_A / RT = b x_B^2 + c x_B^3$$

where $(b + c)/(b + \frac{1}{2}c) = V_A / V_B$, and

$$b + c = (V_A / 2R)[(\Delta E_A^* / V_A)^{1/2} - (\Delta E_B^* / V_B)^{1/2}]^2$$

where ΔE^* is the energy of vaporization of the liquid to the gaseous valence state and V is the molal volume of the liquid. This equation was used (19) to calculate the phase diagrams of molybdenum with all of the Group III and lanthanide metals as well as other metals. Although the regular solution equation works well for solutions of molybdenum and the lanthanides and a

few other metals, internal pressure alone is not adequate to
represent the thermodynamic behavior of metallic solutions. As
mentioned earlier, size differences destabilize structures with
equivalent lattice sites and stabilize those with non-equivalent
lattice sites. However, the major contribution may be electronic when the electronic configurations are different for the
solution components; particularly when a lanthanide, actinide,
or transition metal from the left hand side of the Periodic
Table with vacant d orbitals is mixed with a transition metal
from the right hand side of the Periodic Table which has so many
electrons that some fill d orbitals as non-bonding pairs.

If one considers the variations in thermodynamic stability
and the resulting phase diagram variations as one considers a
wide variety of binary diagrams, one finds that the relative
contributions of size, internal pressure, and electronic factors
vary considerably. To enhance the size factor, let us consider
the interactions of various metals with thorium, uranium and
plutonium. The behavior of metals with thorium is illustrated
in Table 3 in the diagram which separates the metals into four
areas. Because, the alkali metals, the alkaline earth metals
from Ca through radium, Eu, Yb, and the divalent actinides have
internal pressures much lower than that of Th, the internal
pressure difference will predominate and their mutual solubilities will be very small. Because of the small size of the elements of the third period, the size factor will predominate in
greatly reducing the mutual solubilities of Th with the elements
from Sc to Ge. The elements that show large solubilities in
thorium are Y, Zr, Hf, and the lanthanides and actinides excluding those that are divalent. The fourth area to the right of
the transition metals includes those metals with very strong
acid-base interactions that yield very stable intermetallic
phases. Because of the large reduction in thermodynamic activity, the resulting solubilities in thorium are small.

The next diagram in Table 3 shows how the areas shift due
to the lower internal pressure and particularly the smaller size
of Pu, which enhances the solubilities of the third period
metals compared to thorium.

Finally, the distribution in the four groups is illustrated
for uranium. As uranium has a higher internal pressure than
thorium or plutonium, it can dissolve appreciable amounts of the
fifth and sixth transition metals. The mutual solubilities of
the bcc phases of the lanthanides with the bcc phases of Th, U,
or Pu is a particularly good example of the role of internal
pressure. The Engel correlation indicates that these bcc phases
will all have approximately the same number of s and p electrons
and not have non-bonding d electrons pairs; so the effect of
different electronic configurations has been minimized. Bcc

TABLE 3:
Solubilities

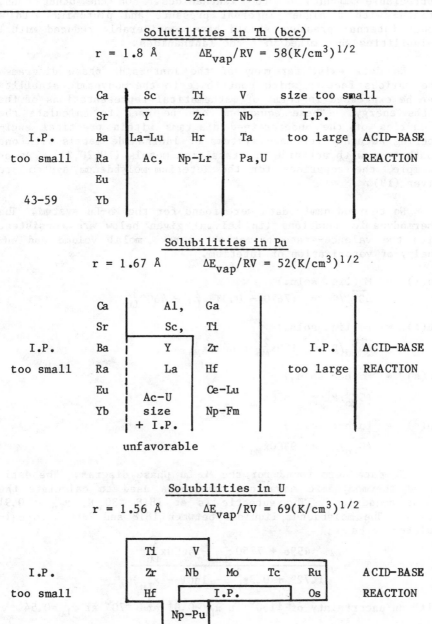

Solutilities in Th (bcc)

$$r = 1.8 \text{ Å} \qquad \Delta E_{vap}/RV = 58(K/cm^3)^{1/2}$$

I.P. too small		Ca	Sc	Ti	V	size too small	
		Sr	Y	Zr	Nb	I.P.	
		Ba	La-Lu	Hr	Ta	too large	ACID-BASE
		Ra	Ac, Np-Lr	Pa,U			REACTION
		Eu					
43-59		Yb					

Solubilities in Pu

$$r = 1.67 \text{ Å} \qquad \Delta E_{vap}/RV = 52(K/cm^3)^{1/2}$$

I.P. too small	Ca	Al,	Ga		ACID-BASE
	Sr	Sc,	Ti		
	Ba	Y	Zr	I.P.	REACTION
	Ra	La	Hf	too large	
	Eu		Ce-Lu		
	Yb	Ac-U size + I.P.	Np-Fm		
		unfavorable			

Solubilities in U

$$r = 1.56 \text{ Å} \qquad \Delta E_{vap}/RV = 69(K/cm^3)^{1/2}$$

I.P. too small	Ti	V				ACID-BASE
	Zr	Nb	Mo	Tc	Ru	REACTION
	Hf	I.P.			Os	
	Np-Pu					

thorium with an intermediate internal pressure that matches the internal pressures of most of the lanthanides should dissolve appreciable amounts of the lanthanides.   On the other hand, uranium with a higher internal pressure and plutonium with a lower internal pressure should have considerably reduced mutual solubilities when mixed with the lanthanides.

No data exist for many of the lanthanide phase diagrams. The various factors which contribute to thermodynamic stability can be expressed in terms of mathematical representations of the Gibbs energy.   These equations can be used to calculate the boundries of the undetermined diagrams within practical engineering accuracy for many systems of lanthanide metals with one another or with actinide or transition metals (19,20,21).   As an example, the equations for the lutetium-molybdenum system are given (19).

No thermodynamic data were found for the Mo-Lu system.   The thermodynamic equations (in kelvin) given below are consistent with the valence-state promotion energy, molal volume and enthalpy of vaporization of lutetium.

$Mo(1) = Mo(liq. soln.)$

$$\Delta \overline{G}^E_{Mo}/R = (7670 - 0.3T)x^2_{Lu} - 4600^3_{Lu}$$

$Lu(1) = Lu(liq. soln.)$

$$\Delta \overline{G}^E_{Lu}/R = 180x^2_{Mo} + 4600x^3_{Mo}$$

$Mo(s) = Mo(bcc soln.)$

$$\Delta \overline{G}^E_{Mo}/R = 7850x^2_{Lu}$$

$Lu(s) = Lu(bcc soln.)$

$$\Delta \overline{G}^E_{Lu}/R = 9300x^2_{Mo}$$

No data were found for the Mo-Lu phase diagram.   The estimated thermodynamic equations have been used to calculate the phase boundries.   The eutectic is at 1818 ±30 K, $x_{Lu} = 0.91$ ±0.03.   The molybdenum liquidus between 1818 and 2890 K is calculated to be

$$T = \frac{4996 + 7750x^2_{Lu} - 4600x^3_{Lu}}{1.729 + 0.3x^2_{Lu} - \ln(1 - x_{Lu})}$$

with an uncertainty of $\pm 150^{\circ}$ at $x_{Lu}=0.91$ and $\pm 70^{\circ}$ at $x_{Lu}=0.54$.

The corresponding solidus is given by

$$x_{Lu} = 7 \times 10^{-5}(2890 - T) - 1.2 \times 10^{-7}(2890 - T)^2$$
$$+ 5.6 \times 10^{-11}(2890 - T)^3$$

with an uncertainty of a factor of 5.

The lutetium liquidus above 1818 K is calculated to be

$$x_{Mo} = 5.88 \times 10^{-4}(1936 - T) + 1.39 \times 10^{-6}(1936 - T)^2$$

with an uncertainty of ±0.01. The lutetium solidus in this temperature range is

$$x_{Mo} = 1.2 \times 10^{-4}(1936 - T)$$

with an uncertainty of a factor of 5.

Below 1818 K the solid-solution phase boundries are calculated to be

$$x_{Mo} = 10^{-4} + 10^{-6}(T - 850) + 4 \times 10^{-9}(T - 850)^2$$
$$+ 10^{-11}(T - 850)^3$$

and

$$x_{Lu} = 10^{-4} + 7 \times 10^{-7}(T - 1000) + 3.5 \times 10^{-9}(T - 1000)^2$$
$$+ 6 \times 10^{-12}(T - 1000)^3$$

The solid solubilities, which are uncertain by a factor of 5, fall to 10^{-4} at 1010 K for lutetium in molybdenum and at 850 K for molybdenum in lutetium.

6. THE ACIDIC PROPERTIES OF THE LANTHANIDE METALS

The previous section has mentioned lanthanides, actinides, and other transition metals with vacant d orbitals acting as Lewis acids toward the basic platinum metals which have non-bonding d electron pairs. Extremely stable compounds are formed (21,22). This effect is best described in terms of the generalized Lewis-acid-base interaction and is usually illustrated by the interaction between BF_3 and NH_3 where BF_3 provides a vacant orbital and NH_3 provides a lone pair of electrons to form a compound of high stability because the valence electrons are used in bonding. If the corresponding reaction of Lu and Pt is considered, Pt metal is fcc and has an electronic configuration close to d^7sp^2. As there are only five d orbitals, four of the d electrons must be paired and are not available for bonding in pure platinum which can only use six of its valence electrons for bonding. This removal of valence electrons from bonding as more electrons are added results in the reduction in enthalpy of sublimation as one moves to the right in the Periodic Table from Re to Au, for example. If platinum is added to lutetium, the non-bonding electrons of the platinum can use the vacant orbitals of lutetium to bond the nuclei together, thus making full use of all of the valence electrons in bonding.

The possibility of strong acid-base interactions when lanthanide compounds are in contact with platinum group metals can produce quite unexpected reactions. As an example, CeS is a refractory that is highly resistant to reduction by strongly electropositive metals. The alkali and alkaline earth metals can be boiled in CeS containers without attack. However, molten platinum will vigorously attack CeS to form $CePt_2$ (23).

Thermodynamic data are available for only a few of the very stable compounds that can be formed, but the general variation of stability can be predicted (22,24). Some of these compounds are among the most stable compounds of any type. Two examples of thermodynamic demonstration of high stability are the determination of $\Delta H^{o}/R = -66\ 000$ kelvin both for the formation of a mole of $HfPt_3$ from the elements (25) and for the formation of a mole of UPd_3 from the elements (24).

One normally considers aqueous H^+ and OH^- as examples of very strong acids and bases. If one were to add 10M HCl to an excess of 10M NaOH at room temperature, the activity of H^+ would be reduced by a factor of 10^{16}. Dissolving Hf or U in an excess of palladium or platinum would reduce the activity of the Hf or U at room temperature by a factor of more than 10^{90}. Reduction of the thermodynamic activity of the lanthanides by the presence of platinum should be almost as large. These strong acid-base reactions profoundly affect the chemistry of the lanthanides and actinides. For example, the oxides are extremely resistant to reduction to the metal. However, the presence of platinum so greatly reduces the activity of the metal that mixtures of lanthanide and actinide oxides with platinum are readily reduced by hydrogen. Bronger and Klemm (26,27) have demonstrated that a variety of lanthanide oxides in contact with platinum can be reduced by hydrogen. Many other unexpected reactions of lanthanide compounds can take place in the presence of platinum group metals.

This work was supported by the Division of
Materials Sciences, Office of Basic Energy
Sciences, U.S. Department of Energy, under
Contract DE-AC03-76SF00098.

APPENDIX 1:

$Ln_2O_3(s)$ Lattice Energy

It is most fortunate that Huber, Holley, Head, Fitzgibbon and their colleagues at Los Alamos have carried out a most thorough evaluation of the enthalpies of formation of lanthanide oxides. The work between 1955 and 1968 was summarized by Holley, Huber, and Baker (28). Several subsequent reviews (29-34) have agreed in acceptance of the original Los Alamos values (28) except for Pr, Sm, Eu, Gd, and Dy. The Gd2O3 value has been corrected (28) for change in the atomic weight. The values for Pr, Sm, Eu, and Dy have been revised by subsequent Los Alamos work (35-37). All ΔH_f^o values for the reaction $2Ln(s)$ + $\frac{3}{2}O_2(g)$ = Ln2O3(s) have been converted to 0 K using $H_{298}^o-H_0^o$ values tabulated by Krause (34). $\Delta H_{0,f}^o/R$ corresponding to the reaction $Ln(s) = Ln^{3+}(g) + 3e^-(g)$ was calculated from the tabulated enthalpies of sublimation of the neutral atoms (2,4) and the ionization potentials (38-40).

The electron affinity (E.A.) of $O(g)$ to $O^=(g)$ is very poorly known, but this does not influence the trend of lattice energies. The enthalpies of formation of $2Ln^{3+}(g) + 6e^-(g)$ plus the enthalpy of formation of 3O(g) minus the enthalpy of formation of Ln2O3(s) yield, by the Born-Haber cycle, the lattice energies of Ln2O3(s) plus 3(E.A.) as tabulated in Table A1.

TABLE A1

	Ln_2O_3 Structure	$Ln_2O_3(s)$ $-\Delta H^o_{0,f}/R$	$Ln^{3+}(g) + 3e^-(g)$ $\Delta H^o_{0,f}/R$	Lattice Energy L/R, plus 3(E.A.)/R
			kilokelvin	
Sc	cub	228.4 ±0.3	556.9 ±0.5	1 431.2
Y	cub	228.2 ±0.3	505. ±2	1 327.
La	hex	215.0 ±0.1	467.5 ±0.6	1 239.0
	cub	(215.) ±0.5		1 239.
Ce	hex	215.2 ±0.4	474.7 ±1	1 253.6
	cub	(215.) ±0.6		1 253.
Pr	hex	216.8 ±0.4	479.7 ±1	1 265.2
	cub	216.8 ±0.4		1 265.2
Nd	hex	216.7 ±0.1	485. ±4	1 276.
	cub	(217.1)±0.5		1 276.
Pm	cub	(219.) ±1.	487. ±5	1 282.
Sm	monocl	218.4 ±0.3	491. ±4	1 289.
	cub	218.9 ±0.4		1 290.
Eu	monocl	198.2 ±0.4	506.9 ±1	1 301.0
	cub	199. ±0.7		1 302.
Gd	monocl	217.9 ±0.5	499. ±2	1 305.
	cub	218.7 ±0.6		1 306.
Tb	cub	223. ±0.9	503. ±2	1 318.
Dy	cub	222.9 ±0.4	504. ±4	1 320.
Ho	cub	225.2 ±0.6	508. ±2	1 330.
Er	cub	227.3 ±0.3	511. ±2	1 338.
Tm	cub	226.3 ±0.7	514.5 ±2	1 344.3
Yb	cub	217.4 ±0.3	523.4 ±0.8	1 353.2
Lu	cub	224.9 ±0.9	518. ±4	1 350.

APPENDIX 2:

Enthalpies of Formation of Gaseous Ln_mO_n Species

TABLE A2

H_0^o/R for m Ln (s) $+ \frac{n}{2} O_2$ (g) $= Ln_mO_n$ (g), kilokelvin

		LnO (g)	Ln2O (g)	Ln2O2 (g)	LnO2 (g)
Sc	hcp	−5.7 ±1	1. ±8	(−58.) ±11	(−48.) ±12
Y	hcp	−4.8 ±1	6. ±7	−59. ±9	(−52.) ±12
La	dhcp	−13.7 ±1	1. ±7	−66. ±9	(−62.) ±11
Ce	αccp	−14.5 ±2	(−2.) ±9	−76. ±8	−63. ±3
Pr	dhcp	−15.7 ±2	(−6.) ±9	(−75.) ±11	(−63.) ±11
Nd	dhcp	−14.3 ±2	(−6.) ±9	(−72.) ±11	−62. ±11
Pm	dhcp	(−14.) ±6	(−8.) ±9	(−71.) ±11	(−60.) ±11
Sm	α	−13.6 ±2	(−12.) ±11	(−70.) ±12	(−58.) ±12
Eu		−4.9 ±2	−15. ±6	−60. ±9	(−44.) ±14
Gd	hcp	−7.5 ±2	7. ±5	−60. ±9	−51. ±9
Tb	hcp	−6.0 ±3	4. ±6	−59. ±9	(−50.) ±11
Dy	hcp	−8.2 ±3	(−2.) ±9	(−63.) ±11	(−51.) ±11
Ho	hcp	−6.1 ±2	0. ±7	−62. ±9	−49. ±9
Er	hcp	−4.4 ±2	(2.) ±9	(−60.) ±11	(−46.) ±11
Tm	hcp	−3.4 ±2	(−1.) ±11	(−57.) ±11	(−44.) ±12
Yb	hcp	0.6 ±3	(−7.) ±14	(−43.) ±14	(−31.) ±16
Lu	hcp	1.4 ±2	1. ±7	(−51.) ±12	(−44.) ±14

All ΔH/R values are given for the formation from the stable crystal structure (2,4) of the metal at 0 K as indicated in the discussion. Recent reviews (29,34,41) of the data for the diatomic oxides show a wide range of values, in part due to divergence of experimental results and in part to the estimation of the contribution of electronic levels to the thermodynamic values. Models which only consider the contribution of ground state will yield entropies which are much too low, as contributions from low-lying electronic levels can be considerable.

Brewer and Rosenblatt (29) have used the electronic partition function for the divalent gaseous ion as an approximation to the value for the gaseous diatomic oxides. As an oxide ion approaches the cation, the electronic levels are split, with some of the low-lying states moving up and some of the higher energy states moving down. For TiO(g) and ZrO(g), where sufficient information on the electronic states of the molecules were available, Brewer and Rosenblatt showed that the use of the free ion over-estimated the electronic contribution of the diatomic molecule. However, for the lanthanides, it is likely that the diatomic electronic partition function may be larger than the ionic value in many instances. It has been suggested that the isoelectronic univalent ion of the next earlier element, e.g. Ce^+ for Pr^{2+}, be used in place of the divalent ion. Calculation of the values around 2000K indicate that many of the univalent ions yield higher electronic partition functions than the divalent ions.

Murad and Hildenbrand (41) have used the model of Smoes et al. (42) which uses a modification of the univalent model, which seems to be a reasonable compromise for incorporating the contributions of excited electronic levels. They have recalculated the previous data and have made new measurements for the monoxides of Gd, Ho, Er, Tm, and Lu. Their recommended values have been accepted. An estimated value is given for PmO. Kordis and Gingerich (43) have given Ln2O values for Sc-La, Eu-Tb, Ho, Yb, and Lu; Ln2O2 values for Sc-Ce, Eu-Tb, and Ho; and LnO2 values for Ce, Nd, Hd, and Ho. Their values have been recalculated to be consistent with the values accepted for the LnO species; the remaining values have been estimated.

APPENDIX 3:

Energies of the Electronic Configurations of the Lanthanide Atoms and Ions

TABLE I: Lanthanum, 10^3 cm^{-1}

Odd terms (lowest level)			Even terms (lowest level)		
	%			%	
			La I		
dsp	71	$^4F_{3/2}$ 13.260	ds^2	85	$^2D_{3/2}$ 0.000
fs^2	53	$^2F_{5/2}$ 15.197	d^{2}s	98	$^4F_{3/2}$ 2.668
s^{2}p	35	$^2P_{1/2}$ 15.22	d^3	98	$^4F_{3/2}$ 12.431
d^{2}p	42	$^4G_{5/2}$ 17.947	fsp	96	$^4F_{3/2}{}^*$ 28.742
fds	47	$^2G_{7/2}{}^*$ 23.221	sp^2		$(^4P_{1/2})$ (31.)±10
fd^2		$^4H_{9/2}{}^*$? 34.715	fdp		$(^4H_{9/2})^*$ (38.)
fp^2		$(^4G_{5/2})$ (49.)±3	f^{2}s		$(^4H_{7/2})$ $(4,\frac{1}{2})$ (52.)±10
			La II (La$^+$)		
fs	96	$^3F_2$ $\frac{5}{2},\frac{1}{2}$ 14.148	d^2	83	$^3F_2$ 0.000
fd	73	$^1G_4{}^*$ $\frac{5}{2},\frac{3}{2}$ 16.599	ds	100	$^3D_1$ 1.895
dp	56	$^1D_2{}^*$ 24.463	s^2	75	$^1S_0$ 7.395
sp	70	$^3P_0$ 27.546	fp	92	$^3G_3$ $\frac{5}{2},\frac{1}{2}$ 35.453
			f^2	98	$^3H_4$ 55.107
			p^2	58	$^1D_2{}^*$ 59.900
			La III (La^{++})		
f		$^2F_{5/2}$ 7.195	d		$^2D_{3/2}$ 0.000
p		$^2P_{1/2}$ 42.015	s		$^2S_{1/2}$ 13.591

Tables I–XV present for all of the lanthanides the energy–level values of the lowest level of the lowest spectroscopic term of the various electronic configurations of the neutral atoms and the singly, doubly, and triply charged ions. The configurations listed are generally restricted to those involving 4f, 5d, 6s, and 6p electrons.

TABLE II: Cerium, 10^3 cm^{-1}

Odd terms (lowest level)				Even terms (lowest level)			
	%				%		
			Ce I				
fds^2	55	$^1G_4$*	0.000	f^2s^2	89	$^3H_4$	4.763
fd^2s	66	$^5H_3$*	2.369	f^2ds	59	$^5I_4$*	12.114
fd^3		$^5I_4$?	12.35	$fdsp$	40	$^3H_4$*	13.514
f^2sp		$(^5I_5)$ 4,1	18.28	fs^2p	14	$(\frac{5}{2},\frac{1}{2})_3$	15.556
f^2dp		$(^5L_6)$	(27.5)	fd^2p	26	$^5I_4$*	19.015
fsp^2		$(^5G_2)$	(31.)±10	f^2d^2		$(^5L_6)$	(25.)±1
f^3s		$(^5I_4)$ $(\frac{9}{2},\frac{1}{2})$	(42.)±8	d^2s^2		$(^3F_2)$	(36.)±5
ds^2p		$(^3F_2)$	(48.)±5	f^2p^2		$(^5I_4)$	(40.)±5
d^2sp		$(^5G_2)$	(50.)±5	d^3s		$(^5F_1)$	(42.)±5
d^3p		$(^5G_2)$	(58.)±5	dsp^2		$(^5F_1)$	(51.)±10
			Ce II (Ce$^+$)				
fd^2	48	$^4H_{7/2}$*	0.000	f^2s	95	$^4H_{7/2}$ 4,$\frac{1}{2}$	3.854
fds	33	$^2G_{9/2}$*	2.382	f^2d	65	$^2H_{9/2}$*	7.012
fs^2	73	$^2F_{5/2}$	9.779	fdp	59	$^2H_{9/2}$*	24.663
f^2p	95	$(^4I_{9/2})$ 4,$\frac{1}{2}$	25.766	fsp	23	$^4F_{5/2}$*	32.139
f^3	97	$^4I_{9/2}$	38.195	d^3	26	$^4F_{3/2}$	32.507
fp^2		$(^4G_{5/2})$	(61.5)±1	d^2s	49	$^4F_{5/2}$*	37.849
dsp		$(^4F_{3/2})$	(73.)±3	ds^2		$^2D_{5/2}$* ?	50.564
d^2p		$(^4G_{5/2})$	(77.)±6				
s^2p		$(^2P_{1/2})$	(92.)±4				

The energies are given in thousands of wavenumbers. Experimental values are given to 0.001 x 10^3 cm^{-1}. Estimated values are given in parentheses to the nearest 0.1 x 10^3 cm^{-1} for accurate

TABLE II (Cont'd): Cerium, 10^3 cm^{-1}

Odd terms (lowest level)				Even terms (lowest level)			
	%				%		

Ce III (Ce^{++})

	%				%		
fd	72	$^1G_4^*$ $\frac{5}{2},\frac{3}{2}$	3.277	f^2	97	$^3H_4$	0.000
fs	99	$^3F_2$ $\frac{5}{2},\frac{1}{2}$	19.236	d^2	95	$^3F_2$	40.440
dp	58	$^3F_2$	92.635	fp	97	($^3G_3$) $\frac{5}{2},\frac{1}{2}$	48.267
sp		($^3P_0$)	(118.)±5	ds	95	$^3D_1$	63.335
gf	100	($^3H_4$)* $^2[\frac{9}{2}]$	122.906	s^2		$^1S_0$	(89.)±5
				p^2		($^3P_0$)	(152.)±9

Ce IV (Ce^{+++})

f		$^2F_{5/2}$	0.000	d		$^2D_{3/2}$	49.737
p		$^2P_{1/2}$	122.585	s		$^2S_{1/2}$	86.602

TABLE III: Praseodymium, 10^3 cm^{-1}

Odd terms (lowest level)				Even terms (lowest level)			
	%				%		

Pr I

	%						
f^{3}s^2	97	$^4I_{9/2}$	0.000	f^{2}ds^2	$^4I_{9/2}^*$		4.432
f^{3}ds	95	$^6L_{11/2}$	8.080	f^{2}d^{2}s	$^6L_{11/2}$		6.714
f^{2}dsp		$^6K_{9/2}$	18.126	f^{3}sp	$^6K_{9/2}$?		13.433
f^{2}s^{2}p		$^4I_{9/2}$?	19.340	f^{2}d^3	($^6L_{11/2}$)		(18.5)±2
f^{3}d^2		($^6M_{13/2}$)	(21.)	f^{3}dp	$^6M_{13/2}$?		22.785
f^{2}d^{2}p		$^6M_{13/2}$?	21.999	f^{2}sp^2	($^6I_{7/2}$)		(38.)±4
f^{3}p^2		($^6K_{9/2}$)	(36.)±3	f^{4}s	($^6I_{7/2}$)		(40.)±6
fd^{2}s^2		($^4H_{7/2}$)*	(51.)±5				

estimates with uncertainty of less than 1 x 10^3 cm^{-1}. When the uncertainty of the estimated value is 1 x 10^3 cm^{-1} or over, some estimate of the uncertainty is given in the tables.

TABLE III (Cont'd): Praseodymium, 10^3 cm^{-1}

Odd terms (lowest level)				Even terms (lowest level)		
	%				%	

Pr II (Pr$^+$)

f^3s	97	$^5I_4\,\frac{9}{2},\frac{1}{2}$	0.000	f^2d^2	$^5L_{6.}$	5.855
f^3d	94	$^5L_6$	3.893	f^2ds	$^5I_4^*$?	7.832
f^2dp		$^5K_5^*$?	30.845	f^2s^2	$^3H_4$	(14.5)±2
f^2sp		$(^5I_4)$	(39.5)±2	f^3p	$^5K_5$	22.675
fd^2s		$(^5H_3)^*$	(55.)±4	f^4	$^5I_4$	(32.)±2
fd^3		$(^5I_4)$	(57.)±4	f^2p^2	$(^5I_4)$	(68.)±2
fds^2		$(^1G_4)^*$	(67.)±3	fd^2p	$(^5K_5)$	(83.)±6
				$fdsp$	$(^3H_4)^*$	(91.)±4

Pr III (Pr^{++})

f^3	97	$^4I_{9/2}$	0.000	f^2d	52	$^2H_{9/2}^*$	12.847
f^2p	82	$^4H_{7/2}^*\,4,\frac{1}{2}$	58.158	f^2s	97	$^4H_{7/2}\,4,\frac{1}{2}$	28.399
fd^2	43	$^4H_{7/2}^*$	60.520	fdp		$(^4H_{9/2})^*$	(114.)±3
fds	62	$^4F_{3/2}^*$	84.410	fsp		$(^4F_{5/2})$	(140.)±5
fs^2		$^2F_{5/2}$	(110.)±5				
fp^2		$(^4G_{5/2})$	(175.)±9				

Pr IV (Pr^{+++})

fd	70	$^1G_4^*\,\frac{5}{2},\frac{3}{2}$	61.171	f^2	97	$^3H_4$	0.000
fs	100	$^3F_2\,\frac{5}{2},\frac{1}{2}$	100.258	fp	98	$^3G_3\,\frac{5}{2},\frac{1}{2}$	136.851
				d^2		$^3F_2$?	139.712

The coupling cases vary considerably among the electronic configurations, but it has been the custom to use LS nomenclature to
designate the levels. When other coupling has been demonstrated
or is expected, JJ or JL designations are also listed. For many

TABLE IV: Neodymium, 10^3 cm^{-1}

Odd terms (lowest level)				Even terms (lowest level)			
	%				%		
Nd I							
f^3ds^2	92	$^5L_6$	6.764	f^4s^2	98	$^5I_4$	0.000
f^3d^2s	97	$^7M_6$	8.800	f^4ds	99	$^7L_5$	8.475
f^4sp		$^7K_4$	13.673	f^3dsp		$^7M_6$	20.272
f^3d^3		$(^7M_6)$	$(22.)\pm2$	f^3s^2p		$(^5K_5)$	(21.4)
f^4dp		$^7M_6$	23.554	f^4d^2		$^7M_6$	21.890
f^5s		$(^7H_2)$	$(40.)\pm5$	f^3d^2p		$^7N_7$	24.856
f^3sp^2		$(^7K_4)$	$(41.)\pm3$	f^4p^2		$(^7K_4)$	$(37.)\pm3$
Nd II (Nd$^+$)							
f^3d^2		$^6M_{13/2}$	8.009	f^4s	98	$^6I_{7/2}$ $4,\frac{1}{2}$	0.000
f^3ds		$^6L_{11/2}$	10.054	f^4d	96	$^6L_{11/2}$	4.438
f^3s^2		$^4I_{9/2}$	$(16.)\pm1$	f^3dp		$^6M_{13/2}$	33.563
f^4p		$^6K_{9/2}$	23.230	f^3sp		$(^6K_{9/2})$	$(42.)\pm1$
f^5		$^6H_{5/2}$	$(33.)\pm3$	f^2d^2s		$(^6L_{11/2})$	$(67.)\pm9$
f^3p^2		$(^6K_{9/2})$	$(71.)\pm4$	f^2d^3		$(^6L_{11/2})$	$(69.)\pm9$
f^2d^2p		$(^6M_{13/2})$	$(95.)\pm9$	f^2ds^2		$(^4K_{11/2})$	$(78.)\pm5$
Nd III (Nd^{++})							
f^3d	88	$^5K_5{}^*$	15.262	f^4	98	$^5I_4$	0.000
f^3s		$(^5I_4)$ $(\frac{9}{2},\frac{1}{2})$	$(30.)\pm2$	f^3p		$(^5K_5)$	$(61.)\pm2$
f^2dp		$(^5L_6)$	$(126.)\pm4$	f^2d^2		$(^5L_6)$	$(71.)\pm5$
f^2sp		$(^5I_4)$	$(151.)\pm7$	f^2ds		$(^5I_4)^*$	$(95.)\pm5$
				f^2s^2		$^3H_4$	$(121.)\pm7$
				f^2p^2		$(^5I_4)$	$(186.)\pm9$

TABLE IV (Cont'd): Neodymium, 10^3 cm^{-1}

Odd terms (lowest level)			Even terms (lowest level)		
	%		Nd IV (Nd^{+++})	%	
f^3	97 $^4I_{9/2}$	0.000	f^2d	$(^4H_{9/2})^*$	(70.)±1
f^2p	$(^4I_{9/2})$ $(4,\frac{1}{2})$(146.)±2		f^2s	$^4H_{7/2}$ $4,\frac{1}{2}$	(100.)±5
fd^2	$(^4H_{7/2})^*$ (158.)±5				

TABLE V: Promethium, 10^3 cm^{-1}

Odd terms (lowest level)			Even terms (lowest level)		
	%		Pm I	%	
f^5s^2	96 $^6H_{5/2}$	0.0	f^4ds^2	$(^6L_{11/2})$	(8.)±1
f^5ds	$(^8K_{7/2})$	(9.4)	f^4d^2s	$(^8M_{11/2})$	(10.5)±1
f^4dsp	$(^8M_{11/2})$	(21.6)±1	f^5sp	$(^8I_{5/2})$	(13.7)
f^4s^2p	$(^6K_{9/2})$	(22.3)±1	f^4d^3	$(^8M_{11/2})$	(24.6)±3
f^5d^2	$(^8L_{9/2})$	(24.)	f^5dp	$(^8L_{9/2})$	(24.7)
f^4d^2p	$(^8N_{13/2})$	(27.5)±1	f^6s	$(^8F_{1/2})$	(34.)±4
f^5p^2	$(^8I_{5/2})$	(38.)±2	f^4sp^2	$(^8K_{7/2})$	(43.)±3

configurations, Hund's rule is expected to hold and the J value given corresponds to Hund's rule. When the levels given in the tables do not correspond to Hund's rule, the symbol is followed by an asterisk. When an experimental value is designated by an asterisk, the value tabulated is the lowest observed level of the configuration, and it is possible that additional observations might find a lower level of the configuration. In general, an attempt was made to estimate the J value corresponding to the lowest expected level. If the lowest observed level does not correspond to the expected J value, it was replaced by an estimated extrapolation to the lowest level if the discrepancy amounts to more than several cm^{-1}. In some instances, the data are too incomplete to make an accurate extrapolation or there may even be doubt about the assignment. In these instances, the level symbol is followed by a question mark. The J value is indicated with the LS designation and is not repeated for the JJ or JL designation.

TABLE V (Cont'd): Promethium, 10^3 cm^{-1}

Odd terms (lowest level)			Even terms (lowest level)		
%				%	

Pm II (Pm$^+$)

f^5s	100	$^7H_2 \frac{5}{2},\frac{1}{2}$	0.000	f^4ds	$(^7L_5)$	$(11.5)\pm2$
f^5d		$^7K_4$	5.332	f^4d^2	$(^7M_6)$	$(12.)\pm3$
f^4dp		$(^7M_6)$	$(36.5)\pm3$	f^4s^2	$^5I_4$	$(17.)\pm3$
f^4sp		$(^7K_4)$	$(42.)\pm3$	f^5p	$(^7I_3)$	(23.4)
f^3d^2s		$(^7M_6)$	$(70.)\pm9$	f^6	$^7F_0$	$(26.)\pm2$
f^3d^3		$(^7M_6)$	$(72.)\pm9$	f^4p^2	$(^7K_4)$	$(71.)\pm4$
f^3ds^2		$(^5L_6)$	$(82.)\pm7$	f^3d^2p	$(^7N_7)$	$(97.)\pm9$

Pm III (Pm^{++})

f^5	$^6H_{5/2}$	0.000	f^4d	$(^6L_{11/2})$	$(16.)\pm2$
f^4p	$(^6K_{9/2})$	$(61.)\pm3$	f^4s	$(^6I_{7/2})$	$(29.5)\pm3$
f^3d^2	$(^6M_{13/2})$	$(74.)\pm6$	f^3dp	$(^6M_{13/2})$	$(130.)\pm5$
f^3ds	$(^6L_{11/2})$	$(97.5)\pm5$	f^3sp	$(^6K_{9/2})$	$(153.)\pm7$
f^3s^2	$^4I_{9/2}$	$(123.)\pm7$			
f^3p^2	$(^6K_{9/2})$	$(190.)\pm9$			

Pm IV (Pm^{+++})

f^3d		$(^5L_6)$	$(73.)\pm2$	f^4	97	$^5I_4$	0.000
f^3s		$(^5I_4) (\frac{9}{2},\frac{1}{2})$	$(110.)\pm5$	f^3p		$(^5K_5)$	$(149.)\pm3$
				f^2d^2		$(^6L_6)$	$(170.)\pm6$

As the odd and even levels are tabulated in separate columns, the superscript ^{o} is omitted for the odd terms.

It is most fortunate that Martin, Zalubas, and Hagan (38) have provided an excellent updated summary of the atomic energy levels of the lanthanides. The experimental values listed in Tables I–XV were taken from their tabulation with a few exceptions noted below. Where experimental values were not available, estimated

TABLE VI: Samarium, 10^3 cm^{-1}

Odd terms (lowest level)				Even terms (lowest level)			
	%				%		
Sm I							
f^6sp	94	$^9G_0$	13.796	f^6s^2	94	$^7F_0$	0.000
f^5ds^2		$^7K_4$	(15.5)	f^6ds	99	$^9H_1$	10.801
f^5d^2s		$(^9L_4)$	(19.3)	f^6d^2		$(^9I_2)$	(26.)±2
f^7s		$(^9S_4)$	(25.)±3	f^5dsp		$(^9L_4)$	(29.2)
f^6dp		$^9I_2$?	25.809	f^5s^2p		$(^7I_3)$	(29.5)±1
f^7d		$(^9D_2)$	(33.)±3	f^5d^2p		$(^9M_5)$	(36.5)±1
f^5d^3		$(^9L_4)$	(34.)±2	f^6p^2		$(^9G_0)$	(39.)±2
Sm II (Sm$^+$)							
f^7		$^8S_{7/2}$	18.289	f^6s	99	$^8F_{1/2}$	0.000
f^5ds		$(^8K_{7/2})$	(19.0)	f^6d	99	$^8H_{3/2}$	7.135
f^5d^2		$(^8L_{9/2})$	(20.5)±1	f^5dp		$(^8L_{9/2})$	(44.)±2
f^6p		$(^8G_{1/2})$? $(0,\frac{1}{2})$	23.352	f^5sp		$(^8I_{5/2})$	(49.)±1
f^5s^2		$^6H_{5/2}$	(24.)±2	f^4d^2s		$(^8M_{11/2})$	(77.)±9
f^5p^2		$(^8I_{5/2})$	(79.)±3	f^4d^3		$(^8M_{11/2})$	(80.)±9
				f^4ds^2		$(^6L_{11/2})$	(89.)±7
Sm III (Sm^{++})							
f^5d		$^7K_4$	24.5	f^6	94	$^7F_0$	0.000
f^5s		$(^7H_2)$	(37.5)±2	f^5p		$(^7I_3)$	(67.5)±2
f^4dp		$(^7M_6)$	(137.)±6	f^4d^2		$(^7M_6)$	(83.)±5
f^4sp		$(^7K_4)$	(160.)±7	f^4ds		$(^7L_5)$	(106.)±5
				f^4s^2		$^5I_4$	(130.)±7
				f^4p^2		$(^7K_4)$	(195.)±9

TABLE VI (Cont'd): Samarium, 10^3 cm^{-1}

Odd terms (lowest level)			Even terms (lowest level)		
%		Sm IV (Sm^{+++})		%	
f^5	96	$^6H_{5/2}$	0.000	f^4d	$(^6L_{11/2})$	(73.)±2
f^4p		$(^6K_{9/2})$	(149.)±5	f^4s	$(^6I_{7/2})$	(102.)±6
f^3d^2		$(^6M_{13/2})$	(172.)±8			

TABLE VII: Europium, 10^3 cm^{-1}

Odd terms (lowest level)			Even terms (lowest level)		
%		Eu I		%	
f^7s^2	98	$^8S_{7/2}$	0.000	f^7sp	92	$^{10}P_{7/2}$	14.068
f^7ds	94	$^{10}D_{5/2}$	12.924	f^6ds^2		$(^8H_{3/2})$	(25.1)
f^7d^2		$(^{10}F_{3/2})$	(28.)±3	f^7dp	94	$^{10}F_{3/2}$	28.520
f^6s^2p		$(^8G_{1/2})$	(39.)±1	f^6d^2s		$(^{10}I_{3/2})$	(30.8)
f^6dsp		$(^{10}I_{3/2})$	(39.)	f^6d^3		$(^{10}I_{3/2})$	(45.5)±1
f^7p^2		$^{10}P_{7/2}$	40.298				
f^6d^2p		$(^{10}K_{5/2})$	(48.)±1				

values were calculated by the methods described with the publication of the earlier version of these tables. (5) The accumulation of data in the last decade has not materially altered the values tabulated earlier. The only major changes have been in the values given for the heavier trivalent ions and for some of the related levels of the divalent ions where the experimental data available earlier were seriously in error resulting in a shift of the values by the Racah plots. (44)

In some instances, several configurations are so intermixed that a clear assignment can not be made. For example, the first d^2p level of LaI at 16857 cm^{-1} is designated (38) as 28% d^2p $^4G^o_{5/2}$ and 24% fs^2 $^2F^o_{5/2}$. It was bypassed to take the $^4G^o_{5/2}$ level at 17947 cm^{-1}, which is designated as 42% d^2p, as the lowest level of the d^2p configuration. In such circumstances, the designation of lowest level may have an uncertainty up to 1000 cm^{-1}. The percent values (38) in the tables correspond to the percent contribution

TABLE VII (Cont'd): Europium, 10^3 cm^{-1}

Odd terms (lowest level)			Even terms (lowest level)		
	%			%	

Eu II (Eu$^+$)

	%				%		
f^7s		$^9S_4$	0.000	f^7p	87	$^9P_{3\,\frac{7}{2},\frac{1}{2}}$	23.774
f^7d	100	$^9D_2$	9.923	f^6ds		$^9H_1$	$(28.5)^{\pm 1}$
f^6dp		$(^9I_2)$	$(54.)^{\pm 2}$	f^6d^2		$(^9I_2)$	$(31.5)^{\pm 2}$
f^6sp		$(^9G_0)$	$(57.)^{\pm 2}$	f^6s^2		$^7F_0$	$(32.5)^{\pm 2}$
f^5d^2s		$(^9L_4)$	$(96.)^{\pm 9}$	f^8		$^7F_6$?	55.652
f^5d^3		$(^9L_4)$	$(98.)^{\pm 9}$	f^6p^2		$(^9G_0)$	$(87.)^{\pm 3}$
f^5ds^2		$(^7K_4)$	$(107.)^{\pm 8}$				

Eu III (Eu^{++})

	%				%		
f^7	97	$^8S_{7/2}$	0.000	f^6d	97	$^8H_{3/2}$	33.856
f^6p	88	$^8G_{1/2}\,0,\frac{1}{2}$	78.982	f^6s	92	$^8F_{1/2}$	46.096
f^5d^2		$(^8L_{9/2})$	$(101.)^{\pm 6}$	f^5dp		$(^8L_{9/2})$	$(156.)^{\pm 8}$
f^5ds		$(^8K_{7/2})$	$(123.)^{\pm 7}$	f^5sp		$(^8I_{5/2})$	$(180.)^{\pm 9}$
f^5s^2		$^6H_{5/2}$	$(147.)^{\pm 8}$				

Eu IV (Eu^{+++})

	%				%		
f^5d		$(^7K_4)$	$(82.)^{\pm 1}$	f^6	93	$^7F_0$	0.000
f^5s		$(^7H_2)$	$(117.)^{\pm 5}$	f^5p		$(^7I_3)$	$(158.)^{\pm 5}$
				f^4d^2		$(^7M_6)$	$(182.)^{\pm 8}$

of the designated LS or JJ symbol, and the total contribution from a given electronic configuration, which can be the sum of several LS designations, can be considerably larger than the tabulated percent value. The lowest CeI fd^3 assignment is given (38) as $^5I_7^o$? at 15022 cm^{-1}. Since the assignment is in qestion, the J = 4 level at 12351 cm^{-1} which was previously selected (5) as the lowest fd^3 level is given in the table. The Nd II f^{3}ds, and f^{3}dp values are revisions by Blaise (48) of the previously tabulated values (38). For SmI, the lowest ds^2 level is given(38) as $^7H_2^o$ at 18076 cm^{-1}. As Hund's Rule is expected to be obeyed here,

TABLE VIII: Gadolinium, 10^3 cm^{-1}

Odd terms (lowest level)				Even terms (lowest level)			
	%				%		
			Gd I				
f^7ds^2	95	$^9D_2$	0.000	f^8s^2		$^7F_6$	10.947
f^7d^2s	98	$^{11}F_2$	6.378	f^7s^2p	70	$^9P_3$	13.434
f^7d^3	98	$^{11}F_2$	22.429	f^7dsp	92	$^{11}F_2$	14.036
f^8sp		$^9D_6^*$ (6,0)	25.658	f^8ds		$^9G_7^*$	24.255
f^7sp^2		$^{11}P_4$	35.561	f^7d^2p	66	$^{11}G_1$	25.069
f^8dp		$(^9H_4)^*$	(41.)±1	f^8d^2		$(^9G_7)^*$	(41.)±3
				f^8p^2		$(^9F_6)^*$	(51.5)±1
			Gd II (Gd$^+$)				
f^7ds	96	$^{10}D_{5/2}$	0.000	f^8s	100	$^8F_{13/2}$	7.992
f^7s^2	93	$^8S_{7/2}$	3.444	f^8d	98	$^8G_{15/2}^*$	18.367
f^7d^2	87	$^{10}F_{3/2}$	4.027	f^7sp	80	$^{10}P_{7/2}$	25.669
f^8p	96	$^8D_{11/2}^*$ $6,\tfrac{1}{2}$	32.595	f^7dp	83	$^{10}F_{3/2}$	25.960
f^7p^2		$(^{10}P_{7/2})$	(56.)±3	f^6d^2s		$(^{10}I_{3/2})$	(77.)±9
f^9		$^6H_{15/2}$	(56.)±2	f^6d^3		$(^{10}I_{3/2})$	(80.)±9
f^6d^2p		$(^{10}K_{5/2})$	(107.)±9	f^6ds^2		$(^8H_{3/2})$	(88.)±7
			Gd III (Gd^{++})				
f^7d	99	$^9D_2$	0.000	f^8		$^7F_6$	2.381
f^7s		$^9S_4$	9.195	f^7p	98	$^9F_3$ $\tfrac{7}{2},\tfrac{1}{2}$	43.020
f^6dp		$(^9I_2)$	(131.)±7	f^6d^2		$(^9I_2)$	(74.)±7
f^6sp		$(^9G_0)$	(155.)±7	f^6ds		$(^9H_1)$	(95.5)±5
				f^6s^2		$^7F_0$	(119.)±7

TABLE VIII (Cont'd): Gadolinium, 10^3 cm^{-1}

Odd terms (lowest level)				Even terms (lowest level)		
%			Gd IV (Gd^{+++})	%		
f^7	97	$^8S_{7/2}$	0.000	f^6d	$^8H_{3/2}$	$(92.)\pm1$
f^6p		$^8G_{1/2}$ $(0,\frac{1}{2})$	$(168.)\pm3$	f^6s	$^8F_{1/2}$	$(126.)\pm4$
f^5d^2		$(^8L_{9/2})$	$(200.)\pm9$			

TABLE IX: Terbium, 10^3 cm^{-1}

Odd terms (lowest level)				Even terms (lowest level)		
%			Tb I	%		
f^9s^2	95	$^6H_{15/2}$	0.000	f^8ds^2 65	$^8G_{13/2}$*	0.286
f^8s^2p		$(^8F_{11/2})$* 6,$\frac{1}{2}$	13.616	f^8d^2s	$^{10}G_{15/2}$*	8.190
f^8dsp		$^{10}H_{13/2}$* ?	14.999	f^9sp	$(^8G_{15/2})$* $\frac{15}{2}$,0 ?	14.888
f^9ds		$(^8G_{15/2})$* ?	15.099	f^8d^3	$(^{10}H_{15/2})$*	$(25.)\pm1$
f^8d^2p		$(^{10}I_{13/2})$* ?	27.399	f^9dp	$(^8G_{15/2})$*	$(32.)\pm1$
f^9d^2		$(^8K_{17/2})$*	$(32.)\pm3$	f^8sp^2	$(^{10}F_{11/2})$*$(6,\frac{1}{2})$	(36.)
f^9p^2		$(^8H_{15/2})$*$(\frac{15}{2},0)$	$(40.8)\pm1$	$f^{10}s$	$(^6I_{17/2})$ $(8,\frac{1}{2})$	$(42.)\pm9$
				$f^7d^2s^2$	$(^{10}F_{3/2})$	$(45.)\pm5$

the value listed in the table is the estimate (5) of 15500 cm^{-1} for ds^2 ^{7}K$_4^O$ which is expected to be confirmed by more complete data to within 500 cm^{-1}.   Similarly, for SmII, the values tabulated (38) for f^{5}ds levels do not extend to the expected Hund's Rule level and an estimated value (5) is given here. On the other hand, the levels listed (38) as f^{6}p down to 21251 cm^{-1} extend to too low an energy and are believed to be misassigned. The expected Hund's Rule level ^{8}G$_{1/2}$ is assigned to the level at 23352 cm^{-1}. For SmIII, the assigned levels (38) do not extend to the expected ^{7}K$_4^O$ level which is estimated (5) to be 24500 cm^{-1}. For EuI, the value given for f^{6}d^{2}s ^{10}H$_{3/2}$ was obtained by extrapolation from the ^{10}I levels of higher J. The value given for f^{6}ds^2

TABLE IX (Cont'd): Terbium, 10^3 cm^{-1}

Odd terms (lowest level)				Even terms (lowest level)		
	%				%	

Tb II (Tb$^+$)

f^9s	94	$^7H_8$ $\frac{15}{2},\frac{1}{2}$	0.000	f^8ds	64 $^9G_7{}^*$	3.235
f^9d	79	$^7H_8{}^*$	11.262	f^8s^2	$^7F_6$	5.898
f^8sp		$(^9F_6)^*$ (6,0)	(28.)±2	f^8d^2	$^9G_7{}^*$?	8.904
f^8dp		$(^9H_6)^*$?	29.913	f^9p	$(^7H_8)$ $\frac{15}{2},\frac{1}{2}$?	25.138
f^7d^2s		$(^{11}F_2)$	(43.)±9	f^{10}	$^5I_8$	(40.)±4
f^7d^3		$(^{11}F_2)$	(48.)±9	f^8p^2	$(^9F_6)^*$ (6,0)	(60.)±5
f^7ds^2		$(^9D_2)$	(53.)±7	f^7d^2p	$(^{11}G_1)$	(72.)±9
				f^7dsp	$(^{11}F_2)$	(78.)±8
				f^7s^2p	$(^9P_3)$	(100.)±9

Tb III (Tb^{++})

f^9	94	$^6H_{15/2}$	0.000	f^8d	73 $^8G_{13/2}{}^*$	8.972
f^7d^2		$(^{10}F_{3/2})$	(43.)±9	f^8s	99 $^8F_{13/2}$	17.676
f^8p	96	$(^8D_{11/2})^*6,\frac{1}{2}$	52.039	f^7dp	$(^{10}F_{3/2})$	(104.)±7
f^7ds		$(^{10}D_{5/2})$	(63.)±5	f^7sp	$(^{10}P_{7/2})$	(125.)±9
f^7s^2		$^8S_{7/2}$	(85.)±7			
f^7p^2		$(^{10}P_{7/2})$	(160.)±9			

Tb IV (Tb^{+++})

f^7d	99	$^9D_2$	51.404	f^8	96 $^7F_6$	0.000
f^7s		$^9S_4$ $\frac{7}{2},\frac{1}{2}$	84.955	f^7p	98 $^9P_3$ $\frac{7}{2},\frac{1}{2}$	127.839
				f^6d^2	$(^9I_2)$	(172.)±9

^{8}H$_{3/2}$ is an estimate (5) which was used in place of the observed ^{8}D value. Similarly for EuII, an estimated value (5) for f^6ds ^{9}H1 was used in place of the observed ^{9}D value. For GdII f^8s ^{8}F$_{13/2}$,

TABLE X: Dysprosium, 10^3 cm^{-1}

Odd terms (lowest level)				Even terms (lowest level)			
	%				%		
			Dy I				
f^9ds^2 81		$^7H_8{}^*$	7.566	$f^{10}s^2$ 94		$^5I_8$	0.000
$f^{10}sp$ 58		$^7H_8$ 8,0	15.567	$f^{10}ds$ 90		$(^7K_9)^*$ $^3[8]$	17.515
f^9d^2s 48		$^9G_8{}^*$	18.473	f^9s^2p 55		$(^7H_7)^*$ $\frac{15}{2},\frac{1}{2}$	20.614
$f^{10}dp$		$(^7L_7)^*$?	34.695	f^9dsp		$(^9K_8)^*$ 8,0 ?	23.031
f^9d^3		$(^9H_9)^*$	(35.)$\pm$2	$f^{10}d^2$		$(^7L_{10})^*$	(35.)$\pm$4
f^9sp^2		$(^9H_7)^*$ $(\frac{15}{2},\frac{1}{2})$	(43.)	f^9d^2p		$(^9L_{10})^*$	(37.)$\pm$2
$f^{11}s$		$(^5I_8)$	(49.)$\pm$9	$f^{10}p^2$		$(^7I_8)^*$ (8,0)	(41.)$\pm$1
			Dy II (Dy$^+$)				
f^9ds 83		$^8H_{17/2}{}^*$	10.594	$f^{10}s$ 95		$^6I_{17/2}$ 8,$\frac{1}{2}$	0.000
f^9s^2 90		$^6H_{15/2}$	12.336	$f^{10}d$ 78		$^6I_{17/2}{}^*$	14.846
f^9d^2 26		$^8G_{13/2}{}^*$	19.492	f^9sp 45		$^8G_{15/2}{}^*$ $\frac{15}{2}$,0	36.212
$f^{10}p$ 22		$^6H_{15/2}{}^*$ (8,$\frac{1}{2}$)	25.192	f^9dp 20		$^8G_{15/2}{}^*$	37.817
f^{11}		$^4I_{15/2}$	(44.)$\pm$2	f^8d^2s		$(^{10}G_{15/2})^*$	(64.)$\pm$9
f^9p^2		$(^8H_{15/2})^*$ $(\frac{15}{2},0)$	(67.)$\pm$3	f^8d^3		$(^{10}H_{15/2})^*$	(70.)$\pm$9
f^8d^2p		$(^{10}I_{9/2})^*$	(93.)$\pm$9	f^8ds^2		$(^8G_{13/2})^*$	(73.)$\pm$7
f^8dsp		$(^{10}H_{9/2})^*$	(98.)$\pm$8				
			Dy III (Dy^{++})				
f^9d		$(^7H_8)^*$	(17.)	f^{10}		$^5I_8$	0.000
f^9s		$(^7H_8)$ $(\frac{15}{2},\frac{1}{2})$	(24.)$\pm$2	f^9p		$(^7H_7)^*$ $(\frac{15}{2},\frac{1}{2})$	(60.)$\pm$2
f^8dp		$(^9H_4)^*$	(125.)$\pm$6	f^8d^2		$(^9G_7)^*$	(64.)$\pm$9
f^8sp		$(^9F_6)^*$ (6,0)	(137.)$\pm$9	f^8ds		$(^9G_7)^*$	(83.)$\pm$5
				f^8s^2		$(^7F_6)$	(104.)$\pm$6
				f^8p^2		$(^9F_6)^*$	(180.)$\pm$9

TABLE X (Cont'd): Disprosium, 10^3 cm^{-1}

Odd terms (lowest level)			Even terms (lowest level)		
	%		Dy IV (Dy^{+++})	%	
f^9	94 $^6H_{15/2}$	0.000	f^8d	$(^8G_{15/2})^*$	(65.)±3
f^8p	$(^8D_{11/2})^*$ $(6,\frac{1}{2})$	(140.)±6	f^8s	$(^8F_{13/2})$	(97.)±5
f^7d^2	$(^{10}F_{3/2})$	(148.)±9			

TABLE XI: Holmium, 10^3 cm^{-1}

Odd terms (lowest level)			Even terms (lowest level)		
	%		Ho I	%	
$f^{11}s^2$	97 $^4I_{15/2}$	0.000	$f^{10}ds^2$	73 $^6I_{17/2}^*$ $8,\frac{3}{2}$	8.379
$f^{10}s^2p$	57 $^6H_{15/2}^*$ $8,\frac{1}{2}$	18.572	$f^{11}sp$	38 $^6H_{15/2}^*$ $\frac{15}{2},0$	15.855
$f^{11}ds$	82 $^6G_{13/2}^*$	18.867	$f^{10}d^2s$	58 $^8G_{15/2}^*$	20.167
$f^{10}dsp$	$(^8L_{15/2})^*$	24.112	$f^{11}dp$	$(^6I_{11/2})^*$	(36.)±1
$f^{10}d^2p$	$(^8M_{21/2})^*$	(39.)±3	$f^{10}d^3$	$(^8L_{19/2})^*$	(39.)±12
$f^{11}d^2$	$(^6L_{19/2})^*$	(40.)±5	$f^{10}sp^2$	$(^8I_{15/2})^*(8,\frac{1}{2})$	(43.5)±2
$f^{11}p^2$	$(^6I_{15/2})^*(\frac{15}{2},0)$	(41.2)	$f^{12}s$	$(^4H_{13/2})$	(47.)±9
			Ho II (Ho$^+$)		
$f^{11}s$	97 $^5I_8$ $\frac{15}{2},\frac{1}{2}$	0.000	$f^{10}s^2$	$(^5I_8)$	(10.)±2
$f^{11}d$	80 $^5G_6^*$ $\frac{15}{2},\frac{3}{2}$	16.282	$f^{10}ds$	$(^7K_9)^*$	(11.5)
$f^{10}sp$	$(^7I_8)^*$ (8,0)	(34.)±2	$f^{10}d^2$	$(^7L_{10})^*$	(22.)±1
$f^{10}dp$	$(^7L_{10})^*$	(39.)±1	$f^{11}p$	95 $(^5I_7)^*$ $\frac{15}{2},\frac{1}{2}$	26.234
f^9d^2s	$(^9G_8)^*$	(72.)±9	f^{12}	$^3H_6$	(46.)±4
f^9ds^2	$(^7H_8)^*$	(79.)±7	$f^{10}p^2$	$(^7I_8)^*$ (8,0)	(66.)±5
f^9d^3	$(^9K_{10})^*$	(79.)±7	f^9dsp	$(^9K_8)^*$	(103.)±8
			f^9d^2p	$(^9L_8)^*$	(103.)±9

TABLE XI (Cont'd): Holmium, 10^3 cm^{-1}

Odd terms (lowest level)				Even terms (lowest level)		
	%				%	

Ho III (Ho^{++})

f^{11}	97	$^4I_{15/2}$	0.000	$f^{10}d$	83 $^6H_{15/2}{}^*$ $8,\frac{3}{2}$	18.033
$f^{10}p$	93	$^6H_{15/2}{}^*$ $8,\frac{1}{2}$	57.498	$f^{10}s$	94 $^6I_{17/2}$ $8,\frac{1}{2}$	21.824
f^9d^2		$(^8K_{17/2})^*$	(72.)±8	f^9dp	$(^8G_{15/2})^*$	(130.)±8
f^9ds		$(^8H_{17/2})^*$	(89.)±5	f^9sp	$(^8H_{15/2})^*(\frac{15}{2},0)$	(143.)±9
f^9s^2		$^6H_{15/2}$	(110.)±8			
f^9p^2		$(^8H_{15/2})^*(\frac{15}{2},0)$	(185.)±9			

Ho IV (Ho^{+++})

f^9d	$(^7H_8)^*$	(73.)±3	f^{10}	93	$^5I_8$	0.000
f^9s	$(^7H_8)$ $(\frac{15}{2},\frac{1}{2})$	(104.)±5	f^9p		$(^7H_7)^*$ $(\frac{15}{2},\frac{1}{2})$	(147.)±5
			f^8d^2		$(^9G_7)^*$	(170.)±9

TABLE XII: Erbium, 10^3 cm^{-1}

Odd terms (lowest level)				Even terms (lowest level)		
	%				%	

Er I

$f^{11}ds^2$	78	$^5G_6{}^*$ $\frac{15}{2},\frac{3}{2}$	7.177	$f^{12}s^2$	99	$^3H_6$	0.000
$f^{12}sp$	71	$^5G_6{}^*$ $6,0$	16.321	$f^{11}s^2p$	95	$(^5I_7)^*$ $\frac{15}{2},\frac{1}{2}$	16.465
$f^{11}d^2s$	85	$^7F_6{}^*$	20.166	$f^{12}ds$	92	$(^5I_5)^*$ $^3[4]$	19.362
$f^{12}dp$		$(^5I_4)^*$ $(6,2)$	(37.)±1	$f^{11}dsp$		$(^7L_6)^*$	22.980
$f^{11}d^3$		$(^7K_7)^*$	(39.)±3	$f^{11}d^2p$		$(^7K_6)^*$	(38.5)±2
$f^{11}sp^2$		$(^7I_7)^*$ $(\frac{15}{2},\frac{1}{2})$	(41.5)±2	$f^{12}p^2$		$(^5H_6)^*$ $(6,0)$	(41.5)
$f^{13}s$		$(^3F_4)$ $(\frac{7}{2},\frac{1}{2})$	(42.)±9	$f^{12}d^2$		$(^5I_4)^*(6,2)$	(43.)±6

the percent value was from Goldschmidt (45). For TbII, an estimate
(5) is given for the f^8sp $^9F_6$ level at 28×10^3 cm^{-1} in place of

TABLE XII (Cont'd): Erbium, 10^3 cm^{-1}

Odd terms (lowest level)			Even terms (lowest level)		
%			%		
		Er II (Er$^+$)			
$f^{11}s^2$	$^4I_{15/2}$	6.825	$f^{12}s$	99 $^4H_{13/2}$ 6,$\frac{1}{2}$	0.000
$f^{11}ds$	$(^6G_{13/2})^*$	10.667	$f^{12}d$	83 $(^4F_{9/2})^*$ 6,$\frac{3}{2}$	16.553
$f^{11}d^2$	$(^6K_{11/2})^*$?	21.820	$f^{11}sp$	$(^6I_{15/2})^*$ $\frac{15}{2}$,0	32.049
$f^{12}p$	$(^4G_{11/2})^*$? (6,$\frac{1}{2}$)	25.592	$f^{11}dp$	$(^6K_{11/2})^*$ ($\frac{15}{2}$,2)	38.439
f^{13}	$^2F_{7/2}$	(43.)±3	$f^{10}d^2s$	$(^8G_{15/2})^*$	(72.)±9
$f^{11}p^2$	$(^6I_{15/2})^*$($\frac{15}{2}$,0)	(65.)±2	$f^{10}ds^2$	$(^6I_{17/2})^*$(8,$\frac{3}{2}$)	(77.)±5
$f^{10}dsp$	$(^8L_{13/2})^*$	(102.)±6	$f^{10}d^3$	$(^8L_{21/2})^*$	(80.)±9
$f^{10}d^2p$	$(^8M_{9/2})^*$	(102.)±9			
		Er III (Er^{++})			
$f^{11}d$	76 $^5G_6^*$ $\frac{15}{2}$,$\frac{3}{2}$	16.976	f^{12}	$^3H_6$	0.000
$f^{11}s$	97 $^5I_8$ $\frac{15}{2}$,$\frac{1}{2}$	19.316	$f^{11}p$	95 $(^5I_7)^*$ $\frac{15}{2}$,$\frac{1}{2}$	55.547
$f^{10}dp$	$(^7L_8)^*$	(130.)±5	$f^{10}d^2$	$(^7L_{10})^*$	(72.)±5
$f^{10}sp$	$(^7I_8)^*$ (8,0)	(140.)±7	$f^{10}ds$	$(^7K_9)^*$	(88.)±3
			$f^{10}s^2$	$^5I_8$	(105.)±5
			$f^{10}p^2$	$(^7I_8)^*$ (8,0)	(180.)±9
		Er IV (Er^{+++})			
f^{11}	97 $^4I_{15/2}$	0.000	$f^{10}d$	$(^6I_{17/2}^*)$	(75.)±3
$f^{10}p$	$(^6I_{15/2})^*$(8,$\frac{1}{2}$)	(147.)±5	$f^{10}s$	$^6I_{17/2}$ (8,$\frac{1}{2}$)	(102.)±5
f^9d^2	$(^8L_{11/2})^*$($\frac{15}{2}$,2)	(180.)±9			

the lowest tabulated (38) J = 7 level at 29913 cm^{-1}. For YbI, the $f^{13}d^2p$ $^5G2$ at 59377 cm^{-1} is from Nir (45). The recent revisions of YbI by Wyart and Camus (46) have been incorporated.

TABLE XIII: Thulium, 10^3 cm^{-1}

Odd terms (lowest level)	%				Even terms (lowest level)	%			
				Tm I					
$f^{13}s^2$		$^2F_{7/2}$		0.000	$f^{12}ds^2$	86	$(^4I_{9/2})^*$	$6,\frac{3}{2}$	13.120
$f^{13}ds$	97	$(^4G_{5/2})^*$	$3[\frac{3}{2}]$	20.407	$f^{13}sp$	78	$(^4F_{7/2})^*$	$\frac{7}{2},0$	16.742
$f^{12}s^2p$	74	$(^4H_{11/2})^*$	$6,\frac{1}{2}$	22.468	$f^{12}d^2s$		$(^6K_{9/2})^*$		(27.0)
$f^{12}dsp$		$(^6K_{9/2})^*$	$6,\frac{3}{2}$	29.309	$f^{14}s$		$^2S_{1/2}$		(33.)±9
$f^{13}p^2$	85	$(^4F_{7/2})^*$	$\frac{7}{2},0$	41.842	$f^{13}dp$	80	$(^4F_{3/2})^*$	$\frac{7}{2},2$	38.319
$f^{13}d^2$		$(^4G_{5/2})^*$	$(3[\frac{5}{2}])$	(44.)±2	$f^{12}d^3$		$(^6K_{11/2})^*$		(48.)±4
$f^{12}d^2p$		$(^6L_{19/2})^*$		(46.)±3	$f^{12}sp^2$		$(^6H_{11/2})^*$	$(6,\frac{1}{2})$	(47.)±2
				Tm II (Tm$^+$)					
$f^{13}s$	100	$^3F_4$	$\frac{7}{2},\frac{1}{2}$	0.000	$f^{12}s^2$	96	$^3H_6$		12.457
$f^{13}d$	91	$^3P_2^*$	$\frac{7}{2},\frac{3}{2}$	17.625	$f^{12}ds$	64	$(^5K_5)^*$	$6,1$	16.567
$f^{12}sp$		$(^5G_6)^*$	$6,0$	38.225	$f^{13}p$	55	$(^3D_3)^*$	$\frac{7}{2},\frac{1}{2}$	25.980
$f^{12}dp$	84	$(^5I_4)^*$	$6,2$	44.838	$f^{12}d^2$	60	$(^5I_4)^*$	$(6,2)$	30.841
$f^{11}d^2s$		$(^7F_6)^*$		(78.)±9	f^{14}		$^1S_0$		(34.)±4
$f^{11}ds^2$		$(^5G_6)^*$	$(\frac{15}{2},\frac{3}{2})$	(83.)±7	$f^{12}p^2$		$(^5H_6)^*$	$(6,0)$	(71.)±5
$f^{11}d^3$		$(^7L_7)^*$		(87.)±9					
				Tm III (Tm^{++})					
f^{13}		$^2F_{7/2}$		0.000	$f^{12}d$	84	$(^4F_{9/2})^*$	$6,\frac{3}{2}$	22.897
$f^{12}p$	96	$^4G_{11/2}^*$	$6,\frac{1}{2}$	62.064	$f^{12}s$	99	$^4H_{13/2}$	$6,\frac{1}{2}$	25.303
$f^{11}d^2$		$(^6L_{11/2})^*$		(85.)±9	$f^{11}dp$		$(^6L_{11/2})^*$	$(\frac{5}{2},2)$	(136.)±7
$f^{11}ds$		$(^6K_{13/2})^*$		(100.)±5	$f^{11}sp$		$(^6I_{15/2})^*$	$(\frac{15}{2},0)$	(143.)±9
$f^{11}s^2$		$(^4I_{15/2})$		(115.)±7					
$f^{11}p^2$		$(^6I_{15/2})^*$	$(\frac{15}{2},0)$	(185.)±9					

TABLE XIII (Cont'd): Thulium, 10^3 cm^{-1}

Tm IV (Tm^{+++})

Odd terms (lowest level)				Even terms (lowest level)			
	%				%		
$f^{11}d$	$(^5G_6)^*$	$(\frac{15}{2},\frac{3}{2})$	(74.)±3	f^{12}	99	$^3H_6$	0.000
$f^{11}s$	$(^5I_8)$	$(\frac{15}{2},\frac{1}{2})$	(106.)±5	$f^{11}p$	$(^5I_7)^*$ $(\frac{15}{2},\frac{1}{2})$		(147.)±5
				$f^{10}d^2$	$(^7L_{10})^*$		(182.)±9

TABLE XIV: Ytterbium, 10^3 cm^{-1}

Yb I

Odd terms (lowest level)				Even terms (lowest level)			
	%				%		
$f^{14}sp$	100	$^3P_0$	17.288	$f^{14}s^2$	99	$^1S_0$	0.000
$f^{13}ds^2$	88	$^3P_2^*$ $\frac{7}{2},\frac{3}{2}$	23.189	$f^{14}ds$	99	$^3D_1$	24.489
$f^{13}d^2s$		$^3D_3^*$	42.384	$f^{13}s^2p$	89	$^3D_3^*$ $\frac{7}{2},\frac{1}{2}$	32.065
$f^{14}dp$	67	$^3F_2$	42.726	$f^{13}dsp$	61	$^5S_2^*$ $\frac{7}{2},\frac{3}{2}$	39.880
$f^{13}sp^2$	$(^5F_3)^*$ $(\frac{7}{2},\frac{1}{2})$		(57.)±2	$f^{14}p^2$		$^3P_0$	42.437
$f^{13}d^3$	$(^5H_5)^*$		(63.)±6	$f^{14}d^2$		$(^3F_2)$?	47.342
				$f^{13}d^2p$		$^5G_2$	59.377

Yb II (Yb$^+$)

Odd terms (lowest level)				Even terms (lowest level)			
$f^{13}s^2$	98	$^2F_{7/2}$	21.419	$f^{14}s$		$^2S_{1/2}$	0.000
$f^{13}ds$	95	$(^4G_{5/2})^*$ $3[\frac{3}{2}]$	26.759	$f^{14}d$		$^2D_{3/2}$	22.961
$f^{14}p$		$^2P_{1/2}$	27.062	$f^{13}sp$	85	$(^4F_{7/2})^*$ $\frac{7}{2},0$	47.912
$f^{13}d^2$	48	$(^4G_{5/2})^*$ $3[\frac{5}{2}]$	45.013	$f^{13}dp$	76	$(^4F_{3/2})^*$ $\frac{7}{2},2$	55.702
$f^{13}p^2$	$(^4F_{7/2})^*$ $(\frac{7}{2},0)$		(81.)±5	$f^{12}d^2s$	$(^6K_{9/2})^*$		(93.)±9
				$f^{12}ds^2$	$(^4H_{9/2})^*$ $(6,\frac{3}{2})$		(97.)±4
				$f^{12}d^3$	$(^6K_{9/2})^*$		(103.)±9

TABLE XIV (Cont'd): Ytterbium, 10^3 cm^{-1}

Odd terms (lowest level)				Even terms (lowest level)		
%					%	
			Yb III (Yb^{++})			
$f^{13}d$	91	$^3P_2^*$ $\frac{7}{2},\frac{3}{2}$	33.386	f^{14}	$^1S_0$	0.000
$f^{12}s$	100	$^3F_4$ $\frac{7}{2},\frac{1}{2}$	34.656	$f^{13}p$ 98	$^3D_3^*$ $\frac{7}{2},\frac{1}{2}$	72.140
$f^{12}dp$		$(^5I_4)^*$ (6,2)	(152.)±4	$f^{12}d^2$	$(^5I_4)^*$ (6,2)	(103.)±3
$f^{12}sp$		$(^5G_6)^*$ (6,0)	(165.)±6	$f^{12}ds$	$(^5I_5)^*$ (6,1)	(106.)±2
				$f^{12}s^2$	$^3H_6$	(130.)±4
				$f^{12}p^2$	$(^5H_6)^*$ (6,0)	(200.)±9

Yb IV (Yb^{+++})

f^{13}		$^2F_{7/2}$	0.000	$f^{12}d$ 85	$^4F_{9/2}^*$ 6,$\frac{3}{2}$	78.529
$f^{12}p$	97	$^4G_{11/2}^*$ 6,$\frac{1}{2}$	152.589	$f^{12}s$ 99	$^4H_{13/2}$ 6,$\frac{1}{2}$	105.979
$f^{11}d^2$		$(^6K_{11/2})^*$	(187.)±9			

TABLE XV: Lutetium, 10^3 cm^{-1}

Odd terms (lowest level)			Even terms (lowest level)		
%				%	
		Lu I			
$f^{14}s^2p$	$^2P_{1/2}$	4.136	$f^{14}ds^2$ 98	$^2D_{3/2}$	0.000
$f^{14}dsp$ 90	$^4F_{3/2}$	17.427	$f^{14}d^2s$ 100	$^4F_{3/2}$	18.851
$f^{14}d^2p$	$^4G_{5/2}$	40.559	$f^{14}sp^2$ 75	$^4P_{1/2}$	32.987
			$f^{14}d^3$	$(^4F_{3/2})$	(51.)±7

TABLE XV (Cont'd): Lutetium, 10^3 cm^{-1}

Odd terms (lowest level)			Even terms (lowest level)		
%			%		

Lu II (Lu$^+$)

$f^{14}sp$	99	$^3P_0$	27.264	$f^{14}s^2$	98	$^1S_0$	0.000
$f^{14}dp$	70	$^3F_2$	41.225	$f^{14}ds$	100	$^3D_1$	11.796
$f^{13}d^2s$		$(^5H_3)^*$	$(75.)\pm 9$	$f^{14}d^2$	97	$^3F_2$	29.407
$f^{13}ds^2$		$(^3F_2)^*$ $(\frac{5}{2},\frac{5}{2})$	$(77.)\pm 4$	$f^{14}p^2$		$(^3P_0)$	$(62.)\pm 5$
$f^{13}d^3$		$(^5H_4)^*$	$(87.)\pm 9$	$f^{13}dsp$		$(^5G_2)^*$ $(\frac{7}{2},\frac{3}{2})$	$(104.)\pm 5$
				$f^{13}d^2p$		$(^5G_2)^*$	$(106.)\pm 9$

Lu III (Lu^{++})

$f^{14}p$		$^2P_{1/2}$	38.401	$f^{14}s$		$^2S_{1/2}$	0.000
$f^{13}d^2$		$(^4G_{5/2})^*$ $(^3[\frac{5}{2}])$	$(73.)\pm 5$	$f^{14}d$		$^2D_{3/2}$	5.708
$f^{13}ds$		$(^4G_{5/2})^*$ $(^3[\frac{3}{2}])$	$(82.)\pm 2$	$f^{13}dp$		$(^4F_{3/2})^*$ $(\frac{7}{2},2)$	$(121.)\pm 5$
$f^{13}s^2$		$^2F_{7/2}$	$(100.)\pm 5$	$f^{13}sp$		$(^4F_{7/2})^*$ $(\frac{7}{2},0)$	$(132.)\pm 9$

Lu IV (Lu^{+++})

$f^{13}d$	83	$(^3P_2)^*$ $\frac{7}{2},\frac{3}{2}$	90.433	f^{14}	100	$^1S_0$	0.000
$f^{13}s$	100	$(^3F_4)$ $\frac{7}{2},\frac{1}{2}$	116.798	$f^{13}p$	98	$(^3D_3)^*$ $\frac{7}{2},\frac{1}{2}$	164.302
				$f^{12}d^2$		$(^5I_4)^*$ $(6,2)$	$(207.)\pm 9$

REFERENCES

1. G. N. Lewis, M. Randall, K. S. Pitzer, L. Brewer, Thermodynamics, Second Ed., McGraw-Hill, N. Y., 1961; Tables 32-3 to 32-5 and Fig. 32-3.

2. R. Hultgren, P. D. Desai, D. T. Hawkins, M. Gleiser, K. K. Kelley, D. D. Wagman, "Selected Values of the Thermodynamic properties of the Elements," American Society for Metals, Metals Park, Ohio, 1973.

3. L. Brewer, J. S. Winn, Faraday Discussions of the Royal Society of Chemistry, Faraday Symposium 14, 126-135 (1980).

4. L. Brewer, "The Cohesive Energies of the Elements," Lawrence Berkeley Laboratory Report LBL-3720 Rev., May 4, 1977.

5. L. Brewer, J. Opt. Soc. Am. 61, 1101-11, 1666-86 (1971).

6. W. Hume-Rothery, The Metallic State, Oxford University Press, Oxford, 1931; Structures of Metals and Alloys, Inst. of Metals, London, 1936.

7. N. Engel, Ingenioeren N101, 1939, M1, 1940; Haandogi Metalläre , Selskabet for Metalforskning, Copenhagen, 1945; Kemisk Maandesblad 30(5), 53; (6), 75; (8), 97; (9), 105; (10), 114 (1949); Powder Met. Bull. 7, 8 (1954); Amer. Soc. Metals, Trans. Quart. 57, 610 (1964); Acta Met. 15, 557 (1967).

8. L. Pauling, Proc. Roy. Soc. London Ser. A 196, 343 (1949); The Nature of the Chemical Bond, Cornell Univ. Press, Ithaca, N. Y., ed. 3, 1960.

9. L. Brewer, Viewpoints of Stability of Metallic Structures, Phase Stability in Metals and Alloys, Ed. P. Rudman, J. Stringer, R. L. Jaffe, McGraw-Hill, New York, 1967, pp. 39-61, 241-49, 344-46, 560-68.

10. A. Kant, S. S. Lin, Menatsch. 103, 757-63 (1972).

11. L. Brewer, Chapter 2, pp. 12-103, High-Strength Materials, Ed. V. R. Zackay, John Wiley, N. Y., 1965.

12. L. Brewer, Science 161, 115-22 (1968).

13. L. Brewer, Thermodynamics and Alloy Behavior of the BCC and FCC Phases of Plutonium and Thorium, Plutonium 1970 and Other Actinides, Ed. W. N. Miner, TMS Nuclear Metallurgy Series, Vol. 17, 650-58 (1970).

14. L. Brewer, Revue de Chimie minerale 11, 616-23 (1974).

15. L. Brewer, J. Nucl. Mater. 51, 2-22 (1974).

16. A. Jayaraman, Handbook on the Physics and Chemistry of Rare Earths, Ed. K. A. Gschneidner, Jr., Le Roy Eyring, North-Holland Publ. Co., N. Y., 1978, pp. 707-747.

17. H. K. Mao, R. M. Hazen, P. M. Bell, J. Wittig, J. Appl. Phys. 52, 4572-74 (1981).

18. J. H. Hildebrand, R. L. Scott, Solubility of Nonelectrolytes, Rainhold, N. Y., Third Ed., 1960; Regular Solutions, Prentice-Hall, Englewood Cliffs, N. J., 1962.

19. L. Brewer, R. H. Lamoreaux, Chapters I and II, pp. 11-356, Molybdenum: Physico-Chemical Properties of Its Compounds and Alloys, Ed. L. Brewer, Atomic Energy Review Special Issue No. 7, International Atomic Energy Agency, Vienna, 1980.

20. L. Brewer, Calculation of Phase Diagrams and Thermochemistry of Alloy Phases, Ed. Y. A. Chang, J. F. Smith, TMS-AIME, 197-206 (1979).

21. L. Brewer, Acta Metall. 15, 553-36 (1967).

22. L. Brewer, P. R. Wengert, Metall. Trans. 4, 83-104, 2674 (1973).

23. E. D. Eastman, L. Brewer. L. A. Bromley, P. W. Gilles, N. L. Lofgren, J. Am. Chem. Soc. 72, 2248-50 (1950).

24. G. Wijbenga, Thermochemical Investigations On Intermetallic UMe Compounds (Me = Ru, Rh, Pd), Ph.D. Thesis, University of Amsterdam, Dec. 1981.

25. V. Srikrishnan, P. J. Ficalora, Metall. Trans. 5, 1471 (1971).

26. W. Bronger, W. Klemm, Z. anorg. allg. Chem. 319 [1-2], 58-81 (1962).

27. W. Bronger, J. Less-Common Metals 12, 63-68 (1967).

28. C. E. Holley, Jr., E. J. Huber, Jr., and F. B. Baker, Progress in the Science and Technology of the Rare Earths, Vol. 3, L. Eyring, Ed., Pergamon Press, Oxford, 1968, pp. 343-433.

29. L. Brewer, G. M. Rosenblatt, "Dissociation energies and Free Energy Functions of Gaseous Monoxides," Advances in High Temperature Chemistry, Vol. 2, 1-83, L. Eyring Ed., Academic Press, N. Y., 1969.

30. D. D. Wagman, W. H. Evans, V. B. Parker, I. Halow, S. M. Bailey, R. H. Schumm, K. L. Churney, "Selected Values of Chemical Thermodynamic Properties," N. B. S. Tech. Note 270-5, U. S. Govt. Printing Office, Washington, D. C., 1971.

31. R. H. Schumm, D. D. Wagman, S. Bailey, W. H. Evans, V. B. Parker, N. B. S. Tech. Note 270-7, U. S. Govt. Printing Office, Washington, D. C., 1973.

32. K. A. Gschneidner, Jr., N. Kippenhan, O. D. McMasters, IS-RIC-6, Rare-Earth Information Center, Inst. for Atomic Res., Ames, Iowa, 1973.

33. L. R. Morss, "Thermochemical Properties of Yttrium, Lanthanum, and the Lanthanide Elements and Ions," Chem. Rev. 76, 827-841 (1976).

34. R. F. Krause, Jr., "Dissociation Energies of the Scandium Group and Rare-Earth Gaseous Monoxides," N. B. S. Rep. NBSIR-600, AFOSR-TR-75-0596, Chapter 9, 123-188, Oct. 1974.

35. G. C. Fitzgibbon, E. J. Huber, Jr., C. E. Holley, Jr. J. Chem. Thermodynam. 4, 349-58 (1972), Rev. Chim. Min. 10, 29-38 (1973).

36. F. B. Baker, G. C. Fitzgibbon, D. Pavone, C. E. Holley, Jr., J. Chem Thermodynam. 4, 621-36 (1972).

37. E. J. Huber, Jr., G. C. Fitzgibbon, C. E. Holley, Jr., J. Chem. Thermodynam. 3, 643-8 (1971).

38. W. C. Martin, R. Zalubas, L. Hagan, "Atomic Energy Levels -- The Rare Earth Elements," NSRDS-NBS 60, U. S. Govt. Printing Office, Washington, D. C., 1978.

39. J. Sugar, C. Corliss, J. Phys. Chem. Ref. Data 9, 473-511 (1980).

40. C. E. Moore, N. B. S. Report NSRDS-NBS 34, U. S. Govt. Printing Office (1970).

41. E. Murad, D. L. Hildenbrand, J. Chem. Phys. 73, 4005-11 (1980).

42. S. Smoes, P. Coppens, C. Bergman, J. Drowart, Trans. Faraday Soc. 65, 682 (1976).

43. J. Kordis, K. A. Gingerich, J. Chem. Phys. 66, 483-91 (1977).

44. P. Camus, J. Phys. (France) 27, 717 (1966).

45. Z. B. Goldschmidt, Chapter 1, Vol. 1, Handbook on the Physics and Chemistry of Rare Earths, Ed. K. A. Geschneidner, Jr., L. Eyring, North-Holland Publishing Co., Amsterdam, 1978.

46. S. Nir, J. Opt. Soc. Am. 60, 354-57 (1970).

47. J. F. Wyart, P. Camus, Physica Scripta 20, 43-59 (1979).

48. J. Blaise, private communication, Sept. 1982.

DISCUSSION

SUCHOW   (question)

Would you mind to elaborate on the electronic structure of α-Ce
from your systematics?

BREWER   (answer)

Yes, I was going to cover that in my next lecture but let me do
it now. Cerium being at the beginning of the lanthanides, the
f-orbitals are not contracted as much as they are later. There is
some contribution to bonding from the f-orbitals, very small but
significant. For the transition metals, where the d-orbitals are
somewhat contracted, if you can push the atoms closer together
you get them to overlap better and greatly improve the cohesion.
The Engel correlation of d^{n-1} s with the bcc (body-centered cubic)
structure and d^{n-2} sp with the hcp (hexagonal close-packed) structure predicts that pressure will stabilize the bcc structure over
the hcp structure for n $\leqslant$ 6 since the structure with the most bonding d electrons will gain the most from the better overlap
of somewhat contracted orbitals. I have had many people write to
ask if this is a violation of thermodynamics to have a metal of
coordination number 8 stabilized by pressure over a close packed
structure. The answer is that even if the bcc metal were not denser than the hcp metal, it would be more compressible as the d
orbitals overlaped better and would become denser. When there are
more than five electrons, the bcc structure with the most d electrons has the fewest bonding d electrons as non-bonding pairs are
formed. So when n $\geqslant$ 8, the order of stabilization by pressure is
fcc > hcp > bcc, just the reverse of the behaviour when n $\leqslant$ 6.
For n = 4 to 11, there has been 100 % confirmation of these predictions. For values of n smaller than 4, the d orbitals have not
extensively contracted and there is not so clear a difference between the overlap of d orbitals and outer-shell orbitals. However,
the 4f orbital of Ce is comparable to the localized 3d orbitals
of Cr to Ni. For all of these metals, pressure improves the overlap resulting in enhanced bonding and removal of magnetism. The
behaviour of cerium is described in more detail in my manuscript.

SUCHOW   (comment)

I am bothered by your use of f electron bonding without promotion.
There is experimental evidence, especially from Russian literature,
where they find large promotion energies. You are probably fami-
liar with the X-ray work where they find promotions of at least
half an electron.

BREWER   (answer)

Well, I'd say it doesn't fit in with the spectroscopic data. This is not an isolated event, we will find a lot of other phenomena that are fit by the same sort of model. For the actinides and transition metals, the role of pressure or decreasing internuclear distance upon the improved inner-shell bonding is a very important factor. The great value in using the spectroscopic data is you're not just making an ad hoc promotion to some vague state. You can say: let's look at the spectroscopic data. Is it feasible? You'll find that in many of these papers, they never looked up the gaseous spectroscopic data to find out whether the proposed promotions were within reach.

SKRIVER   (comment)

One has never believed in promotion. I think your picture is entirely correct. The spectroscopic data really tells that in no way Ce can be tetravalent in the usual sense of the word. Promotion is out.*

SUCHOW   (question)

For Zr and Pt, the electronic structure you showed were not the ground states. I guess you were talking about the configurations in the metals. Also, if the Zr and Pt atoms are paramagnetic, would they become diamagnetic on forming acid-base pair?

BREWER   (answer)

1. Yes, I showed the valence-state configurations corresponding to the metals.
2. No, there would be no magnetism for either the pure metals or the alloy except for a slight paramagnetism due to single electron bonds at the top of the band. The d^3s valence state of Zr has four unpaired electrons but when the atoms combine in the solid, the electrons are paired in electron pair bonds. All of the 4d, 5d or 6d electrons are sufficiently delocalized to form electron pair bonds and no significant magnetism is obtained. The 3d orbitals for the left hand elements through vanadium overlap sufficiently to form bonds. With further increase of nuclear change, some of the orbitals of Cr to Ni will be sufficiently localized to result in localized single electrons that can produce magnetism if the localized orbitals are not filled with non-bonding pairs. The 4f orbitals are so localized that single electron occupancy results in magnetism. The dense low temperature form of Ce is an exception. For Ce, 4f electron pair bonds can form if the nuclei are close enough. Electron pair bonds are formed by the more extended 5f orbitals through plutonium.

---

* However, see Poster No. 4 (G. Netz et al.) - Editor.

SINHA   (question)

You are talking about fractional electron occupancy of orbitals.
What kind of electronic ground state do you expect from fractional orbitals? What is the physical picture?

BREWER   (answer)

Take the case of body-centered cubic Zr which on a fractional basis
would be described as $d^{2.5}sp^{0.5}$ or a mixture of d^3s and d^2sp. You
can describe it in terms of resonance if you like. You can say
that some atoms are in one discrete configuration or other and
then mix them. The fact that we are using fractional configurations is of no particular difficulty for the solid phase. Even
for the gas, you have configuration mixing, but there are some
symmetry limitations.

SINHA   (question)

Your description is an average. That means the electrons may spend
half of the time in the p and half of the time in the d orbital.
What kind of electronic ground state results?

BREWER   (answer)

This is like the business of motion of electrons around an atom.
The electron probability is a constant number at a given position
and the electrons don't jump around.

SINHA   (question)

So you would not subscribe to the ideas of having fractional L
values or effective L values?

BREWER   (answer)

No, I would not, although when L-S coupling is not appropriate,
the L value is not of significance.

GROUP I   (question)

Could you please comment on the Model for intermetallic compounds
proposed by Miedama and compare it to the one you have developed?

BREWER   (answer)

There are a variety of useful models with different input requirements and different ranges of useful accuracy. My model requires inputs of gaseous spectroscopic data, internal pressures,
atomic sizes, and bonding contributions for electrons in different

orbitals. I use the correlation between electronic configuration and crystal structure and take into account generalized Lewis-acid and -base interactions. The Miedama model has fewer inputs and is simpler to apply, but it smoothes out the substantial variations. Thus his model considerably underestimates the strong acid-base interactions in metallic systems.

GROUP I (question)

Could you take one element and show us the available information on the accessible spectroscopic states and briefly how you determine which ones can possibly be used in determining the bonding in the metals?

BREWER (answer)

We can use the energies of the lowest states of each of the electronic configurations of Yb given in Table XIV together with the bonding enthalpies given in reference 9 to illustrate which configurations can contribute to the bonding in Yb metal. The ground state of the gaseous atom, $4f^{14} 6s^2$, is a rare gas configuration with no bonding contribution other than van der Waals bonding. One has to promote to the $f^{14}sp$ configuration at 17288 cm^{-1} to make two electrons available for bonding. The next configuration, $f^{13} ds^2$, has three non-f electrons but only one bonding electron as the filled s^2 subshell is non-bonding. With only one bonding electron, it could not compete with the lower lying $f^{14}sp$ configuration. The next configuration, $f^{14}ds$, has two bonding electrons and will yield a metallic phase almost as stable as that with the $f^{14}sp$ configuration. The next configuration, $f^{13}s^2p$ is not important because it has only one bonding electron. Not until the $f^{13}sp$ configuration at 39880 cm^{-1}, is there a configuration with three bonding electrons. However, the promotion energy is now so large that the additional bonding electron will not offset the extra promotion energy. Thus the configurations with three bonding electrons yield a metal less stable than one with only two bonding electrons. Yttrium is another example without the complication of f electrons. The two lowest configurations, ds^2 and s^2p, have only one bonding electron and would not compete with the next two configurations, trivalent d^2s and dsp. The next higher configurations, d^3, d^2p, sp^2 and dp^2, all have these bonding electrons but they require such large promotion energies that they would not contribute significantly. An even simpler example would be tungsten. The ground state configuration, d^4s^2, with only four bonding electrons cannot compete with the d^5s configuration. The d^4sp configuration is strongly bonded but requires a high promotion. For the formation from the gaseous ground state of a metal phase based on the d^4sp configuration, $\Delta H_o^o/R$ would be about 7000 kelvin higher than for the d^5s configuration which is thus the only configuration of importance for tungsten.

SUCHOW   (question)

This is something I presented to our discussion group earlier today and which we considered at length. The point is that all of
the lanthanides, especially when in water solution, can form 3+
ions and some, as shown by Prof. Klemm yesterday, can form others
as well. The usual explanation is that the electronic structures
tend towards empty, filled and half-filled f sub-shells and, as a
result Eu by holding on to an electron can be 2+, Sm can also be
2+, Ce can be 4+, Pr can be 4+, Nd 4+, Gd only 3+, Tb 4+, Yb 2+,
and Tm probably also 2+. Now, if the criterion is the half-filled
or the empty or filled f-shell, why should Sm stop at 2+ rather
than continuing to 1+? Why should Pr and Tb stop at 4+ rather than
going to 5+, etc.? The answer may not be simple. I have already
spoken to Prof. Klemm privately and he has become sufficiently
interested so that he will attempt calculation in the near future.
Our group offers this problem for open discussion. Perhaps Prof.
Brewer can come up with the answer via reference to his spectral
data.

BREWER   (answer)

It is a mistake to attribute any magic to the empty or half-filled
f-shells. At the beginning of the lanthanides the f orbitals have
not contracted strongly and with a strong enough oxidizing agent,
one can produce Ce^{4+}. As the nuclear charge increases, the f orbitals contract and it is more difficult to remove an f electron to
produce Pr^{4+}. To remove two f electrons to form Pr^{5+} would be
possible only under most extreme oxidizing conditions. As one continues towards Eu, the f orbitals stabilize at a rapid rate and
it becomes quite difficult to promote to an electronic configuration with the non-f electrons or to oxidize to Eu^{3+}. It is easier
to remove an f electron from the f^7 configuration than one might
expect, because of the crystal field effect which controls some
f orbitals below the average distribution. As I showed in the
plot of energies of the gaseous electronic configuration, the
$f^{n-2}s^2$ or $f^{n-2}sp$ configuration drop in energy very rapidly as the
nuclear charge is increased from La to Eu and there is a large
change from Sm to Eu. Thus it is much easier to remove f electrons
from Sm than from Eu. Sm^+ with the f^7s configuration would form a
condensed phase which would be unstable by disproportionation to
Sm metal where three electrons are used in bonding and a higher
oxidation state. Unless there were promotion to f^6ds or some
other configuration with fewer f electrons to allow stronger bonding, using the two non-f electrons in metallic bonding. Due to
electron repulsion, for electron pairs, an f electron added beyond
the f^7 configuration is easier to remove and Tb can be oxidized
to the tetravalent state as for Ce. However, with increasing nuclear charge, the rapid stabilization of the f orbitals quickly
makes it difficult to remove the f electrons to attain the tetra-

valent state for the Tb-Dy-Ho sequence as for the Ce-Pr-Nd sequence. The behaviour in regard to f^{14} configurations is similar to the discussion of the f^7 configuration.

INCLINED W SYSTEMATICS OF THE LANTHANIDES

Shyama P. Sinha

Hahn-Meitner-Institut, Postfach 390128,
D-1000 Berlin 39, Fed. Rep. Germany

ABSTRACT

The concept of the linear variation of properties (P_i) with the orbital angular quantum number (L) of the lanthanides at their ground states is developed ($P_i = w_i L + k_i$). The metallic radii in the lanthanide and actinide series correlate well with the normalized ground state electronic configurations and hence the ground state orbital angular L-values. This L-linearization systematic (Inclined W) allowed the prediction of the following metallic radii: Ce(dhcp, fdsp) 1.85; Eu(hcp, $f^6 dsp$) 1.81; Yb(hcp, $f^{13}dsp$) 1.74; Ac(bcc, sp^2) 1.89 and Np(fcc, $f^4 ds^2$) 1.63 Å. The normalized ionization potentials of the lanthanides from first through fifth produced excellent Inclined W correlation and predictions were made for various energy differences between electronic levels in excellent agreement with the literature values. A model was developed to calculate the sixth ionization potentials of the lanthanides and a comparison with the HF and SCF-HF results was made. Other macro-properties like the M-O bond distances in the C-type M_2O_3, formation constants with various organic carboxylic acids have also been systematized in terms of the L-linearization scheme. In correlating the properties with the L-values care should be exercised that at least all members of a tetrad (if not the whole 1^q series) are in the same oxidation state, possess similar electronic configuration and that no redox reaction has occurred during the progress of the said reaction. An analysis of the electrostatic energies for the f^n configurations in terms of the Racah E^k parameters has been carried out to show the validity of the tetrad classification, the L dependence of the energies and the rhythmic character of the variation of the energies through the series.

S. P. Sinha (ed.), Systematics and the Properties of the Lanthanides, 71–122.
Copyright © 1983 by D. Reidel Publishing Company.

1.   INTRODUCTION

We have heard early this morning that the lanthanide systematics has a long history. I am very happy that Prof. Klemm summarized the early attempts in systematizing the properties of the
lanthanides so elegantly, that I do not need to repeat those. In
his talk we have also witnessed how chemical intuition coupled
with experience have helped Prof. Klemm and his coworkers to arrive
an interpretative synthesis of their results in those days when
instruments had less buttons to push and the digital displays did
not glare at us.

Prof. Klemm has raised an interesting point that the young
scientists often do not consult or read works much older than a
decade or so. I must confess that when in the early seventies I
started thinking about lanthanide systematics, I also did not carefully check the publications of the forties, although references
to Prof. Klemm's early works were known and available to me. Ad-
mitting this guilt, let me also admit that as chemists we often
get stuck to fixed ideas. I believe that nothing has left more im-
pression in our mind than the Periodic Classification of Elements,
correlation of their properties in the series and across the Periodic Table, diagonal relationships etc, during our early training
in the genreal and inorganic chemistry classes. Thus, after gradu-
ation, we were at least endowed with one parameter e.g., the atomic
number (Z) for "free" use. So we happily went on correlating newly
measured elemental, metallic and ionic properties with Z, although
more than often we did not get an "idealized" linear relationship
that we were used to in our elementary chemistry classes. We tried
to disregard or explain away small irregularities in a P_i (any
property) vs. Z plot. Controversies on the interpretation of the
irregularities were not uncommon.

Physicists owe in their part, the idea of the correlation of
properties of atoms and ions with Z, to atomic physics. It was
found that the wavenumbers of the spectral lines of the one-electron
atoms increased in proportion to Z^2. However, for many-electron
atoms this simple relationship is no longer valid, partly because
the energy of the electron undergoing transition does not depend
on the principal quantum number, and partly because it is not proportional to Z^2 but to $(Z-\sigma)^2$ where σ is the screening constant.
The calculated values (SCF) of the screening constant   are found
[1,2] to vary non-linearly through the lanthanide and 3d-element
series causing irregular behaviour of the effective nuclear charge
($Z_{eff}=Z-\sigma$) (Fig. 1,2). As the screening constant (σ) is a variable
quantity, which depends on the electronic configuration and degree
of ionization of the element, Z_{eff} values will vary with the ionicity of the element. Actually, we do not have enough knowledge on
the nature of these variations for the lanthanides at different
stages of ionization. Odiot and Saint-James[3] quote Z_{eff} values
ranging from 19.4 to 23 for Ce to Er in their trivalent state.

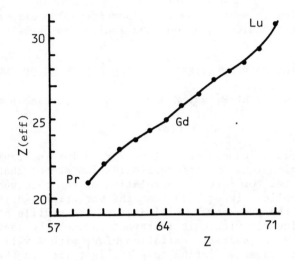

Fig. 1 Variation of Z_{eff} with atomic number (Z) for the neutral lanthanides in the series.

In anyway, it is the valence electrons, lying outside the completely filled orbitals and experiencing the effective nuclear charge, Z_{eff}, that are involved in chemical binding process. Hence, we may ask ourselves a question on the merit of using the atomic number Z (equal to the nuclear charge) as a parameter for correlating properties of the neutral atoms and their ions in different oxidation states with widely varying properties but same Z. On the other hand, Z_{eff} would have been a more logical parameter, if we could have values for different elements at various stages of ionization. However, plots based on Z_{eff} as parameter are uncommon.

Question was also raised during the early development of lanthanide systematics whether the observed irregularities and break near gadolinium for the lanthanide series are simply artifacts. However, careful measurements of the formation constants and other thermodynamic quantities by Prof. Moeller's group in the USA and Prof. Schwarzenbach's group in Switzerland proved beyond doubt that the observed irregularities are real and there is always a break around gadolinium. In Fig. 3, we have collected the data on the complex formation constants ($\log K_1$) for various ligands clearly showing the gadolinium break. Note also that simple ligand like acetate ion (CH_3-COO^-) produces remarkable difference in formation constants within the lanthanide series. We would see later that the profile presented by the acetate ion is more or less a common feature of the simple carboxylate anions and that the Gd(III) aquoion is somewhat unique amongst the lanthanides. We shall return to

Gd(III) aquo ion in my talk on the Fluorescence Spectra and Lifetimes of the Lanthanide Aquo Ions and Their Complexes (Chapter 10).

2.   CORRELATION OF PROPERTIES WITH ATOMIC NUMBERS OF THE ELEMENTS

The inadequacy of P_i vs. Z plots (eq. 1) for the properties P_i

$$P_i = f(Z) \tag{1}$$

of the lanthanides induced me to look for an intrinsic property
of the lanthanide ions in different oxidation states that may be
used as parameter for further correlation of the observed properties. It is implicit in eq. (1) that the variation (increment or
decrement) of the property P_i will dictate the profile of the plot,
as Z varies linearly with unit increment through the lanthanide or
actinide series. A quadratic variation of P_i with Z will result in
a curve, but a linear variation to a straight line in the series.
We would like to recall here and emphasize once again that Z is a
parameter which is independent of the electronic configurations
and ionization stages of the elements. We, however, recognize the
indirect effect of Z (better Z_{eff}) on the ionization potentials
and other properties.

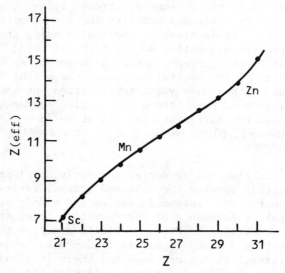

Fig. 2 Variation of Z_{eff} with atomic number (Z) for the 3d-ele-
ments in the series.

Had I at this stage consulted Prof. Klemm's early works, especially the Fig. 5 of his lecture, I may have had some clue on the lanthanide systematics at an early stage. However, during 1973-75 I realized[4] that the ground state total orbital angular quantum numbers (L) of the free trivalent lanthanide ions show periodicity and also generate rhythmics[1]. For the trivalent lanthanides with $4f^n$ configuration we have the following classification of the L values:

L	0	3	5	6
f^q	La	Ce	Pr	Nd
f^{7-q}	Gd	Eu	Sm	Pm
f^{7+q}	Gd	Tb	Dy	Ho
f^{14-q}	Lu	Yb	Tm	Er
q	0	1	2	3

We have later found[5] that the ground state L-values of any pure configuration (1^q) may be calculated using very simple relationship

for $0<q<N/2$ $L=1/2[q(q-N/2)]$ (2)

and for $(N/2+1)<q<N$ $L=1/2[(q-N/2)(q-N)]$ (3)

where q is the number of electrons in the shell (1^q) with $q_{max}=N$ (for s, p, d, f ...; N=2, 6, 10, 14 etc). The variation of the ground state L-values with q of 1^q configurations is shown in Fig. 4.

It is clear from the above equations and from Fig. 4 that the functions have a minimum at N/2, the half-filled shell; and a maxima each at about quarter-filled ($\sim N/4$) and about three quarter-filled shell ($\sim 3/4N$) thus generating a tetrad classification for all configurations from p^q through f^q and may be also for g^q and h^q. The periodicity and rhythmic nature of this variation of L-values are apparent.

Eq. (2) and (3) were tested for generating the ground state L-values for mixed configurations $1_1^q 1_2^n$ (where $1_1 \neq 1_2$) and these are found to hold good provided the ground state follow Hund's rule strictly[5].

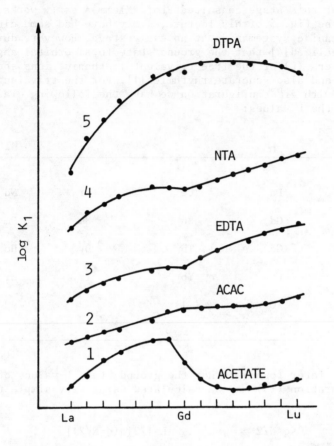

Fig. 3 Variation profiles of the first formation constants ($\log K_1$)
for five representative ligands through the lanthanide series.
Smooth curves are drawn through the points to emphasize the general
profiles.

3.   CORRELATION OF PROPERTIES WITH THE GROUND STATE L-VALUES

It looks like that the ground state L-values depending on the
electronic configurations are intrinsic of the atoms and ions in
their given oxidation states and the L-values could be useful parameters for correlating observed properties. After examining a
variety of properties of the lanthanides, actinides and the d-transition elements, Sinha[1, 2, 4, 6] found that the properties are
linear functions of the ground state L-values

$$P_i = w_i L + k_i \qquad\qquad (i = 1\text{-}4) \qquad\qquad (4)$$

3.1 Analysis of the Metallic Radii of the Lanthanides and the Actinides

First we shall analyze the variation of the metallic radii of the lanthanides and the actinides within the series. The metals are very interesting as they exhibit polymorphism with varying electronic configurations in the condensed phase. The electronic configurations of the gaseous atoms differ considerably from that of the condensed phase. Thus hexagonal lanthanum (dhcp), according to Brewer[7] possesses an electronic configuration of $d^1s^1p^1$ giving a $^4F(L=3)$ electronic ground state in the metallic phase or a $d^{1.5}s^1p^{0.5}$ configuration in its bcc modification. However, the free (gaseous) La atom has a d^1s^2 configuration with $^2D_{11/2}(L=2)$ as ground state. An analysis of the metallic radii of the lanthanides and some actinides according to eq. (4) is presented in Table 1 and a comparison between the classical P_i vs. Z plot and the P_i vs. L plot of eq. (4) is made in Fig. 5. It becomes immediately obvious from the Table and Z-plot that Ce(fcc), Eu(bcc), and

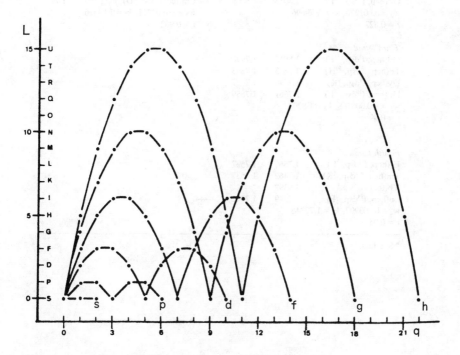

Fig. 4 Periodicity in the variation of the values of L with q for the configurations s^q, p^q, d^q, f^q, g^q, h^q in their ground states[5].

Table 1. Analysis of the Metallic Radii of the Lanthanides and the
Actinides (Å). The Element is followed by the Crystal
Structure, the Electronic Configuration and the $^{2s+1}$L
values used in this Analysis

M	Obs.	Calc.	M	Obs.	Calc.
First tetrad					
La(dhcp, f^0dsp, ^{4}F)	1.8791	1.8788	Ac(bcc)	1.879	*
Ce(fcc)	1.8247	*	Th(bcc, d^3s, ^{5}F)	1.798	1.797
Pr(dhcp, f^2dsp, ^{6}L)	1.8279	1.8297	Pa(bcc, f^1d^3s, ^{6}I)	1.642	1.644
Nd(dhcp, f^3dsp, ^{7}M)	1.8214	1.8200	U(bcc, f^2d^3s, ^{7}L)	1.542	1.541
$w_1 = -0.00988$, $k_1 = 1.90825$			$w_1 = -0.05126$, $k_1 = 1.95116$		
$r = 0.998$			$r = 0.999$		
Second tetrad					
Pm(dhcp, f^4dsp, ^{8}M)	1.811	1.808	Np		*
Sm(hcp, f^5dsp, ^{9}L)	1.8041	1.8072	Pu(fcc, $f^5d^1s^2$, ^{7}K)	1.64	1.65
Eu(bcc)	2.0418	*	Am(fcc, $f^6d^1s^2$, ^{8}H)	1.73	1.71
Gd(hcp, f^7dsp, ^{11}F)	1.8013	1.8008	Cm(fcc, $f^7d^1s^2$, ^{9}D)	1.78	1.79
$w_2 = 0.001276$, $k_2 = 1.79696$			$w_2 = -0.02711$, $k_2 = 1.84316$		
$r = 0.822$			$r = 0.962$		
Third tetrad					
Gd(hcp, f^7dsp, ^{11}F)	1.8013	1.7998			
Tb(hcp, f^8dsp, ^{10}H)	1.7833	1.7863			
Dy(hcp, f^9dsp, ^{9}K)	1.7740	1.7727			
Ho(hcp, f^{10}dsp, ^{8}L)	1.7661	1.7659			
$w_3 = -0.006768$, $k_3 = 1.8201$					
$r = 0.99$					
Fourth tetrad					
Er(hcp, f^{11}dsp, ^{7}L)	1.7566	1.7537			
Tm(hcp, f^{12}dsp, ^{6}K)	1.7462	1.7537			
Yb(fcc)	1.9392	*			
Lu(hcp, f^{14}dsp, ^{4}F)	1.7349	1.7342			
$w_4 = 0.003907$, $k_4 = 1.72246$					
$r = 0.95$					

* See text.

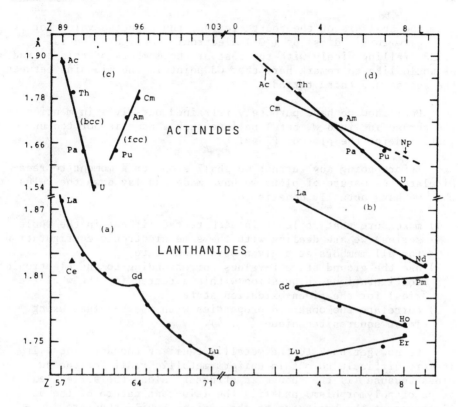

Fig. 5 Plots of the metallic radii of the lanthanides and the
actinides as classical P_i vs. Z and as Inclined W (P_i vs. L).

Yb(fcc) having different crystal structure and probably different
electronic configuration from the rest of the members of their
tetrads fall out of the series. However, if one is trying to correlate the observed metallic radii with Z without paying any attention to the structure, one would be tempted to join La-Ce-Pr-
Nd in the first tetrad creating a cusp around Ce-Pr-Nd region and
explaining it away with elegant theories. The correlation of the
metallic radii with L values allows us to predict the metallic
radius of Ce if it had a dhcp structure (1.849Å) corresponding to
an electronic ground state of $^5I(L=6)$. An increase of 0.025Å in
the metallic radius of Ce for a structural change from fcc to dhcp
is consistent with an increase of ∿0.004Å for the same phase change
of La (La fcc: r = 1.875 Å).

    In the second tetrad Eu behaving as bivalent metal possesses
a bcc structure ($f^7 d^{0.5} s^1 p^{0.5}$). If Eu had the "idealized" configuration $f^6 d^1 s^1 p^1$ with ground state $^{10}I_{3/2}(L=6)$, we would
expect a radius of 1.805Å. Similar prediction on hcp Yb is made

from the analysis of the fourth tetrad: the expected value for Yb(hcp) with $f^{13} d^1 s^1 p^1$ ($^5G_2$) electronic configuration is 1.738Å falling nicely with the rest of the members of this tetrad. We would like to remark here that Gd point is the midpoint between the second and third tetrad.

Until now we have purposely refrained ourselves in mentioning the phrase Inclined W, but I believe such a name is not too unjustified for the plot of P_i vs. L.

Before going any further we shall stop for a moment to recapitulate the mature of plots we have made and lay down the rules. What we have actually done is to

1) make sure that at least in each tetrad (if not in the whole series) we are dealing with the same electronic configuration for all members at a given oxidation state
2) use the ground state L-values corresponding to the electronic configurations in question within a tetrad (or the whole series) for the given oxidation state
3) correlate the observed properties with the L-values using least squares technique (eq. 4).

We now get back to the metallic radii of the actinides. It is more difficult to obtain uniform metallic radii for the actinides because (i) the change in structure along the series (existence of polymorphism) and (ii) the itinerant nature of the 5f electrons at the beginning of the series, while they are localized at the end of the series. Evidences for the participation of the 5f electrons in the lighter actinides (Pa, U, Np, Pu, Am) in forming metallic bonds are now available[8-10].

Within these limitations, we have, none the less, obtained[11] an excellent correlation for the bcc structures of Th, Pa, and U, which allowed us to predict the metallic radius of Ac(bcc) as 1.899Å if it had the expected sp^2(^2p, L=1) ground configuration. But the experimental metallic radius of 1.879Å corresponds to an L-value of 1.4076! This fractional L-value can only be achieved if we consider fractional occupation numbers of electrons involving d-orbitals in Ac. Brewer[7] predicted a ground configuration of $s^1 p^{1.5} d^{0.5}$ for bcc form of Ac. Using the experimental radius and the corresponding L-value of 1.4076 (Fig. 5), Sinha[11] calculated the ground configuration of bcc actinium as $s^1 p^{1.857} d^{0.143}$. If Brewer's prediction on the ground configuration of Ac is correct we obtain[5,11] an L-value of 2.25 which generates a value of 1.836Å for the metallic radius of Ac. This value is undoubtedly very low for Ac. Whatever the "actual" ground state of bcc Ac may be, it is clear that some hybridization involving the d-orbital is operating in this phase of actinium.

   This example of Ac demonstrates the usefulness of our L-linearization scheme or the Inclined W plot. The strength of the Inclined W plot lies not only on the correlation but more on the
deviation from a linear plot. If the property of an element devi-
ates considerably from the linearity within its tetrad, this is
an indication that something extraordinary is happening. Either
the choice of electronic configuration for the given oxidation
state is wrong or it might be that partial oxidation or reduction
of the element has taken place during the experiment. As for example, Ce(III) is easily oxidized to Ce(IV) while Eu(III) has a
tendency to get reduced to Eu(II) in the presence of certain ligands.

3.2   Analysis of the Ionization Potentials of the Lanthanides

   Next, we shall correlate another fundamental quantity, e.g.
the ionization potentials (IP) of the lanthanides using our L-linearization systematics (eq. 4).

   Reasonable linear plots (Fig. 6) for both theoretical[12]
and experimental[13] values of IP_1 amounting to the $f^qs^2 \rightarrow f^qs$ with
q, the number of f-electrons, from Pr to Yb($4f^3-4f^{14}$) with the
exception of La, Ce, Gd(Tb) are obtained. At the beginning (Pr-Eu)
the slopes are almost equal (~ 0.06) and only slight divergence is
seen for the later half of the lanthanide series (IP_1(Expt) =
0.0814 q + 5.1192 eV, r = 0.999 (Tb-Yb); IP_1(HF) = 0.06 q + 4.507 eV,
r = 1.0(Dy-Yb)).

   For elements La, Ce, and Gd the first ionization process,
however, involves transitions of the following type:

   (i) $f^0d^1s^2(^2D) \rightarrow f^0d^2(^3F)$ for LaI

   (ii) $f^1d^1s^2(^1G) \rightarrow f^1d^2(^4H)$ for CeI and

   (iii) $f^7d^1s^2(^9D) \rightarrow f^7d^1s^1(^{10}D)$ for GdI respectively.

It is, however, easy to calculate the energy differences between
the f^qs^2 and f^qs for CeI and GdI using the tables of Brewer[14,15]
i.e., 5.36 and 5.77 eV respectively. In principle the same treatment is also possible for LaI. But the lowest level of fs^2 configuration of LaI ($^2F_{5/2}$) at 1.884 eV above the ground state $^2D_{3/2}$
of ds^2 configuration, is strongly perturbed due to mixing with
levels of other odd configurations. Thus, there is a large uncertainty in estimating the $f^qs^2 \rightarrow f^qs$ energy difference for LaI.
Worden et al[16], who have recently measured the more accurate
IP_1 values for 10 lanthanides by laser spectroscopy, also left
LaI from their least squares plot of $f^qs^2 \rightarrow f^qs$ energy difference
vs. q.

Sinha[6,11,17] has earlier considered LaI, CeI, and PrI to belong to the first tetrad and fixed the energy difference of $(f^2d^1s^2 \rightarrow f^2d^2)$ in PrI at 5.4 eV. He thus obtained[11] a good linear plot (r=0.98) with w_1 = -0.03925 and k_1 = 5.646 (Table 2). With the availability of a more accurate value[16] for CeI (5.5387 eV) from laser spectroscopy, we now reconsider the first tetrad. A slightly better value (5.482 eV) for $f^2d^1s^2 \rightarrow f^2d^2$ energy difference in PrI is obtained by taking the $f^3s^1 \rightarrow f^2d^2$ energy difference in PrII as 0.682 eV and fixing the $f^3s^2 \rightarrow f^2d^1s^2$ difference in PrI at 0.62 eV. The generated set of values for the $f^qd^1s^2 \rightarrow f^qd^2$ process in LaI($f^0d^1s^2$, ^2D), CeI($f^1d^1s^2$, ^1G), and PrI ($f^2d^1s^2$, ^4K) then yield excellent correlation (r=0.999) for this tetrad (Table 3).

Of the other higher ionization potentials of the lanthanides only the second (IP_2) and some of the third (IP_3) potentials have been measured. Sugar[18-20] has, however, calculated and systematized the ionization energies IP_3 to IP_5 for all lanthanides, using a four parameter equation:

$$IP(f^q) = SD(f^q - f^{q-1}5d) + \Delta E(f^{q-1}5d - f^{q-1}6s) + \delta + T \qquad (5)$$

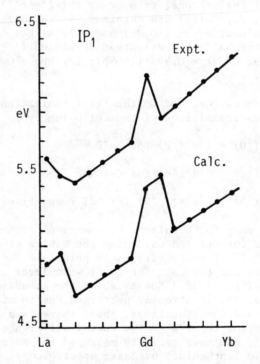

Fig. 6 Comparison of the calculated[12] and the measured[13] first ionization potentials (IP_1) through the lanthanide series.

Table 2. Analysis of the First Ionization Potentials (IP_1) of the Lanthanides

L	M	IP_1 (eV)	Calc.	w_i, k_i, r_i	
2	$La(f^0d^1s^2)$	5.577	5.568	w_1 =	-0.03925
4	$Ce(f^1d^1s^2)$	5.47	5.49	k_1 =	5.646
6	$Pr(f^3s^2)$	5.42	5.41	r_1 =	0.98
6	$Nd(f^4s^2)$	5.49	5.51	w_2 =	-0.02905
5	$Pm(f^5s^2)$	5.55	5.54	k_2 =	5.6867
3	$Sm(f^6s^2)$	5.63	5.60	r_2 =	0.95
0	$Eu(f^7s^2)$	5.67	5.69		
0	$Eu(f^7s^2)$	5.67	5.67	w_3 =	0.04097
2	$Gd(f^7d^1s^2)$	6.14	*	k_3 =	5.6665
5	$Tb(f^9s^2)$	5.85	5.87	r_3 =	0.99
6	$Dy(f^{10}s^2)$	5.93	5.91		
6	$Ho(f^{11}s^2)$	6.02	6.05	w_4 =	-0.03709
5	$Er(f^{12}s^2)$	6.10	6.08	k_4 =	6.2683
3	$Tm(f^{13}s^2)$	6.18	6.16	r_4 =	0.97
0	$Yb(f^{14}s^2)$	6.254	6.268		

* Predicted for Gd IP_1 = 5.79 eV($f^8s^2 \to f^8s$), agreeing with 5.77 eV, the calculated value.

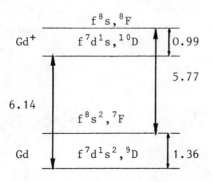

Table 3. Analysis of the Measured First Ionization Potentials
(IP$_1$) of the Lanthanides from Laser Spectroscopy

L	M	IP$_1$(eV)	Calc.	w_i, k_i, r_i
2	La($f^0d^1s^2$)	–	–	
4	Ce($f^1d^1s^2$)	5.5387	–	
6	Pr(f^3s^2)	(5.473)	–	
6	Nd(f^4s^2)	5.525	5.5478	$w_2 = -0.02299$
5	Pm(f^5s^2)	(5.582)	5.5708	$k_2 = 5.6857$
3	Sm(f^6s^2)	5.6437	5.6168	$r_2 = 0.93$
0	Eu(f^7s^2)	5.6704	5.6857	
0	Eu(f^7s^2)	5.6704	5.6675	$w_3 = 0.04281$
2	Gd($f^7d^1s^2$)	6.1502	**	$k_3 = 5.6675$
5	Tb(f^9s^2)	5.8639	5.8815	$r_3 = 0.99$
6	Dy($f^{10}s^2$)	5.9390	5.9243	
6	Ho($f^{11}s^2$)	6.0216	6.0508	$w_4 = -0.03644$
5	Er($f^{12}s^2$)	6.1077	6.0872	$k_4 = 6.2695$
3	Tm($f^{13}s^2$)	6.1844	6.1601	$r_4 = 0.96$
0	Yb($f^{14}s^2$)	6.2539	6.2695	

contd.

Table 3. Contd.

L	M	IP_1 (eV)	Calc.	w_i, k_i, r_i
$f^q d^1 s^2 \rightarrow f^q d^2$				
2	$La(f^0 d^1 s^2)$	5.577	5.5769	$w_1 = -0.01899$
4	$Ce(f^1 d^1 s^2)$	5.5387	5.5389	$k_1 = 5.61487$
7	$Pr(f^2 d^1 s^2)$	5.482	5.4819	$r_1 = 0.999$

** Predicted for Gd IP_1 = 5.7531 eV($f^8 s^2 \rightarrow f^8 s$), agreeing with
5.7839 eV, the calculated value.

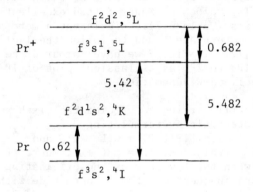

The SD's (system difference) are reasonably well known or may be extrapolated from known values. ΔE is the energy difference between the lowest levels of $f^{q-1}5d$ and $f^{q-1}6s$. 8 represents the stabilization of the ground term of the $f^{q-1}6s$ configuration.
$T = \Delta T + SC$, where ΔT is the energy difference between $f^{q-1}6s$ and $f^{q-1}7s$, and SC is the energy difference between $f^{q-1}7s$ and the ionization limit ($f^{q-1}\infty s$) and is usually obtained from the series calculations[18-20].

A treatment similar to IP_1's may also be applied to the observed IP_2's for the lanthanide series[13] (Table 4). Here also the La, Ce, and Gd data needed readjustment to express the $f^q s^1 \rightarrow f^q$ energy difference. For IP_3, only La and Gd belonging to $f^q d^1$ ground state were readjusted (Table 5). Here, we would like to note that the recently measured[21] more accurate IP_3 value for EuIII e.g., 24.92 eV fitted the second tetrad (Nd-Eu) much better than the previously predicted value of 24.70 eV[19]. The calculated IP_4 and IP_5 values of Sugar fitted excellently the Inclined W systematics (Table 6,7). This is not a mere artifact, as Sugar's four parameters are found to follow the Inclined W systematics closely[17]. In Fig. 7 we have compared the normalized $IP_2(f^q s^1 \rightarrow f^q)$ and IP_3 to IP_5 as a classical plot of IP vs. Z and according to the Inclined W schematics.

Sinha[17] has earlier tried to predict the sixth and some higher ionization potentials by formulating
(i) the series correlation: $IP(\Sigma IP) = w_i L + k_i$ and
(ii) the successive correlation: $IP = m\Delta L + c$, where ΔL is the difference between the L values of the originating and the terminating ions in question. For a lanthanide with f^q configuration the ΔL values can only vary between -3 and $+3$ as the ionization proceeds.

A comparison of the fifth and the sixth potentials calculated by SCF-HF theory based on spherical shell model[12], HF theory[22], Sugar[20] and by the Inclined W systematics[17] (Table 8) is presented in Fig. 8 (a)(b). Individual values of IP_6 for Nd and Er have been calculated earlier by Cowan[22]. These values are in the reasonable agreement with those predicted by the Inclined W schematics (Fig. 8 (b)).

The SCF-HF values of Carlson et al[12] are really too high, probably because these were calculated on the basis of ionization of the 5p electrons. On the contrary, both Sugar's IP_5 and the calculated values of IP_6 using Inclined W schematics follow closely the HF results of Fraga et al[22], who assumed the ionization of the 4f-electrons at these stages of ionization. It is the general feeling of the author that the ionization of the 5p-electrons for most of the lanthanides becomes important from the eighth spectrum onwards.

Table 4: Analysis of the Second Ionization Potentials (IP_2)
of the Lanthanides

L	M	IP_2(eV)	Calc.	w_i,k_i,r_i
3	La(d^2)	11.06	11.08	$w_1 = -0.1607$
5	Ce(f^1d^2)	10.85	10.77	$k_1 = 11.57$
6	Pr(f^3s^1)	10.55	10.61	$r_1 = 0.958$
6	Nd(f^4s^1)	10.72	10.78	$w_2 = -0.08238$
5	Pm(f^5s^1)	10.92	10.87	$k_2 = 11.28$
3	Sm(f^6s^1)	11.07	11.02	$r_2 = 0.969$
0	Eu(f^7s^1)	11.25	11.28	
2	Gd($f^7d^1s^1$)	12.09	12.05	$w_3 = -0.1246$
5	Tb(f^9s^1)	11.52	11.68	$k_3 = 12.3$
6	Dy($f^{10}s^1$)	11.67	11.55	$r_3 = 0.878$
6	Ho($f^{11}s^1$)	11.80	11.84	
5	Er($f^{12}s^1$)	11.93	11.90	$w_4 = -0.0583$
3	Tm($f^{13}s^1$)	12.05	12.02	$k_4 = 12.19$
0	Yb($f^{14}s^1$)	12.17	12.19	$r_4 = 0.972$
$f^qd^2 \rightarrow f^{q+1}$				
3	La(d^2)	11.95	11.5	$w_1 = -0.4089$
5	Ce(f^1d^2)	10.85	11.03	$k_1 = 13.071$
8	Pr(f^2d^2)	9.87	9.80	$r_1 = 0.989$
$f^qs^1 \rightarrow f^q$				
3	La(f^1s^1)	10.20	10.19	$w_1 = 0.112$
5	Ce(f^2s^1)	10.37	10.41	$k_1 = 9.85$
6	Pr(f^3s^1)	10.55	10.52	$r_1 = 0.98$

contd.

Table 4. Contd.

L	M	IP_2(eV)	Calc.	w_i, k_i, r_i
$f^q s^1 \rightarrow f^q$				
3	$Gd(f^8 s^1)$	11.40	11.39	$w_3 = 0.0857$
5	$Tb(f^9 s^1)$	11.52	11.56	$k_3 = 11.13$
6	$Dy(f^{10} s^1)$	11.67	11.65	$r_3 = 0.968$

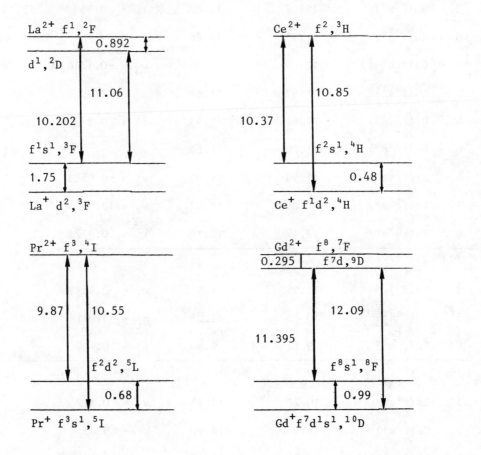

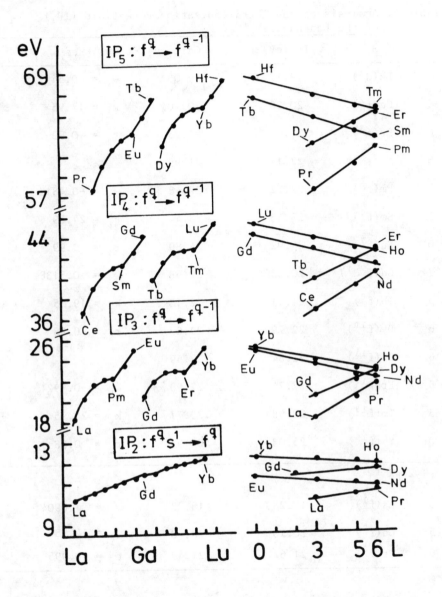

Fig. 7 Comparison of the normalized ionization potentials
$IP_2(f^q s^1 \to f^q)$ and IP_3 to $IP_5(f^q \to f^{q-1})$ as classical plots and
as Inclined W plots.

Table 5. Analysis of the Third Ionization Potentials (IP_3)
         of the Lanthanides

L	M	IP_3 (eV)	Calc.	w_i, k_i, r_i
2	La(d^1)	19.175	19.05	$w_1 = 0.549$
5	Ce(f^2)	20.20	20.70	$k_1 = 17.95$
6	Pr(f^3)	21.62	21.25	$r_1 = 0.93$
6	Nd(f^4)	22.14	21.99	
5	Pm(f^5)	22.32	22.48	$w_2 = -0.4812$
3	Sm(f^6)	23.43	23.44	$k_2 = 24.89$
0	Eu(f^7)	24.92	24.89	$r_2 = 0.99$
2	Gd($f^7 d^1$)	20.63	20.58	$w_3 = 0.5138$
5	Tb(f^9)	21.91	22.12	$k_3 = 19.55$
6	Dy(f^{10})	22.79	22.63	$r_3 = 0.98$
6	Ho(f^{11})	22.84	22.59	
5	Er(f^{12})	22.74	22.98	$w_4 = -0.3921$
3	Tm(f^{13})	23.68	23.77	$k_4 = 24.95$
0	Yb(f^{14})	25.03	24.95	$r_4 = 0.98$
$f^q \to f^{q-1}$				
3	La(f^1)	18.283	18.22	$w_1 = 1.0904$
5	Ce(f^2)	20.20	20.40	$k_1 = 14.95$
6	Pr(f^3)	21.62	21.49	$r_1 = 0.995$

                                                                             contd.

Table 5. Contd.

L	M	IP_3 (eV)	Calc.	w_i, k_i, r_i
$f^q \to f^{q-1}$				
3	Gd(f^8)	20.34	20.32	$w_3 = 0.8121$
5	Tb(f^9)	21.91	21.95	$k_3 = 17.89$
6	Dy(f^{10})	22.79	22.76	$r_1 = 0.999$

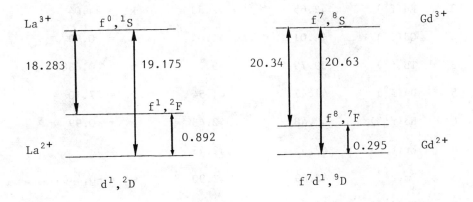

Table 6. Analysis of the Fourth Ionization Potentials (IP_4) of the Lanthanides

L	M	IP_4 (eV)	Calc.	w_i, k_i, r_i
3	$Ce(f^1)$	36.758	36.712	w_1 = 1.2021
5	$Pr(f^2)$	38.98	39.12	k_1 = 33.11
6	$Nd(f^3)$	40.41	40.32	r_1 = 0.99
6	$Pm(f^4)$	41.09	41.02	
5	$Sm(f^5)$	41.37	41.52	w_2 = -0.5038
3	$Eu(f^6)$	42.65	42.53	k_2 = 44.04
0	$Gd(f^7)$	44.01	44.04	r_2 = 0.99
3	$Tb(f^8)$	39.79	39.77	w_3 = 0.8886
5	$Dy(f^9)$	41.47	41.54	k_3 = 37.10
6	$Ho(f^{10})$	42.48	42.43	r_3 = 0.99
6	$Er(f^{11})$	42.65	42.45	
5	$Tm(f^{12})$	42.69	42.90	w_4 = -0.4464
3	$Yb(f^{13})$	43.74	43.79	k_4 = 45.13
0	$Lu(f^{14})$	45.19	45.13	r_4 = 0.99

Table 7. Analysis of the Fifth Ionization Potentials (IP_5) of the Lanthanides

L	M	IP_5 (eV)	Calc.	w_i, k_i, r_i
3	$Pr(f^1)$	57.53	57.47	$w_1 = 1.365$
5	$Nd(f^2)$	60.00	60.20	$k_1 = 53.37$
6	$Pm(f^3)$	61.69	61.56	$r_1 = 0.99$
6	$Sm(f^4)$	62.66	62.67	
5	$Eu(f^5)$	63.23	63.31	$w_2 = -0.6426$
3	$Gd(f^6)$	64.76	64.60	$k_2 = 66.53$
0	$Tb(f^7)$	66.46	66.53	$r_2 = 0.99$
3	$Dy(f^8)$	62.08	62.05	$w_3 = 0.995$
5	$Ho(f^9)$	63.93	64.04	$k_3 = 59.06$
6	$Er(f^{10})$	65.10	65.03	$r_3 = 0.99$
6	$Tm(f^{11})$	65.42	65.26	$w_4 = -0.5137$
5	$Yb(f^{12})$	65.58	65.77	$k_4 = 68.34$
3	$Lu(f^{13})$	66.79	66.79	$r_4 = 0.99$
0	$Hf(f^{14})$	68.375*	68.34	

* Value given by V. Kaufman and J. Sugar, J. Opt. Soc. Amer., <u>66</u>, 1019 (1976). The value of IP_5 for Hf^{4+} predicted from the Tm–Lu tetrad by the Inclined W systematics with $IP_5 = -0.478L + 68.16$, is 68.16 eV agreeing with the measured value of Sugar within 0.3%.

Following the Inclined W technique described earlier[17], it is possible to predict the ionization potentials for several pseudo-lanthanides. Thus the Tm-Lu tetrad for the fifth potentials generated a value of 68.16 eV for IP_5 of Hf. This value (Table 7) is within $\sim$0.3% of that (68.375 eV) given by Kaufman and Sugar[24] recently. Extending the calculations for the sixth IP's, we have obtained the IP_6 values for Hf and Ta, e.g., 88.83 and 90.51 eV, respectively (Table 8). A value of 94.01 eV for IP_6 of Ta is predicted by Kaufman and Sugar[16]. Although the value predicted by the Inclined W systematics for IP_6 of Ta is not as good as that for IP_5 of Hf, it is still close enough to merit a mention. We are presently engaged in analysing the IP_6 values to obtain a better systematic for the lanthanide series.

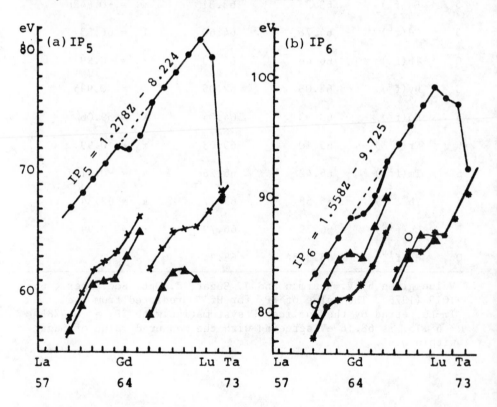

Fig. 8 Comparison of (a) the fifth potentials: Calc. SCF-HF (●) [12], Calc. HF (▲)[22], Sugar (✕)[20] and (b) the sixth potentials: Calc. SCF-HF (●)[12], Calc. HF (▲)[22], Inclined W (corrected values) (✳)[17], Cowan (○)[23].

Table 8. Predicted Sixth Ionization Potentials for the Lanthanides
and Pseudo-Lanthanides from Inclined W Systematics

L	M	IP_6 (eV)	w_i, k_i, r_i
3	$Nd(f^1)$	78.71	$w_1 = 1.0286$
5	$Pm(f^2)$	81.07	$k_1 = 75.7$
6	$Sm(f^3)$	81.72	$r_1 = 0.99$
6	$Eu(f^4)$	81.91	$w_2 = -0.8857$
5	$Gd(f^5)$	82.96	$k_2 = 87.29$
3	$Tb(f^6)$	84.60	$r_2 = 0.999$
0	$Dy(f^7)$	87.29	
3	$Ho(f^8)$	84.17	$w_3 = 0.549$
5	$Er(f^9)$	85.86	$k_3 = 82.67$
6	$Tm(f^{10})$	85.67	$r_3 = 0.91$
6	$Yb(f^{11})$	87.15	$w_4 = -0.586$
5	$Lu(f^{12})$	87.35	$k_4 = 90.51*$
3	$Hf(f^{13})$	88.83	$r_4 = 0.98$

* Note that this is the predicted value of IP_6 for Ta^{5+} (f^{14})
from Inclined W systematics. Kaufman and Sugar[24] predicted
a value of 94.01 eV.

Before closing this section, I would like to say a few words
about the periodic nature of the Inclined W plots. Although the
exact interrelationship between a measured property of the lanthanides and the ground state L-values is not yet fully understood,
the author is willing to argue in the following direction. The
periodic variation of the ground state orbital quantum numbers (L)
within the p^q, d^q, and f^q series is well known[5].

We may express the electrostatic energies of a configuration
in terms of Racah parameters E^k's. For f^q configuration, the electrostatic energy, W_m, is given by the following formulae

for $0<q<7$:

$$W_m(f^q) = 1/2\,[q(q-1)]E^0 + [3(u_1^2 + u_2^2 + u_1u_2 + 5u_1 + 4u_2) \\ - 3/2L(L+1)]E^3 \tag{6}$$

for $q>7$:

$$W_m(f^q) = 1/2\,[q(q-1)]E^0 + 9/2\,(q-2S)E^1 + [3(u_1^2 + u_2^2 + u_1u_2 \\ + 5u_1 + 4u_2) - 3/2L(L+1)]E^3 \tag{7}$$

The values of the coefficients of the individual terms in the above
equations are plotted against q, the number of f-electrons in Fig.9.
It is clearly seen that the [U,L] term (abbreviated for the coefficient of E^3 in the square bracket) is mainly responsible for the
periodicity (tetrad and the Inclined W) in the lanthanide series.
Furthermore, we know that the values of E^1 and E^3 increase through
the series by about 60% and that $E^1 \approx 10E^3$. The strong discontinuity for the coefficient of E^1 at the middle of the series, and
also that for E^3, are noteworthy. The first q dependent term varies monotonically and is often used as scaling parameters. Inclusion of the first order spin orbit coupling would not alter the
general shape of the [U,L] plot, except that q = 1, 6, 7, 8, 13, 14
will be non-degenerate and will not be equal to zero. As Inclined W
systematics makes use of the ground state L-values of the ions, the
contribution of the spin orbit coupling is an important factor and
we are examining this parameter more carefully.

3.3 Correlation of the Lattice Constants and Bond Distances

Prof. Klemm in his lecture mentioned (see Chapter 1) that
the lattice constants of the C-form of lanthanide oxides (M_2O_3)
exhibit only weak discontinuity around Gd[25]. Using recently
measured lattice parameters[26-28], we have calculated the M-O
bond distances using $Ia3(T_R^7)$ space group at two different metal
sites (C_{3i} and C_2). These data are collected in Table 9. Both the
lattice parameters (a) and the average M-O bond distances of the

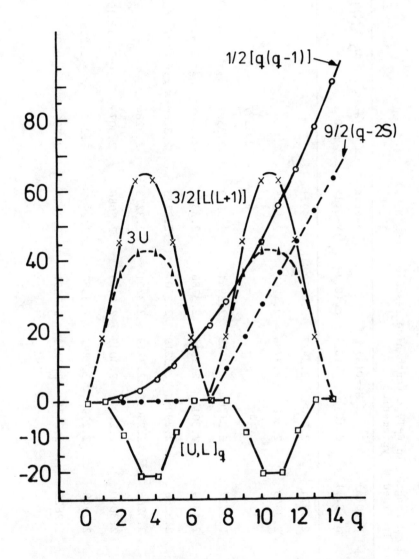

Fig. 9 Plots of the coefficients of E^0, E^1, and E^3 in eq. (6) and (7) against the number (q) of the f-electrons showing periodicity. The [UL]'s are responsible for the tetrad and the Inclined W effects.

Table 9. The M-O Bond Lengths in C-Type Lanthanide Sesquioxides (M_2O_3) and Their Inclined W Systematics

L	M^{3+}	M_1-O(Å) $d_1(C_3i)$	M_2-O(Å)			Average $M-O\langle d\rangle$	Calculated $M-O$	w_i, k_i, r_i
			$d_2(C_2)$	$d_3(C_2)$	$d_4(C_2)$			
5	Pr	2.3974	2.3608	2.3922	2.4379	2.3971		
6	Nd	2.3843	2.3483	2.3786	2.4267	2.3845		
6	Pm	-	-	-	-	-	2.3557	
5	Sm	2.3527	2.3176	2.3473	2.3941	2.3529	2.3507	$w_2= 4.976\times10^{-3}$
3	Eu	2.3369	2.3017	2.3314	2.3786	2.3372	2.3408	$k_2= 2.3258$
0	Gd	2.3266	2.2905	2.3219	2.3701	2.3273	2.3260	$r_2= 0.97$
0	Gd	2.3266	2.2905	2.3219	2.3701	2.3273	2.3284	
3	Tb	2.3090	2.2737	2.3039	2.3479	2.3086	2.3073	$w_3=-7.059\times10^{-3}$
5	Dy	2.2948	2.2599	2.2891	2.3398	2.2959	2.2931	$k_3= 2.3284$
6	Ho	2.2825	2.2474	2.2773	2.3251	2.2831	2.2861	$r_3= 0.99$

contd.

Table 9. Contd.

L	M^{3+}	M_1-O(Å)	M_2-O(Å)			Average	Calculated	w_i, k_i, r_i
		$d_1(C_{3i})$	$d_2(C_2)$	$d_3(C_2)$	$d_4(C_2)$	$M-O\langle d\rangle$	$M-O$	
6	Er	2.2699	2.2351	2.2642	2.3107	2.2700	2.2655	
5	Tm	2.2565	2.2215	2.2519	2.2987	2.2572	2.2602	$w_4=5.295 \times 10^{-3}$
3	Yb	2.2453	2.2107	2.2401	2.2871	2.2458	2.2500	$k_4=2.2338$
0	Lu	2.2360	2.2022	2.2306	2.2758	2.2362	2.2340	$r_4=0.96$

The lattice parameters of Ref.[26] together with the positional parameters from Ref.[29] were used to calculate the M-O bond distances in C_{3i} and C_2 sites. All four distances were averaged and used in the Inclined W correlation. Exclusion of Gd from the third tetrad results in lower correlation coefficient ($r_3=0.97$).
The predicted average M-O bond distance in C-type Pm_2O_3 is 2.356Å.

C-form of the oxides (M_2O_3) exhibit excellent L linearization
(Inclined W) plots (Fig. 10). We would again comment here that
Gd serves as midpoint for second and third tetrad generating what
is known as the symmetric Inclined W plot.

There has been much discussion on the structure of the aquo
lanthanide ions in solution and periodically this theme forms the
subject matter of talking points. While I am willing to subscribe
to the model of the aquo ion as being nonacoordinated (first coordination sphere) for the first half of the series and possibly

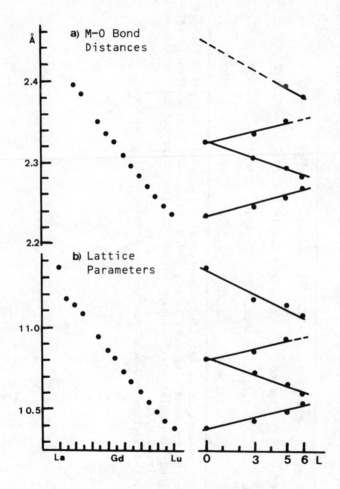

Fig. 10 Plots of (a) the M-O bond distances and (b) the lattice
parameters of the C-form of the lanthanum sesquioxides.

quasi-nonacoordinated (9↔8) for the heavier lanthanides at the
end of series, I believe that the chemical reactions of the lanthanide ions occuring in aqueous solution do not depend on the
exact coordination number of the aquo lanthanides, rather on the
nature of the aquo species, its electronic ground state and its
kinetic and thermodynamic stability.

The situation is, however, different in the solid phase.
Crystals of bromate $[M(OH_2)_9(BrO_3)_3]$ and ethylsulphate $[M(OH_2)_9$
$(C_2H_5SO_4)_3]$ were found to contain nonacoordinated lanthanides
[30-34]. Recently, Albertson and Elding[35] provided accurate
data on the space groups and the lattice constants of all lanthanide bromates $(P6_3/mnc)$ and ethylsulphates $(P6_3/m)$ excepting Pm,
and they found that the members of each series (bromate and ethylsulphate) to be isostructural and to contain nonacoordinated aquo
ion, $[M(OH_2)_9]$, with the water oxygen atoms in tricapped triagonal
arrangement. The lanthanide site symmetry is D_{3h} in bromate and C_{3h}
in ethylsulphate complex. Analysis of the lattice constants (a)
and (c) of the bromates reveals a symmetric Inclined W plot with
high correlation[Table 10]. Actually the exclusion of Gd from the
third tetrad (Tb-Ho) results in lower correlation coefficients.
The predicted lattice parameters for Pm-bromate are a=11.7912 and
c=6.7546Å. The a-parameters of the ethylsulphate also exhibit a
symmetric Inclined W plot but the exclusion of Gd c-parameters in
the analysis of the c-parameters of the third tetrad results in a
better correlation (r = 0.999) than if it is included(r = 0.989).
The predicted lattice parameters for Pm-ethylsulphate are
a=13.9912 and c=7.0936Å.

It is now interesting to compare the slopes (w_i) of the four
tetrads in both compounds. The four slope parameters (w_i) would be
equal or very close to each other, if there is a regular change
in the properties through the series. As is evident from the table,
this is not the case for both compounds. For the cell parameter a,
the slopes(w_i) in the four tetrads of both bromate and ethylsulphate are comparable, while those for the cell parameters c have
larger values for bromate than the ethylsulphate, probably indicating slightly distorted structure with C_{3h} symmetry in ethylsulphate compared to the D_{3h} symmetry in bromate. Another interesting
aspect becomes obvious for the slopes (w_i) of the cell parameter c
which are smaller in ethylsulphate than in bromate. This difference
probably reflects two different types of hydrogen bondings in the
two compounds. According to Albertson and Elding[35], the hydrogen
bonds in ethylsulphate are parallel to the c-direction and thus
causing only very small decrease in the c-parameter through the
lanthanide series (∿0.13Å) while these bonds are perpendicular
to c in the bromates (a decrease of ∿0.22 Å in the series). It
seems interesting that such microcriterion as the direction of
hydrogen bonding plays a role in discriminating the S-state Gd(III)
ion in the third tetrad of the ethylsulphate (Table 10).

Table 10. Analysis of Lattice parameters[35] of Lanthanide
                Bromates and Ethylsulphates

L	M(III)	a(obs)	a(calc)	w_i, k_i, r_i	c(obs)	c(calc)	w_i, k_i, r_i
$M(OH_2)_9$	$(BrO_3)_3$	$(P6_3/mmc)$					
0	La	11.8828	11.8827	$w_1 = -0.00956$	6.8462	6.8477	$w_1 = -0.00979$
3	Ce	11.8523	11.8540	$k_1 = 11.8827$	6.8204	6.8183	$k_1 = 6.84767$
5	Pr	11.8395	11.8349	$r_1 = 0.992$	6.8012	6.7987	$r_1 = 0.994$
6	Nd	11.8224	11.8254		6.7858	6.7889	
6	Pm	-	11.7912	$w_2 = 0.004424$	-	6.7546	$w_2 = 0.00498$
5	Sm	11.7885	11.7868	$k_2 = 11.7646$	6.7511	6.7496	$k_2 = 6.7247$
3	Eu	11.7750	11.7779	$r_2 = 0.975$	6.7371	6.7396	$r_2 = 0.985$
0	Gd	11.7658	11.7646		6.7257	6.7247	
0	Gd	11.7658	11.7665	$w_3 = -0.00526$	6.7257	6.7264	$w_3 = -0.00604$
3	Tb	11.7503	11.7507	$k_3 = 11.7665$	6.7080	6.7083	$k_3 = 6.7264$
5	Dy	11.7453	11.7402	$r_3 = 0.964$	6.7016	6.6962	$r_3 = 0.970$
6	Ho	11.7308	11.7349		6.6859	6.6902	
6	Er	11.7239	11.7215	$w_4 = 0.00432$	6.6733	6.6683	$w_4 = 0.0064$
5	Tm	11.7162	11.7172	$k_4 = 11.6956$	6.6570	6.6619	$k_4 = 6.6299$
3	Yb	11.7056	11.7086	$r_4 = 0.977$	6.6474	6.6491	$r_4 = 0.969$
0	Lu	11.6973	11.6956		6.631	6.6299	

contd.

Table 10: Contd.

L	M(III)	a(obs)	a(calc)	w_i, k_i, r_i	c(obs)	c(calc)	w_i, k_i, r_i
$M(OH_2)_9$ $(C_2H_5SO_4)_3$ $(P6_3/m)$							
0	La	14.1084	14.1093	$w_1 = -0.01323$	7.1491	7.1503	$w_1 = -0.00598$
3	Ce	14.0706	14.0696	$k_1 = 14.1093$	7.1346	7.1323	$k_1 = 7.1503$
5	Pr	14.0454	14.0431	$r_1 = 0.998$	7.1207	7.1204	$r_1 = 0.994$
6	Nd	14.0275	14.0299		7.1130	7.1144	
6	Pm	–	13.9912	$w_2 = 0.00466$	–	7.0936	$w_2 = 0.0033$
5	Sm	13.9884	13.9866	$k_2 = 13.9633$	7.0909	7.0903	$k_2 = 7.0738$
3	Eu	13.9742	13.9772	$r_2 = 0.975$	7.0827	7.0837	$r_2 = 0.995$
0	Gd	13.9645	13.9633		7.0742	7.0738	
0	Gd	13.9645	13.9649	$w_3 = -0.00713$	–	–	
3	Tb	13.9444	13.9435	$k_3 = 13.9649$	7.0645	7.0646	$w_3 = -0.00568$
5	Dy	13.9287	13.9292	$r_3 = 0.9994$	7.0537	7.0533	$k_3 = 7.0817$
6	Ho	13.9221	13.9221		7.0473	7.0476	$r_3 = 0.999$
6	Er	13.9110	13.9091	$w_4 = 0.00368$	7.0371	7.0357	$w_4 = -0.003478$
5	Tm	13.9026	13.9055	$k_4 = 13.887$	7.0308	7.0322	$k_4 = 7.0148$
3	Yb	13.8991	13.8981	$r_4 = 0.979$	7.0247	7.0252	$r_4 = 0.991$
0	Lu	13.8870	13.8870		7.0153	7.0148	

3.4   Analysis of the Formation Constants of the Lanthanides

We have also made a systematic analysis of the formation constants of a series of simple carboxylate complexes of the lanthanides (Table 11). Most ligands are "well behaved" in the sense
that hardly any large anomaly in the trend is observed. However,
I would first of all like to draw your attention to the simplest
carboxylate ligand, the formate. The second tetrad (Pm-Gd) exhibit
a very poor correlation (r=0.55). At the first glance in the Z-plot
(Fig. 11) this anomaly does not become obvious except that there
is a big jump in the value between Sm and Eu. The discrepancy in the
Sm-Eu-Gd region is usually explained by the so-called gadolinium
break and the stabilization of the half filled 4f shell in Gd(III).
However, during the fluorescence lifetime measurements of Eu and
Gd formates (discussed in Chapter 10), we have found quite unexpectedly lower lifetime for Eu-formate complex compared to the
acetate. Formate is known to be a reducing ligand and the poor
correlation may be due to partial reduction of Eu and/or Sm in
formate solution.

The other two deviations are (i) in the second tetrad of
chloroacetate and (ii) in the second and third tetrads of the
iodoacetates (Fig. 12). At the present time we do not have any
convincing explanation for these low correlations. One should be
rather careful with the iodoacetate. This ligand tends to hydrolyze easily. We actually do not know the extent of the influence
of the ligand decomposition on the measured formation constants
for both chloro and iodoacetate.

For the chloro and methoxyacetates, the third tetrad starts
with Tb and not with Gd, i.e., both these ligands generate asymmetric Inclined W plots. Inclusion of Gd in the third tetrad lowers the correlation coefficient significantly for methoxyacetate
(including Gd: r=0.20; excluding Gd: r=0.76) and slightly for
chloroacetate (including Gd: r=0.89; excluding Gd: r=0.93). For
marcaptoacetate, however, inclusion or exclusion of Gd in the
third tetrad produces correlations which are statistically indistinguishable. In the table we have included Gd in the third tetrad,
but in the figure (Fig. 12) we have left it out. The predicted
value for Gd from the analysis of the third tetrad excluding Gd
is log K_1=2.0.

It is, however, becoming necessary to remeasure some of these
formation constants. The w_1 values follow the sequence

propionate>isobutyrate>glycolate>acetate>Cl-acetate>I-acetate>
    0.062   0.057   0.054   0.034   0.031   0.026
HS-acetate>CH$_3$O-acetate>formate
    0.013   0.012   0.008

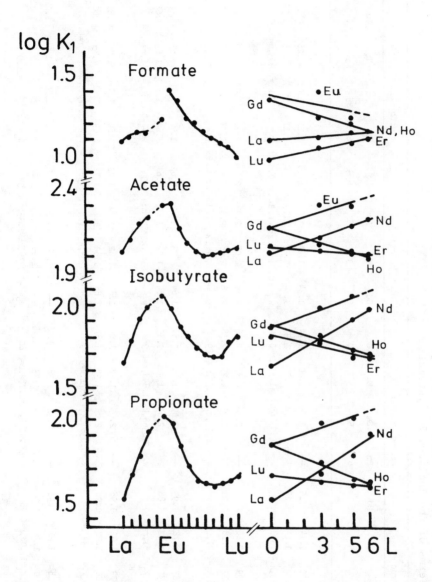

Fig. 11 Comparison of the formation constants ($\log K_1$) of the simple monocarboxylate complexes of the lanthanides.

Table 11. Analysis of the Formation Constants(log K_1) of the Lanthanide Complexes with simple carboxylic Acids in Terms of the Inclined W Systematics

	First Tetrad				Second Tetrad				Third Tetrad				Fourth Tetrad			
M^{3+} =	La	Ce	Pr	Nd	Pm	Sm	Eu	Gd	Gd	Tb	Dy	Ho	Er	Tm	Yb	Lu
L =	0	3	5	6	6	5	3	0	0	3	5	6	6	5	3	0
Formate																
Expt.	1.10	1.12	1.14	1.15	—	1.23	1.40	1.34	1.34	1.24	1.21	1.15	1.11	1.08	1.06	0.99
Calc.	1.10	1.12	1.14	1.15	1.26	1.28	1.32	1.37	1.34	1.25	1.19	1.16	1.11	1.09	1.05	0.99
w_i,k_i,r	$w_1=0.00833,k_1=1.098,r=0.99$				$w_2=-0.01868,k_2=1.373,r=0.55$				$w_3=-0.0295,k_3=1.34,r=0.98$				$w_4=0.019,k_4=0.993,r=0.99$			
Acetate																
Expt.	2.02	2.09	2.18	2.22	—	2.30	2.31	2.16	2.16	2.07	2.03	2.00	2.01	2.02	2.03	2.05
Calc.	2.01	2.11	2.18	2.21	2.36	2.33	2.27	2.18	2.16	2.08	2.03	2.00	2.01	2.02	2.03	2.05
w_i,k_i,r	$w_1=0.0336,k_1=2.01,r=0.99$				$w_2=0.0297,k_2=2.177,r=0.89$				$w_3=-0.0262,k_3=2.157,r=0.99$				$w_4=-0.0064,k_4=2.05,r=0.99$			
Propionate																
Expt.	1.53	1.67	1.78	1.93	—	2.02	1.98	1.84	1.84	1.73	1.63	1.62	1.60	1.61	1.63	1.66
Calc.	1.51	1.70	1.82	1.88	2.07	2.03	1.96	1.85	1.84	1.72	1.65	1.61	1.60	1.61	1.63	1.66
w_i,k_i,r	$w_1=0.0621,k_1=1.51,r=0.97$				$w_2=0.0368,k_2=1.848,r=0.98$				$w_3=-0.0386,k_3=1.84,r=0.99$				$w_4=-0.01,k_4=1.66,r=0.99$			
Isobutyrate																
Expt.	1.64	1.79	1.92	1.98	—	2.05	1.98	1.87	1.87	1.82	1.74	1.70	1.69	1.69	1.78	1.81
Calc.	1.63	1.80	1.92	1.98	2.09	2.05	1.98	1.87	1.88	1.80	1.74	1.71	1.69	1.71	1.75	1.82
w_i,k_i,r	$w_1=0.0569,k_1=1.633,r=0.99$				$w_2=0.0361,k_2=1.871,r=0.99$				$w_3=-0.0283,k_3=1.882,r=0.98$				$w_4=-0.0221,k_4=1.82,r=0.95$			
Chloro-Acetate																
Expt.	1.12	1.18	1.26	1.31	—	1.32	1.38	1.26	—	1.24	1.21	1.15	1.15	1.10	1.04	1.02
Calc.	1.11	1.20	1.26	1.30	1.37	1.35	1.33	1.28	—	1.25	1.19	1.16	1.13	1.11	1.07	1.01
w_i,k_i,r	$w_1=0.0312,k_1=1.108,r=0.98$				$w_2=0.0142,k_2=1.282,r=0.60$				$w_3=-0.0279,k_3=1.33,r=0.93$				$w_4=0.0207,k_4=1.005,r=0.93$			

contd.

Table 11. Contd.

	First Tetrad				Second Tetrad				Third Tetrad				Fourth Tetrad			
$M^{3+}=$	La	Ce	Pr	Nd	Pm	Sm	Eu	Gd	Gd	Tb	Dy	Ho	Er	Tm	Yb	Lu
$L=$	0	3	5	6	6	5	3	0	0	3	5	6	6	5	3	0
Iodo-Acetate																
Expt.	1.18	1.26	1.34	1.32	—	1.43	1.48	1.36	—	1.15	1.11	1.23	1.26	1.18	1.28	1.32
Calc.	1.18	1.26	1.31	1.34	1.48	1.46	1.43	1.38	—	1.13	1.17	1.19	1.22	1.24	1.27	1.32
w_i,k_i,r	$w_1=0.0262,k_1=1.183,r=0.96$				$w_2=0.0161,k_2=1.381,r=0.67$				$w_3=0.02,k_3=1.07,r=0.50$				$w_4=-0.0162,k_4=1.317,r=0.73$			
Methoxy-Acetate																
Expt.	2.03	2.06	2.07	2.11	—	2.13	2.12	2.06	—	2.05	2.05	2.07	2.08	2.08	2.08	2.09
Calc.	2.03	2.06	2.09	2.10	2.15	2.14	2.11	2.07	—	2.05	2.06	2.06	2.08	2.08	2.08	2.09
w_i,k_i,r	$w_1=0.0117,k_1=2.027,r=0.93$				$w_2=0.0145,k_2=2.065,r=0.96$				$w_3=0.0057,k_3=2.03,r=0.76$				$w_4=-0.0017,k_4=2.088,r=0.89$			
Hydroxy-Acetate																
Expt.	2.55	2.70	2.78	2.89	—	2.91	2.94	2.79	2.79	—	2.92	2.99	3.01	3.06	3.13	3.15
Calc.	2.54	2.70	2.81	2.86	2.97	2.94	2.89	2.81	2.79	2.88	2.94	2.98	3.03	3.05	3.10	3.16
w_i,k_i,r	$w_1=0.0533,k_1=2.543,r=0.99$				$w_2=0.0261,k_2=2.81,r=0.83$				$w_3=0.0309,k_3=2.786,r=0.98$				$w_4=-0.0226,k_4=3.167,r=0.93$			
Mercapto-Acetate																
Expt.	1.98	1.99	2.03	2.07	—	2.11	2.07	2.01		1.96	1.93	1.92	1.94	1.98	1.98	2.01
Calc.	1.97	2.01	2.04	2.05	2.13	2.11	2.07	2.01		1.96	1.93	1.92	1.95	1.96	1.98	2.01
w_i,k_i,r	$w_1=0.0141,k_1=1.968,r=0.90$				$w_2=0.02,k_2=2.01,r=0.99$				$w_3=-0.0152,k_3=2.008,r=0.99$				$w_4=-0.0098,k_4=2.012,r=0.90$			

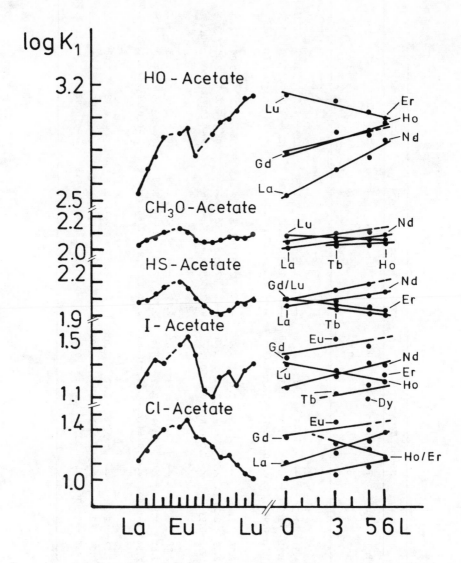

Fig. 12 Comparison of the formation constants (log K₁) of the
mono-substituted monocarboxylate complexes of the lanthanides.

The inductive effect of the CH_3 groups makes the propionate and isobutyrate complexes stronger than the acetate, whereas the electron withdrawing groups like Cl and I make the chloroacetate and iodoacetate complexes weaker than the acetate. It is interesting to see that the hydroxoacetate complexes are somewhat stronger than the acetates. However, tempting it might be to speculate the participation of the OH group in the complex formation, it is the resonance effect that makes hydroxoacetate a ligand comparable to isobutyrate.

We have shown the usefulness of the Inclined W systematics in correlating the properties of the lanthanides and the actinides. But I would like to emphasize once again that extreme caution should be exercised in making sure that the process involved is the same for all lanthanides either in the whole series or at least within each tetrad before attempting the L-linearization (Inclined W) systematics. Caution should also be exercised to see if Gd point belongs to both second and third tetrad (symmetric Inclined W plot) or it belongs to only one of these tetrads by using least squares procedure.

REFERENCES

1. S. P. Sinha, Kemia-Kemi (Helsinki) 5, 238 (1978).

2. S. P. Sinha and A. Bartecki, Inorg. Chim. Acta 31, 77 (1978).

3. S. Odiot and D. Saint-James, J. Phys. Chem. Solids 17, 117 (1960).

4. S. P. Sinha, Helv. Chim. Acta 58, 1978 (1975).

5. S. P. Sinha, Inorg. Chim. Acta 44, L 299 (1980).

6. S. P. Sinha, Struct. Bonding 30, 1 (1976).

7. L. Brewer, Lawrence Barkley Lab. Report, LBL 3720 (1975), see also Chapter 2.

8. H. H. Hill, Plutonium and Other Actinides, Nucl. Met. Series 17, 2 (1970).

9. R. Jullien, E. Galleani d'Agliano and B. Coqblin, Phys. Rev. B 6, 2139 (1972).

10. J. T. Weber, Heavy Element Properties, North-Holland, Amsterdam, 1976, pp. 29-64.

11. S. P. Sinha, Physica 102 B, 25 (1980).

12. T. A. Carlson, C. W. Nestor, N. Wasserman and J. D. Mc Dowell, Oak Ridge National Lab. Report, ORNL - 4562, July 1970.

13. C. E. Moore, Ionization Potentials and Ionization Limits Derived from the Analysis of Optical Spectra, National Bureau of Standard Report, NSRDS-NBS 34, Sept. 1970.

14. L. Brewer, J. Opt. Soc. Amer. 61, 1101 (1971).

15. L. Brewer, J. Opt. Soc. Amer. 61, 1666 (1971).

16. E. F. Worden, R. W. Solarz, J. A. Paisner and J. G. Conway, J. Opt. Soc. Amer. 68, 52 (1978).

17. S. P. Sinha, Inorg. Chim. Acta 27, 253 (1978).

18. J. Sugar and J. Reader, J. Opt. Soc. Amer. 55, 1286 (1965).

19. J. Sugar and J. Reader, J. Chem. Phys. 59, 2083 (1973).

20. J. Sugar, J. Opt. Soc. Amer. 65, 1366 (1975).

21. J. Sugar and N. Spector, J. Opt. Soc. Amer. 64, 1484 (1974).

22. S. Fraga, J. Karwowski and K. M. S. Saxena, Handbook of Atomic Data, Elsevier Publ., Amsterdam, 1976.

23. R. D. Cowan, Nucl. Inst. Methods 110, 173 (1973).

24. V. Kaufman and J. Sugar, J. Opt. Soc. Amer. 66, 1019 (1976).

25. H. Bommer, Z. anorg. allgm. Chem. 241, 273 (1939).

26. R. S. Roth and S. J. Schneider, J. Res. Nat. Bur. Stand. 64 A, 309 (1960).

27. V. B. Glushova and E. K. Keler, Dolkd. Akad. Nauk. SSSR 152. 611 (1963).

28. L. Eyring and N. C. Bainziger, J. Appl. Phys. 33, 428 (1962).

29. G. Gashurov and O. J. Sovers, Acta Cryst. B 26, 938 (1970).

30. S. Singh, Z. Krist. 105, 384 (1944).

31. L. Helmholz, J. Amer. Chem. Soc. 61, 1544 (1939).

32. D. R. Fitzwater and R. E. Rundle, Z. Krist. 112, 362 (1959).

33. E. B. Hùnt, R. E. Rundle and A. J. Stosick, Acta Cryst. $\underline{7}$, 106 (1954).

34. S. K. Sikka, Acta Cryst. $\underline{25\ A}$, 621 (1969).

35. J. Albertson and I. Elding, Acta Cryst. $\underline{33\ B}$, 1460 (1977).

DISCUSSION

KANELLAKOPULOS (question)

How do you calculate the ground state configurations and respective L's? Is really L a good quantum number? As all experimental
data are obtained at room temperature, you have not any idea on
the mixing of J.

SINHA (answer)

For the trivalent lanthanides the ground state configurations without exception are the $4f^n$ configurations and as you know these are
known for surity for a number of years now. We use these configurations and the corresponding free ion L values, which are also
known. In the past, I have developed some simple rules, based on
the number of electrons in the partly filled shell, to calculate
the ground state L values [Inorg. Chim. Acta 44, L299 (1980)]. A
program in BASIC is also available for this calculation [Sinha,
unpublished]. These equations work for all known configurations
provided the ground state follows the Hund's rule strictly. Ground
states for mixed configurations may also be calculated by these
formulae, when the Hund's rule is obeyed.

For the trivalent lanthanides, the ground states are over 98% pure
LS states. Anyway, you start with Russell-Saunders approximations
for all spectroscopic calculations. For the ground states, J mixing is not a problem.

KANELLAKOPULOS (question)

If you plot say, the chemical shift of proton resonance you have
quite different plot. Why is that?

SINHA (answer)

Good that you asked this question. The chemical shifts measured
by any magnetic resonance technique are not expected to follow
the L-correlation systematics presented here. We shall expect the
chemical shifts to follow J values of the ground states more closely. But as J resulted from LS coupling due to spin-orbit interaction, there should be some correlation with L also. I have looked
through this problem in some detail.

Take the case of Eu(III) and Sm(III). The magnetic data even do
not follow the ground state J values (0 and 5/2) as closely as
expected. This is due to the fact that the next spin-orbit state
1 and 7/2 for Eu(III) and Sm(III) respectively are within the
limit of kT, and these states ($^7F_1$ for Eu(III) and $^6H_{7/2}$ for Sm(III))

are well populated around 300° K.

GROUP 1 (question)

Why not use J quantum number for your correlation, and if you did, what form could they take?

SINHA (answer)

We did use the J model, and here is a good plot of the extraction coefficients (log D) against the J quantum numbers for the trivalent lanthanides (Fig. 1). If we plot J versus the atomic numbers of the lanthanides (J = 0-8), the same sequence is followed (Fig. 2). You would, however, notice that now Eu has become the mid point. You would also remark here that we did not take the ground state of Eu(III) to be $^7F_0$. Spectroscopic ground state of Eu is $^7F_0$, but the next state of $^7F_1$ is within the kT. For Eu at room temperature both levels are so well populated that the effective J could be possibly 1. This is why we did not take Eu as 0, instead we took it as 1. You still see that the linear character is preserved within the so called tetrads for the lanthanide series. So J is a parameter which could be used to systematize certain properties which depend on J (magnetic properties).

For the spectroscopic states one may use Prof. Brewer's tables [preceding paper] or C.E. Moore's tables [US Natl. Bur. Std. NSRDS-NBS 34, 1970]. Some J correlations are presented in an article in Struct. Bonding 30, 1 (1976) which was dedicated to Prof. Klemm's 80th. birthday.

NIINISTÖ (question)

The formation constants of Eu with formate and acetate ligands could not be satisfactorily correlated with the Inclined W theory. The case of Eu-formate was explained by partial reduction of trivalent Eu. Is it possible that in the case of the acetate complex there is an abrupt change in complexation (for instance coordination number) which would explain the anomaly, in other words, how reliable are the formation constant data?

SINHA (answer)

Formation constant data are very reliable. I don't know what the authors quoted as reliability factors of their measurements but acetate complexes have been measured many times, not only by potentiometric method but also by spectroscopic technique, and the trend is now well established. All simple carboxylate ligands, show this kind of trend; log K_1 increasing until Eu and then decreasing very drastically. The values which I put up for acetate were from Ingmar Grenthe. He at one time, and also Greg Choppin

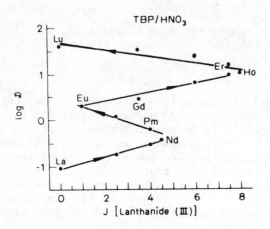

Fig. 1   Plot of the distribution coefficients(D) for the extraction
of the lanthanides with TBP. Note that both Eu and Ho points act
as mid-points between the second/third and the third/fourth tetrads.

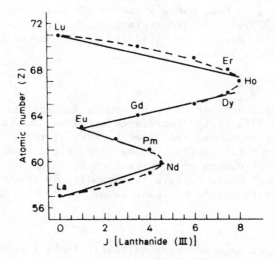

Fig. 2   Variation of the J quantum numbers of the trivalent lanthanides in their ground states with the atomic numbers of the series.
The dotted curve is the observed variation. The straight lines are
drawn to emphasize the variation from linearity.

tried to say something like this: the model they visualized was
that of a solvent separated ion-pair having a lanthanide ion,
water molecule and an acetate ion coming in during the first phase.
As the acetate concentration increased, this cell breaks up and
then the solvent layer covers the acetate complex itself. I do
not know how accurate this model is, because one cannot test this
model by experiments. I would like to add a comment here that the
trend in fluorescence lifetimes of the acetate complexes shows
these to be reasonably strong complex. Moreover, the lifetime data
of Eu, Sm, Gd and Tb aquoions indicate that these ions are kine-
tically different (chapter 10).

KANELLAKOPULOS (comment)

Formation constants have been measured in aqueous systems, and
the problem is that we have no idea of the coordination numbers
of such complexes in solution. Perhaps we observe the irregularities due to different modes of coordination.

SINHA (answer)

The normal classical plots of the formation constants vs. Z do
show different kinds of variation and this was the point of discussion for about the last twenty years, whether the observed irregularities are mere artifacts. However, the experimental measurements, are accurate enough. I think Prof. Thompson, who has done
extensive formation constants measurements, may like to comment
on the accuracy of these measurements. Even if you allow an error
of log 0.1 you would still get a good plot in our systematics as
you have already seen. I guess the question of the coordination
numbers for the lanthanides and their complexes in aqueous solution will be debated for a long time to come.

THOMPSON (comment)

Accuracy is quite good, but the problem is, we have diluted systems.

SINHA (comment)

Oh, very good that you mention of the diluted system. So we know
that the multinuclear complexes are not forming, and we would not
expect a longrange coupling between the lanthanides. Secondly, in
the diluted systems the bonding is still very much ionic bonding,
and the f^n character in the complex is still preserved. What we
actually need in my L-linearization systematics is the concept of
L-effective $[L_{eff} = L_{f.i.} - \Delta L]$. The term ΔL in this equation depends on the internal pressure generated by the ligand atoms for-
ming the complex. However, I have no knowledge how this L_{eff} would
look like in various systems. I might add here that calculations

of L_{eff} are feasible by taking a statistical model for an atom, like that of Thomas-Fermi for example, and carry out pressure dependent calculations of L. I am at present looking into this problem.

KANELLAKOPULOS (question)

You cannot use free ion L values in the complexes of the lanthanides. I also think that the ground state would split due to crystal field and there would be configuration interaction.

SINHA (answer)

As I've mentioned before that our ignorance of the pressure (internal) dependence of ΔL does not allow us to use anything like L_{eff} as a parameter. Thus at present we approximate $L \approx L_{eff}$; that is to say that ΔL is changing by a constant factor and is independent of L. This may or may not be true. Unfortunately, we have to live with this situation until more sophisticated calculations are performed.

To answer the second part of your question let me point out that the crystal field splitting of the ground state in the lanthanide complexes is very small (a maximum of $\sim 200 - 300$ cm^{-1}) and I would

$$[Z] \longrightarrow \{s^x p^y d^q f^n\} \longrightarrow (^{2s+1}L)$$

| At.no. corresponding to the nuclear + ve charges independent of the ionicity of the atom. | Electronic config. dependent on the nature of the ionic state of the atom. Allows the calculation of an electronic ground state. | Electronic ground term depending on the electronic configuration influenced by the environment, crystal field or others. |

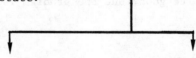

| | Spin orbit coupling negligible compared to the electrostatic interaction preserves the L character of the ground state. | Spin orbit coupling comparable or larger than the electrostatic interaction splits the ground term in the J states: $^{2S+1}L_J \ldots\ldots\ldots ^{2S+1}L_{J^n}$ where the $^{2S+1}L_J$ is the ground state. Environmental effects now operate mostly on J. |

not expect the configuration interaction to show up in the ground
state. This may be the best place to elaborate a bit more on the
concept of L-linearization systematics.
First of all let me give a shcematic appreciation of an ion with
atomic number Z having electronic configuration say $s^x p^y d^q f^n$ for
the ground state. Prof. Brewer has shown earlier how the cohesive
energies may be related to the ground state electronic configuration. I found that Brewer's cohesion energies do fit the L-syste-
matics. This is not at all surprising for me, as the ground state
L's are dependent on the electronic configuration. If I tell you
that I am using an ion of Z = 60, this is not much of an information except that you will gather that I am thinking of Nd at certain stage of ionization. But if you are given the information
that this ion has f^3 configuration, we would immediately know
that Nd is trivalent. If we are now given the information that
this f^3 ion has a ^4I ground term we can already predict many properties, including the nature of the lowest ground state due to
spin orbit splitting. Naively speaking, the ^{2s+1}L or $^{2s+1}L_J$ is a
better description of the ground state of an atom or ion.

RAJNAK (comment)

Perhaps it's a question of terminology, but the configuration interaction as it is normally understood is not changing L. It is
the spin orbit coupling that mixes other terms with different L.

URLAND (question)

I wonder why this works, because mixing by configuration interaction will give some mixed values of L. Couldn't this be the case
in metals?

SINHA (answer)

I think part of your question is already answered by Rajnak's
comment that configuration interactions do not change L. Metal
with a definite electronic configuration also generates a particular ground state having a definite ground state L value. Now
the problem is that if there is a fractional occupation of orbitals which I agree with Prof. Brewer that it might exist in many
metals. It would be very hard for us to calculate with any accuracy the ground state L values for such fractional occupation.
However, even though there is configuration interaction, we are
able to obtain the ground state L values from the ground states
derived by experiments. Here, I must utter a word of caution that
the ground state of the free ions predicted from the spectroscopic
measurements and those in the condensed phase may or may not be
the same. Usually they are different. Hence in selecting the L
values of the metals in the condensed phase extreme caution should
be used. One also needs to take into account the electron distri-

bution of the particular phase under experimental condition (see Table 1 of my paper).

URLAND (question)

How many cases are off?

SINHA (answer)

We haven't found yet any case where it is really off. We have explanations where points (elements) are off the linear plot. If it is off, you better look for the factors that may be operating in that particular system.

NETZ (question)

Do the tetrads give information of the magnetic properties of the rare earth?

SINHA (answer)

It should give some information. Now if you are measuring the magnetic properties of metals or compounds, I would guess the correlation with J-quantum number is more important for the Lanthanides, which in turn depends on the coupling of L and S. So there should be some information available. J also shows a tetrad-classification, but it is not the same type of tetrad that you have when you correlate with L values.

KANELLAKOPULOS (question)

You have shown the correlation of the crystallographic parameter of the lanthanide c-oxides. Why don't you try to use the crystallographic data for the actinide halides and other compounds?

SINHA (answer)

Yes, we have done that and also for the lanthanides. This was presented in Prof. Niinistö's conference that the crystallographic parameters of the chlorides, bromides of the actinides measured by Asprey follow very closely the L systematics presented here [Ref. 1]. But for other correlations I would not be surprised if there are small deviations. In the case of the actinides the jj-coupling might influence our systematics. However, as you would recall that the limited amount of data available for the whole actinide series could be systematized satisfactorily in our systematics.

THOMPSON   (question)

Is there an average distance used for the correlation of the c-oxides?

SINHA   (answer)

Yes, we took the average distance. I had to make some compromise in there. As you know that in the c-oxides there are two lanthanide sites, the C_{3i} site where all six M-O distances are equal and the C_2 site where there are two lots of three different M-O distances. The M-O (C_{3i}) and the average of the three different M-O (C_2) distances exhibit separately the L-linearization. The correlation coefficients for these two plots are statistically indistinguishable from that in the plot of the average M-O distance vs. L.

AMBERGER   (question)

Did you try to plot crystal field parameters $A_1^m <r^1>$ of the lanthanide ions and how good are the correlations ?

SINHA   (answer)

Yes, we have plotted experimental $|A_1^m <r^1>|$'s for the lanthanide ethylsulphates against the L-values of the trivalent lanthanides. Reasonably good linear plots are obtained within the tetrads [see Ref. 6 of my paper for details]. Personally, I am not very satisfied with these plots. There is a lack of data, of course. The four sets of $|A_1^m <r^1>|$ do not vary similarly. You may recall that during the Sixties, we were of opinion that A_1^m's are constant through the lanthanide series. This is not the case. We need accurate data for several systems for the whole lanthanide series to present a general picture. At the moment we are too limited.

URLAND   (comment)

We just began using the angular overlap model for our calculations. We had to use next neighbour, next-next neighbour interactions to explain A_2^o term.

KANELLAKOPULOS   (comment)

I must emphasize that you have localized electrons. If the angular distribution of charge around the central ion changes, there should be an electrostatic (space) correlation between L and the thermodynamic parameters.

RAJNAK   (comment)

The electrostatic interaction, if that's what is diagonal in L,
so you cannot change L by purely electrostatic interaction even
if you're thinking of bonding. If you have some other configuration interaction, you are still going to have the same L.

SKRIVER   (comment)

What you are talking about is the 4f quantum numbers. You can do
this perfectly well in atomic calculations at localized level and
it has nothing to do with bonding.

SINHA   (comment)

I would like to put the key word, configuration interaction, in
its correct perspective. A wavefunction $\Psi_o(L,S,M_L,M_S)$ which has
the correct eigenvalues L and S for a given term may be constructed from suitable linear combinations of the wavefunctions of the
configuration from which the term originates. Other wavefunctions
$\Psi_n(L,S,M_L,M_S)$ may be constructed with the same L and S but from
other configurations. The mixing of Ψ_n's with Ψ_o is called configuration interaction. For a given term (say ground state term)
the best eigenfunction is actually a linear combination of all
such functions

$$\Psi (L,S,M_L,M_S) = \alpha_o\Psi_o + \alpha_1\Psi_1 + \dots \alpha_n\Psi_n$$

As Prof. Brewer has pointed out earlier that for the lanthanides
the energy difference between configurations is very large and
usually larger than the Coulomb interaction. Hence the mixing is
very small indeed, and could be neglected. Actually the inclusion
of configuration interactions results in further screening of the
nucleus.

GROUP 2   (question)

Do you think that the Inclined W has an underlying physical reason
for the correlation of various properties with L? How does a quantum number which is related to the 4f-shell which is rather inac-
tive chemically show up in a formation constant or bond length?
Is there any gas phase data on chemical systems for which the Inclined W works - or any data for solvents other than water?

SINHA   (answer)

This is really a tightly packed one and I shall try to answer as
far as practicable. The magic parameter Z related to the amount of
nuclear positive charge is a good parameter for zeroth approximation. So also is the number of the f electrons in the lanthanide

series (f^n). The screening of the 4f electrons from the nuclear charge is quite effective (see text). We cannot expect Z to correlate all observable properties of atoms and ions and this I have tried to emphasize during my lectures. We need a better, slightly more sophisticated parameter for our correlation. Now, I want to stress once again that Inclined W implies the linear correlation of the properties of the lanthanides (actinides) at any stage of oxidation with the L-values of the ground state electronic configuration of the ions in question. If the plot looks like an Inclined W so much the better, as that's what the first plot I made looked like and hence the name. Inclined W should divide all 1^q configurations into four tetrads [Ref. 4].

I do believe that Inclined W has an underlying physical meaning. The L dependent division in tetrads for all configurations is one. I have also tried to show in Fig. 9 using Eq. (6) and (7) why this periodicity should occur and also that the electrostatic energy depends on the [UL] term of the equations mentioned. I understand your worry about the quantum numbers. When we hear of a "number" our minds tend to focus to a kind of identification tag without telling much about the properties of the material to which the tag is attached. Now, we all know that the Coulomb repulsion between electrons is the dominating interaction operating for the free lanthanide ions. This interaction splits each configuration (f^n) energy levels into terms which are specified by values of total angular momentum quantum number L and total spin quantum number S. Following Hund's rule the ground state has the largest possible value of L. To a first order approximation L and S are still good quantum numbers for the lanthanides, since the overall multiplet plitting due to relativistic effect is much less than the energy separation between the terms.

Let us take the example of the $4f^7$ ion; Gd^{3+}, which I am thinking, (it could well have been the Eu^{2+} with the same electronic configuration), having ground state L = 0. This immediately tells me that it is an S-state ion and there can be no L·S coupling. In a crystal field, the crystal field potential acts only upon the orbital part of the wavefunction, which is non-degenerate $|L=0,M_L=0\rangle$. Hence, according to the first order perturbation model no crystal field splitting is expected. Second order effect is expected to split this ground state of Gd^{3+} but detailed calculations by Wybourne [Phys. Rev. 148, 317 (1966)] were not very successful. Many more properties, like orbital contribution to hyperfine splitting etc. could be predicted for an S-state ion like Gd^{3+} (Eu^{2+}) having L=0. The ground state L gives us the information on the electronic distribution in the ion (atom) of a defined electronic configuration.

Now we shall consider the third ionization potentials amounting to the removal of an f-electron to infinity, $f^q \rightarrow f^{q-1}$ for the

whole lanthanide series. These normalized gaseous ionization potentials were correlated to the ground state L values of the originating ions with a correlation coefficient r = 0.99 [Table 5]
for all tetrads. You probably now see the point. This also answers
your query on the gas phase data. We have only one known case of
the measurements of the ΔH_1 values for lanthanide-ethylenediamine
system in nonaqueous (acetonitile) solvent [J.H. Forsberg and
T. Moeller, Inorg. Chem. 8, 889 (1969)]. These data fits my L-
linearization extremely well [see Ref. 6].

Now I return back to the question of rather inert nature of the
4f electrons. This was the general belief during the Fifties and
the Sixties among the chemists. The nephelauxetic effect (red
shift of absorption band) was thought to be a secondary effect.
Even during the early Sixties it became clear that there could be
appreciable overlap of the ligand orbitals with 4f. Early works
of Ray and his coworkers [Proc. Phys. Soc. 81, 663 (1963); 82, 47
(1963); 86, 1235,1239 (1965); 90, 839 (1967)] positively proved
that a purely ionic description of the lanthanide complexes is
inadequate. In the case of $PrCl_3$ Ray found appreciable overlap between 3p electrons of chlorides and the 4f electrons. The covalency
effect for Pr^{3+} in chloride and bromide was examined by Sinha and
Schmidtke [Mol. Phys. 10, 7 (1965)] and Ellis and Newman [J. Chem.
Phys. 47, 1086 (1967), 47, 4037 (1968)]. It was also found that
small changes in lattice constants produce appreciable effect on
the crystal field for Sm and Er in $LaCl_3$ matrix even when the ligand overlap was neglected [J. Chem. Phys. 50, 1077 (1969)].

The idea of the internal pressure causing a change in electron
distribution in the ground state and hence influencing L is very
attractive. Internal pressure may be generated on the central lanthanide ion by complexing or by changing the lattice (i.e. the
lattice constants, bond distances, etc.). It is my feeling that
for the lanthanide ions the lattice contributions may be significant. We are looking into these matters very carefully. I think
we will be well advised to include both overlap and covalency
effects in our calculations and interpretations of the lanthanide
complexes.

Structures

INORGANIC COMPLEXES OF THE RARE EARTHS

Lauri Niinistö

Department of Chemistry, Helsinki University
of Technology, SF-02150 Espoo 15, Finland

ABSTRACT

Rare earth complexes containing solely inorganic donor
ligands are surveyed with main emphasis on oxygen-donor ligands,
especially the oxoanions. As examples, the sulfato and nitrato
complexes are discussed in more detail. The role of water molecules as ligands and hydrogen bond donors is included in the
discussion. Although solid state structure and properties are
emphasized, a brief comparison to results obtained in solution
is also made.

1.   INTRODUCTION

The early studies on the separation and properties of the
rare earths[1] were largely dealing with inorganic compounds, e.g.
bromates, nitrates and sulfates. Although an enormous amount of
experimental data on the composition and certain properties became
rather soon available for most types of inorganic compounds [1],
their structure and bonding remained unknown until the advent of
modern instrumental techniques.

Many of the classical inorganic compounds of the rare earths
are often referred to as simple or double salts, or when containing
water, as hydrated salts. In spite of the simple stoichiometries
they should be regarded in most cases as complex compounds as
revealed by recent X-ray crystallographic studies.

The present review will attempt to discuss rare earth complexes formed by solely inorganic donor ligands, mainly the

S. P. Sinha (ed.), Systematics and the Properties of the Lanthanides, 125–152.
Copyright © 1983 by D. Reidel Publishing Company.

oxoanions. Although main emphasis will be in the solid state
structure and properties, a comparison to results obtained in
solution will be made whenever possible and relevant. As examples, the sulfato and nitrato complexes will be discussed in more
detail.

2. GENERAL PROPERTIES OF THE RARE EARTH IONS

Several factors contribute to the complex formation of the
rare earth ions and to the properties of the complexes formed.
These include ionic size, oxidation state, bonding to ligands,
coordination number and geometry, thermodynamic and kinetic stability, ligand-ligand repulsions etc. Many of these factors are
related to each other and to the electronic configuration of the
central ion.

In the case of scandium, yttrium and the lanthanoids, the
ionic size, measurable as ionic radii, is often considered to be
the most important factor explaining differencies in a series of
rare earth compounds. In the following ionic radii and some other factors will be briefly discussed.

2.1. Ionic Radii

The first sets of ionic radii were based on a cation coordination number (CN) of six and a fixed value of the anion radius.
The lanthanoid contraction, viz. the gradual decrease of ionic
size from lanthanum to lutetium, was clearly visible in these
values [2].

At the present several more recent sets of ionic radii are
available taking into account the different coordination numbers,
oxidation and spin states as well as the slight variation of anion
radii with CN. Most comprehensive data have been published by
Shannon and Prewitt [3,4] and revised recently by Shannon [5].
These data [5] include the values of effective ionic radii for all
stable trivalent rare earths with coordination numbers of 6,8,
and 9. In addition, values corresponding to CN=10&12 have been
listed for La^{3+} and Ce^{3+} as well as those for 12-coordinated Nd^{3+}
and Sm^{3+}.

On the basis of its effective ionic radius yttrium takes a
position in the middle of the heavier lanthanoids; the values of
6-coordinated Y^{3+} (0.900 Å) and Ho^{3+} (0.901 Å) are equal. On the
other hand, Sc^{3+} (0.745) is in this respect closer to some
trivalent ions of the d block, for example Fe^{3+} (0.645) and Ti^{3+}
(0.670 Å), than to the heavier lanthanoids where the lanthanoid
contraction results in an ionic radius of 0.861 Å for Lu^{3+} [5].

The ionic radii give in general an indication of the expected coordination number in a rare earth complex. This is more apparent in aqueous solutions than in the solid state where the ligands and their coordination modes may result in unexpected coordination numbers. Nevertheless, scandium with its smaller ionic size has a significantly lower average coordination number than the other rare earths, see 3.2.

2.2. Oxidation States

Contrary to some actinoids the oxidation states encountered in lanthanoid compounds in the solid state and especially in the solution are few in number. Standard electrode potentials E^o M(II-III) and E^o M(III-IV) indicate that only Eu^{2+} (-0.35 V), Yb^{2+} (-1.15 V), Sm^{2+} (-1.55 V) and Ce^{4+} (+1.74 V) are sufficiently stable to exist in aqueous solutions [6]. By the use of large complex-forming ligands such as heteropolyanions it has been possible to stabilize to some extent also the tetravalent praseodymium and terbium [7].

From the coordination point of view the tetravalent cerium is interesting due to its high charge and relatively large size which leads to high coordination numbers in the solid compounds and to a resemblance of thorium. In the following only the trivalent state will be dealt with but for comparison an occasional mention to cerium(IV) structures will be made.

2.3. Bonding in Complexes

The properties of the rare earth complexes strongly suggest that the bonding between the metal and the ligands is mainly electrostatic in nature [8,9]. The observed and calculated effects of covalency are very small. For instance, the bonding in contrast to d block elements has no directional characteristics but the spherical symmetry of the central ion is preserved. A closer look reveals that a number of ions (e.g. Sc^{3+}, Y^{3+}, La^{3+}) have a noble-gas configuration and in cases were the 4f orbitals are partly filled they are well shielded from ligand orbital interaction.

The lanthanoids are typical type A acceptors [10] or hard anti-bases (Lewis acids) [11] having strong affinity to oxygen and fluorine. This behaviour is also different from many d block elements.

3.   INORGANIC COMPLEX COMPOUNDS

3.1. Ligands

As noted earlier the lanthanoids as well as scandium and
yttrium are A type acceptors as defined by Ahrland et al. [10]
or hard acids in the Pearson definition [11]. Thus, it is not
surprising that most inorganic complex compounds, especially those
isolated from aqueous solutions, contain oxygen-donor ligands.
Another strong donor atom is fluorine.

In the case of other donor atoms a competition between them
and water molecules usually leads to formation of an aqua complex.
Furthermore, the number of inorganic non-oxygen ligands is limited
and they are usually only monodentate forming rather weak complexes.
The chelate effect which may increase the stability of the complexes formed seem to be limited to organic ligands such as 2,2'-
bipyridyl [12,13].

Apart from halogen-donor ligands, which will be discussed in
more detail below, only a few examples of non-oxygen ligands may
be mentioned. These compounds have been prepared in anhydrous or
nearly anhydrous systems. Thus, rather unstable ammine complexes
of type $LnCl_3 \cdot nNH_3$ have been prepared [14] and the hexacoordinated
isothiocyanate complexes $[Ln(NCS)_6]^{3-}$ stabilized by a large tetra-
n-butylammonium counterion [15]. In the latter case also an X-ray
structural determination has been carried out confirming the
expected but uncommon octahedral coordination for the erbium
Complex[16]. For a review of N-donor ligands, see [13,89].

Rare earth halides, especially the chlorides, have been
extensively studied both in solution and in the solid state.
Fluoride is an exceptional non-oxygen donor ligand as it competes
effectively with water molecules and enters the primary coordination sphere of the cation in aqueous solutions. The stoichiometry of the precipitated fluoride shows it to be nearly anhydrous
while the other halides prepared this way usually contain 6 - 9
water molecules [17]. The removal of water from the halide hydrates is difficult as oxyhalides are easily formed; the iodides
are especially sensitive for decomposition [18].

The bond strengths in halide complexes follow the electronegativity order. The inner vs. outer sphere character may also
be correlated with the basicity of the ligands. Thus, fluoride
forms predominantly inner sphere complexes while already in the
case of the chloride they are of outer sphere type[19].

Table 1 gives examples of some typical halogen- and oxygen-
donor ligands encountered in inorganic rare earth compounds.

Table 1.   Typical ligands in inorganic complex compounds of the
           rare earths.

Ligand type	Examples
X	F^-, Cl^-, Br^-, I^-
XH	OH^-
XH_2	OH_2
XO_2	NO_2^-
XO_3	CO_3^{2-}, NO_3^-, SO_3^{2-}, SeO_3^{2-}, BrO_3^-
XO_4	VO_4^{3-}, NbO_4^{3-}, CrO_4^{2-}, MoO_4^{2-}, WO_4^{2-}, SiO_4^{4-}, PO_4^{3-}, SO_4^{2-}, SeO_4^{2-}, ClO_4^-

Oxygen-donor ligands. As seen from the Table 1 the number
of oxygen-donor ligands is large and the ligands represent different molecular symmetries. In addition, several of these anions
may undergo condensation reactions in solution or in the solid
state thus increasing the number and variety of complexes formed.
For instance, the selenite ion is in aqueous solutions at equilibrium with the hydrogenselenite and diselenite ions and different selenito complexes may be precipitated. An interesting case
is the $PrH_3(SeO_3)_2Se_2O_5$ where the equilibrium has been 'frozen'
in a solid complex. X-ray crystallographic study has shown that
the praseodymium ion is nonacoordinated by six diselenite and
three selenite oxygen atoms[20].

Another group of structures with condensed anions are the
disilicates[21]. An extreme case of condensed anions are
the polyanions and heteropolyanions. They are capable of
forming compounds with the lanthanoids[7]; for instance the
$La_2V_{10}O_{28} \cdot 20H_2O$ has been structurally characterized [22].

A special case is the aqua ligand because it is a neutral
ligand available for coordination in excess under normal preparative conditions employing aqueous solutions. Furthermore, it is
a strong hydrogen bond donor contributing to the stability of the
structures formed. In the next chapter the structures of the
aqua complexes in solution will be discussed in relation to their
solid state structures.

A mixed ligand complexation with both inorganic and organic

ligands is also possible, e.g. the well-known DMSO–nitrato complexes [23] or the crown ether–nitrato compounds [24,25]; these
complexes fall, however, outside the scope of this review.

3.2. Coordination numbers and geometries

A wide range of coordination numbers (CN) extending from
three to twelve has been found in lanthanoid complexes in the
solid state [26]. The lowest coordination numbers, however, are
exceptional and obviously due to steric hindrance caused by
bulky ligands as is the case of the $M[N(Si(NH_3)_6]_3$ complexes [27].

The highly coordinated complexes (CN > 9) are interesting
and several examples can be found in the lanthanoid series. Without going into the rather difficult question as how to define the
coordination number it can be said that CN = 10 and 12 are more
common than previously thought, especially for lanthanum and
cerium. Undecacoordination seemed to be nonexistent for the
trivalent lanthanoids until first such cases were reported in
1978 [28,29]; Table 2 gives a list of all 11-coordinated structures published so far including those with organic ligands and
non-lanthanoid central atoms.

Table 2. Compounds with CN = 11.[2,4]

Compound	Complex	Ref.
$Th_2(OH)_2(NO_3)_6 \cdot 8H_2O$	$[Th(NO_3)_3(H_2O)_3(OH)]_2$	30
$Th(NO_3)_4 \cdot 5H_2O$	$[Th(NO_3)_4(H_2O)_3]$	31,32
$NC_5H_4 \cdot C_5H_4NH\ Ce(NO_3)_4-$	$[Ce(NO_3)_4(H_2O)_2-$	29
$(H_2O)_2(NC_5H_4 \cdot C_5H_4N)$	$(NC_5H_4 \cdot C_5H_4N)]$	
$C_{18}H_{36}O_6N_2Sm_2(NO_3)_6 \cdot H_2O$	$[Sm(NO_3)_5(H_2O)]^{2-}$	33
$La(DAPBAH)^a(NO_3)_3$	$[La(DAPBAH)(NO_3)_3]$	34
$Eu(NO_3)_3C_{10}H_{20}O_5$	$[Eu(NO_3)_3C_{10}H_{20}O_5]$	24
$La(NO_3)_3 \cdot 6H_2O$	$[La(NO_3)_3(H_2O)_5]$	28,35

a 2,6-diacetylpyridinebis(benzoic acid hydrazone)

For each coordination number there are several possibilities for the geometrical arrangement of the ligands leading to the formation of a coordination polyhedron around the lanthanoid ion, cf. Table 3 for CN = 9-12. As the coordination geometries will be discussed in detail elsewhere [36], only two comprehensive surveys will be mentioned here [37,38]. In addition, some examples will be discussed in connection of the nitrato and sulfato structures.

Scandium, yttrium and promethium. The smaller ionic size of scandium together with its tendency to hydrolyze leads into a different structural chemistry [39,40]. Only in a very few cases scandium forms isostructural compounds with other rare earths and coordination numbers exceeding eight are exceptional in inorganic compounds [41].

A survey of 545 crystal structures published in 1970-76, representing 236 different structures, revealed that scandium has an average coordination number of 6.4 while lanthanum has a CN of 8.5 and the smallest lanthanoid lutetium 7.1 [42]. Yttrium had also a value of 7.1 which is unexpectedly low and may be due to

Table 3. High coordination numbers and corresponding coordination
          polyhedra [38,80].

CN	Polyhedron
9	Monocapped square antiprism Tricapped trigonal prism
10	Bicapped square antiprism Tetracapped trigonal prism Tetradecahedron Pentagonal antiprism Pentagonal prism Bicapped square prism
11	Monocapped pentagonal antiprism Decahexahedron Pentacapped trigonal prism
12	Icosahedron Cuboctahedron Hexagonal prism Hexagonal antiprism Bicapped pentagonal prism Anticuboctahedron

the limited data (25 structures). All available information
points out that yttrium should behave like the lanthanoids in the
dysprosium-holmium region in accordance with its ionic radius.

The structural chemistry of promethium, not studied extensively because of its instability, is similar to that of its
adjacent elements, viz. trivalent neodymium and samarium [43].

Structures in solution. The degree and geometry of rare earth
complex formation has been the subject of numerous investigations
by various techniques including X-ray scattering, NMR, MCD and fluorescence spectroscopy as well as ultrasonic, compressibility and
thermodynamic measurements. The results and interpretations seem
to be strongly dependent on the experimental technique used and
sometimes even the same technique has yielded controversial
results. The extensive literature up to 1980 has been recently
summarized and commented by Marcus [44].

If we consider the formation of a lanthanoid complex $LnX_j(H_2O)_z$
[45] by a reaction between the aqua ions $Ln(H_2O)_x$ and $jX(H_2O)_y$
the major questions are: (i) How is X bonded to Ln (inner vs.
outer sphere)?, (ii) What is the coordination number and geometry
for Ln? (iii) How is the CN changed if Ln is changed to another
lanthanoid or yttrium?, and (iv) What is the strength of bonding
as reflected in the rate and mechanism of ligand substitution
reactions?

As none of the experimental techniques mentioned above is
alone capable of answering these questions additional information
must be sought from the structures in the solid state. X-ray
structural information which is available for most inorganic compounds is both useful and necessary when trying to answer at least
the three first questions. For instance, the structures of
lanthanoid bromate and ethylsulfate hydrates contain the
$[Ln(H_2O)_9]^{3+}$ aqua ion thus giving a strong indication for solvate
separated (outer sphere) coordination with a primary hydration
number (coordination number) of nine. Furthermore, as there
exists isostructurality and a smooth variation of unit cell
dimensions from lanthanum to lutetium [46,47] it would be tempting to conclude that the coordination number and geometry remains
constant in the solution.

This conclusion may be valid for the bromates and ethyl-
sulfates but XRD data suggest that in the case of rare earth
chloride solutions there is a change in the coordination number
in solution [48]. The trivalent ions lanthanum through neodymium are 9-coordinated, those between neodymium and terbium
are transitional while the remaining ones are 8-coordinated.
This result is not contradictory to solid state data because the
lanthanoid chloride hydrates form two isostructural series:

the first one with 9-coordinated lanthanoid ions comprises
lanthanum, cerium and praseodymium while the remaining lanthanoids
belong to the second series with 8-coordinated central atoms
[49,50]. It should be noted here that contrary to bromates and
ethyl sulfates the chloride hydrates do not form pure aqua complexes in the solid state but also the chloride ions enter the
primary coordination sphere. The XRD has proven to be a valuable
technique but it requires the use of rather concentrated solutions
(2-4 M) and the results may be affected by this. Nevertheless,
the recent results of Svoronos et al. [51] , using gadolinium as
a fluorescence probe in solids and in dilute (0.1 M) solutions,
indicate that the coordination of gadolinium is similar in the
crystal of $GdCl_3 \cdot 6H_2O$ as in the solution.

3.3. Sulfato complexes

The lanthanoid sulfate hydrates $Ln_2(SO_4)_3 \cdot nH_2O$ have been extensively studied by spectroscopic, thermal and other techniques
[52] but first X-ray diffraction study which unambiquosly established the coordination around the lanthanoid was not made until
1968 [53]. The results on $Ce_2(SO_4)_3 \cdot 9H_2O$ were rather surprising
as they indicated the presence of two crystallographically independent highly-coordinated cerium ions. The first one was 12-
coordinated by exclusively sulfate oxygen atoms while the other
cerium was 9-coordinated by six water molecules and three sulfate
oxygens in a trigonal prismatic arrangement. The remaining three
water molecules were held in the structure by hydrogen bonds.

Subsequent X-ray studies have established the structures
of all stable sulfate hydrates including the anhydrous phase.
In addition to the compounds mentioned in Table 4, a number of
isostructural sulfates [52] and cerium(IV) sulfate tetrahydrate
[61] have been structurally characterized by X-ray methods.
$Ce(SO_4)_2 \cdot 4H_2O$ contains a Ce(IV) ion which is octacoordinated by
an equal number of water and sulfate oxygens.

The sulfate structures consists of polymeric chains or net-
works which are due to the versatile bridging ability of the
polydentate sulfato ligand. By way of an example, the structure
of $Nd_2(SO_4)_3 \cdot 5H_2O$ will be discussed in more detail.

Neodymium sulfate pentahydrate. The monoclinic pentahydrate
series comprise only the largest lanthanoids up to neodymium [52].
On basis of the structural determination the centrosymmetric space
group C2/c was chosen; this is in agreement with an X-ray study of
$Ce_2(SO_4)_3 \cdot 5H_2O$ [62].

The neodymium atom is nine-coordinated by seven sulfate
oxygens and two water molecules in a distorted trigonal prismatic
arrangement (Fig. 1). The coordination polyhedra are joined into

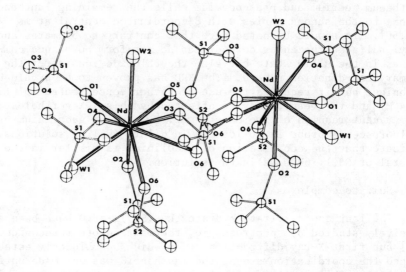

Figure 1.   A perspective view of the coordination abount
            neodymium atoms in the structure of
            $Nd_2(SO_4)_3 \cdot 5H_2O$ [56].

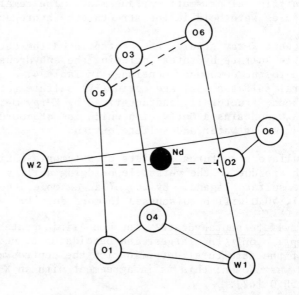

Figure 2.   The distorted tricapped trigonal prismatic
            coordination polyhedron about the Nd atom [56].

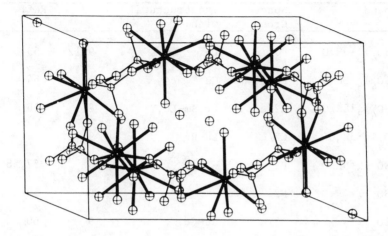

Figure 3.   The unit cell packing of $Nd_2(SO_4)_3 \cdot 5H_2O$ [56].

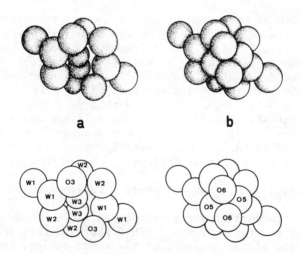

Figure 4.   The environment of the 'noncoordinated'
           water molecule (W3) in the structure of
           $Nd_2(SO_4)_3 \cdot 5H_2O$.   The cavity with the
           water molecules is shown at the left with-
           out the four capping sulfate oxygens
           (05 and 06) [56].

Table 4. Structure types and coordination numbers in the series
$Ln_2(SO_4)_3 \cdot nH_2O$

Compound	Space Group	Isostructural with	CN	Ref.
$Ce_2(SO_4)_3 \cdot 9H_2O$	$P6_3/m$	La	9,12	53
$La_2(SO_4)_3 \cdot 8H_2O$	Pc	Ce	9	54
$Nd_2(SO_4)_3 \cdot 8H_2O$	C2/c	Pr,Sm-Lu,Y	8	55
$Nd_2(SO_4)_3 \cdot 5H_2O$	C2/c	La-Pr	9	56
$Sc_2(SO_4)_3 \cdot 5H_2O$	P1	-	6	57,58
$Ce_2(SO_4)_3 \cdot 4H_2O$	$P2_1/c$	-	9	59
$Nd_2(SO_4)_3$	C2/c	Ce-Gd	9	60

pairs by bridging bidentate sulfate groups; other sulfate groups
join these pairs into a three-dimensional network as seen in Figs.
2 and 3.

One fifth of the water molecules is not coordinated to neodymium but situated in a cavity of the structure by hydrogen
bonds; Cf. Figs. 3 and 4. The different bonding of the water
molecules can be seen in the TG and DTA curves, too; the thermal
stability and dehydration mechanism for the penta- and octahydrates have been recently discussed [63].

'Double sulfates'. Besides the simple sulfates discussed above,
another series of rare earth sulfato compounds is known and extensively studied [52]. The general formula, based on the composition, may be written as follows:

$M_x Ln_y(SO_4)_{(x+3y)/2} \cdot zH_2O$, where M is ammonium or alkali ion

First structural reports became available for $NH_4Sm(SO_4)_2 \cdot$
$4H_2O$ [64,65] and for $K_6Pr_4(SO_4)_9 \cdot 8H_2O$ [66] and they indicated
that, in spite of the presence of the extra cation, the structures were not very much different from those of the 'simple'
sulfates. The subsequent structural studies on a number of compounds have confirmed this [52].

For instance, in $NH_4Sm(SO_4)_2 \cdot 4H_2O$ the samarium ion is nine-
coordinated by six sulfate oxygens and three water molecules; the
coordination polyhedron is an intermediate between a monocapped
square antiprism and a tricapped trigonal prism (Fig. 5). The

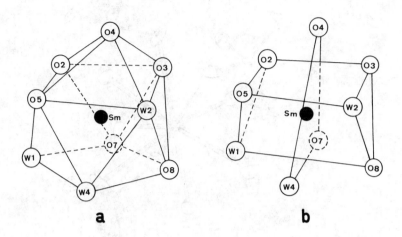

Figure 5.   The coordination polyhedron about the Sm atom
            in the structure of $NH_4Sm(SO_4)_2 \cdot 4H_2O$ [65]: (a)
            monocapped square antiprism; (b) tricapped trigonal
            prism.

polyhedra are joined into infinite chains as seen in Figure 6.
The fourth water molecule and the ammonium ion are located between the chains.

Conclusion.   The coordination of sulfato ligand in rare earth
sulfates is in agreement with its inner sphere character [19,67].
Coordination numbers are generally 8 or 9; the higher ones such
as 10 in $(NH_4)_5La(SO_4)_4$ [68] and 12 in $Ce_2(SO_4)_3 \cdot 9H_2O$ [53] are
exceptional and seem to be limited to lanthanum and cerium compounds only. On the other hand, scandium has an octahedral
coordination in all sulfate structures studied so far [57,58,69];
the same is true for selenate and selenite compounds where a
larger number of structures is known [41].

    The coordination geometries are distorted or intermediate
as expected when two types of ligands, aqua and sulfato, are
present and when the polydentate sulfato ligand has several
bonding possibilities.

3.4. Nitrato complexes

    As with the rare earth sulfates, two main series of nitrato
compounds are known. One representative for each series was
characterized quite early by X-ray methods, namely $Pr(NO_3)_3 \cdot$
$6H_2O$ [70] and $Mg_3Ce_2(NO_3)_{12} \cdot 24H_2O$ [78]. The results indicated

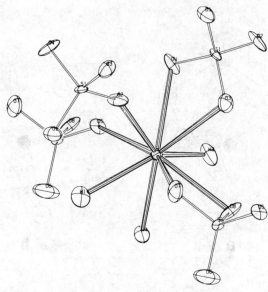

Figure 6a.   A perspective view showing the coordination of
the sulfato and aqua ligands to samarium in the
structure of $NH_4Sm(SO_4)_2 \cdot 4H_2O$.

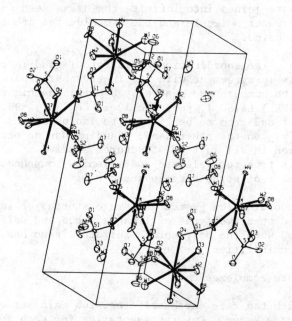

Figure 6b.   The unit cell packing of $NH_4Sm(SO_4)_2 \cdot 4H_2O$.
The 'noncoordinated' water molecules and ammonium
ions are located between the chains.

Table 5. Structural data for nitrato complexes.

Compound	Space Group	Unit Cell $V(Å^3)$	Z	Structure	CN	Ref.
$La(NO_3)_3 \cdot 6H_2O$	$P\bar{1}$	612	2	$[La(O_2NO)_3(H_2O)_5] \cdot H_2O$	11	28,35
$Pr(NO_3)_3 \cdot 6H_2O$	$P\bar{1}$	640	2	$[Pr(O_2NO)_3(H_2O)_4] \cdot 2H_2O$	10	70
$Y(NO_3)_3 \cdot 5H_2O$	$P\bar{1}$	583	2	$[Y(O_2NO)_3(H_2O)_4] \cdot H_2O$	10	71
$Sc(NO_3)_3 \cdot 2N_2O_4$	$P2_1/c$	1291	4	$(NO)_2[Sc(O_2NO)_4(ONO_2)]$	9	72
$K_2Er(NO_3)_5$	$P2_1/c$	1322	4	$K_2[Er(O_2NO)_5]$	10	73
$(NO)_2Ho(NO_3)_5$	$P2_1/c$	1329	4	$(NO)_2[Ho(O_2NO)_5]$	10	74
$K_2La(NO_3)_5 \cdot 2H_2O$	$Fdd2$	3028	8	$K_2[La(O_2NO)_5(H_2O)_2]$	12	75
$(NH_4)_2La(NO_3)_5 \cdot 4H_2O$	$C2/c$	1751	4	$(NH_4)_2[La(O_2NO)_5(H_2O)_2] \cdot 2H_2O$	12	76
$(NH_4)_2La(NO_3)_5 \cdot 3H_2O$	$C2/c$	1697	4	$(NH_4)_2[La(O_2NO)_5(H_2O)_2] \cdot H_2O$	12	76
$K_3Pr_2(NO_3)_9$	$P4_332$	2471	4	$[K_3Pr_2(NO_3)_9]$[a]	12	77
$Mg_3Ce_2(NO_3)_{12} \cdot 24H_2O$	$R\bar{3}$	1209	1	$[Mg(H_2O)_6]_3[Ce(O_2NO)_6]_2 \cdot 6H_2O$	12	78

[a] Polymeric structure

that coordination numbers are unusually high; in these cases 10
and 12, respectively.

Table 5 summarizes the early results as well as more recent
X-ray diffraction studies. Parallel to these investigations a
number of nitrato structures containing organic ligands has been
determined; these compounds are partly included in the two earlier
reviews on the subject [79,17]. The recent review by Eriksson
[80] deals with inorganic nitrate hydrates of the rare earths.

Nitrato ligand, besides being small, often coordinates bi-
dentately with a small 'bite' [81,82]. As seen in the Table 4
bidentate coordination is almost a rule in the case of large rare
earth ions. The only exception is $Sc(NO_3)_3 \cdot 2N_2O_4$ where one
nitrato group is monodentate. In addition, probably steric requirements of the large tetradentate diselenite ion force the
nitrato ligand to be monodentate in the structure of $Ln(Se_2O_5)NO_3 \cdot$
$3H_2O$ (Ln = Pr-Lu,Y) [83,84].

Among the structures listed in the Table 4 two interesting
series may be pointed out: (i) trinitrato hydrates $Ln(NO_3)_3 \cdot nH_2O$
and (ii) pentanitrato series $M_2Ln(NO_3)_5 \cdot nH_2O$; these will be
discussed briefly as examples of the rare earth nitrato structures.

Trinitrato hydrates. Among the several hydrates reported in the
literature [79], the hexa- and pentahydrates can be most easily
prepared. The hexahydrates $Ln(NO_3)_3 \cdot 6H_2O$ form two isostructural
series, first one comprising La and Ce[56,85] and the second one
the remaining lanthanoids including yttrium. The change in the
structure is accompanied by a change in the CN: La is 11-coordi-
nated by three bidentate nitrato groups and five water molecules
(Fig. 7) while Pr has one water molecule less in its coordination
polyhedron. Decacoordination was also observed for $Y(NO_3)_3 \cdot 5H_2O$
[71] but the coordination of the ligands was quite different from
the corresponding hexahydrate [86] as seen in Fig. 8.

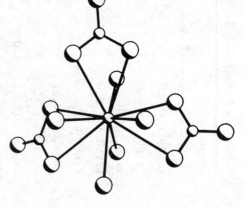

Figure 7. A perspective view
of the 11-coordinated La atom
in the structure of $La(NO_3)_3 \cdot$
$6H_2O$ [35,80].

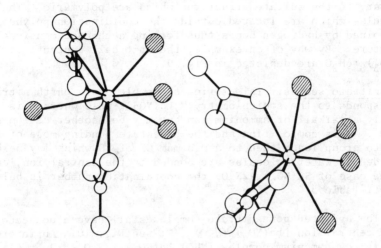

Figure 8. The coordination of nitrato groups and water molecules (shaded atoms) in the complex [Y(NO$_3$)$_3$(H$_2$O)$_4$] [71].
a) Y(NO$_3$)$_3$·5H$_2$O b) Y(NO$_3$)$_3$·6H$_2$O.

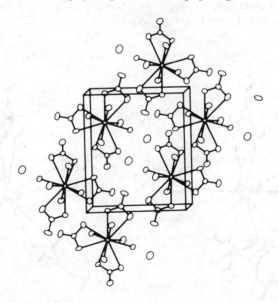

Figure 9. The unit cell packing of La(NO$_3$)$_3$·6H$_2$O showing the [La(NO$_3$)$_3$(H$_2$O)$_5$] complexes and the noncoordinated water molecules [35].

The nitrato complexes are in general discrete molecules contrary to the sulfate structures which are polymeric. The water molecules which are located outside the coordination polyhedra are joined by hydrogen bonds thus forming a three-dimensional structure. By way of an example, the unit cell contents of $La(NO_3)_3 \cdot 6H_2O$ are depicted in Fig. 9.

Pentanitrato series. This series of complex rare earth nitrates corresponds to the stoichiometry $M_2Ln(NO_3)_5 \cdot nH_2O$ where M is usually an alkali or ammonium ion but may be another cation such as NO^+. This composition and the bidentate bonding mode of the nitrato group leads then to a minimum CN of 10 which may be even exceeded if water molecules are bonded to the central ion; only in the case of $Sc(NO_3)_3 \cdot 2N_2O_4$ the coordination number is below 10, cf. Table 5.

The hydrated pentanitrato complexes all have a dodecacoordinated central ion $[La(NO_3)_5(H_2O)_2]^{2-}$; see Fig. 10. An interesting case is the ammonium complex $(NH_4)_2La(NO_3)_5 \cdot nH_2O$ (n=3 or 4) [76]. The two hydrates are structurally closely related as the removal of water does not bring about the collapse of the structure but the coordination geometry and crystal symmetry are preserved. The only significant change is the contraction of the unit cell when there is one water molecule less. The ammonium ions are involved in forming a network of hydrogen bonds with the complex anions and the noncoordinated water molecules.

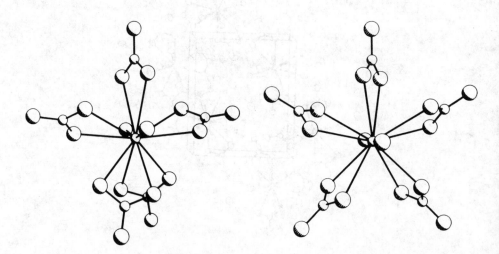

Figure 10.   A perspective view of the $[La(NO_3)_5(H_2O)_2]^{2-}$ complex in the pentanitrato complexes of potassium (left figure) and ammonium (right)[75,76,80].

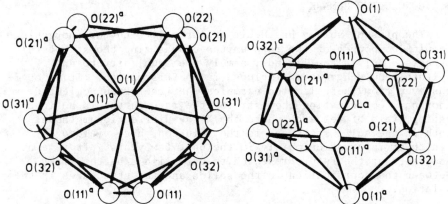

Figure 11.   The coordination polyhedron around the La atom
in the structure of $(NH_4)_2La(NO_3)_5 \cdot nH_2O$ [76].
The left figure shows a view along the capping
water atoms $O(1)$ and $O(1)^a$ and the right one
a view perpendicular to it.

The coordination polyhedron for the potassium [75] and the
ammonium compounds is the icosahedron which is slightly distorted;
in both cases the water molecules occupy the trans positions of
the capping atoms. In the ammonium compounds each of the five
nitrato groups contributes one oxygen atom to each of the two
five-membered rings in the icosahedron (Fig.11) while in the
potassium compound one of the nitrate groups belongs entirely
to one of the rings. It may be noted in this connection that
the symmetry for the dodecacoordinated Ce(IV) in the structure
of $(NH_4)_2Ce(NO_3)_6$ [87] is higher as there are only one type of
ligands present.

Conclusion. A characteristic feature of the rare earth nitrate
structures is the highly coordinated discrete complex molecules
or anions. The coordination numbers often exceed ten especially
in the case of lanthanum and cerium compounds. Even scandium
has unusually high coordination numbers; the same trend has
been reported in organic nitrato structures, for instance in
the urea complexes [88].

The distortions of the coordination polyhedra from the idealized
symmetries seem to be smaller than in the case of the sulfato
complexes. This is quite understandable when comparing the
sizes and bonding modes of the two ligands.

4. CONCLUDING REMARK

The present survey has dealt exclusively with the inorganic
complex compounds of the rare earths emphasizing the structural
aspects. Only a few ligands, namely the aqua, nitrato and
sulfato, could have been included in the discussion. It is obvious even from these limited examples that the often neglected
inorganic coordination chemistry of the rare earths is an interesting field of research where the X-ray diffraction methods,
both in the solid state and in the solution, have yielded a
wealth of information mainly in the past few years. It remains,
however, for the future research to establish a better link
between the coordination in the solids and the structures in the
solution.

NOTES

1. The nomenclature used here follows the recommendations of
    IUPAC (Nomenclature of Inorganic Chemistry, 2nd Edition,
    Butterworths, London 1971). Thus, the rare earths comprise
    Sc, Y and La-Lu while the lanthanoids include the elements
    La-Lu.

2. LaF_3 (and isostructural CeF_3) is sometimes included in the
    11-coordinated compounds. They are not listed here because
    structural determination has shown that two of the La-F
    distances are significantly (>0.35 Å) longer than the
    others (A.K. Cheetham, B.E.F. Fender, H. Fuess, and A.F.
    Wright, Acta Crystallogr., Sect. B., 29, 1093 (1976)).

3. There are earlier reports on the structures of $Nd_2(SO_4)_3 \cdot 8H_2O$
    and $La_2(SO_4)_3 \cdot 9H_2O$ but they do not describe the coordination
    properly or have low accuracy. These reports and the need
    for redetermination have been discussed recently (E. Gebert
    Sherry, J. Solid State Chem., 19, 271 (1976))

4. The structure of tris(tetrahydroborato)tris(tetrahydrofuran)-
    yttrium(III) (B.G. Segal and S.J. Lippard, Inorg. Chem.,17,
    844 (1978)) also contains an 11-coordinated rare earth
    complex where one of the BH_4 groups is bidentate while the
    other two are tridentate.
    The author is grateful for Prof. G.J. Palenik for pointing
    out the existence of this interesting structural information.

REFERENCES

1. Gmelin-Kraut's Handbuch der anorganischen Chemie, Band VI, Abteilung 1-2, Carl Winters Universitätsbuchhandlung, Heidelberg, 1928-32.

2. V.M. Goldschmidt, T. Barth, G. Lunde, and W.H. Zachariasen, Skr. Norske Vidensk. Akad., 1 Mat.-Nat. Kl. No. 2 (1926).

3. R.D. Shannon and C.T. Prewitt, Acta Crystallogr., Sect. B., 25, 925 (1969).

4. R.D. Shannon and C.T. Prewitt, Acta Crystallogr., Sect. B., 26, 1046 (1970).

5. R.D. Shannon, Acta Crystallogr., Sect. A, 32, 751 (1976).

6. L.J. Nugent, MTP Int. Rev. Sci. Inorg. Chem. Ser. 2, 7, 195 (1975).

7. V.I. Spitsyn, Z. Chem., 17, 353 (1977).

8. D.J. Karraker, J. Chem. Ed., 47, 424 (1970).

9. T. Moeller, MTP Int. Rev. Sci. Ser. One, 7, 275 (1972).

10. S. Ahrland, J. Chatt, and N.R. Davies, Quart. Rev., 12, 265 (1958).

11. R.G. Pearson, J. Am. Chem. Soc., 85, 3533 (1963).

12. S.P. Sinha, Complexes of the Rare Earths, Pergamon Press, Oxford, 1966, p. 28.

13. T. Moeller, R.L. Dieck, and J.E. MacDonald, Rev. Chim. Miner., 10, 177 (1973).

14. F. Ephraim and R. Bloch, Chem. Ber., 59, 2692 (1926).

15. J.L. Burmeister and E.A. Deardorff, Inorg. Chim. Acta, 4, 97 (1970).

16. J.L. Martin, L.C. Thompson, L.S. Radonovich, and M.D. Glick, J. Am. Chem. Soc., 90, 4493 (1968).

17. D. Brown, Halides of the Lanthanides and Actinides, Wiley - Interscience, London, 1968.

18. O. Heiniö, M. Leskelä, and L. Niinistö, Acta Chem. Scand. Ser. A., 34, 207 (1980), and references cited therein.

19. G.R. Choppin and S.L. Bertha, J. Inorg. Nucl. Chem., 35
     1309 (1973).

20. M. Koskenlinna and J. Valkonen, Acta Chem. Scand., Ser. A.,
     31,457 (1977).

21. J. Felsche, Struct. Bond. (Berlin), 13, 99 (1973).

22. Yu. N. Safianov, E.A. Kuzmin, and N.V. Belov, Dokl. Akad.
     Nauk SSSR, 242, 603 (1978).

23. D. Brown, MTP Int. Rev. Sci. Inorg. Chem. Ser. 2, 7, 111
     (1975), and references cited therein.

24. J.-C. G. Bünzli, B. Klein, G. Chapuis, and K.J. Schenk,
     Inorg. Chem., 21, 808 (1982).

25. G. Bombieri, G. de Paoli, and F. Benetello, J. Inorg.
     Nucl. Chem., 42, 1417 (1980).

26. S.P. Sinha, Struct. Bond. (Berlin), 25, 69 (1976).

27. R.A. Andersen, D.H. Templeton, and A. Zalkin, Inorg. Chem.,
     17, 2317 (1978), and references cited therein.

28. B. Eriksson, L.O. Larsson, and L. Niinistö, J.C.S. Chem.
     Comm., 616 (1978).

29. M. Bukowska - Strzyzewska and A. Tosik, Inorg. Chim. Acta,
     30, 189 (1978).

30. G. Johansson, Acta Chem. Scand., 22, 389 (1968).

31. T. Ueki, A. Zalkin, D.H. Templeton, Acta Crystallogr., 20,
     836 (1966).

32. J.C. Taylor, M.H. Mueller, R.L. Hitterman, Acta Crystallogr.,
     20, 842 (1966).

33. J.H. Burns, Inorg. Chem., 18,3044 (1979).

34. J.E. Thomas, R.C. Palenik, and G.J. Palenik, Inorg. Chim. Acta
     37, L459 (1979).
35. B. Eriksson, L.O. Larsson, L. Niinistö, and J. Valkonen,
     Inorg. Chem., 19, 1207 (1980).

36. G.J. Palenik, Chapter 5 in this volume.

37. M.G.B. Drew, Coord. Chem. Revs., 24, 179 (1977).

38. M.C. Favas and D.K. Kepert, Progr. Inorg. Chem., 28, 309
     (1981).

39. G.A. Melson and R.W. Stotz, Coord. Chem. Revs., 7, 233 (1971).

40. G.T. Horowitz (ed.), Scandium, Academic Press, New York,
     1975, Chapters 6 and 8.

41. J. Valkonen, Ann. Acad. Sci. Fenn. Ser. A II, 188, (1979).

42. M. Leskelä and L. Niinistö, Unpublished results.

43. F. Weigel, Chem. Z., 102, 339 (1978).

44. Y. Marcus, In Gmelin Handbook of Inorganic Chemistry, Sc, Y,
     La-Lu, Vol. D3, Springer-Verlag, Berlin, 1982, p. 1.

45. I. Grenthe, Kem.-Kemi, 5, 234 (1978).

46. J. Albertson and I. Elding, Acta Crystallogr., Sect. B., 33,
     1460 (1977).

47. J. Albertsson, Kem-Kemi, 5, 229 (1978).

48. A. Habenschuss and F.H. Spedding, J. Chem. Phys., 73, 442
     (1980).

49. E.J. Peterson, E.I. Onstott, and R.B. van Dreele, Acta
     Crystallogr., Sect. B, 35, 805 (1979), and references cited
     therein.

50. A. Habenschuss and F.H. Spedding, Cryst. Struct. Commun.,
     7, 535 (1978) and references cited therein.

51. D.R. Svoronos, E. Antic-Fidancev, M. Lemaitre-Blaise, and
     P. Caro, Nouv. J. Chim., In the press.

52. Gmelins Handbuch der anorganischen Chemie, Sc, Y, La-Lu,
     Vol. C8, Springer-Verlag, Berlin, 1981.

53. A. Dereigne and G. Pannetier, Bull. Soc. Chim. Fr., 174
     (1968).

54. L.A. Aslanov, I.S.A. Farag, V.M. Ionov, and M.A. Porai-
     Koshits, Zh. Fiz. Khim., 47 2172 (1973).

55. L.A. Aslanov, V.B. Rybakov, V.M. Ionov, M.A. Porai-Koshits,
     and V.I. Ivanov, Dokl. Akad. Nauk. SSSR, 204, 1122 (1972).

56. L.O. Larsson, S. Linderbrandt, L. Niinistö, and U. Skoglund,
     Suom. Kemistil. B, 46, 314 (1973).

57. L. Niinistö, J. Valkonen, and L.O. Larsson, Finn. Chem. Lett.,
     45 (1975).

58. J. Valkonen, L. Niinistö, B. Eriksson, L.O. Larsson and
     U. Skoglund, Acta Chem. Scand, Ser. A., 29, 866 (1975).

59. A. Dereigne, J.-M. Manoli, G. Pannetier, and P. Herpin,
     Bull. Soc. Fr. Minéral. Cristallogr., 95, 269 (1972).

60. S.P. Sirotinkin, V.A. Efremov, L.M. Kovba, and A.N. Pokrovskii,
     Kristallografiya, 22, 1272 (1977).

61. O. Lindgren, Acta Chem. Scand., Ser. A., 31, 453 (1977).

62. I.S.A. Farag, L.A., Aslanov, V.M. Ionov, and M.A. Porai-
     Koshits, Zh. Fiz. Khim. 47, 1056 (1973).

63. L. Niinistö, P. Saikkonen and R. Sonninen, The Rare Earths
     in Modern Science and Technology, G.J. McCarthy, J.J. Rhyne
     and H.B. Silber (eds.), Plenum Press, New York, 1982. In the
     press.

64. L. Niinistö, Chem. Commun. Univ. Stockholm, 13, (1973).

65. B. Eriksson, L.O. Larsson, L. Niinistö, and U. Skoglund,
     Inorg. Chem., 13, 290 (1974).

66. I.S.A. Farag, L.A. Aslanov, M.A. Porai-Koshits, M.B.
     Varfolomeev, V.M. Ionov, and L.D. Iskhakova, Zh. Fiz. Khim.
     47, 1056 (1973).

67. C.F. Hale and F.H. Spedding, J. Phys. Chem., 76, 1887 (1972).

68. L. Niinistö, J. Toivonen and J. Valkonen, Finn. Chem. Lett.,
     87, (1980).

69. R.G. Sizova, A.A. Voronkov, and N.V. Belov, Sov. Phys. Dokl.,
     19, 472 (1975).

70. I.M. Rumanova, G.F. Volodina, and N.V. Belov, Sov. Phys.
     Cryst., 9, 545 (1965).

71. B. Eriksson, Acta Chem. Scand., Ser. A, 36, 186 (1982).

72. C.C. Addison, A.J. Greenwood, M.J. Haley, and N. Logan,
     J.C.S. Chem. Comm., 580 (1978).

73. E.G. Sherry, J. Inorg. Nucl. Chem., 40, 257 (1980).

74. G.E. Toogood and C. Chieh, Can. J. Chem., 53, 831 (1975).

75. B. Eriksson, L.O. Larsson, L. Niinistö and J. Valkonen,
     Acta Chem. Scand., Ser. A., 34, 567 (1980).

76. B. Eriksson, L.O. Larsson, and L. Niinistö, Acta Chem.
     Scand., Ser. A, 36, 465 (1982).

77. W.T. Carnall, S. Siegel, J.R. Ferraro, B. Tani, and E.
     Gebert, Inorg. Chem., 12, 560 (1973).

78. A. Zalkin, J.D. Forrester, and D.H. Templeton, J. Chem.
     Phys., 39, 2881 (1963).

79. Gmelins Handbuch der anorganischen Chemie, Sc, Y, La-Lu,
     Vol. C2, Springer-Verlag, Berlin 1974.

80. B. Eriksson, Chem. Commun. Univ. Stockholm, 3, (1982).

81. C.C. Addison, N. Logan, S.C. Wallwork, and C.D. Garner,
     Quart. Rev., 25, 289 (1971).

82. A. Leclaire, J. Solid State Chem., 28, 235 (1979).

83. L. Niinistö, J. Valkonen, and P. Ylinen, Inorg. Nucl. Chem.
     Lett., 16, 13 (1980).

84. J. Valkonen and P. Ylinen, Acta Crystallogr., Sect. B, 35,
     2378 (1979).

85. N. Milinski, B. Ribár, and M. Sararic, Cryst. Struct.
     Commun., 9, 473 (1980).

86. B. Ribár, N. Milinski, and Ž. Budovalčev, Cryst. Struct.
     Commun., 9, 203 (1980).

87. T.A. Beineke and J. Delgandio, Inorg. Chem., 7, 715 (1968).

88. V.I. Kuskov, E.N. Treushnikov, and N.V. Belov, Sov. Phys.
     Crystallogr.,23, 677 (1978).

89. J.H. Forsberg, Coord. Chem. Rev., 10, 195 (1973).

DISCUSSION

URLAND (question)

This is regarding one of the figures you showed, e.g., the dependence of the average coordination number of the rare earth series.
There Y actually tops out. You were telling us that this may be
due to the number of available examples but on the other hand you
had extremely few examples for Tm. Is there any connection between
availability of f-orbitals in Tm in contrast to Y?

NIINISTÖ (answer)

It is very difficult for me to answer, because we have not analyzed the data in such context. As I have mentioned when presenting the Table that it is difficult to see the reason for the
anomaly. I am very cautious because we did not analyze in detail,
such as what factors are involved, but we intend to do that in
the near future.

ASCENSO (question)

Do you expect that results by X-ray scattering technique could be
different from other methods because you need to use concentrated
solutions whereas with other methods like NMR and luminescence
you are able to investigate dilute solutions.

NIINISTÖ (answer)

Yes, I quite agree that this is one of the drawbacks of this X-ray
scattering method, one has to use concentrated solutions and the
results cannot be directly translated to the dilute systems. There
are, however, some ways to do this, for instance one can study
how the structures change when the solutions are diluted in a
stepwise manner but there is also a lower concentration limit
where one can go down to.

COUTURE (comment)

I'll speak of the work we have done with Katheryn Rajnak on the
optical absorption spectroscopy of rare earth chlorides in frozen
aqueous solutions which are very dilute. Our results may add something to the result that has been shown by X-ray. We have for the
chloride, the bromide and the perchlorate exactly the same spectra,
and we conclude that the nearest neighbours of the rare earth ions
are only water molecules. If you take the nitrate for instance,
there is a completely different spectrum. The rare earth ions
have energy levels which are split by the crystal field. The number of components we get shows that we have in frozen solutions
only one site for the rare earth. Some lines are very faint; it

means that there are optical selection rules arising from a high symmetry. We did an energy level calculation; it is not easy with frozen solution since we cannot have polarization nor Zeeman effect results so this energy level calculation started from comparison to similar spectra of crystals. The conclusion is: The Nd^{3+} is ninefold coordinated; the spectra are very similar to those of the ethylsulfate so a D_{3h} symmetry explains all the levels and we think by the analogy with crystal that the polyhedron is a tricapped triangular prism. For Er^{3+}, the comparison with crystals shows us that the spectra compare with those of hydrated chloride where Er^{3+} ions have an eightfold coordination. We have shown that in hexahydrated chloride, the surroundings of the rare earth, although there are two chlorides and six oxygens, have nearly the shape of a square antiprism. So we did the calculation for the crystal with D_{4d} symmetry for the selection rules, the interpretation of the levels and of the Zeeman effect. The results were very good. With the crystal results as a guide we were also able to analyze the Er^{3+} frozen solution data in D_{4d} symmetry and concluded that the surroundings of Er^{3+} consist of 8 water molecular arranged in the shape of a square antiprism. We can also add that these results could be extended to liquid aqueous solutions: there is a great similarity of the absorption spectra of Nd^{3+} in frozen solution and liquid solution.

SINHA   (comments)

You have so little intensities on those "faint" lines, I wonder if any sensible calculation could be carried out. Secondly, species observed in frozen solution does not mean that the same species is present at room temperature. Thirdly in frozen solutions (at liquid N_2 temperature) often crystallites are present which may show spectra with relatively high symmetry. Furthermore, for Nd^{3+} and Er^{3+} with half-integral J values all symmetries except cubic would result in $(J + 1/2)$ splitting. This may hinder the assignment of site symmetry for these two ions especially as you say that you do not have Zeeman data for the frozen solutions.

COUTURE   (answer)

For Nd^{3+} solutions we have done spectra at several different temperatures. If we compare the spectra we see that as the temperature is increased from 4 K to 200 K and then to room temperature, the low-temperature bands broaden but the frequencies and the distribution of intensities remain the same. This indicates that the surroundings are unchanged.

SINHA   (question)

But what about Er^{3+}?

COUTURE   (answer)

We did not do that study for Er^{3+}.

KANELLAKOPULOS   (question)

Have you tried to run the spectra in deuterated water?

COUTURE   (answer)

Yes, it is exactly the same.

SINHA   (comment)

Recent X-ray studies of Habenschuss and Spedding (J. Chem. Phys.
70, 3758 (1979)) showed an average coordination number of 8.9
(H_2O) with Nd-O (water) distance of 2.51 Å. An extensive neutron
diffraction work on the Nd aquo ion is in progress at Oak Ridge
National Laboratory which may bring some new insight (This work
is now published: A.H. Narten and R.L. Hahn, Science 217, 1249
(1982); where the authors conclusively proved the presence of a
well-defined first hydration sphere of water molecules. Each Nd^{3+}
ion is surrounded by 8.5 oxygen atoms (D_2O) at a distance of
2.48 Å - Editor).

MÖLLER   (question)

Among the most common minerals are the feldspars, e.g., $Na[AlSi_3O_8]$,
$Ca[Al_2Si_2O_8]$. Why is $La[Al_3SiO_8]$ is unknown? Ln-silicates are,
however, known. Did somebody search for Ln-aluminosilicate?
Do you have any information if compounds like $NaLn[Al_2Si_2O_8]_2$ might
exist?

Niinistö   (answer)

If I remember it correctly that some studies on the system Ln_2O_3-
Al_2O_3-SiO_2 have been reported and these indicate the existence of
lanthanoid aluminosilicates. The structural information on these
phases is, however, scarce.

It seems thus possible to prepare lanthanoid aluminosilicate but
whether or not structures such as $Ln[Al_2Si_3O_8]$ and $NaLn[Al_2Si_3O_8]_2$
are stable, is a different question, which I am afraid I cannot
answer conclusively at present.

STRUCTURAL CHEMISTRY OF THE LANTHANIDES

Gus J. Palenik

Department of Chemistry, University of Florida
Gainesville, Florida, USA 32611

ABSTRACT

The structural chemistry of the lanthanides, yttrium, and
scandium is reviewed, using the Cambridge Crystallographic Data
Centre file. In addition, relevant complexes with only inorganic
ligands have been included. Organolanthanides are divided into
the cyclopentadienides, σ-bonded compounds, and the cyclooctatetraene complexes. A brief discussion of the covalent contribution to the bonding in organolanthanides is included. The coordination complexes are organized in terms of their coordination
numbers. Lower coordination numbers 3 to 7 appear to be rare and
directly related to steric effects. The coordination numbers of
8 and 9 are the "favored" ones for these elements. A number of
examples for the coordination numbers 10, 11 and 12 are discussed. Relatively long lanthanide-nitrogen bond distances are a
recurring feature in the structural chemistry of these elements.
A more general effect may be long bond lengths to "soft" donor
atoms.

INTRODUCTION

The structural chemistry of the lanthanides is dominated by
the regular decrease from La to Yb in the size of either the atom
or ion.   This decrease can be seen in the values of the crystal
radii for the lanthanides, scandium, and yttrim which are presented in Table 1. The radii are from the tabulation by
Shannon [1] and include values for different coordination numbers
and oxidation states.   The radius for Y(III) for CN-VI is very
close to that of Ho(III) and is only very slightly larger for the

153

S. P. Sinha (ed.), Systematics and the Properties of the Lanthanides, 153–210.
Copyright © 1983 by D. Reidel Publishing Company.

Table 1. Crystal Radii of the Lanthanides, Scandium, and Yttrium

Ion	La^{3+}	Ce^{3+}	Pr^{3+}	Nd^{3+}	Pm^{3+}	Sm^{3+}	Eu^{3+}
CN-VI	1.172	1.15	1.13	1.123	1.11	1.098	1.087
CN-VIII	1.300	1.283	1.266	1.249	1.233	1.219	1.206
CN-IX	1.356	1.336	1.319	1.303	1.284	1.272	1.260

Ion	Gd^{3+}	Tb^{3+}	Dy^{3+}	Ho^{3+}	Er^{3+}	Tm^{3+}	Yb^{3+}
CN-VI	1.078	1.063	1.052	1.041	1.030	1.020	1.008
CN-VIII	1.193	1.180	1.167	1.155	1.144	1.134	1.125
CN-IX	1.247	1.235	1.223	1.212	1.202	1.192	1.182

Ion	Lu^{3+}	Sc^{3+}	Y^{3+}	Ce^{4+}	Pr^{4+}	Tb^{4+}	Sm^{2+}
CN-VI	1.001	0.885	1.040	1.01	0.99	0.90	
CN-VIII	1.117	1.010	1.159	1.11	1.10	1.02	1.41
CN-IX	1.172		1.215				1.46

Ion	Eu^{2+}	Yb^{2+}
CN-VI	1.31	1.16
CN-VIII	1.39	1.28
CN-IX	1.44	

higher coordination numbers. Therefore, we shall follow the common practice of including yttrium in our discussion of the lanthanides. However, the size of Sc(III) is very much less than that of Yb(III), and one rarely finds Sc(III) complexes with a CN greater than six. Consequently, there is much less justification on a size basis for including Sc(III) in our discussions. Nevertheless, the chemistry of Sc(III), Y(III), and the lanthanides is so similar that Sc(III) will be included in our discussion.

A second variation in the structural chemistry of the lanthanides is found in the existence of oxidation states of two and four for certain elements. The radii for the M(IV) ions of Ce, Pr and Tb, and the M(II) ions of Sm, Eu and Yb have been included in Table 1. However, complexes with these oxidation numbers are not common. The lack of organic complexes of the M(IV) ions is not surprising when one considers that the standard reduction potential is very high for all three ions. All of the M(II) ions are easily oxidized to the M(III) state, even by water. However, complexes of the M(II) state for Sm, Eu and Yb should exist in nonaqueous solvents in the absence of oxygen, but this area has not been pursued actively.

In any discussion of the structural aspects of coordination compounds, two terms, coordination number and coordination polyhedron, invariably arise. In spite of the wide usage of these terms, there is no general agreement on how to define either term rigorously. The problem of defining the CN can be particularly difficult in low symmetry materials where there are a number of near neighbors with only slightly longer distances. While these problems have been discussed recently [2,3], there is no generally accepted method for calculating the CN in some complex phases.

The problem of determining the CN of lanthanide complexes arises most frequently in organometallic compounds and nitrato complexes. In cyclopentadienyl compounds where the metal atom is equidistant from all five carbon atoms of the ring, should the ring be considered as occupying five coordination sites? This is an important question if one attempts to retrieve structural data on the basis of CN. The Cambridge Crystallographic Data File considers a cyclopentadienyl ring as occupying only one coordination site although this decision is currently under review [4,5]. To avoid the problem of defining the CN in organometallic compounds, we shall discuss these compounds separately.

A less controversial problem arises with bidentate nitrate coordination. If the two M-O distances are approximately equal, should the nitrate ion be considered as occupying only one coordination site? While there are examples where the discussion of the coordination polyhedra might be simplified by considering the nitrate ion as occupying only one site, we will not follow that procedure in our discussion. In other words, the number of sites occupied by the nitrate ion will be dependent only on the M-O distances.

The X-ray experiment provides, in most cases, an accurate description of the positions of the various donor atoms around the central metal atom or ion. A discussion of the results can involve an analysis of the M-L distances, the L-M-L' angles, the L•••L distances, and the arrangement of L donors around the central atom. The latter point is usually considered in terms of some idealized geometry and is frequently of some chemical interest. However, when we consider coordination numbers greater than six, the number of possible polyhedra increases rapidly. For example, there are 34 nonisomorphic convex polyhedra with seven vertices [6] so that the choice of a coordination polyhedron for a seven coordinate complex is not straightforward. In fact the merit of assigning polyhedra to cases with CN of 10 and up has been raised in a recent review of high coordination complexes [7]. Nevertheless, the desire to assign a coordination polyhedron to a particular complex is strong and should not be discouraged. In this case, the criteria for selecting a polyhedron should be considered.

There are two general methods currently in use for assigning coordination polyhedra, and these have been discussed to some extent [7]. One method in wide use is to fit the n-donor atoms to an idealized polyhedron and to determine the rms average separation of the two sets. This method was proposed by Dollase [8] although other workers had developed similar computer programs [9,10]. In the Dollase program, the idealized polyhedron can be translated, rotated, and, if consistent with the symmetry of the polyhedron, expanded or contracted. In addition one has a choice of fitting the observed polyhedron with varying M-L distances or of normalizing the various M-L distances to a unit sphere. The normalization of the M-L lengths has been followed by Drew [7,11] in his two review articles. However, one can argue that different bond lengths are an important part of the description of polyhedra and should not be ignored.

The second method used in assigning a coordination polyhedron involves the use of the so-called δ-angles. The δ-angles are the angles formed by the normals between adjacent faces of the polyhedron. This method was first proposed by Porai-Koshits and Aslanov [12] and extended by Muetterties and Guggenberger [13-15]. The rms difference between the δ-angles for an ideal polyhedron and the observed δ-angles can be used to determine the best fit. Two problems arise with the use of δ-angles which make the assignment not always straightforward. The question whether to calculate δ-angles using the observed M-L distances or after normalizing to an M-L distance of 1 Å still remains. In addition the assignment of vertices to an ideal polyhedron can be very difficult if there are large distortions from ideality. This problem arose in determining the coordination polyhedron in $Er(DAPSAH)(H_2O)_2(NO_3)^{2+}$, where the rms values were 11.88 and 12.59, for the TTP and MSAP, respectively [16].

One of the non-trivial problems in assigning a polyhedron by either method is the derivation of an ideal model. Kepert, in a series of articles [17-20], has calculated the idealized polyhedron for a CN of 6-12, excluding 11. He considered a variety of constraints imposed by chelating ligands as well as varying the coefficient in the repulsion expression. Nevertheless, there still remain problems because of the large number of polyhedra theoretically possible for any CN greater than 8. For example, in the case of CN 10, there are a number of 1-6-3 polyhedra [21] which differ by the orientation of the lower three vertices relative to the planar hexagon. Energy minimization using the points on a sphere model reduces these to two limiting polyhedra. A tetracapped trigonal prism results when the starting model is not completely staggered and is of slightly lower energy. However, with the completely staggered model, the higher energy form retains the 1-6-3 symmtery after minimization of the energy. The latter 1-6-3 polyhedra is an important model in the case of com-

plexes with an approximately planar hexadentate ligand [22]. The
large number of polyhedra available for high coordination numbers, together with the constraints imposed by multidentate ligands, make the assignment of coordination polyhedra almost an exercise in futility. Therefore, in this review we will not dwell
extensively on the exact nature of the coordination polyhedron
but will attempt to concentrate mainly on the CN and the factors
which influence the final CN and to some extent the polyhedron.

STRUCTURES OF ORGANOLANTHANIDES

    The organolanthanides, with three exceptions [23-25], all
involve π-type ligands. Furthermore, again with only three
exceptions [26-28], the complexes all involve the trivalent
state. For the present discussion, we will divide the organolanthanides into five groups: tris(cyclopentadienyl), tris(cyclopentadienyl)•base, bis(cyclopentadienyl), σ-bonded compounds, and
cyclooctatetraene complexes. This division is somewhat similar
to that used in a recent review of organolanthanides [29] and is
chosen to emphasize the structural similarities.

    The first organolanthanides to have been studied in any
detail were the LnCp$_3$ compounds. At present, there are three
reported structural studies of LnCp$_3$ compounds [30-32] and one of
scandium [33]. The first LnCp$_3$ compound to be studied by X-ray
diffraction techniques was SmCp$_3$. The space group is Pbcm with
a = 14.23, b = 17.40, and c = 9.73 Å and eight molecules per
cell [30]. The Pr, Pm, Gd and Tb derivatives were reported to be
isomorphous with the SmCp$_3$ compound [34]. However, the existence
of a second modification in which the b axis was approximately
halved was also noted. Furthermore, TmCp$_3$ was reported to crystallize in the space group Pc2$_1$n or Pcmn with a = 13.82,
b = 8.59, and c = 19.98 Å in which b is approximately halved and
c is doubled relative to the SmCp$_3$ cell. The disordered structure of SmCp$_3$ was thought to be derived from a twinned crystal [35] so that Nd(MeCp)$_3$ was synthesized and studied [31]. The
Nd(MeCp)$_3$ is probably the best example of a LnCp$_3$ compound, and
the tetrameric unit is shown in Figure 1. Each Nd atom is bonded
in an η^5 fashion to three MeCp rings and to one carbon atom of a
MeCp ring in another Nd(MeCp$_3$) unit. The 20 Nd-C distances to
the nonbridging MeCp rings average 2.769(31) Å, while the 10 Nd-C
distances to the bridging ring average 2.841(42) Å. While the
difference between these Nd-C distances is not statistically significant, longer distances to the bridging ring are reasonable.
In addition each Nd has a long Nd-C bond, 2.990(7) and
2.978(7) Å, to the bridging ring from the other unit. The Nd-C
interactions are somewhat similar to those reported for SmCp$_3$ and
presumably for the other LnCp$_3$ compounds, at least through Tb.
However, since TmCp$_3$ is not isomorphous with the lighter LnCp$_3$

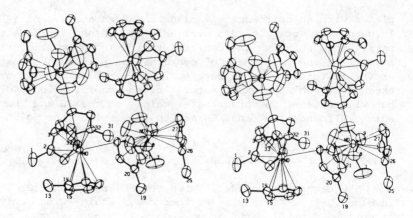

Figure 1.   Stereoview of the tetrameric unit of the Nd(MeCp)₃
            structure [21].

derivatives, the structure may not be the same.   In fact $ScCp_3$,
which crystallizes in the space group $Pbc2_1$ with a = 12.881,
b = 8.954, and c = 9.925 Å, has a different structure as shown in
Figure 2.   The Sc atom is bonded to only two Cp rings in an η^5

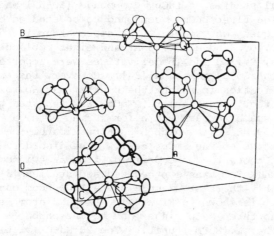

Figure 2.   Packing and structure of $ScCp_3$ [33].

fashion, with an average Sc-C distance of 2.494(22) Å.   Two Sc-C
bonds of 2.629(4) and 2.519(4) Å to Cp rings which appear to
bridge the $ScCp_2$ unit complete the coordination sphere around the
Sc atom [33].   If we assume that a Cp ring with six electrons in
three molecular orbitals occupies three coordination sites, then

the CN in ScCp₃ is only 8, compared to 11 in Nd(MeCp)₃. A decrease in CN with a decrease in size is reasonable and could explain the fact that TmCp₃ is not isomorphous with the lighter lanthanides. The bonding in the LnCp₃ and ScCp₃ compounds is believed to be ionic, and comparisons with the CaCp₂ [36] and InCp₃ [37] structures are frequently made to support this view. However, the ScCp₃ structure is quoted as giving direct evidence for covalency in the Sc-C bond so that the matter is not completely settled and will be discussed in more detail below.

Finally, we note that if edge interactions are restricted as in Sm(ind)₃, the metal is bonded in a η^5-fashion to all three C₅ rings with no additional Sm-C bonds, as shown in Figure 3 [32].

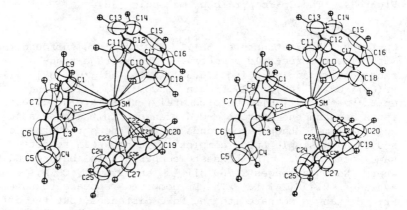

Figure 3. A stereoview of Sm(ind)₃ [32].

The Sm-C distances range from 2.678 to 2.872 Å, with an average of 2.744(48) Å. The Sm-C distances are longer than would be expected by comparison with the Nd-C distances in Nd(MeCp)₃ after corrections for the change in the radii as well as in CN. Unfortunately, the esds on the M-C distances are large so that the difference is not significant. However, longer Sm-C distances would be in agreement with increased steric effects in a Ln(ind)₃ versus a LnCp₃ compound.

The synthesis of the LnCp₃ compounds usually results in the isolation of a LnCp₃·base complex which is decomposed to yield the anhydrous species. The formation of LnCp₃·base compounds is reasonable in view of the structures of the LnCp₃ species which usually bond to a fourth ring in some way. The crystal structures of four of these adducts have been reported [38-41] and offer an interesting comparison with the LnCp₃ compounds. Unfortunately, the four compounds studied involve Ce, Pr, Gd and Yb and are not

directly comparable to any of the LnCp₃ structures. A typical
LnCp₃•base compound is shown in Figure 4 which illustrates the

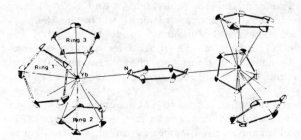

Figure 4. The $Cp_3Yb \cdot pyr \cdot YbCp_3$ molecule [38].

$Cp_3Yb \cdot pyr \cdot YbCp_3$ structure [38]. The Cp rings surround the metal
with almost three-fold symmetry. The average ring-center-metal-
ring-center angle being 116.3° in $Ce(ind)_3 \cdot py$ [39], 118.9° in
$Pr(Cp)_3 \cdot CHIN$ [40], 117.5° in $GdCp_3 \cdot THF$ [41], and 117.9° in
$Cp_3Yb \cdot pyr \cdot YbCp_3$ [38]. Unfortunately, there are no direct comparisons between the average M-C distances in the base adducts versus the simple LnCp₃ compounds because the structures of a base
adduct and simple LnCp₃ complex have not been reported for the
same metal ion. The closest comparison is between $Nd(MeCp)_3$,
where Nd-C averages 2.769(31) Å, and $Pr(Cp)_3 \cdot CHIN$, where Pr-C
averages 2.778(24) Å. The Pr compound appears to have longer
Pr-C distances compared to the Nd-C distances, but the difference
is less than would be expected. A somewhat similar situation
exists with $GdCp_3 \cdot THF$ [Gd-C average 2.738(25) Å] compared to
$SmInd_3$ [Sm-C is 2.744(48) Å] where the expected difference should
be about 0.03 Å. However, the differences are not significant in
either comparison because of the relatively large esds on the
average M-C distances. The esds in the simple LnCp₃ are about
twice those of the base adduct, which may be related to increased
libration of the Cp rings in the LnCp₃ compounds. Low temperature studies of a LnCp₃ and LnCp₃•base with the same Ln are
needed to decide whether there are significant differences in the
Ln-C distances in the two cases.

There has been a great deal of interest recently in the synthesis and structure of compounds containing a Cp_2Ln fragment.
The chemistry of the various Cp_2Ln compounds is fascinating
because of the way in which the coordination sphere around the Ln
is completed. These compounds can be classified into two groups,
the simple monomeric compounds and the dimeric derivatives.

The simplest compounds are the $Cp_2LnX \cdot solvent$ monomers (see
Figure 5) where the solvent is usually THF. The reported

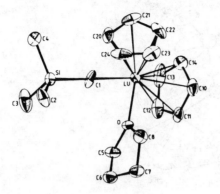

Figure 5.   The Cp₂Lu[CH₂Si(CH₃)₃]•THF molecule [43].

examples are Cp₂Lu(t-butyl)THF [42], Cp₂Lu[CH₂Si(CH₃)₃]•THF [43], and (PMCp)₂Yb[Co(CO)₄]•THF [44].   There are   more complex   mono- mers (see Figure 6) where the coordination sphere is completed by

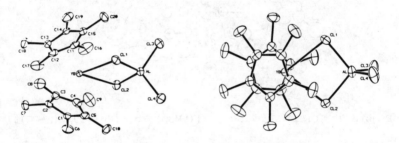

Figure 6.   Two   views   of   the   (PMCp)₂YbCl₂AlCl₂   molecule.   The right-hand   view   is perpendicular   to   the   "YbCl₂Al" plane [45].

bridging groups to a terminal group.   These derivatives are (PMCp)₂YbI₂Li(ether)₂,   (PMCp)₂YbCl₂Li(ether)₂,   (PMCp)₂YbCl₂AlCl₂ and   [φ₂(CH₃)SiCp]₂YbCl₂Li(ether)₂   [45],   Cp₂Yb(μ-CH₃)₂Al(CH₃)₂ [46],   and Cp₂Y(μ-CH₃)₂Al(CH₃)₂ [47].   The dimers can be halogen bridged such as [(MeCp)₂YbCl]₂ [48]   and   [Cp₂ScCl]₂ [49], methyl bridged [Cp₂YCH₃]₂ [50]   and [Cp₂YbCH₃]₂ [50],   acetylene bridged [Cp₂ErC≡CC(CH₃)₃]₂ [51] (see Figure 7),   or with a metal carbonyl bridge [(PMCp)₂Yb]₂Fe₃(CO)₁₁ [52] (see Figure 8).   A surprisingly large number of the above are Cp₂Yb derivatives; therefore,   the dimensions in the Cp₂YbL₂ compounds are summarized   in Table 2. The Yb-C distances to the Cp rings vary over a surprisingly small range (2.573 to 2.634 Å),   considering the   variety of   the other two ligands and   the substitution on the Cp rings.   In fact   the

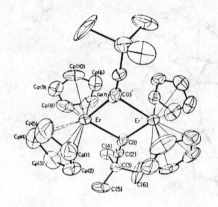

Figure 7. A view of the Cp₂ErC≡CC(CH₃)₃ dimer [51].

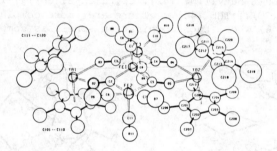

Figure 8. ORTEP drawing of [(PMCp)₂Yb]₂[Fe₃(CO)₇(μ-CO)₄] [52].

Table 2. Summary of distances (in Å) in Cp₂YbL₂ compounds.

Compound	Ref	Yb-Cp	Other Distances
Cp₂Yb(μ-CH₃)₂Al(CH₃)₂	46	2.609(30)	2.608, 2.561 to C
Cp₂Yb(μ-CH₃)₂YbCp₂	50	2.613(13)	2.535, 2.485 to C
[(PMCp)₂Yb]₂Fe₃(CO)₁₁	52	2.573(13)	2.243(5) to O
[(PMCp)₂Yb](THF)Co(CO)₄	44	2.596	2.335 to O(THF)
(PMCp)₂Yb(μ-Cl)₂AlCl₂	45	2.584(17)	2.756 to Cl
[(MeCp)₂YbCl]₂	48	2.585(24)	2.637 to Cl
(PMCp)₂Yb(μ-Cl)₂Li(ether)₂	45	2.611(14)	2.595 to Cl
[φ₂(CH₃)SiCp]₂Yb(μ-Cl)₂Li(ether)₂	45	2.634(10)	2.593 to Cl
(PMCp)₂Yb(μ-I)₂Li(ether)₂	45	2.624	3.027 to I

Yb-Cl distances of 2.593 to 2.756 Å show a larger variation than
the Yb-C distances.   The reasons for the variations in the Yb-Cl
distances are not obvious but may be related to the differences
in the electronegativity of the metal atom bonded to the bridging
chlorine atoms.   The synthesis and structure of other derivatives
of the type $Cp_2Yb(\mu-X)_2MX_2$ would be very useful in determining
whether the Yb-Cl distances are a function of the electronegativity of M.   The M-C distances will be discussed in more detail
below.

     The   remaining Cp_2LnL_2 compounds, with the exception of
$[Cp_2ScCl]_2$ [49], all contain a Ln-C σ-bond, either bridging or
terminal.   Therefore, a discussion of the Ln-C σ-bond is appropriate at this point.   A comparison of $Cp_2Lu(t-butyl)\cdot THF$ [42]
and $Cp_2Lu[CH_2Si(CH_3)_3]\cdot THF$ [43] is particularly interesting since
the Lu-O(THF) distances are the same [2.31(2) and 2.29 Å] but the
Lu-C distances of 2.47(2) and 2.38 Å differ markedly.   The longer
distances in the t-butyl derivative can be explained in terms of
steric effects.   Support for the steric arguments can be found in
the structure of $Cp_2Y(\mu-CH_3)_2Al(CH_3)_2$ [47].   The Y-C(bridge) distance, avg 2.585 Å, is corrected by the difference between the
Al-C(terminal) and Al-C(bridge) distances of 0.16 Å [53] to give
an estimated Y-C(nonbridge) distance of 2.42 Å.   A similar calculation using the data for $[Cp_2Y(\mu-CH_3)]_2$ [50] gives a Y-C distance of 2.39 Å.   Since Y is about 0.04 Å larger than Lu, we can
estimate that a Lu-C(nonbridge) · distance should be about 2.35 to
2.38 Å.   In addition the structure of $[Cp_2Yb(\mu-CH_3)]_2$ [50] gives
a Yb-C distance of 2.35 and a Lu-C distance of 2.34 Å.   However,
the structure of $Cp_2Yb(\mu-CH_3)_2Al(CH_3)_2$ [46], where Yb-C(bridge)
is a rather long (2.59 Å) and the difference in the Al-C bond is
only 0.13 Å, gives Lu-C a distance of 2.45 Å.   Furthermore, the
structural data for $Li(THF)_4^+Lu(2,6-dimethylphenyl)_4^-$ [25] [Lu-C
is 2.452(33) Å] and for $Yb[CH(SiMe_3)_2]_3Cl^-$ [23] [Yb-C is 2.379 Å]
predict a longer Lu-C bond. However, the lower coordination number in both compounds suggests that steric effects are extremely
important in both compounds, and therefore the Lu-C bond distance is longer than would be predicted.   Steric problems
must also be important in a comparison of the dimensions of
$[(tmen)Li(\mu-CH_3)_2]_3Er$ [24] (see Figure 9) with those in
$Cp_2Er[C\equiv C-C(CH_3)_3]_2$ [51] where the Er-C distances are 2.57 and
2.445 Å, respectively.   Since Lu^{3+} is about 0.03 Å smaller than
Er^{3+}, we would estimate Lu-C as 2.54 and 2.45 Å, respectively,
after the difference in hybridization has been taken into
account. We see that the various compounds give an Lu-C distance
ranging from 2.35 to 2.54 Å which appears to be too large a
spread.   The reasons for the large variation in the Lu-C bond
length is easily ascribed to steric effects, usually difficult to
prove.   However, the low CN of 4 and 6 for the lanthanides is
usually achieved only via steric hindrance so that the long bond
lengths may be understandable. We must conclude that a Lu-C bond

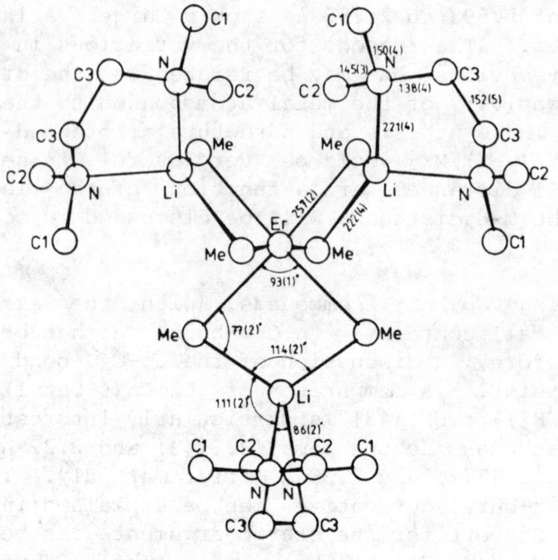

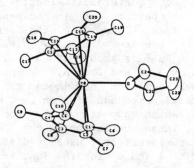

Figure 9.   The structure of [tmenLi(μ-CH₃)₂]₃Er [24].

for CN-8 is most likely about 2.36 Å.   However, additional structural data on organolanthanides containing a Ln-C bond would be
most helpful in resolving the question of steric influences on
the Ln-C bond length.

In concluding our survey of the Ln-Cp complexes, we should
mention the three examples of Ln²⁺ derivatives that have been
reported: (PMCp)₂Yb(THF)   [26],   [(CH₃)₃SiC₅H₄]₂Yb(THF)₂ [27],
and (MeCp)₂Yb(THF) [28].   Both the PMCp and (CH₃)₃SiCp derivatives are monomeric (see Figure 10), and the Yb-O distances
(2.412 and 2.404 Å, respectively) are equal.   By comparison of Cp₂Lu(THF)(t-butyl) [42]   (Lu-O, 2.300 Å)   and

Figure 10.   ORTEP drawing of (PMCp)₂YbTHF [26].

Cp$_2$Lu[CH$_2$Si(CH$_3$)$_3$]THF [43] (Lu-O, 2.29 Å), we predict that Yb^{3+}-O would be about 2.30-2.31 Å, which is in reasonable agreement with the Yb-O distance of 2.335(2) Å in (PMCp)$_2$Yb[Co(CO)$_4$]THF [44]. Therefore, we predict that Yb^{2+} will be about 0.08 to 0.10 Å larger than Yb^{3+}. Since the Yb^{3+}-C (Cp ring) distances varied from 2.573 to 2.634 Å, we would expect Yb^{2+}-C to vary from 2.65 to 2.73 Å. Indeed, the Yb-C distance in (PMCp)$_2$Yb(THF) averages 2.663(15) Å and in [(CH$_3$)$_3$SiC$_5$H$_4$]$_2$Yb(THF)$_2$ is 2.748 Å. There is some evidence that the (CH$_3$)$_3$Si group on a Cp ring lengthens the M-C distances by about 0.03 Å, which apparently is the case in this compound. In contrast in (MeCp)$_2$Yb(THF) [28] the Yb-C distances (avg 2.85 Å) and the Yb-O bond (2.53 Å) are both longer than expected. However, the compound is polymeric in the solid state (see Figure 11), and the Yb-C distances to the nonbridging

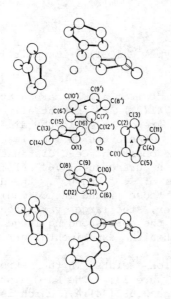

Figure 11. A plot of the structure of (MeCp)$_2$Yb·THF [28].

Cp ring of 2.76 Å are normal, while the Yb-C distances to the bridging Cp ring are long [2.89(5) Å], as might be expected. The difference between the M-C distances to bridging and nonbridging η^5-Cp rings is 0.11 Å in CaCp$_2$ [36] and 0.08 Å in Nd(MeCp)$_3$ [31]. Therefore, we see that there are three examples where the distances to bridging Cp rings are longer than to nonbridging Cp rings, which is reasonable in terms of sharing the fixed electron density on the Cp rings.

The last group of organolanthanides all contain the COT
ligand: K(diglyme)Ce(COT)$_2$ [54], [Ce(COT)(THF)$_2$(μ-Cl)]$_2$ [55],
and Nd(COT)(THF)$_2$Nd(COT)$_2$ [56]. The Ce(COT)$_2^-$ is shown in
Figure 12 and consists of a Ce^{3+} ion between two planar COT

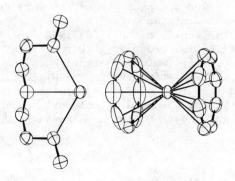

Figure 12. A view of the K(diglyme)$^+$Ce(COT)$_2^-$ ion pair [54].

rings. The Ce^{3+} ion is equidistant from all 16 carbon atoms with
an average Ce-C distance of 2.741(21) Å. Essentially the same
Ce-C distances [avg 2.710(14) Å] are observed in the dimeric species [Ce(COT)(THF)$_2$(μ-Cl)]$_2$ [55], where the Ce ion is bonded to
only one COT ring. Although the observed Ce-C distances can be
derived from the U-C distances in uranocene (2.647 Å) [57] and
the difference between the radii of Ce^{3+} and U^{4+}, the distances
are significantly shorter than those found in Ce(ind)$_3$py [39]
[avg Ce-C, 2.859(94) Å]. However, the large variation in the
Ce-C distances in Ce(ind)$_3$py and the corresponding large esd make
the two Ce-C distances equal within the experimental error.

A somewhat similar situation exists in the structure of
Nd(COT)(THF)$_2$Nd(COT)$_2$ shown in Figure 13. The Nd-C distances to
the nonbridging COT rings average 2.682(34) Å, which is shorter
than the 2.769(31) Å found for the average Nd-C distances to nonbridging Cp in Nd(MeCp)$_3$ [31]. Unfortunately, the difference of
0.087 Å is not statistically significant because of the relatively large esds. However, we see that in all three COT complexes
the average M-C distances are shorter than in the corresponding
Cp case. Simple electrostatic arguments would have predicted
that the interaction between a Ce^{3+} ion and the COT ion would
be stronger than with the mononegative Cp ion which could
explain the shorter distances. Alternatively, the shorter M-C
distances in the COT complexes could be related to the larger
diameter of the COT ring relative to the Cp ring which would
require shorter M-C distances for an equivalent overlap. In
essence shorter distances in the COT complexes can be interpreted

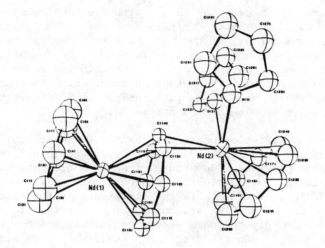

Figure 13.   A drawing of the Nd(COT)$_2$Nd(COT)(THF)$_2$ molecule [56].

in terms of either an electrostatic or covalent model for the
bonding in these complexes.

The question of whether there is a covalent contribution to
the bonding in the organolanthanides is controversial. Since the
electronegativities of the lanthanides are low, approximately
1.1, we would predict that the covalent contribution would be
small. As has been noted [58], one of the difficulties is developing a definition of covalent bonding which is experimentally
verifiable.   Another problem is the question of whether the 4f
orbitals are involved in the bonding to any large extent.   Since
these points have been discussed in detail elsewhere [58,59], no
attempt will be made to review all the evidence relating to 4f
orbital participation or covalent bonding.   However, some comments on the proposed criteria for covalent bonding seem appropriate.

The two criteria which have been proposed to distinguish between ionic and covalent bonding were the geometry of the compounds and the regularity of the bond lengths [58].   The suggestion that "The geometries of ionic compounds tend to be irregular..." which "...is in marked contrast to the regular, directional bonds which typify covalent compounds" does not appear to
be valid.   One need only consider the various forms of the element sulfur or the existence of various sulfur-fluorine compounds
which give a variety of shapes depending, in essence, only on the
number of fluorines bonded to the central sulfur atom to realize
that covalent bonding can be extremely varied for any elements
other than those of the first row Li to F.   Furthermore, a

comparison of either $CaCp_2$ or $InCp_3$ with the $LnCp_3$ compounds may also not be valid. The 4d orbitals have just been filled in the case of In, and the use of the 5d orbitals in bonding may not be energetically favorable. In the In case, the formation of a tetrahedral hybrid using 5s and 5p orbitals is reasonable and prevents the formation of a π-type interaction. In summary the use of geometrical variations to identify covalent versus ionic bonding does not appear to be justifiable.

The second criterion postulates that "... ionic radii can be used to predict bond lengths" while "...predominantly covalent compounds show pronounced departures from such predictions." The distance between two covalently bonded atoms, r_{AB}, is given by $r_A + r_B - 0.09\Delta_{AB}$, where Δ_{AB} is the difference in the electronegativity of A and B. If we are considering a series of elements, A, with approximately the same electronegativity which are bonded to the same atom, then $r_{AB} \approx r_A + k$, where k is a constant. In the case of the lanthanides where crystal field effects would be minimal, we would expect that the covalent radius would be related to the ionic radius by an expression r(covalent) = k_1r(ionic) + k_2, where k_1 and k_2 are constants. Consequently, we would write $r_{AB} \approx k_1$r(ionic) + k + k_2 which is a linear relationship between r_{AB} and r(ionic). Figure 14 gives a plot of the M-C distance versus r(ionic) for the 3d-metallocenes, MCp_2, and some organolanthanides, $LnCp_3$ and $LnCp_2$ derivatives.

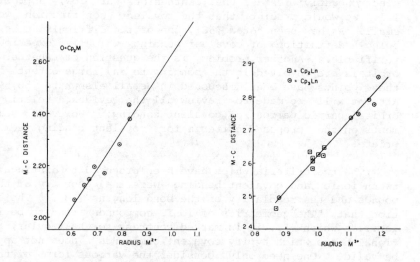

Figure 14. A plot of the M-C distance versus the ionic radius. The left-hand figure is for MCp_2 complexes, where M is a first row transition element. The right-hand figure is for the $LnCp_3$ and $LnCp_2X$ complexes.

The lanthanides give a straight line with a slope of 0.971 and a correlation coefficient of 0.97. A total of 7 $LnCp_3$ and 8 $LnCp_2$ structures were used in constructing the plot. The MCp_2 plot was drawn using the electron diffraction data given in Table II of reference 58. In the MCp_2 case the slope is 1.49 and the correlation coefficient is 0.92, in contrast to the values of 1.18 and 0.83 found for the X-ray data. The question, of course, is how do these two plots relate to the question of covalent bonding in the lanthanides. Certainly, the poorer fit in the MCp_2 case is an indication of a reasonable covalent contribution to the bond in the 3d metallocenes. Furthermore, the large deviation of the slope from 1.00 is another indication that a strictly ionic model is not applicable in the 3d-metallocenes. The linear relationship found in the lanthanide case, $LnCp_3$, together with a slope near unity suggests that the covalent contribution may be small. However, we should bear in mind the fact that if the covalent radii of the lanthanides differ by a constant amount from the ionic radii, a linear relationship between the M-C distance and M^{3+} would still be valid. In summary we see that attempting to determine the covalent contribution to chemical bonding using structural criteria is not straightforward.

COORDINATION COMPOUNDS OF THE LANTHANIDES

A search of the January 1982 CCDC Bibliographic File gave 266 entries containing either a lanthanide, scandium, or yttrium together with an organic residue of some kind. The present discussion will emphasize these compounds, but selected inorganic species will be included. The distribution of these entries among the various elements is shown in Table 3.

Table 3. The distribution of the number of entries in the January 1982 release of the CCDC Bibliographic File versus the element.

Element	La	Ce	Pr	Nd	Pm	Sm	Eu	Gd	Tb
Entries	28	28	22	40	0	11	23	10	2

Element	Dy	Ho	Er	Tm	Yb	Lu	Sc	Y
Entries	7	7	21	0	25	9	15	18

The distribution represents the availability of the element, the interest in other physical properties, and, of course, the ease with which a particular complex crystallizes. Furthermore, if we exclude the organolanthanides because of the difficulties in defining the CN in these compounds, we find the distribution versus coordination number given in Table 4.

Table 4.   The distribution of the coordination compounds of the
           lanthanides, Sc and Y, excluding the organolanthanides,
           as a function of CN.

CN	3	4	5	6	7	8	9	10	11	12
Number	4	1	1	19	20	62	70	23	4	7

The preponderance of structures with a CN of 8 and 9 illustrates the preference of the lanthanides for these two CN.   In
fact the $Ln(H_2O)_9{}^{3+}$ ion is well known for the elements La to Yb,
which also demonstrates the tendency of these ions to form complexes with a CN of 8 or 9.

We shall discuss the structures of the lanthanides in terms
of their CN rather than with regard to a particular type of
ligand or donor atom. This approach emphasizes the factors which
influence the final CN of the lanthanide complex, especially in
cases other than the usual 8 and 9 coordinate complexes.

LANTHANIDE COMPLEXES WITH A CN OF 3 TO 7

If we assume that a CN of 8 or 9 is the usual one found for
lanthanide complexes, then complexes with a CN of 3 to 7 can be
considered "abnormal".   As we shall see, steric effects are the
primary mechanism for synthesizing complexes with a low CN.

The classical example of using steric factors to limit the
CN was provided with the tmsa ligand.   All of the reported
lanthanide complexes with a CN of 3, 4 or 5 involve the tmsa
ligand [60-63].   A view of the Nd(tmsa)₃ molecule is shown in
Figure 15.   As the M-N distance decreases from Nd [2.29(2) Å] to

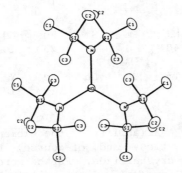

Figure 15.   A view down the c axis of Nd(tmsa)₃ [62].

Eu [2.259(9) Å] to Yb [2.158(13) Å] to Sc [2.047(6) Å], the N-Si
distance increases from 1.70(1) Å to 1.702(3) Å to 1.720(4) Å to
1.751(2) Å.   In addition, the Si-N-Si angle decreases from
126.4(9)° to 126.0(6)° to 122.1(8)° to 121.6(4)°.   The above
changes in distances and angles suggest increased steric effects
in going from Nd to Sc as would be expected.

The Ln(tmsa)$_3$ complexes react readily with tppo to form
four-coordinate species, Ln(tmsa)$_3$tppo (Figure 16), and an unusual five-coordinate complex, [La(tmsa)$_2$tppo]$_2$•O$_2$ (Figure 17) [63].

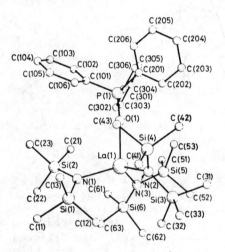

Figure 16.   The four-coordinate complex La(tmsa)$_3$tppo [63].

The latter is the only known five-coordinate lanthanide complex.
The isoelectronic carbanion (Me$_3$Si)$_2$CH$^-$ has also been used to
prepare the four-coordinate species Yb[CH(SiMe$_3$)$_2$]$_3$Cl$^-$, which has
been discussed earlier [23].   We see that the basic principle is
that steric effects can be used to restrict dramatically the
number of coordinating groups even with the relatively large lanthanides.   Presumably, more conventional bulky ligands could be
synthesized to yield other lanthanide complexes with a low CN,
although there appears to be only limited activity in this area
at present.

There are 13 lanthanide [64-74], 5 scandium [75-79], and one
yttrium [80] complexes, as well as one organolanthanide [24] compound with a CN of 6.   Since scandium is quite a bit smaller than
yttrium and the lanthanides, a CN of 6 is not unusual.   If we
exclude the ScCl$_3$(THF)$_3$ complex [77], the remaining 30 Sc-O
distances range from 2.014 to 2.119 Å with an average of

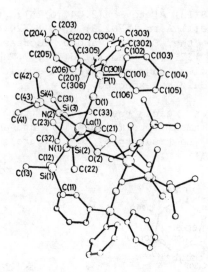

Figure 17. A view of the [La(tmsa)₂tppo]₂•O₂ complex. Note the
            unusual five-coordination of the La ion [63].

2.075(28) Å.   There appear to be small differences in Sc-O bond
lengths as a function of the ligand, but there are too few
reported structures of Sc complexes to allow further speculation.
However, the structure of mer-ScCl₃(THF)₃ appears to have significantly longer Sc-O distances as well as a difference between
the Sc-O trans to Sc-Cl [2.236(8) Å] and Sc-O trans to Sc-O
[2.147(7) and 2.164(7) Å].   This so-called trans effect has been
noted in other structures [81,82] and appears to be relatively
common [82].

     Of the 14 reported lanthanide and yttrium structures with a
CN of 6, 7 involve either Cl or S as donor atoms [65-69], and
6 involve bulky ligands [70-74,80].   The fact that large donor
atoms or bulky ligands restrict the CN of the lanthanides is not
surprising in view of our earlier discussions of CN 3 to 5.   One
unusual structure is that of [(C₄H₉)₄N]₃Er(NCS)₆ [64], where the
Er ion is surrounded octahedrally by the N-bonded anions.

     The Ln-S distances in the complexes of Pr, Sm, Dy and
Lu [66,67] are very close to the values estimated from the radii
of Ln³⁺ for CN-VI and S²⁻.   The complexes are distorted from an
idealized octahedral or trigonal prismatic geometry, with the Ln
complex being almost midway between the two polyhedra.   However,
the distortions are not easily rationalized as a function of the
size of the lanthanide, but steric interactions between the
ligands appear to be important [67].

The three chloro complexes [65,68,69] have Ln-Cl distances which are in agreement with the predictions from the ionic radii. In two of the cases other ligands are present in the coordination sphere but do not appear to affect the Ln-Cl distances. There is a long Nd-Cl distance, 3.05 Å, in $Nd_6(i-OPr)_{17}Cl$, but in this case the Cl atom is at the center of a trigonal prism of Nd atoms and is equidistant to all six Nd ions [74].

The six structures involving bulky ligands all have oxygen donors in the coordination sphere. The Ln-O distances in four cases [71-73,80] are about 0.10 Å shorter than the values predicted from the ionic radii of Ln^{3+} and O^{2-}. The assumption of an O^{2-} in an organic ligand is probably not valid so that the observed shortening is reasonable. On the other hand, the Eu-O distances of 2.638 and 2.756 Å in $Eu(tmsa)_2(dme)_2$ [70] are longer than the predicted value of 2.57 Å. Unfortunately, whether the lengthening results from steric factors or reflects the poor donor ability of an ether oxygen cannot be decided at present. However, the observed Eu-O distance is also longer than would be predicted from the Ln-O distance in the various organolanthanides, vide supra. In contrast the Nd-O distance of 2.05 Å in $Nd_6(i-OPr)_{17}Cl$ [74] is very much shorter than the predicted value of 2.38 Å. In summary we see that the Ln-O distances are not simply the sum of the ionic radii for Ln and O but are dependent on the nature of the oxygen donor.

The $Er(NCS)_6^{3-}$ anion is rather unusual since the Er ion is octahedral but the NCS^- ion is not expected to be sterically demanding. Furthermore, the Er-N distance of 2.34(2) Å is about 0.14 Å longer than the Er-O found in Er complexes with a CN of 6. Unfortunately, the paucity of experimental data and details precludes an extensive discussion of this complex. Certainly the reasons for a CN of only six are not obvious with the information at hand, and a complete report on this interesting structure would be most welcome.

A CN of 7 is not favored for any element in the periodic table. Nevertheless, complexes with a CN of 7 are known for most elements and are usually the result of steric constraints imposed by the ligand. A similar situation exists in the case of the large lanthanide ions where 15 [83-97] of the 21 structures with a CN of 7 [83-103] involve the addition of a base molecule to a tris(β-diketone)lanthanide. In most cases [83-93] the β-diketone has very bulky substituents; however, with the heavier lanthanides base·tris(2,4-pentanedionato)lanthanide(III) complexes can be formed [94-97]. A typical structure is shown in Figure 18. The majority of these structures have been discussed in two reviews on seven-coordination [11,19] and will not be discussed extensively. All the base·tris(β-diketone)lanthanides have either a capped octahedron or capped trigonal prism geometry. The M-O

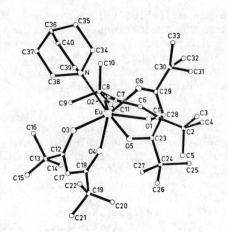

Figure 18.   The (dpm)₃Eu(quin) molecule is an example of a seven-
coordinate complex [91].

distances with two exceptions [84,89] follow the changes expected
from the changes in the radius of the Ln(III) ion.   In one case,
(fod)₃Lu(H₂O) [84], the Lu-O distances average 2.17(9) Å which is
less than the predicted value of 2.24 Å [85], but the accuracy of
the values precludes any discussion.   In the second case the Pr-O
distances of 2.356(33) Å also appear to be slightly shorter than
the predicted value of 2.39 Å, but the difference is not significant.   In summary the Ln-O(β-diketone) distances in these seven-
coordinate   complexes   follow   the   trends   predicted   from   other
structural studies.

The bases in the base•tris(β-diketone)lanthanide(III) complexes are all oxygen donors with two exceptions [85,91].   The
Ln-O(base) distances are usually longer than the Ln-O(β-diketone)
distances by about 0.07 to 0.10 Å.   One exception is the
(dpm)₃Pr(tppo) complex [89] where the Pr-O(tppo) distance is
2.349 Å compared   to   an   average   Pr-O(dpm)   distance   of
2.356(33) Å.   Furthermore, the Pr-O(dpm) distances are slightly
shorter than the predicted value of 2.39 Å. However, an explanation of these observations is not obvious at the present time.
One problem with the majority of these studies is the relatively
low accuracy of the reported distances.

The two nitrogen base adducts, (dpm)₃Lu(3-Mepy) [85] and
(dpm)₃Eu(quin) [91], are particularly interesting because of the
apparently long Ln-N distances of Lu-N of 2.492(8) Å and Eu-N of
2.603(7) Å.   The difference between these two distances is
approximately the difference in the ionic radii of the two ions.
However, in both cases the distances are about 0.25 Å longer than

the corresponding Ln-O distance, and the large difference is unexpected on the basis of the similar sizes of O and N. The reason for the lengthening of the Ln-N bond is not obvious since both O and N are considered "hard" donor atoms. A correspondingly longer Er-N bond was also reported in the Er(NCS)$_6{}^{3-}$ ion [64], as noted above, so that the elongation of the Ln-N bond may be a general phenomenon independent of the coordination number.

The remaining four lanthanide [98-102] and one scandium structures [103] with CN of 7 incorporate a variety of ligands but apparently all have a PBP geometry. The Er(dmp)$_7$(ClO$_4$)$_3$ complex (Figure 19) is one of the few species known which contains

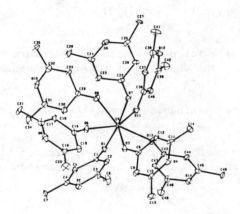

Figure 19.   The Er(dmp)$_7{}^{3+}$ ion is one of the few species with seven identical ligands around a central atom [101].

seven identical ligands [101]. The Er-O (axial) distances, average 2.230(10) Å, are slightly shorter than the Er-O (equatorial) distances of 2.300(29) Å, and the difference is possibly significant. Since the corresponding lanthanum compound, La(dmp)$_8$(ClO$_4$)$_3$, is octacoordinate [104], the seven coordination found in the Er complex must be a result of the lanthanide contraction. A somewhat similar effect is seen in the complex Sm[S$_2$P(OEt)$_2$]$_2$(tppo)$_3{}^+$S$_2$P(OEt)$_2{}^-$ which is seven-coordinate [99] while the lighter lanthanide complexes are eight-coordinate. In essence the formation of seven-coordinate lanthanide complexes appears to be related to the ionic size of the lanthanide ion and the steric requirements of the ligand.

The seven-coordinate scandium complex Sc(dapsc)(H$_2$O)$_2{}^{3+}$ shown in Figure 20 illustrates an alternative approach to preparing complexes with a CN of 7 [101]. A planar pentadentate ligand

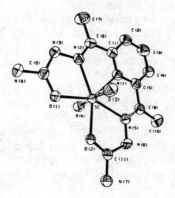

Figure 20.   The Sc(dapsc)(H$_2$O)$_2$$^{3+}$ ion [103].

with a suitable cavity size forces the metal ion to form PBP complexes.   To date that is the only   reported seven-coordinate   Sc
complex.

The seven-coordinate   lanthanide   complexes   demonstrate the
importance of   steric factors in   determining the final coordination number of even the large lanthanide ions.   The use of bulky
organic ligands   appears to be   a simple way of   restricting   the
coordination number, although specially designed ligands are also
useful.

LANTHANIDE COMPLEXES WITH A CN OF 8

Approximately two-thirds of the lanthanide coordination complexes with reported   structures   have a   CN of   either   8 or 9.
There is little doubt   that these are the   "favored"   coordination
numbers for lanthanide ions. However, an extensive discussion of
complexes with a CN of 8 or 9 will not be given   since a majority
of the structures   have been discussed   in recent reviews of high
coordination numbers [7,18,20].

For a CN of 8   the Dod and SAP   are the two polyhedra   which
are found most frequently.   Cubic coordination is rarely found in
metal complexes.   Indeed, a majority of the lanthanide complexes
with a CN of 8 have either the Dod or SAP arrangement, although a
few examples with some   distorted   intermediate   geometries have
been reported.

There are only four reported lanthanide complexes with eight
identical ligands [104-106].   The La(dmp)$_8$$^{3+}$ ion has a SAP geometry [104] and, as noted earlier, the Er(dmp)$_7$$^{3+}$ ion has a PBP
arrangement.   These   two   ions   illustrate   the   effect   of   the

lanthanide contraction on the CN.   However, the situation with
aqua complexes is more complicated since both octa- and nonaaqualanthanide ions can be isolated.   Although the $Ln(H_2O)_9{}^{3+}$ ions
appear to be the   usual case, the structure of $Gd(H_2O)_8{}^{3+}$ was
reported recently to have a Dod arrangement [105].   There
appear to be two different Gd-OH$_2$ distances, 2.451(10) Å and
2.354(10) Å, which correspond to the A and B sites in the
Dod [7].

The structures of two $Ln(py-NO)_8{}^{3+}$ ions, Ln = La and Nd (see
Figure 21) have been reported [106].   The La derivative has C$_2$

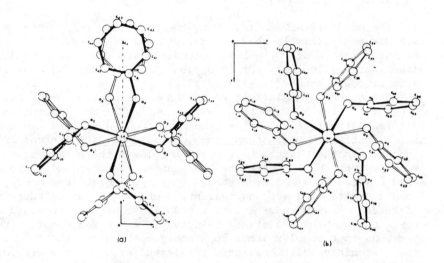

Figure 21.   Structures of the $La(PyO)_8{}^{3+}$ (a) and $Nd(PyO)_8{}^{3+}$ (b)
              ions viewed along the respective crystal c
              axes [106].

symmetry and the La-O distances are equal, with an average of
2.497(3) Å. The less accurate values in $La(dmp)_8{}^{3+}$ are 2.48(8) Å.
The structure was initially reported to be distorted from a SAP
toward a cubic arrangement. However, an analysis of the δ angles
suggests that the geometry is intermediate between the SAP and
Dod arrangements, a more common result.   The $Nd(py-NO)_8{}^{3+}$ cation
has a SAP arrangement but with two possibly different Nd-O distances of 2.431(11) Å and 2.382(19) Å.   The former is close to
the value expected from the La results, while the latter distance
appears to be a bit shorter. Unfortunately, the esds on the Nd-O
distances are large enough that the difference is only possibly
significant and so preclude any speculation on the results.

Approximately one-third of the lanthanide complexes with a CN of 8 involve a β-diketone either as a tetrakis(β-diketone)lanthanide anion [107-115] or as a bis adduct of a neutral tris(β-diketone)lanthanide complex [116-130]. The geometry of virtually all of these complexes has been discussed [7,18,20] and will not be considered in this review. One interesting feature of these β-diketone complexes is that the Ln-O distance appears to be independent of the β-diketone and of whether the complex is a tetrakis(β-diketone)lanthanide anion or a tris(β-diketone)lanthanide bisbase adduct. In a few cases [111,112,114,124] there appear to be differences in the bond lengths to the A and B sites in the Dod, but more frequently the bond lengths are equal within the rather large esds.

The bond distances involving the bases in the complexes of Ln(β-diketone)₃(base)₂ show several unusual features. With only one exception [123], the base molecules are equivalent and are either water [116-121], other oxygen donors [124,128], or aromatic nitrogen bases [125-127,129,130]. In the case of the oxygen donors the Ln-O distance to the base is approximately 0.10 Å longer than the Ln-O to the β-diketone. The one exception is Nd(ttfac)₃(tppo)₂ [124] where the Nd-O(tppo) distances are equal and slightly shorter than the average Nd-O(β-diketone) bond lengths. In the Nd(ttfac)₃(tppo)₂ complexes there appear to be two different Nd-O(β-diketone) distances corresponding to the A and B sites in a Dod even though the tppo molecules occupy an A and a B site and are equivalent. One might argue that the β-diketone oxygens have a slight negative charge which would decrease the Ln-O bond distance relative to a neutral donor. However, the tppo complex above apparently contradicts that hypothesis. Certainly additional structural studies of higher precision would be helpful in explaining these observations.

A second feature of the Ln(β-diketone)₂(base)₂ complexes is the very long Ln-N bond in all cases. The difference between the Ln-N bond length and the average Ln-O(β-diketone) distance ranges from 0.25 to 0.32 Å for the aromatic nitrogen bases and 0.18 Å for the amine nitrogen base [123]. The large difference between neutral O and N donors is somewhat surprising although a similar observation was made in the case of monobase adducts with a CN of 7, vide supra. One possible explanation is that aromatic nitrogen bases are not "hard" bases but borderline in character [131]. Consequently, in bonding to the "hard" lanthanide ions the "hard" oxygen donors are favored relative to the borderline nitrogen bases and form the shorter and stronger bonds.

Lanthanide complexes with a CN of 8 which do not have a β-diketone ligand can be divided into several different groups. There are a number of tetrakisligand complexes [132-139] several of which are sulfur ligands [135-139], although the majority of

the complexes involve carboxylic acid groups [140-160]. Finally,
there are a number of complexes containing a variety of different
ligands [161-169].

One of the most unusual tetrakis complexes is the
La(bipyO$_2$)$_4$$^{3+}$ ion (Figure 22) which was reported to have a cubic

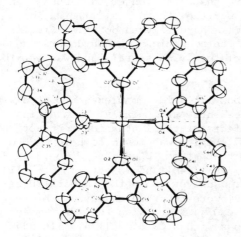

Figure 22.   The La(bipyO$_2$)$_4$$^{3+}$ complex viewed along the pseudo-
             fourfold   axis to illustrate the cubic coordina-
             tion [132].

arrangement of oxygen atoms around the La(III) ion [132]. In
fact only two other examples of cubic coordination have been
reported, and both of these involve actinide ions [18].   The
authors have suggested that a SAP model appeared possible for the
cation and that the cubic arrangement may arise because of better
packing in the solid state.   An examination of salts involving
other anions might show deviations toward the SAP geometry which
would support the packing arguments.

The PF$_6$$^-$ salts of Sm(tmma)$_4$$^{3+}$ and Er(tmma)$_4$$^{3+}$ are not isomorphous, but both cations have a SAP geometry [134].   An extra
ligand is present in the La(tmma)$_5$$^{3+}$ salt which was not surprising in view of the large size of the La(III) ion.   However, in
the La(III) salt one of the tmma ligands is only monodentate so
that the final CN is 9 [170].   While the increase in CN is not
unexpected in going from Sm(III) to La(III), the dangers of
extrapolating only on the basis of stoichiometry are evident.

There are five tetrakisligand complexes which have eight
"soft" sulfur donors bonded to a Ln ion [136-139], as well as the

La(deops)$_3$(tppo)$_2$ complex which has six sulfur donors [99]. The
Ln-S distances in the tetrakis(sulfur ligand) complexes are about
0.5 Å longer than the corresponding Ln-O distances in the
tetrakis(β-diketone) complexes. Furthermore, in La(deops)$_3$(tppo)$_2$
the La-S distances average 3.054(42) Å compared to the La-O
average distances of 2.439 Å, a difference of 0.62 Å.   However,
the S atom is only about 0.3 to 0.4 Å larger than an O atom.
Consequently, we believe that the longer Ln-S bonds indicate a
weaker interaction between the "hard" La(III) ion and a "soft"
sulfur donor compared to a "hard" oxygen donor.   Similar lengthening of the Ln-N bonds to aromatic nitrogen donors was observed
in other lanthanide complexes (vide supra) and was attributed to
a similar effect.

Of the various carboxylic acid complexes reported with a CN
of 8, the edta complexes are particularly interesting because
they illustrate the effects of the lanthanide contraction on the
CN.   The La(edta)(H$_2$O)$_3^-$ complex was found to have a CN of 9, and
the prediction was made that from Tb on the CN would be 8 [171].
Indeed, the Er(edta)(H$_2$O)$_2^-$ [140] and Yb(edta)(H$_2$O)$_2^-$ [141] were
found to have a CN of 8.   However, both the Sm(edta)(H$_2$O)$_3^-$ [172]
and Tb(edta)(H$_2$O)$_3^-$ [173] were reported to have a CN of 9.   More
recently, the Ln(edta)(H$_2$O)$_3^-$ salts with Ln = Pr, Sm and Gd were
reported to be isomorphous with a CN of 9 [174].   In addition the
authors were unable to obtain the orthorhombic modification of
the Dy salt reported earlier [141] and raised some questions
regarding the final refinement of the Dy salt.   Nevertheless, the
available data support the hypothesis of a change in the CN of
the edta complexes from 9 to 8 as the end of the lanthanide
series is approached.   Where exactly the change occurs remains to
be determined, but certainly the change occurs between the three
elements Dy, Ho or Er.

A somewhat similar situation exists with the lanthanide
complexes of nta which appear to form three distinct crystallographic groups.   The heavier lanthanides from Dy to Ln have a CN
of 8 as shown in Dy(nta)(H$_2$O)$_2$·2H$_2$O [143].   For the heavier lanthanides, the stoichiometry is Ln(nta)(H$_2$O)$_2$·H$_2$O, Ln = Pr-Tb, and
the crystal structure of the Nd [142] and Pr [176] complexes have
shown a CN of 9.   In all three structures the nitrogen atom of
the nta ligand is coordinated, but the Ln-N distance is about
0.2 Å longer than the corresponding Ln-O bond distance. Crystals
of the third group have Ln = La or Ce but were of such poor
quality that they could not be characterized.   However, on the
basis of studies with edta and related ligands, vide infra, we
might expect a CN of 10 for the La and Ce complexes with nta.   In
any case the nta complexes provide another example of how the CN
decreases because of the lanthanide contraction.

The last series of complexes with a CN of 8 which we will
discuss are the complexes with either nicotinic acid or isonico-
tinic acid [157-160].   While the Sm, Eu, Ho and Er complexes all
have a CN of 8, the La complexes of both nic and isonic are difficult to classify.   In the La complexes there are 8 short La-O
distances, 2.424 to 2.604 Å in La(isonic)₃ and 2.462 to 2.646 Å
in La(nic)₃.   There is also one long La-O bond, 2.969 Å in
La(isonic)₃ [159] and 2.919 Å in La(nic)₃ [160].   In both cases
the long La-O distance involves a bridging carboxyl group oxygen
atoms.   The long La-O distance appears to be a bonding interaction and is in agreement with our expectations that the CN of
La complexes are usually higher than for the heavier lanthanides.

Of the structures involving the Ce(IV) ion, three have a CN
of 8 [107,108,133,168].   Since the radius for Ce(IV) is approximately the same as that of Lu(III), we might expect that very
high coordination numbers would not be quite as common as with
the larger Ce(III).   The Ce-O distances average 2.324(23) Å in
the two Ce(acac)₄ complexes [107,108] and 2.360(2) Å in the
Ce(cat)₄ structure [133].   While these two values are not significantly different, they are both longer than the average Ce-O
distance of 2.211(4) Å reported for a Schiff base complex
Ce(htfmd)₂ (Figure 23) [168].   In the latter complex the usual

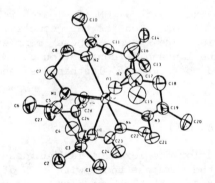

Figure 23.   A drawing of the Ce(htfmd)₂ molecule, an example of
             an eight-coordinate Ce(IV) complex [168].

long Ln-N bond distance, average 2.623(6) Å, was also reported.
A comparison of the three structures suggests that there is a
correlation between the Ce-O distance and the presence of nitrogen donors in the coordination sphere. Further studies of Ce(IV)
complexes would be useful in establishing whether the shortening
of the Ce-O distances in the presence of nitrogen atoms is a general occurrence or a peculiarity of that particular ligand system.

The NdPc₂ is an unusual lanthanide complex with a CN of 8 [169]. The sandwich-like structure is shown in Figure 24. The

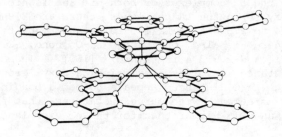

Figure 24. The molecular structure of NdPc₂ [169].

eight Nd-N distances vary from 2.39 to 2.49 Å, although specific details were not given. These distances are approximately the same as the Nd-O distances found in other Nd complexes with a CN of 8 rather than the much longer Ln-N distances which are usually found. In addition the location of a lone hydrogen could not be established conclusively. The unusual nature of the LnPc₂ complex warrants further crystallographic studies.

The lanthanide complexes with a CN of 8 are relatively common and have mainly a SAP or a Dod as the geometry. A few examples of other geometries are known. These complexes also illustrate the effect of the lanthanide contraction on CN and the weaker interactions formed between the lanthanides and "soft" donor atoms.

LANTHANIDE COMPLEXES WITH A CN OF 9.

The number of coordination complexes of lanthanides with a CN of 9 is approximately the same as those with a CN of 8. As noted earlier, the complexes with a CN of 8 or 9 account for approximately two-thirds of the reported lanthanide structures.

For a CN of 9, the TTP and MSAP are the two polyhedra which are observed most frequently in coordination complexes. Two recent reviews on nine and higher coordination numbers [7,20] have focussed on the questions of coordination geometry so that this review will not deal extensively with the geometrical aspects of CN-9.

A broad overview of the complexes with a CN of 9 does not provide an easy breakdown into various ligand types. The nona-aqua complexes [176-179] are some of the very few complexes with nine identical ligands. There is a group of tris(tridentate ligand)Ln complexes with a CN of 9 [180-191] which have both oxygen and nitrogen donor atoms. However, by far the largest group of complexes with a CN of 9 contains oxygen donors usually with a carboxylic acid ligand. The ligands include acetates [192-198], hydroxyacetates [199-201], malonates [54,202-204], fumarate [204], oxalates [205-211], and iminodiacetates [212-215]. In addition there are a small number of trisligand trisanionlanthanide complexes [216-223] of which the majority have nitrate as the anion [216-221]. Finally, there are a few complexes which are not easily classified in any of the above groups [224-228], as well as the edta and nta complexes discussed earlier in connection with eight-coordination [141,171-175].

The nonaaqualanthanide ions, $Ln(H_2O)_9^{3+}$, are among the very few complexes with nine identical ligands and have been extensively discussed [20]. With ethyl sulphate as the anion, the cations are all TTP with symmetry C_{3h} and are isomorphous for La-Lu and Y. The most interesting feature is that the Ln-O to the prism atoms are all shorter than Ln-O to the capping atoms. Furthermore, the prismatic distances change in agreement with the expected lanthanide contraction, 0.15 Å from Pr to Yb, but the capping distances decrease about half as much, 0.074 Å from Pr to Yb. These observations are in agreement with theoretical calculations which indicate increased repulsive forces on the capping atoms relative to the prismatic ones [20].

Three different ligands, terpy [180], dipic [181-184], and odac [185-190], are found in the reported tris(tridentate ligand)Ln complexes. With one exception [189], the complexes form discrete molecules of the mer isomer of the TPP in the solid state. The terpy complex has both typical and rather long Eu-N bond distances. However, the lack of atomic coordinates or individual bond lengths precludes further discussion. In both the dipic and odac complexes the capping atoms in the TPP are about 0.1 Å longer than the other donor atoms, similar to the situation found in the nonaaqualanthanide ions. In the odac complexes the difference is the same in the Nd complex, Nd-O is 2.523 and 2.428 Å, and in the Yb complex, Yb-O is 2.431 and 2.339 Å [188], although the decrease is only about two-thirds of the expected 0.12 Å. In the dipic complexes the differences from Nd (Nd-N of 2.581 and Nd-O of 2.485 Å [18]) to Yb (Yb-N of 2.512 and Yb-O of 2.365 Å [183]) are not constant although the Ln-O distances follow the lanthanide contraction. Whether the difference in the behavior of the dipic and odac complexes is related to the rigidity of the ligands or the presence of the nitrogen donor in dipic cannot be decided upon at this time. However, the experimental evidence certainly supports the conclusion that the capping atoms in a TTP behave differently under the influence of the lanthanide contraction than do the prismatic donor atoms.

The structures of complexes involving carboxylic acid groups are invariably polymeric in the solid state. The bridging groups can be water or a carboxylic acid group. The net result is that any discussion of distances or geometries is complicated by the polymeric nature of the structure. Consequently, we will not consider these complexes in our discussion.

There is an interesting group of complexes with a CN of 9 formed by the addition of three base molecules to a tris(bidentate anion)Ln. The majority of the complexes are tris(nitrato)tris(ligand)Ln complexes [216-221]. There are five possible isomers for a TTP complex with stoichiometry Ln(bidentate)$_3$(unidentate)$_3$. However, the isomer in which the bidentate spans the three parallel edges requires a larger ligand bite [20] than is possible. Therefore, with the nitrate ion the prediction is that there will be two different Ln-O(nitrate) distances, the accuracy of the various determinations is low and possible differences are obscured by experimental errors. However, the average Ln-O of the donor molecules and Ln-O to the nitrates decrease as predicted by the lanthanide contraction. In contrast, as noted earlier, the capping versus prismatic atom distances usually changed at different rates. More accurate values would be very useful in determining whether the various predictions are indeed valid.

Two tris(anion)tris(ligand)Ln complexes with β-diketones as the anions have been studied [222,223]. Surprisingly, these two adducts have a CN of 9. Our previous discussions with β-diketone complexes with a CN of 7 or 8 suggested that bulky β-diketones would restrict the CN. The structure of Pr(facam)$_3$(DMF)$_3$ was the first example of a β-diketone complex with a CN of 9 [222]. The 1:1 binuclear complex of Co(acac)$_3$ and Eu(fod)$_3$ is the second example of a tris(β-diketone)Ln complex with a CN of 9 [223]. The Eu(fod)$_3$ case is somewhat easier to rationalize since the coordination involves a triangular face of the Co(acac)$_3$ complex to form one face of a TTP. The fod ligands have one oxygen capping and one on the corner. The capping Eu-O distances are slightly but not significantly longer than the noncapping Eu-O distances. The existence of these two complexes with a CN of 9 reinforces the fact that generalizations on the species in solution can be risky.

The tris(isonich)triaquasamarium(III) cation belongs in the general class of tris(bidentate)tris(unidentate)Ln complexes which we have been discussing. Although the authors suggest that the cation was a TTP with capping water molecules [224], the cation is best described as a TTP with the nitrogen donors as capping atoms [20]. The Sm-N distances are long, avg 2.66(47) Å, as has been observed in virtually all the lanthanide complexes with nitrogen donors. Since the capping atoms are usually at longer distances (see above), placing of the nitrogen donors in these sites is reasonable. We might predict that the synthesis of a TTP with a bidentate ligand spanning the three parallel edges would require not only a normalized ligand bite of about 1.3 Å [20], but a bidentate ligand having two oxygen donors as well.

The last four complexes with a CN of 9 are difficult to classify [225-228]. In one case there are eight Pr-O distances ranging from 2.383 to 2.568 Å, with the ninth Pr-O distance being a very long 2.895 Å [225]. A very long Nd-S bond of 3.15(3) Å is found in $Nd(tdac)(H_2O)_4^{2+}$ [228], which is polymeric in the solid state. While the determination is not very accurate, the long Nd-S bond is in agreement with the observation that bonds of Ln to second row ion donor atoms are longer than expected from size considerations alone. The $Pr(terpy)(H_2O)_5Cl^{2+}$ ion [227] also illustrates the lengthening of bonds from Ln to "soft" donor atoms. The distances Pr-N of 2.635 and 2.625 Å and Pr-Cl of 2.876(2) Å are all long relative to either the Pr-O distances in the cation (avg 2.513 Å) or to Pr-O distances in other nonacoordinate species.

We see that complexes with a CN of 9 usually have a TTP arrangement of donors around a lanthanide ion. In many cases the Ln-capping atom distances are longer than to the prismatic atoms, in agreement with theoretical predictions. Finally, we see that lanthanide distances to "soft" nitrogen bases or to second row ion donor atoms appear to be longer, and presumably weaker, than would be predicted from the radii alone.

LANTHANIDE COMPLEXES WITH A CN OF 10, 11 OR 12

As the CN increases, the number of possible polyhedra also increases, and the question of molecular geometry is less meaningful. Furthermore, the number of examples of complexes with a CN of 10 or higher decreases as the CN increases. The result is that a discussion of the characteristic polyhedra of CN 10 and higher is not extremely useful. The shapes and interconvertability of the three most common polyhedra for CN 10, BSAP, BCDod and 46Dod have been discussed [7,20] and will not be considered here.

A common feature of the majority of the complexes with a CN of 10 and higher is the presence of a bidentate ligand with a small bite. The nitrate ion is by far the most common anionic bidentate ligand found in complexes with a high CN, although carbonates and acetates have also been found. The fact that a ligand with a small bite leads to complexes with a high CN is not surprising since steric problems are an important consideration in achieving a high CN.

A survey of lanthanide complexes with a CN of 10 reveals a small group which contains a multidentate ligand or ligands with only water molecules completing the coordination sphere [229-234]. However, the remainder of the complexes with a CN of 10 have at least one anion in the coordination sphere [235-257], and in only one case [257] is the anion not a bidentate ligand with a small "bite."

A protonated edta complex, La(Hedta)(H$_2$O)$_4$, with a CN of 10 has been reported [229]. In contrast the tetranegative edta ligand forms complexes with a CN of 9 with La down to 8 with Er. When the related trinegative ligand heedta is used, a dimer with a CN of 10 has been found [230-232]. The two ligands edta and heedta have virtually the same steric requirements and differ mainly in the number of ionizable protons. Consequently, the final CN of lanthanide complexes with ligands of the edta and heedta type are dependent upon the charge on the ligand, in essence an electroneutrality effect.

As the number of donor atoms in the ligand decreases from six in edta or heedta to five and four, bisligand lanthanide complexes have been reported [233,234]. The cation Ce(dapsc)$_2$$^{3+}$ has two pentadentate ligands coordinated to the Ce atom and no water or anions in the coordination sphere [234] (see Figure 25).

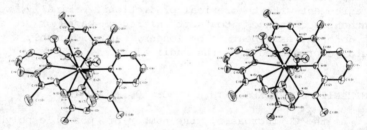

Figure 25.   A stereoview of the ten-coordinate Ce(dapsc)$_2$$^{3+}$ ion [234].

The complex was prepared in an aqueous solution so that the absence of coordinated water was surprising. Furthermore, the planar ligand must twist from a possible bicapped cubic arrangement to achieve the observed BSAP. The use of multidentate ligands to prepare complexes with a high CN is an area of current activity (vide infra).

The majority of the complexes with a CN of 10 have at least one anion in the coordination sphere and, in a few cases, five [235-240]. The pentanitrato complexes of Ce [235], Eu [236], Ho [237], and Er [238] have all been reported. The Ln-O

distances decrease from Ce to Er as expected although there are
some indications that the range of Ln-O distances increases from
Ce [2.553(18) to 2.591 Å] to Er [2.392(7) to 2.493 Å].   Whether
the increase in the range is a steric effect related to the
smaller size of Er relative to Ce or just reflects the accuracy
of the determinations cannot be decided at this point.

There are   two reported   pentacarbonatocerium(IV)   complexes [239,240], a tricarbonatocerium(IV) complex [241], as well
as a tetranitratocerium(IV) complex [244].   These complexes are
among the few examples of Ce(IV) complexes which have been
reported.

The tri(bidentate anion)Ln complexes with a CN of 10 have
coordinated nitrates [251-256] with only one exception, the
Nd(phen)₂(sal)₂ complex [257].   With six of the coordination
sites occupied by bidentate nitrate ions, the remaining four
sites are filled by bipy [252-254], dmso [255,256], or a quadri-
dentate Schiff base [251].   The latter is one of the few Schiff
base complexes of the lanthanides that has been prepared and
characterized by diffraction techniques.   One of the interesting
features of the Gd complex is the difference between the Gd-N
bond distances to the pyridine nitrogen (2.576 Å) and to the
Schiff base nitrogen (2.506 Å).   The latter value is very close
to the average Gd-O(nitrate) distance of 2.502(34) Å. There is an
apparent differentiation by the Gd atoms between the two different nitrogen atoms.   Whether the difference is a steric effect or
is related to the donor properties of the two nitrogen atoms is
difficult to determine.   In Tb(NO₃)₃(bipy)₂ [252] the Tb-O and
Tb-N distances are shorter by about the expected amount compared
to the La-O and La-N distances in La(NO₃)(bipy)₂ [253].   (Tb-O is
2.47 Å, La-O is 2.599 Å, Tb-N is 2.50 Å, and La-N is 2.656 Å.)
However, there is a large difference between Ln-O and Ln-N in the
two complexes similar to that observed in the Gd complex above.
Unfortunately, the low accuracy of the Tb(NO₃)₃(bipy)₂ results
obscures any significance in these differences.

The Ln(NO₃)₃(dmso)₄ complexes observed for La [255] and
Nd [256] provide another example of the effects of the lanthanide
contraction.   While the lighter lanthanides have a CN of 10 with
4 dmso groups, the heavier lanthanides have only 3 dmso groups
and a CN of 9 [218-220]. However, of more interest is the observation that in all five complexes the Ln-O distances to the dmso
group are shorter than to the nitrate anion. We can compare La-O
of 2.475 versus 2.653 Å, Nd-O of 2.385 versus 2.661 Å, Er-O of
2.270 versus 2.465 Å, Yb-O of 2.24 versus 2.43 Å, and Ln-O of
2.253 versus 2.462 Å.   This observation is directly opposite to
the bipy complexes where the Ln-N ligand distances were longer
than the Ln-O distances.   Since the differences in bond lengths
were interpreted in terms of a hard-soft interaction, the

reasonable conclusion is that dmso is harder than pyridine. However, steric requirements for dmso may be less than for bipy because of the two hydrogens adjacent to the nitrogen atom. Further structural studies on both dmso and bipy complexes could help to explain the observed differences.

The remaining complexes with a CN of 10 all have a macrocyclic ligand of either a cryptate [245,246] or a polyether type [247-250]. Both the cryptate complexes have a 2.2.2 cryptate coordinated to either a Sm [245] or Eu [246] ion with either a bidentate NO_3^- [245] or ClO_4^- [246] completing the coordination sphere. The separation of the strands of the crypt to allow a bidentate ion to coordinate has been rationalized on the basis of the small size of the ion relative to the size of the cavity.

The polyether complexes include a Eu 12-crown-4 complex [247], two 18-crown-6 complexes [248,249], and a dithio 18-crown-6 complex with La [250]. The differences in the Ln-O(polyether) and Ln-O(anion) distances are difficult to rationalize. The Ln-O polyether distances appear to be longer in general than Ln-O to the coordinated anions or water. However, in two cases [248,250] the Ln-O distance to a bidentate ClO_4^- ion are longer than the Ln-O polyether distances. When the ClO_4^- is monodentate [248,250], the Ln-O distances to the anion are shorter than to the polyether. Finally, the La-S distance (avg 3.038 Å) are longer than the La-O distances (avg 2.610 Å), as expected. The difference is close to the difference between the radius of O^{2-} and S^{2-}. The La-S distances in this case do not appear to be quite as long as in other Ln-S complexes, vide supra. Hopefully, as more structural data on lanthanide complexes with polyethers and other macrocycles become available, the observed variations in bond distances will become more explicable.

There are only 9 examples [245,258-265] of lanthanide and yttrium complexes with a CN of 11. Because of the rarity of this CN, discussions of CN 11 were omitted from recent reviews on complexes with high coordination numbers [7,20,266]. Although 10 different polyhedra have been postulated for CN 11 [267], the question of the favored polyhedron has not been explored extensively. Calculations using the points on a sphere model have indicated that the DHH is favored for values of n $\leq$ 6, but that for n $\geq$ 7 the MCPA is slightly more stable. However, the difference in energy between the DHH and MCPA is less than 3%. In all cases the PCTP is less stable [268]. Another study stated that the MCPA was the favored polyhedron without any restrictions [264], although the MCPA was said to be an unattractive choice for polyhedral boranes [269]. Consequently, we see that the choice of a favored polyhedron for CN 11 is by no means settled.

With one exception [258], complexes with a CN of 11 all contain bidentate nitrate ions which means that distorted polyhedra
will be the rule, not the exception.   Therefore, the description
of a complex in terms of a particular polyhedron is not only difficult but possibly futile.   In view of these problems we shall
minimize our discussion of the polyhedra for CN 11.

One of the first examples of CN 11,   and certainly a unique
one, was the $Y(BH_4)_3(THF)_3$ complex [258].   There are two tridentate and one bidentate BH_4^- ions coordinated to the Y^{3+} ion,
which, together with three THF molecules, give a CN of 11. This
is the only example of a complex with CN 11 which does not contain nitrate ions coordinated to the metal ion.   There are some
lanthanide complexes with identical stoichiometries which presumably will also have a CN of 11.   One interesting question is
whether the larger La^{3+} ion will have all three BH_4^- ions as tridentate, giving a final CN of 12.

The remaining complexes with a CN of 11 can be divided into
two groups, depending on the ligands surrounding the metal ion.
The first group has only bidentate nitrate ions and only monodentate ligands in the coordination sphere [245,259-262].   The three
complexes in the second group have three bidentate nitrate ions
and a pentadentate ligand coordinated to the metal ion.

The complexes containing only bidentate nitrates and monodentate ligands can have either three [259,260], four [261,262],
or five [245] bidentate nitrates coordinated to the metal ion.
The complexes $Ln(NO_3)_3(H_2O)_5$, Ln = La [259] or Ce [260], are not
found in the Cambridge File but have been included because there
are so few examples of 11-coordination.   In both complexes there
are three different sets of Ln-O distances which are not related
to whether the O atoms are from the nitrate ions or water molecules.   There are two long distances, four intermediate, and five
short distances. The two long distances average La-O of 2.860(21)
and Ce-O of 2.862(27) Å.   The four intermediate distances average
La-O 2.678(18) and Ce-O 2.646(15) Å, while the five short distances average La-O 2.568(34) and Ce-O 2.537(37) Å.   The differences between the three sets are significant although the differences do not appear to be related to the location of the atoms on
the coordination polyhedron.   Although similar effects are found
in the other complexes with a CN of 11, the differences do not
appear to be as significant as in these two complexes.

The Sm-O distances in the $Sm(NO_3)_5(H_2O)^-$ ion can also be
divided into two long, four intermediate, and five short.   The
averages are 2.641(38), 2.556(21), and 2.504(11) Å, respectively [245]. In this case the differences are possibly significant.
A comparison of differences between the La-O, Ce-O, and Sm-O
distances in the three complexes reveals that the long and inter-

mediate distances have decreased by more than the expected lanthanide contraction while the short distances have decreased by less than that amount. Furthermore, the five short values are slightly longer than the average Sm-O in the 10-coordinate cation, 2.496(39) Å [245]. One possible explanation is a simple steric argument that a fixed nitrate bite does not allow 11 oxygens to coordinate to the metal ion at the same distance. However, the existence of $Ln(NO_3)_6^{3-}$ ions, vide infra, with identical Ln-O distances tends to rule out that rationalization.

The two complexes $Ln(NO_3)_4(H_2O)_2(4,4'-bipy)$, Ln = Ce [261] and Ln = Nd [262], present an even more confusing set of distances. The Ce-N distance of 2.832(4) Å is very much longer than the Ce-O distances which range from 2.501(4) to 2.769(4) Å. As noted above, longer Ln-N(pyridine) bonds are very common in lanthanide complexes. However, in the Nd case the Nd-N distance of 2.767(12) Å is even slightly shorter than the longest Nd-O bond, range 2.434(11) to 2.780(11) Å. While the Ce-O distances appear to fall into four groups, [2.753(23), 2.673(11), 2.601(21) and 2.501(4)], a similar division in the isomorphous Nd complex is not apparent. Furthermore, the difference between the radius of Ce(III) and Nd(III) is only about 0.03 Å so that the large difference reported for these two complexes is difficult to understand.

The three remaining complexes with a CN of 11 all have a pentadentate ligand and three bidentate nitrates [263-265]. The geometry is very similar in the two complexes involving a noncyclic pentadentate ligand. The five donor atoms of the ligand are approximately planar with one bidentate nitrate above and one below the ligand plane. The third nitrate occupies the open end of the ligand with one oxygen above and one below the ligand plane (Figure 26). The La(dapbah)(NO_3)_3 complex was classified as a DHH on the basis of an analysis of the δ-angels [263]. In contrast the La(teg)(NO_3)_3 complex was described in terms of a hexagonal basal plane with the two nitrates in an approximately bisphenodral position [265] despite the similarities in the

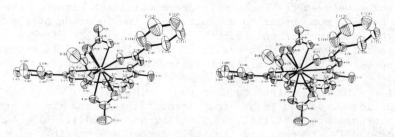

Figure 26.   Stereoview of the eleven-coordinate La(dapbah)(NO_3)_3 molecule [263].

arrangement of ligands in the two complexes. However, the large
deviations of the atoms from the hexagonal plane make the
description questionable. The macrocyclic 15-crown-5 complex was
described in terms of a MCTP [264]. This description is reasonable since the cyclic 15-crown-5 ligand has no open end to be
spanned by a nitrate ion, and the three nitrate ions all coordinate to the metal on the same side of the 15-crown-5 plane.

The metal-to-ligand distances in the three complexes are
very irregular which appears to be typical of CN-11. In the
La(dapbah)(NO$_3$)$_3$ case [263] the La-N distances are much longer
than the La-O bond lengths, as is usually observed in lanthanide
complexes. The La-O(dapbah) distances are the shortest bond
lengths in the complex [2.560(3) and 2.507(3) Å]. The corresponding La-O(nitrate) distances range from 2.563(3) to
2.648(3) Å, with an average of 2.617(29) Å. The spread in the
La-O(nitrate) distances in La(teg)(NO$_3$)$_3$ [265] is even larger,
2.59(1) to 2.79(1) Å. The average La-O(nitrate) distance is
2.64(8) Å, but the validity of averaging the values is questionable and is reflected in the large esd. Similarly, there
are large differences in the La-O(ligand) [265] and
Eu-O(ligand) [264] distances in the two complexes where in both
cases the ligand has five oxygen donors. The conclusion would
appear to be that the distances in complexes with a CN of 11 are
not regular regardless of the nature of the groups in the coordination sphere.

The icosahedron is one of the five regular Platonic solids
and is generally accepted as the favored polyhedron for a CN of
12 [7,20,266,269]. Although there are more examples of complexes
with a CN of 12 compared to a CN of 11, the CN of 12 is still not
common. The Ln(NO$_2$)$_6^{n-}$ ion is one of the most frequent examples
of CN 12 so that we have included several examples [270-273] of
inorganic salts which would not be found in the Cambridge File.
With two exceptions, Pr(nap)$_6^{3+}$ [274] and La(crypt)(NO$_3$)$_2^+$ [275],
the other examples of CN-12 are the type La(NO$_3$)$_3$L [221,276-279],
where L is a hexadentate ligand.

One of the few examples of a Ce(IV) complex is the
well-known salt (NH$_4$)$_4$Ce(NO$_3$)$_6$ which contains the Ce(NO$_3$)$_6^{2-}$
ion [270]. The ion has a symmetry very close to T$_h$, with an
average Ce-O distance of 2.502(15) Å. The Ce-O distance is
0.133 Å shorter than in the Ce(NO$_3$)$_6^{3-}$ ion which is close to the
expected difference. The ions Ln(NO$_3$)$_6^{3-}$ with Ln = La [271,275],
Ce [272], Pr [273], and Ne [245] have all been determined with
various cations. The Ln-O distances (La 2.658(20) [275], Ce
2.635(29) [272], and Nd 2.602(43) Å [245]) decrease by about the
predicted amount. However, the neutron diffraction results for
La [2.670(28) Å] are slightly longer, perhaps reflecting a difference due to the radiation rather than any chemical effects.

The Pr case [2.633(78) Å] is somewhat more complicated since the stoichiometry $K_3Pr_2(NO_3)_9$ requires nitrate ions to bridge the Pr ions.   Strictly speaking,   the $Pr(NO_3)_6^{3-}$ unit does not exist in the solid,   and there appear to be two types of Pr-O distances (2.631 and 2.745 Å versus 2.577 and 2.580 Å).   Consequently, averaging the four values is not correct but does show that the average Pr-O distance is only slightly longer than would be predicted.

The $Pr(nap)_6^{3+}$ ion is very close to an icosahedron (Figure 27) in spite of the small bite of the nap ligand.   The

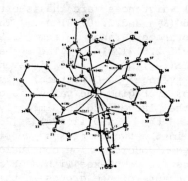

Figure 27.   The twelve-coordinate ion $Pr(nap)_6^{3+}$ [274].

most interesting features are the long Pr-N bonds [average 2.748(9) Å] to the aromatic nitrogen atoms. Long Ln-N bonds have been a recurrent phenomenon of the structural chemistry of the lanthanides, although a satisfactory explanation is not available at present.   A steric effect due to the hydrogens on the carbon adjacent to the nitrogen is unlikely.   A Ln-N distance of 2.60 Å and an Ln•••H distance of 3.44 Å would be calculated assuming an ideal geometry while any bends of the C-H bond could increase the distance, if required.   Although the "soft-hard" rational appears to be the most satisfactory explanation, additional structural data would be useful.

The `La(NO_3)_2(crypt)^+` ion has only two bidentate nitrates in the coordination sphere [275] compared to the three found in the remaining complexes with a CN of 12.   The anion in the crys-tal is $La(NO_3)_6^{3-}$ so that an internal comparison of the La-O dis-tances is possible.   As noted, the La-O distances in the anion average 2.658(20) Å, and the La-O(nitrate) in the cations average 2.678(13) and 2.668(28) Å.   Consequently, the La-O(nitrate) dis-tances are not significantly different in the anion and cation. The La-O(crypt) distances are slightly longer as might be

expected [2.705(18) and 2.708(38) Å], but the large esd's make
the differences not significant. However, the La-N bonds are
long [2.814 and 2.839(11) Å] as usual. Probably the most unusual
feature is the fact that the crypt is sufficiently flexible to
open and allow two nitrates to coordinate to the lanthanum ion.

Another unusual lanthanum complex is formed via a template
synthesis involving lanthanum nitrate, diacetylpyridine and
ethylenediamine,   La(NO$_3$)$_3$(DAPEND) [276].   The ligand is an
approximately planar hexadentate ligand with the La ion approximately equidistant from 6 nitrogens [avg 2.728(30) Å], with one
nitrate above and two below the plane [La-O avg 2.732(29) Å].
This molecule is one of the few cases where Ln-N distances are
not longer than Ln-O distances.   Also the La-O distances are
possibly significantly longer than in the La(NO$_3$)$_6$$^{3-}$ ion [275],
although the various values are all very similar. The fact that
in the macrocyclic ligand the La-N distances are not unusually
long would argue for steric effects in the other complexes with
long Ln-N bonds.

The four remaining complexes with a CN of 12 all have a
stoichiometry Ln(NO$_3$)$_3$L, where L is a cyclic polyether with six
oxygens [221,277-279].   The Nd(NO$_3$)$_3$(18-crown-6) complex was
determined independently by two groups, and the results are identical [277,278]. The Nd-O(nitrate) distances (2.587(12) Å [277]
and 2.602(14) Å [278]) are equal and identical to the value
observed in the Nd(NO$_3$)$_6$$^{3-}$ ion (2.602(43) Å [245]).   Similarly,
the La-O(nitrate) distances in the La(NO$_3$)$_3$(18-crown-6) complex
(avg 2.664(14) Å [221]) are equal to those in the La(NO$_3$)$_6$$^{3-}$ ion
and other La complexes with a CN of 12.   One of the interesting
features which is found in all three 18-crown-6 complexes is a
puckering of the 18-crown-6 to make three different sets of trans
Ln-O(polyether) distances. The two long trans Ln-O distances are
not significantly different in the La (2.776(6) Å [221]) and Nd
complexes (2.780(12) [278] and 2.756(20) Å [277]). The decrease
in the metal radius is more apparent in the two shorter
Ln-O distances (2.644(24) in La [221], 2.615(24) [278], and
2.594(26) Å [277] in the Nd complexes).   Presumably, the puckering of the ring toward one coordinated nitrate is necessary to
permit two additional nitrates to coordinate to the metal ion.
Unfortunately, the individual distances in the dicyclohexyl-18-
crown-6 complex with La(NO$_3$)$_3$ [279] are not available since the
folding of the polyether in this case could be different.

ACKNOWLEDGEMENTS

I would like to thank the following for their permission to
reproduce the Figures used in the text: The American Chemical
Society for Figures 1-4, 6-9, 10, 11, 12, 15, 21-24, and 27;   The

Royal Society of Chemistry for Figures 11, 16, and 17; Acta Chemica Scandinavica for Figure 18; Verlag Chemie Gmdh for Figures 8 and 9; Elsevier Science Publications for Figures 20, 25, and 26; and Crystal Structure Communications for Figure 19. I would also like to thank Laura E. Griggs for her efforts in typing the manuscript, my wife Ruth for her diligence in proof-reading, and finally, Professor S. Sinha for his extraordinary patience and encouragement in producing this manuscript.

REFERENCES

1.   R. D. Shannon, Acta Cryst., A32, 751 (1976).
2.   M. O'Keefe, Acta Cryst., A35, 772 (1979).
3.   F. L. Carter, Acta Cryst., B34, 2962 (1978).
4.   Cambridge Crystallographic Data Centre User Manual, 2nd edi-
      tion (1978).
5.   D. G. Watson, private communication.
6.   D. Britton and J. D. Dunitz, Acta Cryst., A29, 362 (1973).
7.   M. G. B. Drew, Coord. Chem. Revs., 24, 179 (1977).
8.   W. A. Dollase, Acta Cryst., A30, 513 (1974).
9.   C. J. Fritchie, Jr., Acta Cryst., B31, 802 (1975).
10.   S. C. Nyburg, Acta Cryst., B30, 251 (1974).
11.   M. G. B. Drew, Prog. Inorg. Chem., 23, 67 (1977).
12.   M. A. Porai-Koshits and L. A. Aslanov, J. Struct. Chem., 13,
      244 (1972).
13.   E. L. Muetterties, Tetrahedron, 30, 1600 (1974).
14.   E. L. Muetterties and L. J. Guggenberger, J. Am. Chem. Soc.,
      96, 1748 (1974).
15.   L. J. Guggenberger and E. L. Muetterties, J. Am. Chem. Soc.,
      98, 7221 (1976).
16.   S. W. Gaines and G. J. Palenik, unpublished results.
17.   D. L. Kepert, Prog. Inorg. Chem., 23, 1 (1977).
18.   D. L. Kepert, Prog. Inorg. Chem., 24, 179 (1978).
19.   D. L. Kepert, Prog. Inorg. Chem., 25, 41 (1979).
20.   M. C. Favas and D. L. Kepert, Prog. Inorg. Chem., 28, 309
      (1981).
21.   J. B. Casey, W. J. Evans, and W. H. Powell, Inorg. Chem.,
      20, 3556 (1981) discuss a means of numbering boron polyhedra
      which is applicable to other polyhedra. In addition the
      series by R. B. King, Inorg. Chem., 20, 363 (1981) and ref-
      erences therein also describes nomenclature for polyhedra.
22.   H. Aghabozorg, V. Lynch, R. C. Palenik, and G. J. Palenik,
      unpublished results.
23.   J. L. Atwood, W. E. Hunter, R. D. Rogers, J. Holton,
      J. McMeeking, R. Pearce, and M. F. Lappert, J. C. S. Chem.
      Comm., 140 (1978).
24.   N. Schumann, J. Pickardt, and N. Bruncks, Angew. Chem. Int.
      Ed. Engl., 20, 120 (1981).

25. S. A. Cotton, F. A. Hart, M. B. Hursthouse, and A. J. Welch, J. C. S. Chem. Comm., 1225 (1972).

26. T. D. Tilley, R. A. Andersen, B. Spencer, H. Ruben, A. Zalkin, and D. H. Templeton, Inorg. Chem., 19, 2999 (1980).

27. M. F. Lappert, P. I. W. Yarrow, J. L. Atwood, R. Shakir, and J. Holton, J. C. S. Chem. Comm., 987 (1980).

28. H. A. Zinnen, J. J. Pluth, and W. J. Evans, J. C. S. Chem. Comm., 810 (1980).

29. T. J. Marks, Prog. Inorg. Chem., 24, 51 (1978).

30. C.-H. Wong, T.-Y. Lee, and Y.-T. Lee, Acta Cryst., B25, 2580 (1969).

31. J. H. Burns, W. H. Baldwin, and F. H. Fink, Inorg. Chem., 13, 1916 (1974).

32. J. L. Atwood, J. H. Burns, and P. G. Laubereau, J. Am. Chem. Soc., 95, 1830 (1973).

33. J. L. Atwood and K. D. Smith, J. Am. Chem. Soc., 95, 1488 (1973).

34. P. G. Laubereau and J. H. Burns, Inorg. Chem., 9, 1091 (1970).

35. Reference 6 in reference 31 above.

36. R. Zerger and G. Stucky, J. Organometal. Chem., 80, 7 (1974).

37. F. W. B. Einstein, M. M. Gilbert, and D. G. Tuck, Inorg. Chem., 11, 2832 (1972).

38. E. C. Baker and K. N. Raymond, Inorg. Chem., 16, 2710 (1977).

39. A. Zazzetta and A. Greco, Acta Cryst., B35, 457 (1979).

40. J. H. Burns and W. H. Baldwin, J. Organometal. Chem., 120, 361 (1976).

41. R. D. Rogers, R. V. Bynum, and J. L. Atwood, J. Organometal. Chem., 192, 65 (1980).

42. W. J. Evans, A. L. Wayda, W. E. Hunter, and J. L. Atwood, J. C. S. Chem. Comm., 292 (1981).

43. H. Schuman, W. Genthe, and N. Bruncks, Angew. Chem. Int. Ed. Engl., 20, 119 (1981).

44. T. D. Tilley and R. A. Andersen, J. C. S. Chem. Comm., 985 (1981).

45. P. L. Watson, J. F. Whitney, and R. L. Harlow, Inorg. Chem., 20, 3271 (1981).

46. J. Holton, M. F. Lappert, D. G. H. Ballard, R. Pearce, J. L. Atwood, and W. E. Hunter, J. C. S. Dalton, 45 (1979).

47. G. R. Scollary, Aust. J. Chem., 31, 411 (1978).

48. E. C. Baker, L. D. Brown, and K. N. Raymond, Inorg. Chem., 14, 1376 (1975).

49. J. L. Atwood and K. D. Smith, J. C. S. Dalton, 2487 (1973).

50. J. Holton, M. F. Lappert, D. G. H. Ballard, R. Pearce, J. L. Atwood, and W. E. Hunter, J. C. S. Dalton, 54 (1979).

51. J. L. Atwood, W. E. Hunter, A. L. Wayda, and W. J. Evans, Inorg. Chem., 20, 4115 (1981).

52. T. D. Tilley and R. A. Andersen, J. Am. Chem. Soc., <u>104</u>, 1772 (1982).

53. J. C. Huffman and W. E. Streib, J. C. S. Chem. Comm., 911 (1971) found Al-CH$_3$ (terminal of 1.952 Å and Al-CH$_3$ (bridge) of 2.124 Å or a difference of 0.172 Å which is in agreement with our value.

54. K. O. Hodgson and K. N. Raymond, Inorg. Chem., <u>11</u>, 3030 (1972).

55. K. O. Hodgson and K. N. Raymond, Inorg. Chem., <u>11</u>, 171 (1972).

56. C. W. DeKock, S. R. Ely, T. E. Hopkins, and M. A. Brault, Inorg. Chem., <u>17</u>, 625 (1978).

57. A. Avdeef, K. N. Raymond, K. O. Hodgson, and A. Zalkin, Inorg. Chem., <u>11</u>, 1083 (1972).

58. K. N. Raymond and C. W. Eigenbrot, Jr., Accts. Chem. Res., <u>13</u>, 276 (1980).

59. E. C. Baker, G. W. Halstead, and K. N. Raymond, Struct. Bonding (Berlin), <u>25</u>, 23 (1976).

60. J. S. Ghotra, M. B. Hursthouse, and A. J. Welch, J. C. S. Chem. Comm., 669 (1973).

61. P. G. Eller, D. C. Bradley, M. B. Hursthouse, and D. W. Meek, Coord. Chem. Rev., <u>24</u>, 1 (1977).

62. R. A. Andersen, D. H. Templeton, and A. Zalkin, Inorg. Chem., <u>17</u>, 2317 (1978).

63. D. C. Bradley, J. S. Ghotra, F. A. Hart, M. B. Hursthouse, and P. R. Raithby, J. C. S. Dalton, 1166 (1977).

64. J. L. Martin, L. C. Thompson, L. J. Radonovich, and M. D. Glick, J. Am. Chem. Soc., <u>90</u>, 4493 (1968).

65. J. G. H. DuPreez, H. E. Rohwer, J. F. DeWet, and M. R. Caira, Inorg. Chim. Acta, <u>26</u>, L59 (1978).

66. Y. Meseri, A. A. Pinkerton, and G. Chapuis, J. C. S. Dalton, 725 (1977).

67. A. A. Pinkerton and D. Schwarzenbach, J. C. S. Dalton, 1300 (1980).

68. L. J. Radonovich and M. D. Glick, J. Inorg. Nucl. Chem., <u>35</u>, 2745 (1973).

69. R. J. Majeste, D. Chriss, and L. M. Trefonas, Inorg. Chem., <u>16</u>, 188 (1977).

70. T. D. Tilley, A. Zalkin, R. A. Andersen, and D. H. Templeton, Inorg. Chem., <u>20</u>, 551 (1981).

71. S. Onuma, H. Inoue, and S. Shibata, Bull. Chem. Soc. Japan, <u>49</u>, 644 (1976).

72. J. P. R. deVilliers and J. C. A. Boyens, Acta Cryst., <u>B28</u>, 2335 (1972).

73. L. A. Aslanov, V. M. Ionov, and S. S. Sotman, Kristallografiya, <u>21</u>, 1200 (1976).

74. R. A. Andersen, D. H. Templeton, and A. Zalkin, Inorg. Chem., <u>17</u>, 1962 (1978).

75. T. J. Anderson, M. A. Neuman, and G. A. Melson, Inorg. Chem., <u>12</u>, 927 (1973).

76. E. Hansson, Acta Chem. Scand., 27, 2841 (1973).
77. J. L. Atwood and K. D. Smith, J. C. S. Dalton, 921 (1974).
78. T. J. Anderson, M. A. Neuman, and G. A. Melson, Inorg. Chem., 13, 158 (1974).
79. M. K. Gusejnova, A. S. Antsyshkina, and M. A. Porai-Koshits, Zh. Struckt. Khim., 9, 1040 (1968).
80. R. W. Baker and J. W. Jeffery, J. C. S. Dalton, 229 (1974).
81. R. Restivo and G. J. Palenik, J.C.S. Dalton, 341 (1972).
82. T. G. Appleton, H. C. Clark, and L. E. Manzer, Coord. Chem. Rev. 10, 335 (1973).
83. A. F. Kirby and R. A. Palmer, Inorg. Chem., 20, 1030 (1981).
84. J. C. A. Boeyens and J. P. R. deVilliers, J. Cryst. Mol. Struct., 1, 297 (1971).
85. S. J. S. Wasson, D. E. Sands, and W. F. Wagner, Inorg. Chem., 12, 187 (1973).
86. R. M. Wing, J. J. Uebel, and K. K. Andersen, J. Am. Chem. Soc., 95, 6046 (1973).
87. J. C. A. Boeyens and J. P. R. deVilliers, J. Cryst. Mol. Struct., 2, 197 (1972).
88. A. Zalkin, D. H. Templeton, and D. G. Karraker, Inorg. Chem., 8, 2680 (1969).
89. L. A. Aslanov, V. M. Ionov, V. B. Ribakov, E. F. Korytnii, and L. I. Martynenko, Koord. Khim., 4, 1427 (1978).
90. C. S. Erasmus and J. C. A. Boeyens, Acta Cryst., B26, 1843 (1970).
91. E. Bye, Acta Chem. Scand., A28, 731 (1974).
92. C. S. Erasmus and J. C. A. Beoyens, J. Cryst. Mol. Struct., 1, 83 (1971).
93. F. A. Cotton and P. Legzdins, Inorg. Chem., 7, 1777 (1968).
94. E. D. Watkins, II, J. A. Cunningham, T. E. Phillips, II, D. E. Sands, and W. F. Wagner, Inorg. Chem., 8, 29 (1969).
95. M. F. Richardson, P. W. R. Corfield, D. E. Sands, and R. E. Sievers, Inorg. Chem., 9, 1632 (1970).
96. E. F. Korytnii, L. A. Aslanov, M. A. Porai-Koshits, and O. M. Petrukhin, Zh. Strukt. Khim., 11, 311 (1970).
97. J. A. Cunningham, D. E. Sands, W. F. Wagner, and M. F. Richardson, Inorg. Chem., 8, 22 (1969).
98. M. C. Mattos, E. Surcouf, and J.-P. Mornon, Acta Cryst., B33, 1855 (1977).
99. A. A. Pinkerton and D. Schwarzenbach, J. C. S. Dalton, 2466 (1976).
100. D. B. Dell'Amico, F. Calderazzo, F. Marchetti, and G. Perego, J. C. S. Chem. Comm., 1103 (1979).
101. C. C. Bisi, A. Coda, and V. Tazzoli, Cryst. Struct. Comm., 10, 703 (1981).
102. C. C. Bisi, M. Gorio, E. Cannillo, A. Coda, and V. Tazzoli, Acta Cryst., A31, S134 (1975).
103. D. D. McRitchie, R. C. Palenik, and G. J. Palenik, Inorg. Chim. Acta, 20, L27 (1976).

104. C. C. Bisi, A. D. Giusta, A. Coda, and V. Tazzoli, Cryst.
      Struct. Comm., 3, 381 (1974).
105. M. Bukowska-Strzyzewska and A. Tosik, Acta Cryst., B38,
      265 (1982).
106. A. R. Al-Karaghouli and J. S. Wood, Inorg. Chem., 18, 1177
      (1979).
107. H. Titze, Acta Chem. Scand., A28, 1079 (1974).
108. H. Titze, Acta Chem. Scand., A23, 399 (1969).
109. M. J. Bennett, F. A. Cotton, P. Legzdins, and
      S. T. Lippard, Inorg. Chem., 7, 1770 (1968).
110. J. H. Burns and M. D. Danford, Inorg. Chem., 8, 1780
      (1969).
111. L. A. Butman, L. A. Aslanov, and M. A. Porai-Koshits, Zh.
      Struct. Khim., 11, 46 (1970).
112. A. L. Il'Inskii, M. A. Porai-Koshits, L. A. Aslanov, and
      P. I. Lazarev, Zh. Struct. Khim., 13, 277 (1972).
113. R. A. Lalancette, M. Cefola, W. C. Hamilton, and
      S. J. LaPlaca, Inorg. Chem., 6, 2127 (1967).
114. J. G. Leipoldt, L. D. C. Bok, S. S. Basson,
      A. E. Laubscher, and J. S. Van Vollenhoven, J. Inorg. Nucl.
      Chem., 39, 301 (1977).
115. A. T. McPhail and P.-S. W. Tschang, J. C. S. Dalton, 1165
      (1974).
116. T. Phillips, D. E. Sands, and W. F. Wagner, Inorg. Chem.,
      7, 2295 (1968).
117. L. A. Aslanov, M. A. Porai-Koshits, and
      M. O. Dekaprilevich, Zh. Strukt. Khim., 12, 370 (1971).
118. A. L. Il'Inskii, L. A. Aslanov, V. I. Ivanov,
      A. D. Khalilov, and O. M. Petrukhin, Zh. Strukt. Khim., 10,
      285 (1969).
119. J. G. White, Inorg. Chim. Acta, 16, 159 (1976).
120. L. A. Aslanov, E. F. Korytnyi, and M. A. Porai-Koshits, Zh.
      Struckt. Khim., 12, 661 (1971).
121. J. A. Cunningham, D. E. Sands, and W. F. Wagner, Inorg.
      Chem., 6, 499 (1967).
122. J. P. R. deVilliers and J. C. A. Boeyens, Acta Cryst., B27,
      692 (1971).
123. J. G. Leipoldt, L. D. C. Bok, S. S. Basson,
      J. S. Van Vollenhoven, and A. E. Laubscher, J. Inorg. Nucl.
      Chem., 38, 2241 (1976).
124. J. G. Leipoldt, L. D. C. Bok, A. E. Laubscher, and
      S. S. Basson, J. Inorg. Nucl. Chem., 37, 2477 (1975).
125. W. H. Watson, R. J. Williams, and N. R. Stemple, J. Inorg.
      Nucl. Chem., 34, 501 (1972).
126. W. DeW. Horrocks, Jr., J. P. Sipe, III, and J. R. Luber,
      J. Am. Chem. Soc., 93, 5258 (1971).
127. R. E. Cramer and K. Seff, Acta Cryst., B28, 3281 (1972).
128. J. A. Cunningham and R. E. Sievers, Inorg. Chem., 19, 595
      (1980).

129. J. G. Leipoldt,   L. D. C. Bok,   S. S. Basson,   and
     A. E. Laubscher, J. Inorg. Nucl. Chem., 38, 1477 (1976).
130. E. F. Korytnyi,   N. G. Dzyubenko,   L. A. Aslanov,   and
     L. I. Martynenko, Russ. J. Inorg. Chem., 26, 39 (1981).
131. See for example "Inorganic Chemistry" by J. E. Huheey,
     2nd edition, Harper and Row, New York, 1978, page 276 ff
     and references therein.
132. A. R. Al-Karaghouli, R. O. Day, and J. S. Wood, Inorg.
     Chem., 17, 3702 (1978).
133. S. R. Sofen, S. R. Cooper, and K. N. Raymond, Inorg. Chem.,
     18, 1611 (1978).
134. E. E. Castellano and R. W. Becker, Acta Cryst., B37, 61
     (1981).
135. D. Brown, D. G. Holah, and C. E. F. Rickard, J. Chem. Soc.
     (A), 786 (1970).
136. M. Ciampolini, N. Nardi, P. Colamarino, and P. Orioli,
     J. C. S. Dalton, 379 (1977).
137. A. A. Pinkerton and D. Schwarzenbach, J. C. S. Dalton,
     2464 (1976).
138. A. A. Pinkerton and D. Schwarzenbach, J. C. S. Dalton,
     1470 (1981).
139. S. Spiliadis, A. A. Pinkerton, and D. Schwarzenbach, J. C.
     S. Dalton, 1809 (1982).
140. T. V. Filippova,   T. N. Polynova,   A. L. Il'Inskii,
     M. A. Porai-Koshits,   and L. I. Martynenko,   Zh. Strukt.
     Khim., 18, 1127 (1977).
141. L. R. Nassimbeni, M. R. W. Wright, J. C. Van Niekerk, and
     P. A. McCallum, Acta Cryst., B35, 1341 (1979).
142. K. F. Belyaeva, M. A. Porai-Koshits, N. D. Mitrofanova, and
     L. I. Martynenko, Zh. Strukt. Khim., 9, 541 (1968).
143. L. L. Martin and R. A. Jacobson, Inorg. Chem., 11, 2789
     (1972).
144. Y. M. Nesterova, M. A. Porai-Koshits, N. D. Mitrofanova,
     and E. D. Filippova, Russ. J. Inorg. Chem., 25, 1109
     (1980).
145. I. N. Polyakova, T. N. Polynova, M. A. Porai-Koshits, and
     N. D. Mitrofanova, Zh. Strukt. Khim., 19, 766 (1978).
146. I. Grenthe, Acta Chem. Scand., 23, 1253 (1969).
147. I. Grenthe, Acta Chem. Scand., 25, 3721 (1971).
148. L. M. Dikareva, A. S. Antsyshkina, M. A. Porai-Koshits,
     V. N. Ostrikova, I. V. Arkhangel'skii, and A. Z. Zamanov,
     Dokl. Akad. Nauk Az. SSR, 34, 41 (1978).
149. H.-R. Wenk, Z. Krist, 154, 137 (1981).
150. A.S. Antsyshkina, M.A. Porai-Koshits, I.V. Arkhangel'skii,
     and L.A. Butman, Koord. Khim., 2, 565 (1976).
151. L. A. Aslanov,   V. M. Ionov,   V. B. Ribakov,   and
     I. D. Kiekbaev, Koord. Khim., 4, 1598 (1978).
152. S. P. Bone, D. B. Sowerby, and R. D. Verma, J. C. S.
     Dalton, 1544 (1978).
153. E. Hansson, Acta Chem. Scand., 27, 823 (1973).

154. E. Hansson, Acta Chem. Scand., 27, 2827 (1973).
155. E. Hansson, Acta Chem. Scand., 26, 1337 (1972).
156. M. S. Khiyalov, I. R. Amiraslanov, Kh. S. Mamedov, and
      E. M. Movsumov, Koord. Khim., 7, 445 (1981).
157. L. A. Aslanov, I. D. Kiekbaev, I. K. Abdul'Minev, and
      M. A. Porai-Koshits, Krist., 19, 170 (1974).
158. I. K. Abdul'Minev, L. A. Aslanov, M. A. Porai-Kcshits, and
      R. A. Chupakhina, Zh. Strukt. Khim., 14, 383 (1973).
159. J. Kay, J. W. Moore, and M. D. Glick, Inorg. Chem., 11,
      2818 (1972).
160. J. W. Moore, M. D. Glick, and W. A. Baker, Jr., J. Am.
      Chem. Soc., 94, 1858 (1972).
161. A. R. Davis and F. W. B. Einstein, Inorg. Chem., 13, 1880
      (1974).
162. T. J. Anderson, M. A. Neuman, and G. A. Melson, Inorg.
      Chem., 13, 1884 (1974).
163. B. Barlic, L. Golic, and F. Lazarini, Cryst. Struct. Comm.,
      3, 407 (1974).
164. V. I. Kuskov, F. N. Treushnikov, and N. V. Belov, Krist.,
      23, 1196 (1978).
165. E. Baraniak, R. S. L. Bruce, H. C. Freeman, N. J. Hair, and
      J. James, Inorg. Chem., 15, 2226 (1976).
166. V. B. Ribakov, L. A. Aslanov, and M. A. Porai-Koshits,
      Koord. Khim., 5, 1723 (1979).
167. V. I. Ponomarenko, E. N. Kurkutova, M. A. Porai-Koshits,
      L. A. Aslanov, and K. Soulaymankulov, Dokl. Akad. Nauk.
      SSSR, 228, 360 (1976).
168. J. H. Timmons, J. W. L. Martin, A. E. Martell, P. Rudolf,
      A. Clearfield, J. H. Arner, S. J. Loeb, and C. J. Willis,
      Inorg. Chem., 19, 3554 (1980).
169. K. Kasuga, M. Tsutsui, R. C. Petterson, K. Tatsumi,
      N. Van Opdenbosch, G. Pepe, and E. F. Meyer, Jr., J. Am.
      Chem. Soc., 102, 4835 (1980).
170. E. E. Castellano and R. W. Becker, Acta Cryst., B37, 1998
      (1981).
171. J. L. Hoard, B. Lee, and M. D. Lind, J. Am. Chem. Soc., 87,
      1612 (1965).
172. T. F. Koetzle, Acta Cryst., A31, S22 (1975).
173. B. Lee, Diss. Abst., B28, 84 (1967).
174. L. K. Templeton, D. H. Templeton, A. Zalkin, and
      H. W. Ruben, Acta Cryst., B38, 2155 (1982).
175. L. L. Martin and R. A. Jacobsen, Inorg. Chem., 11, 2785
      (1972).
176. J. Albertsson and I. Edding, Acta Cryst., B33, 1460 (1977).
177. D. R. Fitzwater and R. E. Rundle, Z. Krist., 112, 363
      (1959).
178. C. R. Hubbard, C. O. Quicksall, and R. A. Jacobson, Acta
      Cryst., B30, 2613 (1974).
179. R. W. Broach, J. M. Williams, G. P. Felcher, and
      D. G. Hinks, Acta Cryst., B35, 2317 (1979).

180. G. H. Frost, F. A. Hart, C. Heath, and M. B. Hursthouse, J. Chem. Soc. (D), 1421 (1969).
181. J. Albertsson, Acta Chem. scand., 26, 1023 (1972).
182. J. Albertsson, Acta Chem. Scand., 26, 1005 (1972).
183. J. Albertsson, Acta Chem. Scand., 26, 985 (1972).
184. J. Albertsson, Acta Chem. Scand., 24, 1213 (1970).
185. J. Albertsson and I. Elding, Acta Chem. Scand., A31, 21 (1977).
186. I. Elding, Acta Chem. Scand., A30, 649 (1976).
187. J. Albertsson and I. Elding, Acta Cryst., B32, 3066 (1976).
188. J. Albertsson, Acta Chem. Scand., 24, 3527 (1970).
189. I. Elding, Acta Chem. Scand., A31, 75 (1977).
190. J. Albertsson, Acta Chem. Scand., 22, 1563 (1968).
191. N.-G. Vannerberg and J. Albertsson, Acta Chem. Scand., 19, 1760 (1965).
192. G. G. Sadikov, G. A. Kukina, and M. A. Porai-Koshits, Zh. Strukt. Khim., 8, 551 (1967).
193. G. G. Sadikov and G. A. Kukina, Zh. Struckt. Khim., 9, 145 (1968).
194. G. V. Romanenko, N. V. Podberezskaya, and V. V. Bakakin, Dokl. Akad. Nauk SSSR, 248, 1337 (1979).
195. M. C. Favas, D. L. Kepert, B. W. Skelton, and A. H. White, J. C. S. Dalton, 454, 186 (1980).
196. J. W. Bats, R. Kalus, and H. Fuess, Acta Cryst. B35, 1225 (1979).
197. L. A. Aslanov, I. K. Abdul'Minev, M. A. Porai-Koshits, and V. I. Ivanov. Dokl. Akad. Nauk SSSR, 205, 343 (1972).
198. L. A. Aslanov, V. M. Ionov, and I. D. Kiekbaev, Koord. Khim., 2, 1674 (1976).
199. I. Grenthe, Acta Chem. Scand., 26, 1479 (1972).
200. I. Grenthe, Acta Chem. Scand., 23, 1752 (1969).
201. I. Grenthe, Acta Chem. Scand., 25, 3347 (1971).
202. E. Hansson, Acta Chem. Scand., 27, 2813 (1973).
203. E. Hansson, Acta Chem. Scand., 27, 2441 (1973).
204. E. Hansson and C. Thornqwist, Acta Chem. Scand., A29, 927 (1975).
205. W. Ollendorf and F. Weigel, Inorg. Nucl. Chem. Lett., 5, 263 (1969).
206. E. Hansson, Acta Chem. Scand., 27, 2852 (1973).
207. E. Hansson, Acta Chem. Scand., 24 2969 (1970).
208. E. Hansson and J. Albertsson, Acta Chem. Scand., 22, 1682 (1968).
209. H. Steinfink and G. D. Brunton, Inorg. Chem., 9, 2112 (1970).
210. G. D. Brunton and C. K. Johnson, J. Chem. Phys., 62, 3797 (1975).
211. T. R. R. McDonald and D. J. M. Spink, Acta Cryst., 23, 944 (1967).
212. J. Albertsson and A. Oskarsson, Acta Chem. Scand., A28, 347 (1974).

213. A. Oskarsson, Acta Chem. Scand., 25 1206 (1971).

214. J. Albertsson and A. Oskarsson, Acta Chem. Scand., 22, 1700
      (1968).

215. I. N. Polyakova,   T. A. Senina,   T. N. Polynova,
      M. A. Porai-Koshits, and N. D. Mitrofanova, Zh. Strukt.
      Khim., 18, 1128 (1977).

216. K. K. Bhandary, H. Manohar, and K. Venkatesan, Acta Cryst.,
      B32, 861 (1976).

217. C. Chieh, G. E. Toogood, T. D. Boyle, and C. M. Burgess,
      Acta Cryst., B32, 1008 (1976).

218. L. A. Aslanov, L. I. Soleva, and M. A. Porai-Koshits, Zh.
      Strukt. Khim., 13, 1101 (1972).

219. K. K. Bhandary, H. Manohar, and K. Venkatesan, J. C. S.
      Dalton, 288 (1975).

220. L. A. Aslanov, L. I. Soleva, and M. A. Porai-Koshits, Zh.
      Strukt. Khim. 14, 1064 (1973).

221. J. D. J. Backer-Dirks,   J. E. Cooke,   A. M. R. Galas,
      J. S. Ghotra, C. J. Gray, F. A. Hart, and M. B. Hursthouse,
      J. C. S. Dalton, 2191 (1980).

222. J. A. Cunningham and R. E. Sievers, J. Am. Chem. Soc., 97,
      1586 (1975).

223. L. F. Lindoy, H. C. Lip, H. W. Louie, M. G. B. Drew, and
      M. J. Hudson, J. C. S. Chem. Comm., 778 (1977).

224. L. B. Zinner,   D. E. Crotty,   T. J. Anderson,   and
      M. D. Glick, Inorg. Chem., 18, 2045 (1979).

225. L. A. Aslanov, I. K. Abdul'Minev, and M. A. Porai-Koshits,
      Zh. Strukt. Khim., 13, 468 (1972).

226. J. H. Burns and W. H. Baldwin,   Inorg. Chem., 16, 289
      (1977).

227. L. J. Radonovich and M. D. Glick, Inorg. Chem., 10, 1463
      (1971).

228. T. Malmborg and A. Oskarsson, Acta Chem. Scand., 27, 2923
      (1973).

229. M. D. Lind, B. Lee, and J. L. Hoard. J. Am. Chem. Soc. 87,
      1611 (1965).

230. C. C. Fuller, P. D. Murphy, and R. A. Jacobson,   Cryst.
      Struct. Comm. 8, 9 (1979).

231. Ya. M. Nesterova,   A. L. Il'Inskii,   N. N. Anan'Eva,
      M. A. Porai-Koshits, T. N. Polynova, and N. D. Mitrofanova,
      Zh. Strukt. Khim., 19, 543 (1978).

232. C. C. Fuller, D. K. Molzahn, and R. A. Jacobson, Inorg.
      Chem., 17, 2138 (1978).

233. H. B. Kerfoot, G. R. Choppin, and T. J. Kistenmacher,
      Inorg. Chem., 18, 787 (1979).

234. J. E. Thomas and G. J. Palenik, Inorg. Chim. Acta, 44, L303
      (1980).

235. A. R. Al-Karaghouli and J. S. Wood, J. C. S. Dalton, 2318
      (1973).

236. J.-C. Bünzli, B. Klein, G. Chapuis, and K. J. Schenk,
      J. Inorg. Nucl. Chem., 42, 1307 (1980).

237. G. E. Toogood and C. Chieh, Can. J. Chem., 53, 831 (1975).

238. E. G. Sherry, J. Inorg. Nucl. Chem., 40, 257 (1978).

239. S. Voliotis, A. Rimsky, and J. Faucherre, Acta Cryst., B31, 2607 (1975).

240. S. Voiliotis and A. Rimsky. Acta Cryst. B31, 2620 (1975).

241. L. A. Butman, V. I. Sokol, and M. A. Porai-Koshits, Koord. Khim., 2, 265 (1976).

242. G. G. Sadikov, G. A. Kukina, and M. A. Porai-Koshits, Zh. Strukt. Khim., 12, 859 (1971).

243. P. I. Lazarev, L. A. Aslanov, and M. A. Porai-Koshits, Koord. Khim., 1, 706 (1975).

244. M.-Ul-Haque, C. N. Caughlan, F. A. Hart, and R. Van Nice, Inorg. Chem., 10, 115 (1971).

245. J. H. Burns, Inorg. Chem., 18, 3044 (1979).

246. M. Ciampolini, P. Dapporto, and N. Nardi, J. C. S. Dalton, 974 (1979).

247. J.-C. Bünzli, B. Klein, D. Wessner, and N. W. Alcock, Inorg. Chim. Acta, 59, 269 (1982).

248. M. Ciampolini, N. Nardi, R. Cini, S. Mangani, and P. Orioli, J. C. S. Dalton, 1983 (1979).

249. J.-C. Bünzli, B. Klein, D. Wessner, K. J. Schenk, G. Chapuis, G. Bombieri, and G. de Paoli, Inorg. Chim. Acta, 54, L43 (1981).

250. M. Ciampolini, C. Mealli, and N. Nardi, J. C. S. Dalton, 376 (1980).

251. G. D. Smith, C. N. Caughlan, M.-Ul-Haque, and F. A. Hart, Inorg. Chem., 12, 2684 (1973).

252. D. S. Moss and S. P. Sinha, Z. Phys. Chem., 63, 190 (1969).

253. A. R. Al-Karaghouli and J. S. Wood, Inorg. Chem., 11, 2293 (1972).

254. V. B. Kravchenko, Zh. Strukt. Khim., 13, 345 (1972).

255. K. K. Bhandary and H. Manohar, Acta Cryst., B29, 1093 (1973).

256. L. A. Aslanov, L. I. Soleva, M. A. Porai-Koshits, and S. S. Goukhberg, Zh. Strukt. Khim., 13, 655 (1972).

257. L. A. Aslanov, I. K. Abdul'Minev, and M. A. Porai-Koshits, Zh. Fiz. Khim., 47, 601 (1973).

258. B. G. Segal and S. J. Lippard, Inorg. Chem., 17, 844 (1978).

259. B. Eriksson, L. O. Larsson, and L. Niinistö, Inorg. Chem., 19, 1207 (1980).

260. M. Milinski, B. Ribar, and M. Satarie, Cryst. Struct. Comm., 9, 473 (1980).

261. M. Bukowska-Strzyzewska and A. Tosik, Inorg. Chim. Acta, 30, 189 (1978).

262. K. Al-Rasoul and T. J. R. Weakley, Inorg. Chim. Acta, 60, 191 (1982).

263. J. F. Thomas, R. C. Palenik, and G. J. Palenik, Inorg. Chim. Acta, 37, L459 (1979).

264. J.-C. G. Bünzli, B. Klein, G. Chapuis, and K. J. Schenk,
      Inorg. Chem., 21, 808 (1982).
265. U. Casellato, G. Tomat, P. DiBernardo, and R. Graziani,
      Inorg. Chim. Acta, 61, 181 (1982).
266. S. P. Sinha, Struct. Bonding (Berlin), 25, 69 (1971).
267. R. B. King, J. Am. Chem. Soc., 92, 6460 (1970).
268. R. Gopal, J. S. Rutherford, and B. E. Robertson, J. Solid
      State Chem., 32, 29 (1980).
269. E. L. Muetterties, Rec. Chem. Prog., 31, 51 (1970).
270. T. A. Beineke and J. Delgaudio, Inorg. Chem., 7, 715
      (1968).
271. M. B. Anderson, G. T. Jenkin, and J. W. White, Acta Cryst.,
      B33, 3933 (1977).
272. A. Zalkin, J. D. Forrester, and D. H. Templeton, J. Chem.
      Phys., 39, 2881 (1963).
273. W. T. Carnall, S. Siegel, J. R. Ferraro, B. Tani, and
      E. Gerbert, Inorg. Chem., 12, 560 (1973).
274. A. Clearfield, R. Gopal, and R. W. Olsen, Inorg. Chem., 16,
      911 (1977).
275. F. A. Hart, M. B. Hursthouse, K. M. A. Malik, and
      S. Moorhouse, J. C. S. Chem. Comm., 549 (1978).
276. J. D. J. Backer-Dirks, C. J. Gray, F. A. Hart,
      M. B. Hursthouse, and B. C. Schoop, J. C. S. Chem. Comm.,
      774 (1979).
277. G. Bombieri, G. DePaoli, F. Benetollo, and A. Cassol,
      J. Inorg. Nucl. Chem., 42, 1417 (1980).
278. J.-C. G. Bünzli, B. Klein, and D. Wessner, Inorg. Chim.
      Acta, 44, 147 (1980).
279. M. E. Harman, F. A. Hart, M. B. Hursthouse, G. P. Moss, and
      P. R. Raithby, J. C. S. Chem. Comm., 396 (1976).

LIST OF ABBREVIATIONS

acac	acetylacetonate
base	any Lewis base
BCDod	bicapped dodecahedron
bipy	2,2'-bipyridyl
4,4'-bipy	4,4'-bipyridyl
bipyO$_2$	2,2'-bipyridyl-N,N'-dioxide
BSAP	bicapped square antiprism
cat	catecholate
CHIN	cyclohexylisonitrile
CN	coordination number
COT	cyclooctatetraene anion, $C_8H_8^{2-}$
Cp	cyclopentadienide, $C_5H_5^-$
12-crown-4	1,4,7,10-tetraoxacyclododecane
15-crown-5	1,4,7,10,13-pentaoxacyclopentadecane
18-crown-6	1,4,7,10,13,16-hexaoxacyclooctadecane

crypt	4,7,13,13,21,24-hexaoxa-1,10-diazabicyclo[8.8.8]-hexacosane
dapbah	2,6-diacetylpyridinebis(benzoic acid hydrazone)
DAPEND	2,7,13,18-tetramethyl-3,6,14,17,23,24-hexaazatri-cyclo[17,3,1,1^{8,12}]tetracosa-1(23),2,6,8,10, 12(24),13,17,19,21-decaene
dapsah	2,6-diacetylpyridinebis(salicylic acid hydrazone)
dapsc	2,6-diacetylpyridine disemicarbazone
dedtc	diethyldithiocarbamate
deops	O,O'-diethylphosphorodithioate
DHH	decahexahedron
dipic	2,6-pyridinedicarboxylate
dme	1,2-dimethoxyethane
DMF	dimethylformamide
dmp	2,6-dimethyl-4-pyrone
dmso	dimethylsulfoxide
Dod	dodecahedron
46Dod	4A,6B-expanded dodecahedron
dpdm	dibenzoylmethane
dpm	2,2,6,6-tetramethylheptane-3,5-dionato
edta	ethylenediaminetetraacetate
esd	estimated standard deviation
Et	ethyl
facam	3-trifluoroacetyl-d-camphor
fod	1,1,1,2,2,3,3-heptafluoro-7,7-dimethyloctane-4,6-dionato
heedta	hydroxyethylethylenediaminetriacetate
htfmd	1,1,1,12,12,12-hexafluoro-2,11-bis(trifluorometh-yl)-4,9-dimethyl-2,11-diolato-5,8-diazadodeca-4,8-diene(2-)
i-OPr	isopropoxide, $C_3H_7O^-$
ind	indenyl
isonic	isonicotinate
isonich	isonicotinic acid hydrazide
L	any ligand
Ln	any lanthanide, yttrium or scandium
M	any metal atom
MCPA	monocapped pentagonal antiprism
Me	methyl
MeCp	methylcyclopentadienide, $CH_3C_5H_4^-$
3-Mepy	3-methylpyridine
MSAP	monocapped square antiprism
nap	1,8-naphthyridine
nic	nicotinate
nta	nitrilotriacetate
odac	oxydiacetate
Pc	phthalocyaninato
PCTP	pentacapped trigonal prism
PMCp	pentamethylcyclopentadienide, $C_5(CH_3)_5^-$
phen	1,10-phenanthroline

py	pyridine
py-NO	pyridine-N-oxide
pyr	pyrazine
quin	quinuclidine
rms	root-mean-square
sal	salicylato
SAP	square antiprism
tdac	thiodiacetate
teg	tetraethyleneglycol
terpy	2,6-di(2,'2"-pyridyl)pyridine
THF	tetrahydrofuran
tmen	tetramethylethylenediamine
tmma	tetramethylmalonamide
tmsa	bis(trimethylsilyl)amido
tmu	tetramethylurea
tppo	triphenylphosphine oxide
ttfac	thenoyltrifluoroacetonate
TTP	tricapped trigonal prism
X	any Lewis base anion

DISCUSSION

GROUP I   (question)

Are there any discrete complexes of the rare earths in which the
coordination polyhedron is a cube or approximately a cube?
What ligand features might one expect to be most important in
forcing the rare earths to adopt this coordination polyhedron?

PALENIK   (answer)

The square antiprism is favored over the cube for eight-coordination because of decreased steric repulsions between the donor
atoms. Consequently, cubic coordination will occur only when the
steric requirements of the ligand prevent the formation of either
the square antiprism or the dodecahedron. The steric requirements
of the ligand are believed to be responsible for the cubic coordination found in $U(bipyridine)_4$ and $Ba(phenanthroline)_2(H_2O)_4^{2+}$.

A second factor in cubic coordination is the formation of an appropriate hybrid orbital with cubic symmetry. Presumably, f-orbital
participation is required to form eight hybrid orbitals with cubic
symmetries. Under these circumstances, the formation of cubic coordination might be expected to be restricted to the actinide elements. Indeed the majority of the examples $[N(C_2H_5)_4]_4 U(NCS)_8$,
Na_3PaF_8, and $U(bipyridine)_4$ involve an actinide as the central
atom.

For the lanthanides only La $(2,2'$-bipyridine-N-oxide$)_4$ has been
shown by X-ray studies to have cubic coordination. The La (pyri-
dine-N-oxide)$_8^{3+}$ ion is distorted toward cubic symmetry but is best
described as a square antiprism. The existence of one example of
cubic coordination in the lanthanides suggests that other cubic
complexes could be prepared.

Two possible tactics are envisioned for synthesizing cubic coordinated lanthanide complexes: steric factors in the ligand or a cubic ligand. Steric factors in a ligand might be easier to control
but the predictions are much more difficult. For example, the
2-methylpyridine-N-oxide or 2,6-dimethylpyridine-N-oxide ligands
might have additional steric requirements relative to the unsub-
stituted ligand to form cubic coordination about a lanthanide. On
the other hand increasing the steric requirements may prevent the
formation of an octakis (ligand) lanthanide complex.

The construction of a ligand with the appropriate cubic symmetry
is a much more difficult synthetic task. One possibility would be
to link two porphyrin rings into a ligand with cubic symmetry. The
chain length between the two rings must be chosen so that a twist
to a square antiprism could not occur. Professor Collman at Stan-

ford has synthesized some bridged porphyrins which might be suitable for cubic coordination with the lanthanides.

PALENIK (question)

Can I ask you a question, which I do not know if it is fair, but,
did the structure of hexathiocyanato-Er(III) complex ever got refined beyond the 12 percent?

THOMPSON (answer)

I don't know why the hexaisothiocyanatoerbate(III) structure was
never completed by Milton Glick. I think it was, but I don't know
why it was not published. All of the lanthanide ions can be prepared with that formulation, except Ce and La. The Er compound is
simple to make but if you try to make the La compound, it is fairly difficult and we were not able to get single crystals. The chemical analysis indicates it has four tertbutylammonium cations
and seven thiocyanate anions. Diphenyl sulfoxide is another relatively good ligand and although the early literature contains compounds with six ligands with perchlorate as the anion, Serra and I
showed a number of years ago that with perchlorate and hexafluoro-
phosphate only the compounds containing seven ligands can be isolated. Here again, there is no structural work done on any of
this, although the spectra are completely consistent with seven
coordination for the diphenyl sulfoxide. The other comment one
might make about the structures is that we all have a tendency to
generalize from the structures that are available. Unfortunately,
the structures that are available are probably not completely typical, since the compounds were made for specific reasons. There
is a reason why there are 17 or whatever 7-coordinated compounds
that contain the ß-diketonates or that sort of thing. So its really
pretty hard to talk about what kind of structural characteristics
one really does have. It has been one of my questions as it has
obviously been one of yours that if you have monodente ligands,
what really does happen. Do you have six coordinations at one end
and seven or eight at the other?

PALENIK (comment)

It was one of the interesting structures and it would have been
nice to have the refined structure. I agree with you about gene-
ralizing about a few pieces of data. But on the other hand, if you
never generalize from what you have, they will never point you in
a direction of things that you want to do. For example, there are
so few of the 3, 4 and 5 coordinated complexes to start with and
they all involve silylamine. It would be nice to see if you can
make other ones. I think one of the reasons that there are so many
of the 7-coordinated acetylacetonates is because of the tremendous
interest in these compounds with respect to their laser activities.

SINHA   (comment)

Monodented ligands can induce different coordination numbers for
the lanthanides. The trifluorides are 9 coordinated, whereas you
can make heptafluoride and Prof. Hoppe and Rödder [Z. Anorg. All-
gem. Chem. 312, 277 (1961)] made earlier the hepta-fluorides
(Na$_3$CeF$_7$) which are similar in structure to the trisodium-uranium-
heptafluoride. A complete structure was never made, but they
assure us that the heptafluoride anion possesses a pentagonal
bipyramid.

MÖLLER   (question)

If I understood you correctly, you spoke of bulky ligands that you
need to get these structures. In geochemistry we observe that the
rare earth elements try to substitute Ca in its octahedral positions, so they have the coordination number of 6 and obviously in
these compounds the coordination number 6 is a preferred one. You
can have it up to 9 but if you compare the amount of rare earth
elements that you can put in it, say in calcite that has the coordination number 6 and aragonite has the coordination number 9,
there is not much difference. Could you comment on that?

PALENIK   (answer)

I think you have a problem with charge balance with these things
as well because you are replacing a 2+ ion by a 3+ ion. In the
case of calcite, I would suspect that there are constraints because of the carbonate anion. You would have to change the local
symmetry somewhat to incorporate the lanthanide, but you could not
balance off the attractive forces versus the repulsion. If you
try to add too much you just disrupt the whole structure. I think
roughly that is why you will be limited to the amount, no matter
what.

MÖLLER   (question)

Is it possible to make all these funny structures with Ca as well,
can you substitute rare earth structures by Ca?

PALENIK   (answer)

The answer to that would be no. For two reasons: one of them is
the Ca ion has a radius of about 1 Å and the second one is the
charge balance problem. There is, however, a large number of Ca
compounds that have coordination number 7 in which you have seven
waters tucked around the Ca, these are very common. I think it is
an important problem, because if you are using lanthanides for
NMR-probes, you really want to know the coordination chemistry of
the lanthanides relative to the 2+ ions, because this is the one

they normally substitute. There is some interesting work with zeolites as I remember. One can prepare some zeolites with Eu substitution where they substitute the cavities fairly freely.

SUCHOW   (question)

Your work employing three different ions of the same size in seeking different coordination numbers is interesting but it seems to me that it may be faulty because for one thing, mercury is likely to have a very different electronegativity from the lanthanides and actinides you are using, and more importantly perhaps, Hg does not have 4f orbitals available for bonding.

PALENIK   (answer)

Well, it's true that you don't get much bonding from f-orbitals in the case of the lanthanides or in the case of Hg. I think that everybody will agree that a change in the size of an ion, at constant charge, would lead to an increase in the coordination number. The question is what sort of experiment can you design to show the effect of charge on coordination number. I think intuitively you'd say that if you increase the charge with a fixed size, you would expect the coordination number to increase, if for no other reason than Pauling's electroneutrality principle. But I don't know very many cases where you can actually design an experiment to show this. I grant that there may be other variables but at least we can look at a system and see if it does change according to what we expect. There are a couple of other possible series. If you start looking at electronegativity you should also look at changes in ionization potentials. We have done an In derivative, and In(III) is slightly smaller than the Y but In(III) forms a 7 coordinated species while Y(III) forms a 9 coordinated species. Here again it may be a question of size; Y is just big enough that it cannot go to 7, or it may be something to do with the electronegativity, or with the ionization potentials but it's still a question which I think is worth asking.

Electronic and Magnetic Properties

ELECTRONIC STRUCTURE AND COHESION IN THE RARE EARTH METALS

Hans L. Skriver

Risø National Laboratory
DK-4000 Roskilde, Denmark

ABSTRACT

We review the variation in a few selected physical properties within the series of the rare earth metals, and discuss to what extent the observed trends may be explained in terms of electronic structure theory.

1. INTRODUCTION

The lanthanide metals are the series of 15 elements appearing between La and Lu in the extended sixth row of the periodic table (Fig. 1). It is customary to add Sc and Y to the lanthanide series, and then refer to the whole group as the rare earth metals. The lanthanides themselves are often also referred to as the 4f-metals because each new electron which is added as one proceeds from La to Lu enters into the 4f-shell. Furthermore, since the 4f-shell is located inside the shell of the 5d6s-conduction states the nature of the latter changes little as a function of atomic number. Hence the rare earths exhibit great similarities in many of their physical properties, and it is mainly only when the 4f-electrons are directly involved that they do show distinctly different behaviour.

As an example of this state of affairs let us consider the case of the atomic radius[1], plotted as a function of atomic number in Fig. 2. The first observation we make is that the atomic radii of the trivalent lanthanides, i.e. Eu and Yb excluded, within ± 4% are equal to that of Gd, and this is what we mean by similar pro-

213

S. P. Sinha (ed.), Systematics and the Properties of the Lanthanides, 213–254.
Copyright © 1983 by D. Reidel Publishing Company.

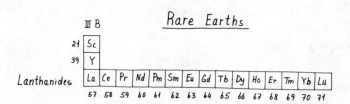

Fig. 1 The rare earth metals, their atomic numbers and chemical symbols.

perties. On the other hand, if one looks more carefully one finds that S decreases monotonically by 8% from La to Lu. This is known as the lanthanide contraction. Furthermore, one may discern two linear trends, Pr - Sm and Gd - Lu, from which La, Ce, Gd and Lu deviate. It turns out that the lanthanide contraction, the deviations from the linear trends, and the special properties of Ce, e.g. the γ → α transition, to a varying degree are caused by the 4f-electrons. Hence, it is the presence of the 4f-electrons, even in the core, which separates one lanthanide metal from the next.

In the present series of lectures we shall try to indicate to what extent some selected properties of the rare earth metals may be understood in terms of their electronic structure. We shall concern ourselves with two kinds of properties: Those which may be described purely in terms of the one-electron energy levels, i.e. the electronic energy-band structure, and those which require the total energies both in the atomic and the metallic state. As it turns out, the first kind includes properties such as lattice spacing, compressibility, and crystal structure and involves mainly the 5d6s-conduction electrons. In contrast the second kind, which includes properties such as the cohesive energy and the valence state of the metal, involves directly the 4f-electrons.

2. ONE-ELECTRON THEORY

We shall base our one-electron description on the density functional approach suggested by Hohenberg, Kohn, and Sham [2,3]. These authors considered the Hamiltonian (atomic Rydberg units)[2]

$$H = T + U + V$$

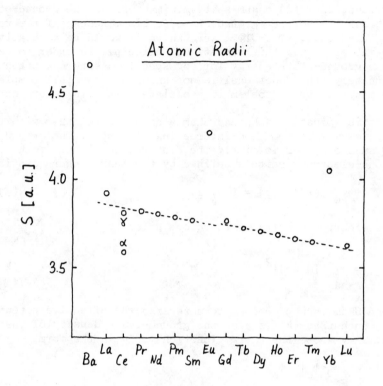

Fig. 2 The atomic radii[1] for the lanthanide metals. The data is taken from Ref. [1] and the unit of length is the Bohr radius[2] .

$$= \sum_i^M (-\nabla_i^2) + \tfrac{1}{2} \sum_{i \neq j}^M \frac{2}{r_{ij}} + \sum_i^M v_{ext}(\underline{r}_i) \qquad (1)$$

of a system of M interacting electrons moving in some fixed external potential, v_{ext}. Here T is the kinetic energy, U the electron-electron Coulomb repulsion and V the interaction with the external potential, which includes the electrostatic interaction with the fixed nuclei.

Firstly Hohenberg and Kohn showed that the external potential is a unique functional of the electron density, $n(\underline{r})$, and hence that the ground state, Φ, and the energy functional

$$\langle\Phi|H|\Phi\rangle = \langle\Phi|T+U|\Phi\rangle + \int v_{ext}(\underline{r})n(\underline{r})d\underline{r} \qquad (2)$$

are unique functionals of $n(\underline{r})$. Secondly, they showed that the
energy functional (2) assumes its minimum value, the ground-state
energy, for the correct ground-state density. Hence, if the uni-
versal functional, $\langle\Phi|T+U|\Phi\rangle$, were known it would be a relatively
simple matter to use this variational principle in order to determine the ground-state energy for any specified external potential.
Unfortunately, the functional is not known and the full complexity
of the many-electron problem is associated with its determination.

In this situation Kohn and Sham considered together with the
real system a system of non-interacting electrons. The ground state,
Φ_s, of such a system is simply the Slater determinant of the lowest-
lying one-electron orbitals defined by the Schrödinger equation

$$[-\nabla^2 + v_{eff}(\underline{r})]\psi_j(\underline{r}) = E_j\psi_j(\underline{r}) \qquad (3)$$

and the electron density is therefore given by

$$n(\underline{r}) = \sum_j^{occ.} |\psi_j(\underline{r})|^2 \qquad (4)$$

Kohn and Sham then proceeded to determine the effective potential,
$v_{eff}(\underline{r})$, such that (4) is also the ground-state density of the real
system. They wrote the energy functional (2) in the form

$$\langle\Phi|H|\Phi\rangle = \langle\Phi_s|T|\Phi_s\rangle$$
$$+\int\{\tfrac{1}{2}\frac{2n(\underline{r}')}{|\underline{r}-\underline{r}'|} d\underline{r}' + v_{ext}(\underline{r}) + \varepsilon_{xc}[n(\underline{r})]\}n(\underline{r})d\underline{r} \qquad (5)$$

where the first term is the kinetic energy of the non-interacting
system and the second term is the well-known Hartree term. It is
these two terms which presumably play an important role also for
the interacting system. The remainder has been grouped into the
fourth so-called exchange-correlation term which therefore describes
the difference between the true kinetic energy and the kinetic energy of the non-interacting system plus the difference between the
true interaction energy and that given by the Hartree term.

The energy functional (5) may now be minimized with respect to
the density, $n(\underline{r})$, to give an effective single-particle Schrödinger
equation of the form (3) with the effective potential

$$v_{eff}(\underline{r}) = \int \frac{2n(\underline{r}')}{|\underline{r}-\underline{r}'|} \, d\underline{r}' + v_{ext}(\underline{r}) + v_{xc}(n(\underline{r})) \qquad (6)$$

Here, the first term is the Hartree potential, the second term is the external potential which includes the Coulomb attraction from the nuclei, and the final term is the exchange-correlation potential. The latter is given by

$$v_{xc}(\underline{r}) = d[n\epsilon_{xc}(n)]/dn$$

$$\equiv \mu_{xc}[n(\underline{r})] \qquad (7)$$

where μ_{xc} is a function of the (local) electron density at $\underline{r}$. Useful estimates of ϵ_{xc} and μ_{xc} have been obtained from calculations for a homogeneous electron gas of density $n(\underline{r})$ by Hedin, Lundqvist, von Barth, and Gunnarsson [4-6].

It follows from (6) that the effective potential which enters the one-electron Schrödinger equation (3) depends upon the electron density which in turn is determined by (3). Hence, when we wish to find the ground state of the system we must solve equations (3,4,6) self-consistently in the spirit of the classical Hartree or Hartree-Fock methods. To this end we need energy band theory which allows us to solve the effective, one-electron Schrödinger equation (3) for an infinite single-crystal.

3. ENERGY BAND THEORY

The theory of energy bands [7] is concerned with the solution of the one-electron Schrödinger equation (3) in the situation where the effective potential (6) has the translational symmetry of a periodic crystal lattice. In this case the eigenfunctions, $\psi_j(\underline{k})$, and the eigenvalues, $E_j(\underline{k})$, are functions of a vector, $\underline{k}$, in reciprocal space, and one may characterize all the electron states by k-vectors lying inside the so-called Brillouin zone plus the band index, j, defined such that $E_j(\underline{k}) \leq E_{j+1}(\underline{k})$. The eigenvalues, $E_j(\underline{k})$, constitute the energy band structure.

We shall use the Atomic Sphere Approximation (ASA) of Andersen [8-10] both to calculate and to describe the energy bands of the rare earths. In this approximation one surrounds each atom in the crystal lattice by an atomic sphere[1] of radius S (see Fig. 3). Inside a single of these overlapping spheres the potential is assumed to be spherically symmetric, and the solutions of Schrödingers differential equation at any prescribed energy, E, may be taken as the partial waves

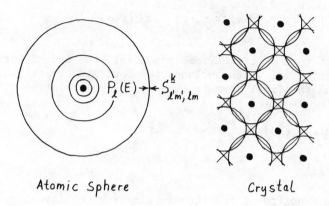

Fig. 3 The atomic sphere approximation. Shown on the right is a
square crystal lattice with its atomic spheres, and on the left
a single atomic Wigner-Seitz sphere. In the ASA (cf. equation (13))
the energy band problem may be approximated by a boundary-value
problem: The surrounding crystal lattice imposes a $\underline{k}$-dependent
and ℓ-quantum number destroying boundary condition carried by the
structure constants, $S^{k}_{\ell'm';\ell m}$, on the solution inside any atomic
sphere the information about which is carried by the potential
functions, $P_{\ell}(E)$.

$$\phi_{\ell m}(E,\underline{r}) = \phi_{\ell}(E,r)\ i^{\ell}Y^{m}_{\ell}(\hat{r}) \tag{8}$$

where the radial function, $\phi_{\ell}(E,r)$, is a solution of the appropriate radial Schrödinger equation. Based on the partial waves one
defines the energy dependent orbitals

$$\chi_{\ell m}(\underline{r}) = i^{\ell}Y^{m}_{\ell}(\hat{r})$$

$$\times \begin{cases} \phi_{\ell}(E,r) + p_{\ell}(E)(r/S)^{\ell} & r \leq S \\ (r/S)^{-\ell-1} & r \geq S \end{cases} \tag{9}$$

which are continuous and differentiable if

$$p_{\ell}(E) = \frac{D_{\ell}(E)+\ell+1}{D_{\ell}(E)-\ell} \tag{10}$$

where the logarithmic derivative function is defined by

$$D_\ell(E) = S\phi'_\ell(E,S)/\phi_\ell(E,S) \tag{11}$$

The orbitals (9) are usually referred to as Muffin-Tin Orbitals (MTO) although a better name in the present context would be atomic sphere orbitals. The tails of these orbitals are the fields from a multipole of order 2^ℓ, and when centred at $\underline{R}$ they may be expanded around the origin in angular momenta, i.e. in terms of functions of the form $i^\ell Y_\ell^m(\hat{r})r^\ell$. The expansion coefficients are the so-called canonical structure constants $S^{\underline{k}}_{\ell'm',\ell m}$ [8-10] which are independent of the scale of the lattice and of the potential, and which may therefore be calculated once and for all for a given crystal structure. One now places a linear combination of the Muffin-Tin orbitals, multiplied by the appropriate Bloch factor, in each atomic sphere throughout the crystal, i.e.

$$\Psi(\underline{k},E,\underline{r}) = \sum_{\ell m} A^{\underline{k}}_{j;\ell m} \sum_{\underline{R}} e^{i\underline{k}\cdot\underline{R}} \chi_{\ell m}(E,\underline{r}-\underline{R}) \tag{12}$$

and seek to determine the coefficients $A^{\underline{k}}_{j;\ell m}$ such that (12) is a wavefunction for the entire crystal. Since the partial wave (8) is already a solution to the Schrödinger equation inside the atomic sphere at the chosen energy, any linear combination of $\phi_{\ell m}$, in particular the one which appears through (12) and (9), is a solution. Hence, the condition that (12) is a solution for the entire crystal is that inside the sphere at the origin the sum of the tails from all the other spheres interfere destructively with the terms proportional to $i^\ell Y_\ell^m(\hat{r})p_\ell(E) (r/s)^\ell$ from the MTO at the origin. This tail cancellation condition leads to the set of homogeneous equations

$$\sum_{\ell m} (S^{\underline{k}}_{\ell'm',\ell m} - P_\ell(E)\delta_{\ell'\ell}\delta_{m'm})A^{\underline{k}}_{j;\ell m} = 0 \tag{13}$$

for all $\ell'm'$ and with the potential functions defined by

$$P_\ell(E) = 2(2\ell+1) \frac{D_\ell(E)+\ell+1}{D_\ell(E)-\ell} \tag{14}$$

i.e. $2(2\ell+1)$ times the p-function which appears in (9) and (10). The ASA equations (13) have non-trivial solutions only for those values of $\underline{k}$ and E for which the determinant of the matrix

$S-P\delta_{\ell'\ell}\delta_{m'm}$ vanishes, and a direct way to determine the energy bands of a given crystal would be to trace the roots of this determinant as a function of the k-vector. In actual calculations we prefer to use the Linear Muffin-Tin Orbitals (LMTO) method which is obtained by the variational principle using energy independent MTO's [9], but since the ASA equations and the LMTO method are mathematically equivalent we shall use the conceptually simpler ASA equations to describe the calculated energy bands.

In the ASA-equations the crystal structure and $\underline{k}$ dependent part, $S^{k}_{\ell'm',\ell m}$, is separated from the potential and volume dependent part, $P_{\ell}(E)$. Furthermore, the potential function, $P_{\ell}(E)$, does not depend on the m-quantum number owing to the spherical symmetry of the potential, and one may therefore transform (13) to a representation where each sub-block, $S^{k}_{\ell m',\ell m}$, with $\ell' = \ell$ is diagonal without changing the potential function, obtaining the $2\ell+1$ canonical sub-bands, $S^{k}_{\ell i}$. Examples of such canonical bands may be found in Ref. [10]. If we now neglect structure constants with $\ell' \neq \ell$ we have diagonalized (13) provided that

$$S^{k}_{\ell i} = P_{\ell}(E) \tag{15}$$

which is an implicit equation for the unhybridized ℓ-energy band. It represents a monotonic mapping of the kind shown in Fig. 4 of the canonical bands onto an energy scale.

The potential function is an increasing tan-like function of the energy (cf. Fig. 4), and within the n'th period of $P_{\ell}(E)$ the energy dependence may be parametrized by a few potential parameters, e.g.

$$P_{\ell}(E) = \mu_{\ell}S^{2}(E-C_{\ell}) \tag{16}$$

Here C_{ℓ} is the centre of the ℓ-band, and

$$\mu_{\ell} = [S^{3}\phi^{2}_{\ell}(C_{\ell},S]^{-1} \tag{17}$$

is the intrinsic mass of the ℓ-band. The latter is unity for free electrons and inversely proportional to the ℓ-band width. If we insert the parametrized form of $P_{\ell}(E)$ into (15) we find

$$E_{n\ell i}(\underline{k}) = C_{\ell} + [\mu_{\ell}S^{2}]^{-1}S^{k}_{\ell i} \tag{18}$$

which is an explicit equation for the $n\ell$-band in terms of the

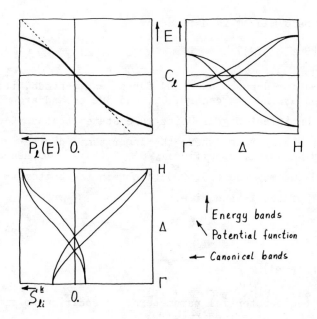

Fig. 4 The formation of energy bands viewed as a non-linear scaling of the canonical bands according to (15). The simple linear approximation, eq. (16), to the potential function is shown by a broken line.

canonical band. It shows that the $n\ell$-energy band is obtained by placing the ℓ-canonical band at C_ℓ and scaling it by $1/\mu S^2$.

A somewhat more accurate parametrization may be obtained if one introduces an extra "distortion" parameter γ_ℓ. In that case the energy function analogous to (18) and inverse to $P_\ell(E)$ is (cf. Ref. [10])

$$E_\ell(P) = C_\ell + \frac{1}{\mu_\ell S^2} \frac{P_\ell}{1-\gamma_\ell P_\ell} \tag{19}$$

or

$$E_\ell(P) = V_\ell + \frac{2(2\ell+1)^2 (2\ell+3)}{\tau_\ell S^2} \frac{\gamma_\ell}{1-\gamma_\ell P_\ell} \tag{20}$$

where the extra band mass τ is defined by

$$\tau_\ell = (2\ell+3)[S^3\phi_\ell^2(V_\ell,S)]^{-1} \tag{21}$$

We shall use (19) for d-bands and (20) for s- and p-bands, i.e. we expand the d-band around its centre, C_ℓ, and the s-band around its bottom, V_ℓ. The empirical Wigner-Seitz rule states that an ℓ-band is formed in the energy range where the logarithmic derivative function (11) is negative, specifically $0 > D_\ell(E) > -\infty$. Hence, by means of (14) one may use (19, 20) to obtain the bottom, B_ℓ, and the top, A_ℓ, of the ℓ-band, i.e.

$$B_\ell = E_\ell(P_\ell(D_\ell = 0)) \tag{22}$$

$$A_\ell = E_\ell(P_\ell(D_\ell = \infty)) \tag{23}$$

in terms of the basic potential parameters. In addition we find that the centre, C_ℓ, is defined by

$$C_\ell = E_\ell(P_\ell(D_\ell = -\ell-1)) \tag{24}$$

Inspection shows that for the s-band the bottom, B_s, coinsides with the square-well pseudopotential, V_s.

In order to avoid misunderstanding we stress that the approximate theory outlined in the present section is used for interpretation purposes only. The complete band calculations which we shall proceed to present are calculated by the accurate LMTO method.

4.   THE ELECTRONIC STRUCTURE

We have performed self-consistent energy band calculations for all the rare earth metals by means of the LMTO method including the so-called combined correction terms [9]. Exchange and correlation was included in the parametrized form given by von Barth and Hedin [5], and relativistic effects were included except spin-orbit splitting which was neglected. The appropriate cores, Ar for Sc, Kr for Y, and Xe for the lanthanides, were obtained

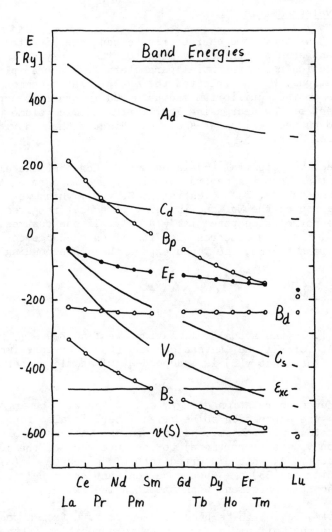

Fig. 5 Energy bands for the trivalent lanthanides evaluated at the atomic radius, S = 3.75 a.u. Plotted are: the potential, v(S), and the exchange correlation density, ε_{xc}, at the sphere radius plus the square-well pseudopotential, V_ℓ, the bottom, B_ℓ, the centre, C_ℓ, and the top, A_ℓ, of the relevant bands together with the Fermi level, E_F.

from atomic calculations and frozen during the self-consistent iterations for the conduction states which included s-, p-, and d-partial waves.

4.1 The Trivalent Metals

In Fig. 5 we show some characteristic band energies for the trivalent lanthanides obtained by means of the theory of the preceding section using potential parameters from the complete LMTO calculations. We have plotted the energies at a common atomic radius close to the equilibrium radius of Gd in order to separate out the effect of the decreasing equilibrium atomic volume (cf. Fig. 2) through the series. We shall now discuss the behaviour of the quantities in Fig. 5.

The effective potential (6) is adjusted such that the electro-static potential from the neutral atomic sphere is zero at the surface of the sphere. The potential at S, v(S), which we shall loosely refer to as the Muffin-Tin zero, is therefore given by the exchange-correlation potential (7) only. If one assumes, like Wigner-Seitz, that the electron digs itself an exchange-correlation hole on its parent atom, one would expect the corresponding poten-tial to be

$$v(s) = - \frac{2Z_{ws}}{S} \qquad\qquad (25)$$

where $Z_{ws} = 1$. In the actual calculations we find $Z_{ws} = 1.12$, i.e. close to one, for all the lanthanides and (25) therefore immediately explains the fact that v(S) is independent of atomic number (cf. Fig. 5).

The exchange correlation energy, ε_{xc}, which will be important for the pressure expression to be given later, is also found (Fig. 5) to be independent of atomic number. In the well-known Slater approximation v_{xc} is proportional to the cube root of the local electron density, n, which inserted into (7) gives $\varepsilon_{xc} = (3/4)v_{xc}$. Hence according to (25) ε_{xc} should be constant and equal to $(3/4)$ v(S) in reasonable agreement with the data in Fig. 5.

The bottom, B_S, of the s-band is seen to move from a position 0.25 Ry above the Muffin-Tin zero in La to a position in the vici-nity of the Muffin-Tin zero in Lu. This is in contrast to the naïve belief that the bottom of the conduction band is tied to the Muffin-Tin zero. The drop in B_s has been explained for the 4d-metals by Pettifor [11] who observed that the bottom of the s-band followed closely the so-called Fröhlich-Bardeen expression, i.e.

$$B_s = - \frac{3Z_o}{\tau_s S} [1- (\frac{S_o}{S})^2] \qquad\qquad (26)$$

where we have introduced the band mass τ_s (cf. (21)). In this

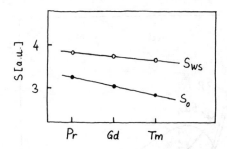

Fig. 6 The experimentally observed equilibrium radius, S_{ws}, and the "core radius", S_0, obtained from a fit to the Fröhlich-Bardeen expression (26).

theory S_0 represents the extent of some core region from which the 6s-states are excluded by orthogonality requirements. If we assume $Z_0 = 1.45$ we find the behaviour of S_0 shown in Fig. 6. It follows from the figure that at a fixed atomic radius ($S = 3.75$ a.u.) the region in space which the s-electrons are allowed to occupy increases as we move from La to Lu, and hence that the s-band moves down in energy as observed in Fig. 5. The relevance of the values for S_0 may be judged from Fig. 7 where one naïvely would estimate the extent of the 5s-core to be 2. - 3. a.u. The unoccupied 6p-band in Fig. 5 follows the same trend as the 6s-band described above, and for the same reason: The space available increases with increasing atomic number.

The decrease in core-size may be explained by the phenomenon of incomplete screening. As one goes from one element to the next the extra electron which is added to the system enters the 4f-orbital. Now the 4f-orbital is concentrated (cf. Fig. 7) inside the 5s-, 5p-, 5d-, and 6s-orbitals, and seen from these outer orbitals the extra proton which is also added (to the nucleus) is almost screened out. The screening, however, is not complete, and the 5s- and 5p-electrons therefore see an effective Coulomb attraction which increases with atomic number thereby shifting the bulk of corresponding orbitals towards the interior of the atomic sphere. In addition to the reduction in orbital extent the increased Coulomb attraction also causes the orbital energies to decrease relative to those of the preceding element.

The screening mechanism is also operational for the 5d-states, but since these are concentrated towards the exterior of the atomic sphere they are less affected than the 5s- and 5p-orbitals. Hence the centre of the 5d-band (cf. Fig. 5) falls only slightly through the series. One might even argue that since the orbitals correspond-

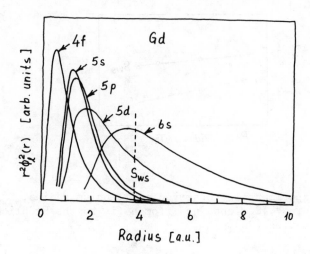

Fig. 7 Radial electron densities from atomic calculations for Gadolinium.

ing to the bottom of the d-band are those which are most concentrated at the exterior of the atomic sphere they should be least affected by an extra electron in the 4f-orbitals. This may explain the almost constant value of B_d in Fig. 5. In contrast the energy, A_d, of the antibonding orbitals at the top of the d-band should show a more marked variation with atomic number as indeed it does (Fig. 5).

The bandwidths of the conduction bands are also affected by the incomplete screening. As the orbitals are being pulled towards the interior of the atomic sphere the amplitude at the sphere-surface, and hence the bandwidth, decreases. This is reflected by the increase in the band masses, τ_s and μ_d shown in Fig. 8, with atomic number since they are inversely proportional to the amplitude (cf. (16, 21)).

The essence of Fig. 5 is that, as a function of atomic number the position of the 6s-band falls relative to the Fermi level and to the position of the 5d-band. As a consequence the number of s-electrons increases through the series at the expense of the number of d-electrons as may be seen from Fig. 9. The finite p-occupation numbers in Fig. 9 does not reflect a genuine p-occupation but rather the $\ell=1$ component of the tails from the neighbouring sites. As we shall see in the following it is this shift in the character of the conduction states which is behind many of the trends observed in the physical properties of the lanthanide metals.

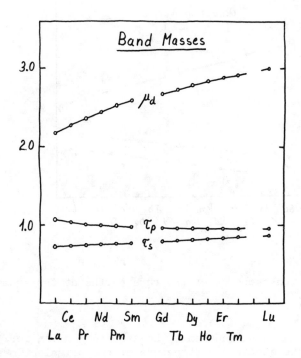

Fig. 8 Band masses for the 6s-, 6p-, and 5d-bands in the lanthanides at the atomic radius, S = 3.75 a.u.

4.2 The Divalent Metals

The results of the energy band calculations for the divalent
metals in the lanthanide series, where we include Ba for comparison, are shown in Fig. 10. As before we plot the results at a
common atomic radius close to the equilibrium radius of Yb in
order to avoid the effect of the decreasing equilibrium radius
(cf. Fig. 2). The trends seen in Fig. 10 are very similar to those
exhibited by the trivalent metals in Figs. 5, 8, 9, i.e. we observe
a decrease in the position of the 6s-band and a depletion of the
5d-band as a function of atomic number. The main difference between
the trivalent and the divalent lanthanides is, of course, that the
latter have approximately one d-electron less than the former which
in connection with the fall of the s-band and the relative stability in the position of the d-band results in an essentially
empty 5d-band in Yb. An analysis of v(S) and B_S in Fig. 10 in
terms of (25) and (26) gives results similar to those for the

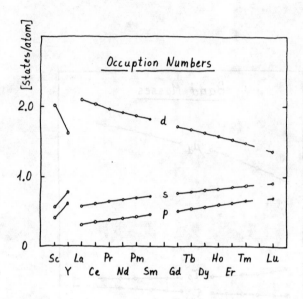

Fig. 9 Occupation numbers for the rare earths at the atomic
radius, S = 3.75 a.u.

trivalent lanthanides, and hence the arguments given for the
variation in the electronic structure of the trivalent metals can
also be applied to the divalent metals.

5. PRESSURE CALCULATIONS

We shall briefly introduce the so-called pressure relation
[10, 12, 13] which may be used to calculate the electronic pressure once the self-consistent energy band problem has been solved.
From the local density theory in section 2 we note that when self-
consistency is reached the kinetic energy which enters the energy
functional (5) may be obtained from (3) in the form

$$\langle \Phi_s | T | \Phi_s \rangle = \sum_j^{occ.} \int \psi_j^* (-\nabla^2) \psi_j \, d\underline{r}$$

$$= \sum_j^{occ.} E_j - \int V_{eff}(\underline{r}) n(\underline{r}) d\underline{r}$$

(27)

The energy functional (5) which is the total energy of the elec-

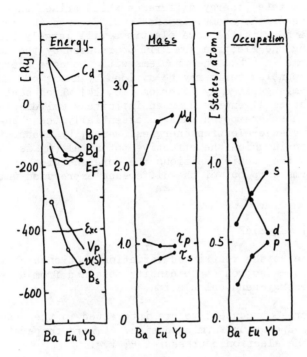

Fig. 10 Energy bands for the divalent metals Ba, Eu, and Yb
evaluated at the atomic radius S = 4.10 a.u., left panel, corresponding band masses, central panel, and the calculated occupation
numbers, right panel. An explanation of the symbols are given in
the captions for Figs. 5, 8, 9.

tronic system and which we rename U may now be written

$$U = \sum_{j\underline{k}}^{\text{occ.}} E_j(\underline{k}) - \tfrac{1}{2} \int_S \frac{2}{|\underline{r}-\underline{r}'|} n(r')n(r) \, d\underline{r}d\underline{r}'$$

$$-\int_S (\mu_{xc}[n(r)] - \epsilon_{xc}[n(r)])n(r)d\underline{r}$$

(28)

where we have used (6). In addition we have introduced the atomic
sphere approximation, added the nuclear-nuclear repulsion, and
made use of the fact that the spheres are neutral in the ASA for
the case of one atom per cell. The total energy (28) is simply
the sum of the one-electron energies corrected once for the Hartree
and exchange-correlation terms which enters twice in the one-
electron sum through the effective potential (6).

    Usually we are not interested in the total energy itself but
rather in the total energy difference which arises when we change
the positions of the atoms. In such a difference the large energies associated with the core electrons will largely cancel, and
the remaining one-electron sum will be reduced by the double-
counting terms in (28). The problems with the core energies and
the double-counting terms may be avoided by means of the so-called
force theorem suggested by Andersen [10, 14] which states that the
change in the total energy upon an infinitessimal displacement of
frozen atomic sphere potentials is essentially the change in the
sum of the valence-electron energies. Hence, by means of a restricted variation in which the effective potential (6) is shifted
rigidly with the atomic positions the core energies and the double-
counting terms drop out of the difference expression which we write

$$dU = \delta \int^{E_F} EN(E)dE \qquad\qquad\qquad (29)$$

Here δ indicates the restricted variation mentioned above, and N
(E) is the one-electron state density obtained from a self-
consistent energy-band calculation.

    In the present context we are interested in the change in the
total energy upon a uniform expansion. At zero temperature this is
related to the electronic pressure, P, i.e.

$$P = - dU/d\Omega \qquad\qquad\qquad (30)$$

where Ω is the volume of the primitive cell. Therefore, by performing self-consistent energy-band calculations at a series of
atomic volumes one may map out the zero-temperature equation of
state, $P(\Omega)$, determine the equilibrium volume, Ω_{eq}, by

$$P(\Omega_{eq}) = 0 \qquad\qquad\qquad (31)$$

or alternatively the equilibrium radius $S = (3\Omega_{eq}/4\pi)^{1/3}$, and the
bulk modulus

$$B = - dP/d\ln\Omega|\Omega_{eq} \qquad\qquad\qquad (32)$$

In the actual calculations we use a scheme based on equations
(77) and (78) in Ref. [10], but for the purpose of interpretation
we shall now give a simple derivation of a less accurate expression.
We write the band energy in the form

$$E = C(S) + \frac{E-C}{W} (n) \times W(S) \qquad\qquad\qquad (33)$$

where the brackets indicate that the centre, C, and the bandwidth, W, depend on the atomic volume only while the combination (E-C)/W depends only on the occupation number. The sum of the valence one-electron energies is therefore

$$\int^{E_F} EN(E)dE = C(S)n + W(S) \int^{E_F} \frac{E-C}{W}(n)dn \tag{34}$$

The differentiation with respect to volume as required by (30) may now be performed directly, and we have the ℓ-partial pressure

$$3P_\ell \Omega = n_\ell \left(-\frac{\delta C_\ell}{\delta \ell nS} \right) + n_\ell (\bar{E}_\ell - C_\ell) \frac{\delta \ell n \mu_\ell S^2}{\delta \ell nS} \tag{35}$$

where we have used $dU = - Pd\Omega = - 3P\Omega d\ell nS$ and where

$$\bar{E}_\ell = n_\ell^{-1} \int^{E_F} EN(E)dE \tag{36}$$

is the centre of gravity of the occupied part of the ℓ-band. Similarly, we find by expansion around V_ℓ

$$3P_\ell \Omega = n_\ell \left(-\frac{\delta V_\ell}{\delta \ell nS} \right) + n_\ell (\bar{E}_\ell - V_\ell) \frac{\delta \ell n \tau_\ell S^2}{\delta \ell nS} \tag{37}$$

which one may use for s- and p-bands while (35) is appropriate for d-bands. The expressions for the volume derivatives of the potential parameters appearing in (35, 37) may be found in Ref. [10]. They are

$$\frac{\delta C}{\delta \ell nS} = - 2 \frac{C - \epsilon_{xc}}{\mu}$$

$$\frac{\delta V}{\delta \ell nS} = - (2\ell+3) \frac{V - \epsilon_{xc}}{\tau} \tag{38}$$

$$\frac{\delta \ell n \mu S^2}{\delta \ell nS} = (2\ell+1) + \frac{2}{\mu} - \frac{2(C - \epsilon_{xc})S^2}{D_\upsilon + \ell + 1}$$

$$\frac{\delta \ell n \tau S^2}{\delta \ell nS} = -(2\ell+1) + \frac{2\ell+3}{\tau} - \frac{2(V - \epsilon_{xc})S^2}{D_\upsilon - \ell}$$

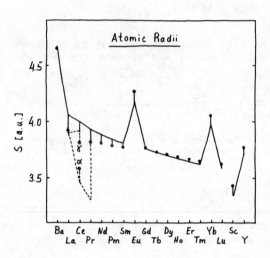

Fig. 11 The equilibrium atomic radii for the rare earth metals.
The full line indicates the calculation including s-, p- and d-
partial waves only while the broken line indicates that f-partial
waves have also been included. The full circles indicate the
experimental radii.

where the extra potential parameter D_ν is the logarithmic derivative of $\partial\phi_\ell(E,S)/\partial E$.

5.1 The Atomic Volume (Radius)

We have calculated the electronic pressure (30) over a volume
range sufficiently large to determine the equilibrium atomic volume
(cf. (31)) for all the rare earths. Since the volume is relatively
independent of the crystal structure all self-consistent calculations have been carried out assuming the closely packed fcc
structure. The results are shown in Fig. 11, and a comparison with
the experimentally observed values shows that the calculated radii
are typically 3% too large for the light lanthanides but almost
correct for the heavy lanthanides. As a result, the lanthanide

contraction is somewhat overestimated by the calculation. However, as seen in Fig. 11, the errors in the beginning of the series may partly be corrected if the 4f-electrons are included in the valence-state calculations. For La the reason is that the onuccupied 4-f band is so close to the Fermi level that it modifies the occupied bands through hybridization, and in the case of α-Ce there is one itinerant, i.e. band-like, 4f-electron which gives extra binding thus helping to reduce the volume below that given by the spd-electrons alone. In γ-Ce the 4f-electron is localized, i.e. it does not contribute to the binding, and for that reason the equilibrium radius is increased over that of a α-Ce. We shall later return to a calculation of the γ-α transition in Ce. If we include the 4f-electrons in the valence-state calculations in Pr [15] the 4f-contribution to the binding becomes so large and the calculated radius so small (cf. Fig. 11) as to be unacceptable. Hence, beyond Ce one should treat the 4f-states as localized, as we all know.

5.2 The Bulk Modulus

The bulk modulus (32) which is the inverse of the compressibility may be obtained from the calculated equation of state, $P(\Omega)$, by a numerical differentiation. The results of such a procedure are shown and compared with the experimental values in Fig. 12. Apart from the low bulk modulus in α-Ce which is connected with the itinerant 4f-electron in this material, we note that the experimental bulk modulus for the trivalent rare earths varies only from 30 GPa in Pr to 50 Gpa in Lu. This is a very moderate variation compared to that found in the remainder of the 6th row, i. e. the 5d-transition series, where the bulk modulus changes by more than an order of magnitude from La to Os.

The calculated bulk modulii are on the average 30% higher than the measured values, an error which is not untypical for the present type of calculations (cf. Fig. 5.16 in Ref. [10]). On the other hand, the calculation does describe the observed trends correctly, i.e. the moderate increase for the trivalent metals, the depression of B in Ce caused by the 4f-electron, and the relative low values for the divalent metals. Hence, it may be used to explain these trends in terms of the changes in the electronic structure.

5.3 The Partial Pressures

We shall now discuss the individual s-, p-, and d-contributions to the calculated electronic pressure in the light of the expressions (35, 37, 38) and using the band structure calculations presented earlier. Thereby we will be able to understand the

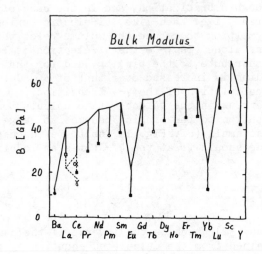

Fig. 12 Bulk modulii for the rare earth metals. The full line
indicates the calculation including s-, p-, and d-partial waves
only while the broken line indicates that f-partial waves have
also been included. The bulk modulii in the former calculations
are evaluated at the calculated equilibrium radii and those in
the latter at the measured radii. The full squares are the "best"
experimental values at low temperature [16], the open squares are
experimental values at room temperature [16], the full circle is
the experimental value from Ref. [17] and the open circle is the
value estimated in Ref. [17].

variation in the atomic volume and the bulk modulus through the
lanthanide series in terms of the electronic structure of the
lanthanide metals.

Of the three contributions to the total pressure shown in Fig.
13 the s- and p-pressures are positive while the d-pressure is
negative. We therefore refer to the s- and p-pressures as repulsive
and to the d-pressure as bonding, and use the loose expression
that the d-electrons bind the crystal together at equilibrium.
The signs of the individual contributions may be understood if we

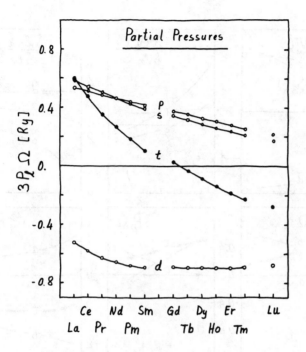

Fig. 13 Partial s-, p-, d- and total, t, pressures for the trivalent lanthanides at the common atomic radius S=3.75 a.u. obtained from the LMTO band calculations.

note that in the present case the s- (and p-) orbitals have a negative curvature at the atomic radius (cf. Fig. 7), and that the logarithmic derivative function for fixed radius is a de-creasing function of energy. When we therefore decrease the volume we must increase the energy in order to maintain a volume independent boundary condition, e.g. logarithmic derivative, at the atomic radius as prescribed by the ASA equations (13) (cf. also Fig. 3 and equation (10)). Hence, the derivative $-\delta E/\delta \ln S$, which is essentially the pressure per ℓ-electron (cf. (29, 30)), is positive for s- (and p-) electrons and, because the d-orbitals have a positive curvature at the atomic radius, negative for d-electrons.

According to (35, 37) there are two contributions to the individual pressures, the bandposition term proportional to $\delta C/\delta \ln S$ or $\delta V/\delta \ln S$ and the bandwidth term proportional to $\delta \ln \tau S^2/\delta \ln S$ or $\delta \ln \mu S^2/\delta \ln S$. In Lu $V < \varepsilon_{xc}$ (cf. $B_S = V_S$ and V_p in Fig. 5) for s-

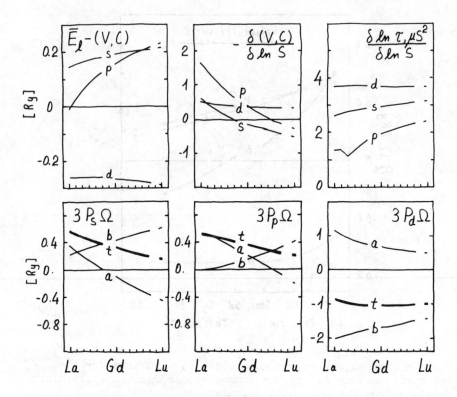

Fig. 14 Plots of the quantities which enter the approximate pressure expressions (35, 37), upper panels, and a decomposition of the
partial pressures, labelled t, into bandposition terms, labelled a,
and bandwidth terms, labelled b, lower panels.

and p-states. The bandposition term therefore provides an attractive pressure, labelled a in Fig. 14, which is the correction due
to exchange and correlation to the repulsive bandwidth term proportional to the "kinetic" energy ($\bar{E}-V$). As we move to the left
from Lu, V increases above ε_{xc} owing to the relative increase in
the core size. Hence, according to (38) $\delta V/\delta \ln S$ changes sign and
the bandposition term becomes repulsive. The bandwidth term,
labelled b in Fig. 14, on the other hand, decreases from Lu to La
as a result of the increase in the core size which make n, $\bar{E}-V$,
and $\delta \ln TS^2/\delta \ln S$ go down. In total the s- and p-electrons provide
strongly repulsive pressures which are responsible for the incom-
pressibility of the rare earth metals.

For d-electrons the centre, C, is well above ε_{xc} (cf. Fig. 5) and the bandposition term , labelled a in Fig. 14, is therefore repulsive. As we move to the left of Lu it becomes even more so because C rises relative to ε_{xc}. The attraction is provided by the bandwidth term, labelled b in Fig. 14, which is negative because $\bar{E}$ is below the centre. This term becomes more bonding from Lu to La because the number of d-electrons increases (cf. Fig. 9). In total, the d-contribution is bonding and remains relatively independent of atomic number with a shallow minimum near Gd.

A comparison between the lower panels of Fig. 14 and the results of the LMTO calculation, Fig. 13, shows that the approximate expressions (35, 37) provide almost exact estimates of the s- and p-pressures. Hence, we conclude that the s- and p-pressures decrease from La to Lu because V falls with respect to ε_{xc}, the variation being essentially that of the bandposition term. The magnitude of the d-contribution, however, is overestimated some 30%, but the trend is correctly described. We therefore conclude that the initial decrease in the d-pressure from Lu to Gd is caused by the fact that the centre falls with respect to ε_{xc} thereby modifying the bonding bandwidth term which varies as the occupation number, i.e. falls in magnitude from La to Lu.

The total pressure in Fig. 13 and the calculated equilibrium atomic radius show a very similar variation with atomic number as indeed they should if the bulk modulus did not vary too much through the series. Hence, one may conclude that the lanthanide contraction is caused by the contraction of the core and the resulting fall in the position of the s- and p-energy bands. Also, the error in the calculated equilibrium radius in the beginning of the lanthanide series seems to be caused by too high lying d-bands.

The calculated bulk modulus depends sensitively on the atomic radius. Evaluated at a common radius, say that of Gd, it would start out high in La and decrease towards Lu. The reason is that the space available for the sp-electrons between the core and the atomic sphere increases from La to Lu. Therefore, the sp-electrons in the beginning of the series will respond more strongly to changes in the volume through the bandposition term than do those of higher atomic number, and the bulk modulus decreases through the series. If we evaluate the bulk modulus at the calculated equilibrium radius the space available for the sp-electrons is still larger in La than in Lu but it has decreased to the extent that the bulk modulus in La now falls below that in Lu. In Ref. [18] it was shown that La has a relatively low bulk modulus because s-electrons are transferred into the d-band under pressure, and that the importance of this softening increases the closer one is to the point where the s-band is completely depleted. In the present case La is closer to having an empty s-band than Lu,

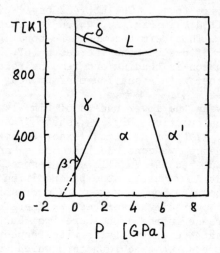

.Fig. 15 PT-phase diagram for Cerium. The γ-α phase line has been
extrapolated to zero temperature and negative pressure.

and hence it is consistent with the above observations that the
bulk modulus increases through the series of lanthanide metals.

5.4 4f-delocalization, γ-α Ce

Cerium is one of the most intriguing metals in the lanthanide
series. At low temperature it forms in the fcc structure, the α-
phase in Fig. 15, with an atomic volume which is about 20% lower
than the volumes of the high temperature β-, γ-, and δ-phases. The
α-phase is therefore also referred to as the collapsed fcc-phase.
At room temperature Ce undergoes a pressure induced isostructural
γ-α transition in which the volume collapses by about 15%. In the
pressure range above 6 PGa Ce attains an orthorhombic structure,
α' in Fig. 15, which is also found in the actinide metal Uranium.

Several explanations have been proposed for the γ-α transition
in Ce. Here, we shall restrict ourselves to show to what extent
band theory supports the viewpoint taken by Johansson [19] namely
that the α to γ transition in Ce is a Mott metal to insulator
transition within the 4f-shell. We have, to this end, plotted
several pressure-volume curves, i.e. equations of state, on Fig.
16. From these plots we note that the calculation which treats the
4f-electrons on the same footing as the spd-electrons, marked spdf

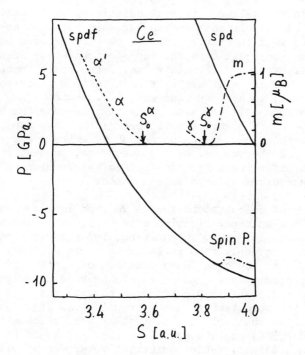

Fig. 16 Calculated zero temperature pressure-volume curve for Cerium, full lines, and measured [20] room temperature equation of state, broken lines. The dot-dash curve indicates the effect of spin-polarization. S_0^α is the low temperature experimental equilibrium radius in the α-phase and S_0^γ is the measured room temperature equilibrium radius in the γ-phase.

on the figure, gives a fairly good description of the measured α,α' equation of state. We find furthermore from the calculation that the 4f-band is occupied by about one electron. Hence, band theory leads us to conclude that in the low-volume α-phase of Ce the 4f-electron is itinerant, i.e. on the metallic side of the Mott transition. Next we note from Fig. 16 that the calculation which treats the 4f-electron as a core electron gives the best description of the γ-phase. Hence, we conclude that in the high-volume γ-phase the 4f-electron is localized, i.e. on the insulating side of the Mott transition.

There is nothing in the above straight-forward application of band theory which indicates at which point on the calculated equations of state the itinerant to localized transition will take place. In order to resolve this question we shall allow for a spin-polarization in the valence state calculations. It is well-known from energy band calculations on the magnetic metals that a spin-polarized state will be stable when the overlap between neighbouring orbitals becomes so small that the one-electron state density at the Fermi level exceeds a certain value. Hence, as a function of increasing volume one will see a transition from a non-polarized, i.e. non-magnetic, to a-spin-polarized, i.e. magnetic ground state. In the calculations [21] one also finds that because the spin-polarized state has a higher kinetic energy than the non-polarized state the spin-polarized electrons set up an extra pressure which further increases the volume. As a consequence, on has, in most cases, a first order phase transition from a low-volume non-polarized state to a high-volume spin-polarized state.

In the present calculations for Ce we find in agreement with Glötzel [22] that spin-polarization sets in close to the measured equilibrium radius for γ-Ce. In addition, the polarization, m in Fig. 16, rises so steeply that it leads to a van der Waal's loop around - 8.5 GPa in the calculated equation of state, and there-fore to a first order phase transition of the kind described above. The calculated - 8.5 GPa should be compared with the - 1.0 GPa obtained from the zero temperature extrapolation, Fig. 15, of the γ-α phase line. We shall point out, that similar calculations for the 3d-monoxides [14] and the actinides [23, 24] leads to a similar picture and considerably better quantitative agreement with exper-iments.

The picture of Ce which emerges from the band calculations is that in the low-volume α-phase the 4f-states are itinerant, i.e. band-like, while in the high volume phases they are localized, i.e. core-like, and that the volume change involved in the γ-α transition arises because itinerant electrons contribute to the binding in contrast to localized electrons. Although the α-phase nominally has four valence electrons and the γ-phase only three, the γ-α transition is not a valence transition in the usual sense which would involve a transition between the configurations $f^n(ds)^3$ and $f^{n-1}(ds)^4$. Rather, the fourth valence electron in α-Ce resides in the 4f-band which has different binding-properties than the sd-band, and for that reason the atomic radius of α-Ce cannot be obtained by an extrapolation from Hf metal where all four valence electrons reside in the sd-band.

The γ-α transition in Ce is found experimentally [25] to have an analogue in the phase III to phase IV transition in Pr. From calcu-lations [15] we find again that a spin-polarization similar to that found in Ce sets in at about the volume where the III - IV tran-sition is observed. Hence, the onset of spin-polarization may be used to mark the volume at which the 4f-states become localized.

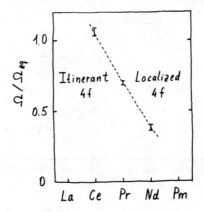

Fig. 17 Itinerant versus localized behaviour of the 4f-electrons.
Plotted is the relative volume at which the spin-polarization
leads to an anormaly (van der Waal's loop for Ce and Pr) in the
calculated equation of state.

Judged from Fig. 17 it will be extremely difficult to delocalize
the 4f-electrons beyond Pr under static conditions since this
would require more than two-fold compression.

6. THE CRYSTAL STRUCTURES

The rare earth metals are famous for the appearance of the
crystal structure sequence hcp → Sm-type → dhcp → fcc. This
sequence is realized both as a function of decreasing atomic
number, i.e. going from Lu to La, and as a function of pressure.
Thus, if one applies pressure to, say, hcp-Gd one will eventually
obtain the fcc structure. The observed regularity in the crystal
structures was used by Johansson and Rosengren [26] to construct
out of the individual phase diagrams a generalized phase diagram
for the lanthanide metals. It follows from the generalized phase
diagram that under pressure each individual element looks like
the preceding and lighter elements. Johansson and Rosengren
rationalized this behaviour by reference to the ratic between
the atomic radius S_{ws} and some core radius S_I. They observed that
although S_{ws} decreases through the series (cf. Fig. 2) the ratio
S_{ws}/S_I increased much as S_{ws}/S_O does in Fig. 6. Hence, when pressure is applied to a given element S_{ws} decreases but S_I remains
constant thereby simulating a lighter element. Subsequently,

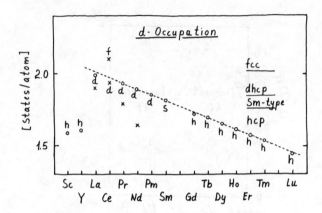

Fig. 18 d-occupation numbers at the experimentally observed equilibrium atomic radius (cf. Fig. 2). The crosses show the d-occupation when the 4f-electrons are treated as valence electrons. Indicated are also the experimentally observed crystal structures: f for fcc, d for dhcp, S for Sm-type, and h for hcp, and to the right in the figure the empirical d-occupation numbers which are found to separate the crystal structures.

Duthie and Pettifor [27] showed that the ratio S_{WS}/S_I was intimately connected with the relative occupancy of the s- and d-orbitals as we discussed in section 4, and they established by means of unhybridized ASA calculations (cf. section 3) a correlation between the crystal structures and the calculated d-occupation numbers.

6.1 The Trivalent Metals

Duthie and Pettifor [27] used canonical band theory to calculate structural energy differences for the trivalent lanthanides, and they showed that as the d-band occupancy increases one goes through the crystal structure sequence hcp → Sm-type → dhcp → fcc. However, the quantitative agreement with experiment was not perfect, and in order to obtain Lu in the hcp-structure they had to argue that the inclusion sd-hybridization would remove 0.4. d-electrons in Lu but not in La so that Lu would have effectively 1 d-electron less than La. The present fully hybridized calculations do not support this conjecture, the difference between La and Lu in Fig. 18 is 0.54 d-electrons, and instead we believe that it is the calculation of structural energy differences by

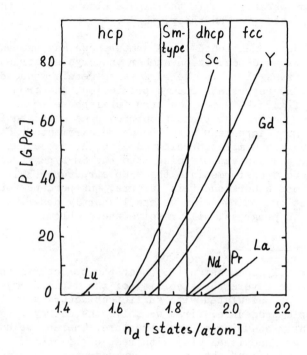

►Fig. 19 The calculated electronic pressure, P, as a function of
the d-occupation number, n_d. The vertical lines indicate the
d-occupancy at which structural phase transitions are expected to
occur.

means of canonical band theory which although qualitatively correct
is not sufficiently accurate to give quantitative agreement with
experiments in the present case.

Our d-occupation numbers are shown in Fig. 18, and it may
readily be seen that the number of d-electrons may be used to
separate the four crystal structures of the rare earths. Since we
have not yet performed total energy calculations of the kind we
shall present in the next section when discussing the divalent
rare earths, we have used the experimentally known pressure-induced
structural phase transitions in La (dhcp → fcc), Sm (Sm-type →
dhcp), and Gd (hcp → Sm-type) to fix three d-occupation numbers
which will separate the crystal structures. We find n_d = 2.0, 1.85,
and 1.75 for the three transitions, respectively, and these numbers

are indicated to the right in Fig. 18. One notes that Sc and Y
in this picture behave as heavy lanthanides, and that the empirical correlation established in the figure also explains the two
low-pressure phases (fcc and dhcp) in Ce.

Given the above-mentioned three numbers one may use the
correlation between d-occupation number and crystal structure to
predict phase transitions in the rare earth metals induced by
pressure. Fig. 19 represents such a prediction, and from the
curves in the figure at which the structural phase transitions
are expected to occur. The transition pressures obtained for Y
and Gd are in good agreement with recent measurements [28, 29].
However, for Sc the simple correlation with the d-occupancy seems
to break down, and instead of going into the Sm-type structure
around 20 GPa Sc transforms into the β-Np structure [30]. At
present we do not understand this failure, but perhaps total
energy calculations of the kind we shall shortly describe, and
which are now in progress, will resolve the problem.

6.2 The Divalent Metals

It has recently been shown [31] that the structural energy
differences which govern the stability of the various crystal
structures may be calculated with sufficient accuracy by the
following simple procedure. First we solve the energy band problem
self-consistently assuming an fcc structure. Thereby we have minimized the energy functional U{n} with respect to changes in the
electron density, n, and obtained the ground-state density n^{sc}_{fcc}.
Then we construct a trial density, n^{tr}_{bcc}, by positioning frozen,
self-consistent fcc-potnetials in a bcc geometry, solve the
corresponding one-electron Schrödinger equation (3), construct
the kinetic energy (27), and thereby evaluate the energy functional
(28) for the trial density. The structural energy difference is
now simply

$$\Delta(bcc-fcc) = U\{n^{tr}_{bcc}\} - U\{n^{sc}_{fcc}\} \tag{39}$$

In complete analogy with the force theorem [10, 14] the use of
rigidly shifted potentials ensures that the chemical shifts in the
core-electron energies do not enter (39) and that at the same time
the double counting terms cancel. Hence, we are led to consider
only the sum of the one-electron energies for the valence electrons,
i.e.

$$\Delta(bcc-fcc) = \int^{E_F} EN^{tr}_{bcc}(E)dE - \int^{E_F} EN^{sc}_{fcc}(E)dE \tag{40}$$

Within the atomic sphere approximation all that is required, in

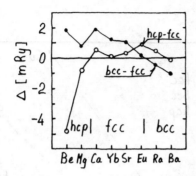

Fig. 20 Structural energy differences relative to the fcc phase calculated at the experimentally observed equilibrium radii for the divalent metals arranged according to d-occupation.

addition to the fcc-calculation which we have already performed, is to calculate the energy bands in, say, the bcc structure, and subtract according to (40).

The structural energy differences for a series of divalent metals obtained by the above-mentioned procedure are shown in Fig. 20. It follows from the figure that at low temperature Be and Mg should form in hcp structure, Ca and Sr in the fcc structure, and Ra and Ba in the bcc structure in complete agreement with experiments. For Yb the calculations predict the fcc phase at low temperature in contrast to the observed hcp structure. However, it follows from Fig. 20 that the stability of the fcc phase against the hcp phase is extremely small, and in fact an increase in the atomic radius of only 0.3% will stabilize the hcp structure. Hence, it appears that the anomalous low temperature hcp phase observed in Yb is simply a consequence of the particular electronic structure of the metal. For Eu we find that the fcc phase is marginally more stable than the experimentally observed bcc phase. However, already at the calculated equilibrium radius is bcc the stable phase, and hence a consistent application of the theory would predict the correct low temperature phase in Eu. One should also note that we have neglected the contribution to the total energy from the zero point motion which may be comparable to the calculated Δ(bcc-fcc) for Eu, and probably tip the balance in favour of the bcc structure.

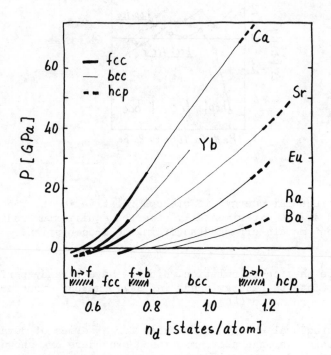

Fig. 21 Calculated pressure and stable crystal structures as functions of the d-occupation number.

Viewed on a pressure scale the above-mentioned uncertainties are extremely small and Fig. 21 may therefore be used to predict structural phase transitions with reasonable confidence. We would like to point out that neither experimental evidence nor adjustable parameters of any kind have been used to construct Fig. 21. It is therefore very satisfactory to find that those transitions which have been verified so far by experiments are in good agreement with the predictions one can make from the figure.

It follows from Fig. 21 that the divalent alkaline earth metals have a crystal structure sequence of their own, i.e. hcp → fcc → bcc → hcp, as a function of d-occupation number, and that the divalent lanthanides, Eu and Yb, are part of that sequence. One might therefore expect Eu and Yb to transform into the hcp structure at high pressures. However, before this happens the two metals

will have the changed valence [26], and instead they are expected
to form in one of the phases observed within the series of trivalent lanthanides (cf. Fig. 18).

7. POST SCRIPTUM

Since the present notes were conceived as a very personal
view of the electronic structure of the rare earth metals based
on new, unpublished energy band calculations it is not felt
appropriate to include in these notes, although it will be included in the lectures, the theory of Johansson and Rosengren
[26, 32-34] for the cohesive energy and the valence state of
the rare earths. For the same reason, we have also not included
references to all the previous energy band calculations performed
on the rare earth metals. However, they may be found in the excellent reviews by Dimmock [35], Freeman [36], and Liu [37].

NOTES

1. In a monoatomic solid the atomic radius, S, is defined as the
   radius of a sphere, the atomic sphere, which has the volume
   of the primitive crystallographic cell. It is directly proportional to the 12-coordinated metallic radius, R_m, i.e.
   $S = (3/4\pi)^{1/3} 2^{5/6} R_m$. For a given crystal structure S is also
   proportional to the lattice parameter.

2. In atomic units (a.u.) the unit of length is the Bohr radius
   equal to 0.5292 Å and the unit of energy is the Rydberg (Ry)
   equal to 13.6 eV.

REFERENCES

1. B.J. Beaudry and K.A. Gschneidner, "Preparation and basic properties of the Rare Earth Metals" in "Handbook on the Physics
   and Chemistry of Rare Earths", vol. 1, ed. by Karl A. Gschneidner
   and Le Roy Eyring (North Holland, Amsterdam, 1978).

2. P. Hohenberg and W. Kohn, Phys. Rev. 136, B 864 (1964).

3. W. Kohn and L.J. Sham, Phys, Rev. 140, A 1133 (1965).

4. L. Hedin and B.I. Lundqvist, J. Phys. C4, 2064 (1971).

5. U. von Barth and L. Hedin, J. Phys. C5, 1629 (1972).

6. O. Gunnarsson and B.I. Lundqvist, Phys. Rev. B13, 4274 (1976).

7. J. Callaway, "Energy Band Theory" (Academic Press, New York and London 1964).

8. O.K. Andersen, Solid State Commun, 13, 133 (1973).

9. O.K. Andersen, Phys. Rev. B12, 3060 (1975).

10. A.R. Mackintosh and O.K. Andersen, "The Electronic Structure of Transition Metals" in "Electrons at the Fermi Surface" ed. by M. Springford (Cambridge University Press, Cambridge 1980).

11. D.G. Pettifor, J. Phys. F7, 613 (1977).

12. R.M. Nieminen and C.H. Hodges, J. Phys. F6, 573 (1976).

13. D.G. Pettifor, Commun. Phys. 1, 141 (1976).

14. O.K. Andersen, H.L. Skriver, H. Nohl and B. Johansson, Pure & Appl. Chem. 52, 93 (1979).

15. H.L. Skriver, "Electronic Transitions in Praseodymium under Pressure", in "Physics of Solids under High Pressure" ed. by J.S. Schilling and R.N. Shelton (North Holland, Amsterdam 1981).

16. T.E. Schott, "Elastic and Mechanical Properties" in "Handbook on the Physics and Chemistry of Rare Earths", vol. 1, ed. by Karl. A. Geschneidner and Le Roy Eyring (North Holland, Amsterdam 1978).

17. K.A. Geschneidner, "Physical Properties and Interrelationship of Metallic and Semimetallic Elements" in "Solid State Physics", vol. 16, ed. by F. Seitz and D. Turnbull (Academic Press, New York 1964).

18. A.K. McMahan, H.L. Skriver and B. Johansson, Phys. Rev. B23, 5016 (1981).

19. B. Johansson, Phil. Mag. 30, 469 (1974).

20. W.H. Zachariasen and F.H. Ellinger, Los Alamos Report LA-6251 (1976), unpublished.

21. O.K. Andersen, J. Madsen, U.K. Poulsen, O. Jepsen, and J. Kollar, Physica 86-88B, 249 (1977).

22. D. Glötzel, J. Phys. F8, L163 (1978).

23. H.L. Skriver, O.K. Andersen, and B. Johansson, Phys. Rev. Lett. <u>41</u>, 42 (1978).

24. H.L. Skriver, O.K. Andersen, and B. Johansson, Phys. Rev. Lett. <u>44</u>, 1230 (1980).

25. J. Wittig, Z. Physik B-Condensed Matter <u>38</u>, 11 (1980).

26. B. Johansson and A. Rosengren, Phys. Rev. <u>B11</u>, 2836 (1975).

27. J.C. Duthie and D.G. Pettifor, Phys. Rev. Lett. <u>38</u>, 564 (1977).

28. Y.K. Vohra, H. Olijnik, W. Grosshans, and W.B. Holzapfel, Phys. Rev. Lett. <u>47</u>, 1065 (1981).

29. W.A. Grosshans, Y.K. Vohra, and W.B. Holzapfel, to be published.

30. W. Grosshans, Y.K. Vohra, and W.B. Holzapfel, J. Magn. Magn. Mater, to be published.

31. H.L. Skriver and B. Johansson, to be published.

32. B. Johansson, J. Phys. <u>F4</u>, L169 (1974).

33. B. Johansson, Phys. Rev. <u>B20</u>, 1315 (1979).

34. B. Johansson, Phys. Rev. <u>B19</u>, 6615 (1979).

35. J.O. Dimmock, "The Calculation of Electronic Energy Bands by the Augmented Plane Wave Method" in "<u>Solid State Physics</u>", vol 26, ed. by H. Ehrenreich, F. Seitz, D. Turnbull (Academic Press, New York 1971).

36. A.J. Freeman, "Energy Band Structure, Indirect Exchange Inter-actions and Magnetic Ordering" in "<u>Magnetic Properties of Rare Earth Metals</u>", ed. by R.J. Elliot (Plenum Press, London and New York 1972).

37. S.H. Liu, "Electronic of Rare Earth Metals" in "<u>Handbook on the Physics and Chemistry of Rare Earths</u> ", ed. by K.A. Gschneidner, L.R. Eyring (North Holland, Amsterdam 1978).

DISCUSSION

BREWER   (question)

What temperature were these calculations carried out?

SKRIVER   (answer)

At zero degree.

GROUP I   (question)

What are the approximations, what is the accuracy of the various
properties calcuated compared to experiment?

SKRIVER   (answer)

The accuracy is about 1-3 % for bulk modulus, for compressibility
10-20 %.

DELONG   (question)

Can you predict the occurrence or stability of the α-U structure
for rare earths?

SKRIVER   (answer)

This may or may not be within the limit of the accuracy of the
methods applied to the alkaline earths metals including Eu and
Yb. There are band calculations for α-U structures in lanthanides.

DELONG   (question)

What is the feasibility of a "finite" contribution of a Ce^{2+} state
to the ground state properties of a metallic Ce compound (or even
elemental say α-Ce)? There may be some X-ray evidence for the
contribution of Ce^{2+} character in some metallic materials.

SKRIVER   (answer)

The spectroscopic data and the known systematics of cohesive energies show that Ce metal has a $4f^1 5d 6s^2$ configuration. If the
promotion energy for 2+/3+ is very large you are going to keep
the divalent state for Ce: this energy is too small and I think
Ce metal is trivalent.

BREWER   (comment)

In the use of expressions like divalent and trivalent, we have to
distinguish
(i)   metallic phase: here how many non-f-electrons we have and
(ii) compounds: like in oxides or sulphides, we talk about how
      many electrons we have transferred. There are clearly diva-
      lent Ce compounds available, CeS, where Ce has given up two
      electrons to sulphur. But these are different types of
      interactions.

SKRIVER   (comment)

You should call it a configuration rather than valence state.

ROSSAT-MIGNOD   (comment)

In CeS, Ce is trivalent, it is a metal.

BREWER   (comment)

I wish to comment on one topic mentioned by Prof. Skriver in his
lecture that might have caused some confusion. Unfortunately Prof.
Skriver had to leave early, but I have discussed this with him
and he agreed that it is important to clarify his reference to
the relationship between structure and the number of d electrons
for transition metals.

To introduce students to the behaviour of transition metals at
high temperatures, I show them the distribution of crystal structures at the melting points. There are three groups. At the left
hand side, only the body-centered cubic (bcc) structure is found
for K to Fe with 0 to 6.5 3d electrons per atom (e/a), for Rb to
Mo with 0 to 5 4d e/a, and for Cs to W with 0 to 5 5d e/a. The
stability of the bcc structure is thus independent of the number
of d electrons over a wide range. Also for the lanthanides to Ho,
and the actinides to Pu the solid structure at the melting point
is bcc. The next group of hcp structures includes Tc, Re, Ru, and
Os. Finally, we have the ccp metals of the ninth, tenth, and
eleventh groups of the transition metals. For the transition metals of group 6 through 11, only one structure is stable at all
temperatures except for Mn, Fe, and Co. These metals have configurations with non-bonding electron pairs that can compete with
more promoted configurations with fewer d non-bonding pairs because of the poor bonding of the 3d electrons. The metals of groups
2 to 4 transform to close-packed structures at lower temperatures
because the $d^{n-1}s$ and $d^{n-2}sp$ configurations are of almost equal
stability.

For the formation of bcc and hcp structures the difference in
$\Delta H_0/R$ is less than 0.7 kilokelvin for Ca to Ra, Sc to La and on
to Yb, Ac to Pu, and Ti to Hf. For the fifth and six groups, the
bcc structure becomes more stable than the hcp structure by 3-11
kilokelvin as increased nuclear charge lowers the energy of the
d orbitals. From the gaseous spectroscopic values and bonding
enthalpies of the different types of electrons, one can calculate
$\Delta H_0^0/R$ for the bcc and hcp structures where they are not the stable
phases. Taking $\Delta H_0^0/R$ of the bcc structure as zero, the values of
$\Delta H_0^0/R$ in kilokelvin for one ppm-atom of bcc structure transforming to hcp or ccp structures are given below for groups six to
eleven. For structures that are unstable between 0^0 K and the
melting point at pressures around 1 atm, the calculated values at
0^0 K are given in parentheses.

	Cr	Mn	Fe	Co	Ni	Cu
hcp	(+5.7)	(+0.2)	(+0.5)	-0.9	(0)	(+2.)
ccp		-0.4	+0.7	-0.85	-6.4	-19.9

	Mo	Tc	Ru	Rh	Pd	As
hcp	(+6.1)	-3	-1.2			
ccp				-2.3	-3.9	-6.

	W	Re	Os	Ir	Pt	Au
hcp	(+11.)	-1.1	-2.7	(-1.5)	(-1.5)	(+2.)
ccp				-2.	-5.4	-9.5

If we examine the stability of different electronic configurations,
we find for both normal and transition metals in pure metallic
phases or alloys that the bcc is found for a wide range of d electron concentrations, but the number of outer shell s,p electrons
is restricted to 1 - 1.5 e/a. Similarly the hcp structure is
found for d electron concentrations from 0 to 2 and 5 to 7 e/a,
when the d^{n-2}sp configuration is low enough in energy to yield a
stable solid. Thus the outer electron concentration of 2 e/a controls the occurrence of the hcp structure.

The main group of stable ccp structures is found for Mn to Ag,
Rh to Cu, and Ir to Au where d electrons are promoted to p orbitals to reduce the number of non-bonding d electron pairs, thus
the s,p concentration must be higher than for the hcp metals and
is expected in the range 2.5 to 3 in accordance with the values
of ccp Al, Ga, In, and their alloys.

Thus, it is clear that it is the number of outer shell p electrons
that controls the long-range order for both normal and transition
metals with a wide range of d electron concentrations possible in
each structure. There is normally one s e/a for all structures.
With less than 0.5 p e/a, the bcc structure is obtained. With
0.7 - 1.1 p e/a, the hcp structure is obtained. With 1.5 - 2 p e/a,
the ccp structure is obtained. Finally, with three p e/a, the

diamond structure is obtained. The Engel extension of the Hume-Rothery Rule to transition metals depends upon substantial contraction of the d orbitals compared to the outer-shell s,p orbitals. Thus the correlation may not be applicable to the beginning of each transition period and the d electrons may play a role in establishing the crystal structures of the alkaline earth and third group metals.

GROUP I & II   (question to the whole audience)

a) Are there any phase studies of mixed lanthanides, e.g., La/Gd or Sc/La?
b) What structure do these mixtures adopt?
c) Has Professor Skriver done any calculations on these mixtures?
d) What is the structure of commercial mischmetal? Is mischmetal less reactive than the pure metals and if so, why?

SKRIVER   (answer)

No calculation has yet been carried out in mixed lanthanide systems.

SINHA   (answer)

As far as I recall there is at least one study done on the phase diagram of La-Gd [see R. Valetta, Dissert. Abs. 20, 3539 (1960)]. The high temperature phases of La and Gd are completely soluble. However, an intermediate phase with approximate composition $LaGd_3$ has been observed. This phase ($LaGd_3$) is isostructural with Sm, with a = 3.667 and c = 26.482 Å. A slightly older but a very good compilation of phase diagrams of the binary lanthanide-metal, lanthanide-lanthanide systems was made by C.E. Lundin, in The Rare Earths (Eds. F.H. Spedding and A.H. Daane), John Wiley, New York, 1961, pp. 224-385.

Mischmetal is a mixture of usually the cerium group of light lanthanides and varies widely in composition depending on the source and the material that is reduced to prepare it. Mischmetal prepared from monazite is relatively high in Nd and Pr, whereas that from bastnasite is rich in La and depleted in Nd and Pr. Ce is also present at a very high concentration ($\sim$ 48-56 %). Mischmetal has an average analysis of Ce 49, La 24, Nd 19, Pr 6, Sm 2, Gd 0.5 percent. Thus I would presume the structure of mischmetal would be dictated by the major components present. If one consults the Darken-Gurry plots (electronegativity of elements vs. radii) for α and γ Ce, it becomes quite clear that γ-Ce would dissolve Mg, Ca, In, Ac, Sc, Yb, Th, and possibly Eu in addition to other trivalent lanthanides. But for α-Ce the points for lighter lanthanides (La, Pr, Nd) lie just outside the boundary line. Thus some reduction of solubility of these elements in α-Ce is expected at

equilibrium. However, La lowers the transition temperature of Ce
and hence the fractionation of La between α-Ce and γ-Ce becomes
difficult.

I do not think that a good sample of mischmetal to be any more or
less reactive than pure metals or say La-Ce alloy. In the early
days samples of lanthanides metals were found to be very active,
igniting at a relatively low temperature. However, this phenomenon was closely connected with the purity and the particle size.
The lighter lanthanides are usually attacked by moisture more rapidly than the heavier ones. However, if heated above room temperature all lanthanides start oxidizing at a relatively low temperature and the rate increases tremendously at higher temperature.

In general lanthanide rich alloys exhibit very much the same corro-
sion resistance as the parent lanthanides. However, the lighter
flint, which is an alloy of mischmetal with Fe, is rather pyro-
phoric.

MAGNETIC PROPERTIES OF LANTHANIDE COMPOUNDS

J. Rossat-Mignod

Laboratoire de Diffraction Neutronique
Département de Recherche Fondamentale
Centre d'Etudes Nucléaires de Grenoble
85 X - 38041 Grenoble Cédex, France

ABSTRACT

In this paper we will not give an exhaustive review of the
magnetic properties of lanthanide compounds, but from a few
examples we want to emphasize the characteristic properties encountered in such compounds. After a brief survey of the crystal
field theory and the interaction mechanisms, we will present
successively the problems related to the magnetic anisotropy, the
nature of the magnetic ordering, the dipolar ordering, the anomalous rare earth compounds, the investigation of magnetic phase
diagrams and to the determination of the exchange and crystal
field parameters.

1.   INTRODUCTION

The aim of this lecture is not to give a complete review of
the magnetic properties of the different kinds of rare earth
compounds. It would be a too tedious task in regard with the
numerous compounds which can be synthetized with rare earth elements. We want only to present a few examples, in order to illustrate the large variety of magnetic properties which have been
encountered among rare earth metals, intermetallics, semimetals,
semiconductors and ionic compounds. Their main features will be
underlined.

To get a general view on the magnetic properties of rare
earth compounds   a   reading of the Handbook on the Physics and
Chemistry of Rare Earths [1] is recommanded. With a series of
four volumes it represents the more recent review on the subject.

255

S. P. Sinha (ed.), Systematics and the Properties of the Lanthanides, 255–310.
Copyright © 1983 by D. Reidel Publishing Company.

However, on rare earth metals and their alloys, a complete and up to date description has been given by B. Coqblin [2].

Among the transition elements the main characteristic of rare earth ions is the small spatial extension of the 4f shell. Therefore in compounds, the overlap between two 4f orbitals centered on two neighbouring rare earth ions is very weak because the radius of the 4f shell is about ten times smaller than interatomic distances. So in both ionic and metallic compounds the rare earth atoms can be treated as a collection of independent ions on a periodic lattice. The 4f electrons being well localized, in any compound the rare earth ions present a well defined valence. Usually they are in a trivalent state except anomalous rare earths which can be divalent (tetravalent) or in an intermediate state. Then the interactions between rare earth ions will be introduced as a small perturbation which is quite different from 3d alloys or compounds. The understanding of the magnetic properties, which are very often complex in lanthanide compounds, needs a precise knowledge of the nature of the system without interactions and the interaction mechanisms which are involved. By taking some examples among intermetallic and ionic compounds the influence of the crystalline electric field (CEF) and the symmetry of the rare earth atom site will be shown. The nature of the interaction mechanisms and the symmetry of the crystallographic cell is important for determining the nature of the magnetic ordering. The case of ionic compounds in which the interaction of dipolar type is predominant will be examined by considering the rare earth gallium garnets. A special attention will be devoted to the presentation of the properties of anomalous rare earth compounds. The magnetization processes and the studies of magnetic phase diagrams will be described in a few cases to show how they can yield informations to the understanding of the magnetic properties. Finally we will underline how inelastic neutron scattering experiments can give a direct determination of the microscopic parameters involved in the magnetic properties.

2.   IONIC DESCRIPTION OF THE 4F ELECTRONS

2.1. Free ion

In rare earth ions, 4f electrons are in an inner shell characterized by a radius $\sqrt{\langle r^2 \rangle} \cong 0.35$ Å. This value is very small in comparison with the radius of the $5s^2 5p^6$ close shell (~ 1 Å) and the distance between RE-ions which is in fact about 10 time larger. So the 4f electrons interact only weakly with the electrons of the surrounding atoms. Then the intra-atomic interactions are much stronger than the inter-atomic ones and even in metallic compounds the width of the 4f band is perfectly negligible because the overlap between different 4f wave functions is

extremely small. Therefore whatever the compound, ionic or
metallic, an ionic picture can be adopted to describe RE atoms.
First the energy levels of an isolated ion must be computed [3].
The results are well known and we recall here only the main
points :
i) The most important contribution to the energy comes from the
central potential which gives the $4f^n$ contribution. The $4f^n$
configuration, where the number n of 4f electrons is integer, is
stable for normal rare earths which are mainly in a trivalent
state.
ii) The Coulomb and exchange interactions between the 4f electrons
lift the degeneracy of the $4f^n$ configuration. The resulting levels can be ordered according to the values of the quantum
numbers L and S. In pratice, we deal only with the levels of L
and S values given by Hund's rules.
iii) The spin-orbit coupling is large and lifts the degeneracy
of the levels inside the multiplet L, S. The resulting levels are
ordered according to the values of the total angular momentum
$\vec{J} = \vec{L} + \vec{S}$, the ground multiplet has J = L - S for the light elements and J = L + S for the heavy ones. Except in the cases of
Sm^{3+} and Eu^{3+} only the groud multiplet $|f^n, \alpha LSJ>$ is populated
at usual temperatures. Within this multiplet we can define the
magnetic moment in μ_B by $\vec{\mu} = \vec{L} + 2\vec{S} = g_J\vec{J}$ where g_J is the Lande
factor.

    Then the susceptibility of a free ion is given by :

$$\chi_0 = \frac{(g_J\mu_B)^2 J(J+1)}{3kT} \tag{1}$$

The effective magnetic moments $\mu_{eff} = g_J\sqrt{J(J+1)}$ (table 1) are close
to the experimental values obtained at high temperatures, as can
be seen in figure 1, except for Sm^{3+} and Eu^{3+} where the excited
multiplets must be taken into account to give a Van Vleck contribution. These results indicate that the energy levels of a free
R^{3+} ion are only slightly modified when it is placed in a crystal.

    Thus the assumptions of an ionic character, i.e. an integer
number of 4f electrons, the validity of Hund's rules (J, L, S) and
the smallness of the crystal field are a good starting point for
the study of the magnetic properties of normal rare earth
compounds.

Table 1. Some ionic properties of the trivalent lanthanides.

	L	S	J	g_J	$g_J\sqrt{J(J+1)}$	$g_J J$	$\alpha \times 10^2$	$\beta \times 10^4$	$\gamma \times 10^6$
Ce^{3+}	3	1/2	5/2	6/7	2.54	2.14	-5.72	63.5	0
Pr^{3+}	5	1	4	4/5	3.58	3.2	-2.10	-7.35	61.0
Nd^{3+}	6	3/2	9/2	8/11	3.62	3.27	-0.643	-2.91	-38.0
Sm^{3+}	5	5/2	5/2	2/7	0.84	0.71	4.13	25.0	0
Eu^{3+}	3	3	0	0	0	0			
Gd^{3+}	0	7/2	7/2	2	7.94	7			
Tb^{3+}	3	3	6	3/2	9.7	9	-1.01	1.22	-1.12
Dy^{3+}	5	5/2	15/2	4/3	10.6	10	-0.635	-0.592	1.03
Ho^{3+}	6	2	8	5/4	10.6	10	-0.222	-0.333	-1.30
Er^{3+}	6	3/2	15/2	6/5	9.6	9	0.254	0.444	2.07
Tm^{3+}	5	1	6	7/6	7.6	7	1.01	1.63	-5.60
Yb^{3+}	3	1/2	7/2	8/7	4.5	4	3.17	-17.3	148.0

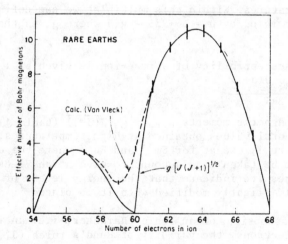

Fig. 1 Effective magnetic moment of trivalent lanthanides.

2.2. Crystal field description

Each RE ions, placed in a crystal, is submitted to an inhomogeneous electrostatic potential coming from the other ions and the conduction electrons in the case of metallic compounds. This contribution to the energy can be divided into a one-ion contribution which corresponds to the crystal electric field (CEF) term and a two-ion contribution which accounts for the dipolar or multipolar interactions between magnetic ions. This last term will be considered in the next section. Here we will consider only a RE ion in a crystal without interaction. The Hamiltonian of a R^{3+} ion containing n 4f electrons can be written as:

$$H = H_o + H_c \qquad (2)$$

H_o is the free ion part and H_c is the CEF contribution which can be parametrized by a tensorial expansion [4] :

$$H_c = \sum_{i=1}^{n} \sum_{k,q} B_{kq} t_q^{(k)}(i) = \sum_{k,q} B_{kq} T_q^{(k)} \qquad (3)$$

k = 2, 4, 6 for 4f electrons. The parameters B_{kq} contain many contributions : simple electrostatic term, overlap, configuration mixing, covalency, etc...

Within a given $|4f^n, \alpha LSJ>$ multiplet, H_c can be expressed by using the Steven's operator equivalent technique [5] :

$$H_c = \sum_{\ell,m} V_\ell^m \theta_\ell O_\ell^m \qquad (4)$$

where $V_\ell^m = A_\ell^m <r_\ell^\ell>$ are the CEF parameters determined from experimental data, $<r^\ell>$ is the ℓth moment of the distribution of the 4f electron cloud [6] and θ_ℓ are the reduced matrix elements of the Stevens operators θ_ℓ^m ($\theta_\ell = \alpha, \beta, \gamma$ for $\ell = 2, 4, 6$). The number of terms in (4) will depend on the symmetry of the RE-site. As example for a cubic symmetry with a quantization axis along a [100] axis the CEF Hamiltonian depends only on two parameters.

$$H_c = V_4^0 \beta_J (O_4^0 + 5O_4^4) + V_6^0 \gamma_J (O_6^0 - 21O_6^4) \qquad (5)$$

According to Lea, Leask and Wolf [7] eq. (5) can be rewritten as :

$$H_c = W\{x \frac{O_4}{F(4)} + (1 - |x|) \frac{O_6}{F(6)}\} \qquad (6)$$

with $Wx = V_4^0 \beta_J F(4)$; $W(1 - |x|) = V_6^0 \gamma_J F(6)$ and $|x| \leq 1$ (7)

Within the ground multiplet (L, S, J), which is the only one of interest in the study of magnetic properties, the value of the coefficient α, β and γ are reported in table 1.

The energy splitting of the ground multiplet $|J, M>$ by the crystal field will depend on both the RE ion and the compound. Usually it is of the order of a few 10^{-2} of Kelvin, thus the CEF is a very important contribution for determining the magnetic anisotropy.

For Kramers ions (odd number of 4f electron), the CEF levels are only doublets except in cubic symmetry where quartets (Γ_8) can occur. For non-Kramers ions the CEF levels are only singlets when the site symmetry is low (C_2, C_3, D_2, D_{2v}), whereas for the other groups both singlets and doublets exist and even triplets in cubic symmetry.

The number of CEF levels is determined only by the symmetry, it is given by the reduction of the representation D_J upon the irreducible representations of the point group of the RE-site.

$$D_J = \sum_\nu p_\nu \Gamma^{(\nu)} \text{ with } \sum_\nu p_\nu d_\nu = 2J + 1 \tag{8}$$

where d_ν is the dimension of $\Gamma^{(\nu)}$. Such decompositions are tabulated in [8].

2.3. Determination of the crystal field parameters

By measuring physical quantities such as the specific heat, the susceptibility or the magnetization in high magnetic field, the CEF-parameters can be determined by fitting the experimental data to the calculated values.

The specific heat per RE-ion is given by

$$\frac{C}{k} = \frac{1}{Z} \sum_{i=1}^{2J+1} (\beta\delta_i)^2 e^{-\beta\delta_i} - \frac{1}{Z^2} (\sum_{i=1}^{2J+1} \beta\delta_i \, e^{-\beta\delta_i})^2 \tag{9}$$

δ_i are the energies of the CEF-levels, $\beta = \frac{1}{kT}$ and Z is the partition function $Z = \sum_i e^{-\beta\delta_i}$.

The susceptibility per RE-ion, for a magnetic field applied along the direction H_α can be written as :

$$\chi_\alpha T = \frac{\mu_B^2 g_J^2}{k} \frac{1}{Z} \sum_i e^{-\beta\delta_i} \left(|M_{ii}^\alpha|^2 + 2kT \sum_j \frac{|M_{ij}^\alpha|^2}{E_j - E_i} \right) \tag{10}$$

where the matrix elements M_{ij}^α are defined as $M_{ij}^\alpha = \langle i|J^\alpha|j\rangle$ with $\alpha = x$, y or z. In the case of an uniaxial symmetry of the RE-site, the anisotropy of the paramagnetic Curie temperature, determined at high temperature, is directly related to the second order CEF-parameter V_2^0 by the relation :

$$\theta_\perp - \theta_{//} = \frac{3(2J-1)(2J+3)}{10 \, k} \alpha V_2^0 \tag{11}$$

These indirect methods are indeed sensitive only when a few CEF-levels are involved (Ce^{3+}, Sm^{3+}, Yb^{3+}). However a direct determination of the energies of the CEF-levels as spectroscopic methods does give more accurate CEF parameters.

In ionic compounds, the optical absorption and fluorescence techniques are well known. In particular to get the CEF-levels of the ground multiplet, fluorescence spectra recorded at low temperature are very useful because the transitions originate from a single excited level. An example is given in figure 2 which shows the spectrum corresponding to the $^5D_4 \rightarrow {}^6F_6$ transitions in Y_2O_2S : Tb^{3+} [9]. In oxysulfides RE-ions are located in a site of C_{3v} symmetry, then the $J = 6$ multiplet of Tb^{3+} is split into 5 singlets and 4 doublets, i.e. in 9 levels which can be clearly seen in figure 2.

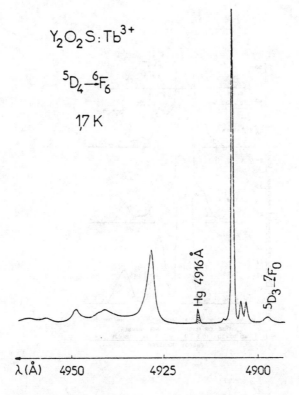

Fig. 2 Fluorescence spectrum of Y_2O_2S : Tb^{3+} at $T = 1.7$ K associated with the $^5D_4 \rightarrow {}^6F_6$ transition.

However, optical methods cannot be used for metallic compounds ; in that case only the neutron spectroscopy can give some
results. This technique is analogous far infrared spectroscopy
since the energy of the incident neutrons can vary from 10 K to
1000 K. An example is given in figure 3 which shows the spectrum
obtained with the intermetallic compound ErRh [10] which has a
cubic structure of CsCl-type. This method is particularly convenient for a cubic symmetry because only two CEF-parameters have
to be determined. For a lower symmetry the problem becomes more
complex because the number of transitions which can be seen are
not, very often, large enough to get a set of CEF parameters
without any ambiguity. Therefore it is needed to combine both
direct and indirect methods. Indeed more information can be
obtained by using a single crystal because selection rules then
be used for the determination of the irreducible representations
associated to the CEF levels.

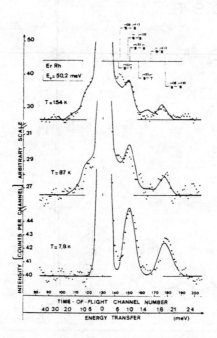

Fig. 3 Neutron time of flight spectra of the compound ErRh.

3.   INTERACTION BETWEEN RARE EARTH IONS

The interaction mechanism between two localized spins, in RE-compounds, cannot be the famous Heisenberg-type direct exchange interaction - $2I_{mn} \vec{S}_m \cdot \vec{S}_n$ because the exchange integral is too small. This exchange integral is due to the Coulomb interaction which is expected to originate from the overlap of two atomic 4f-orbitals on different sites and this f - f overlapping is negligible.

Therefore all interactions must be indirect ones, via either through the p-anion orbitals or the conduction electrons, depending on the ionic or metallic character of the compound.

3.1. Metallic compounds

In metallic compounds it is well known that a strong exchange interaction exists between the localized 4f and the conduction electrons. As the conduction electrons have mainly a d-character, this interaction corresponds indeed to a d-f intra-atomic exchange mechanism which can be expressed as [11] :

$$H_{df} = - I_{df} \vec{S}_d \cdot \vec{S}_f + \sum_{K=1}^{4} \left[A_K - B_K (1/2 + \frac{1}{S_f} \vec{S}_d \cdot \vec{S}_f) \right] U_d^k U_f^k \qquad (12)$$

Usually only the scalar product part is taken into account because conduction electrons are considered to have a S-like character. However for light rare earths as Ce, Pr or Nd the anisotropic part may give a larger contribution. However in the following we will restrict ourselves to the isotropic part

$$H_{df} = - I_{df} \vec{S}_d \cdot \vec{S}_f = - I_{df} (g_J - 1) \vec{S}_d \cdot \vec{J}_f \qquad (13)$$

Equation (13) gives a ferromagnetic aligment of the spins of conduction and 4f electrons ($I_{df} \sim 0.1$ eV) but the polarization of the conduction electrons can be either parallel or antiparallel to the 4f magnetic moment ($\vec{\mu} = g_J \vec{J}$) depending on whether g_J is smaller (light RE) or larger (heavy RE) than unity (see table 1).

Such an interaction leads, in second order perturbation theory, to the famous RKKY (Ruderman - Kittel - Kasuya - Yosida) indirect interaction [12]

$$H_{mn} = J(\vec{R}_m - \vec{R}_n) \vec{S}_m \cdot \vec{S}_n = (g_J - 1)^2 J(\vec{R}_m - \vec{R}_n) \vec{J}_m \cdot \vec{J}_n \qquad (14)$$

$J(\vec{R}_m - \vec{R}_n)$ is a long range and oscillating function $F(2k_F|\vec{R}_m - \vec{R}_n|)$, such as $F(x) = \dfrac{x\cos x - \sin x}{x^4}$, which may give rise to very complex magnetic structures as will be seen in section 5. This interaction is stronger for RE ions of the middle of the RE series, as Gd^{3+}, because they have the largest spin value. The exchange integral, then, follows the famous de Gennes law in $(g_J - 1)^2$ [2].

In metallic compounds we can have a mixing between f and conduction electrons, this d-f mixing will be important when the energy Δ to promote a f-electron into the d-band is low. This term of the order of $|V_{df}|^2/\Delta$ (~ 0.01 eV) gives an antiferromagnetic coupling ; but is less important than the d-f exchange interaction (0.1 eV) except in anomalous RE compounds where the 4f level is close to the Fermi energy, as we will see in section 7.

3.2   Semiconductors and insulators

In this case there are no conduction electrons and then only the mixing term remains. Indeed f-electrons can mix either with the p-anion orbitals or the d (or s) cation orbitals of the neighbouring RE ion. In both cases the interactions are of the indirect type, the former one is very often referred as the superexchange mechanism.

The usual superexchange mechanism, as described by Anderson [13] for 3d compounds, corresponds to a virtual transfer of a 4f-electron via the p-state of the anion from one site to another one, and then produces a pair of ions, R^{2+} and R^{4+} for a R^{3+} ion or R^+ and R^{3+} for a R^{2+} ion. However in Eu chalcogenides this mechanism was found by Kasuya [14] to be an order of magnitude too small to account for the experimental data because of the small overlap between the p and 4f states as can be shown in figure 4.

Therefore Kasuya [14] proposed two other possible mechanisms. The first one is superexchange via the d-f exchange interaction. It corresponds to a virtual transfer of an anion p-electron to the 5d (or 6s) state of the neighbouring R^{3+} cation which aligns the 4f spin through the large d-f (or s-f) exchange interaction. This p-d mixing mechanism is more important than the p-f one because the overlap between p and 5d states is larger. This mechanism leads also to an antiferromagnetic coupling which explains quite well the negative coupling between n.n.n. in EuSe or EuTe for the 180 degree configuration.

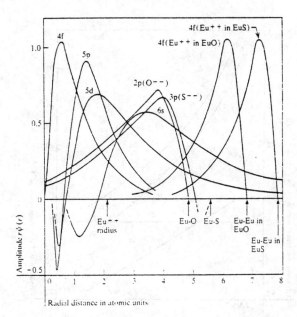

Fig. 4   Atomic wave functions for Eu^{2+}, O^{2-} and S^{2-}. Lattice distances between nearest neighbour Eu and chalcogen ions are indicated by arrows. Wave functions are drawn with the respective origins at these points to show the overlap of anion-cation and cation-cation functions.

However, a cross term between the Anderson (4f → p → 5d transfers) and the d-f exchange mechanism may overcome the above mechanism because it leads to a ferromagnetic coupling, this mechanism can explain why the n.n.n. exchange integral in EuO is positive (figure 5).

We shall now describe the indirect exchange mechanism in which the cation wave functions have the more important role whereas the anion p-bands are relatively unimportant. Since the overlap between cation wave functions is more significant between n.n. cations, the main contribution to this mechanism will be for nearest neighbour interactions. This process corresponds to a virtual transfer of a f-electron to a 5d-state which then experiences the usual d-f exchange interaction. This transfer can be induced only by a phonon excitation (optical phonons), so the exchange constant I_1 has the same sign as I_{df}, i.e. is positive and is a decreasing function when the temperature is cooled down. Indeed in EuO Kasuya [14] has shown that this mechanism yields too weak interaction.

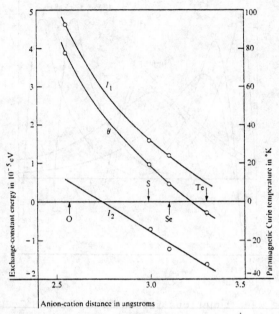

Fig. 5 The exchange constants I_1 and I_2 and the paramagnetic Curie temperature θ of Eu chalcogenides as a function of the anion-cation distance.

Another way to transfer a 4f electron on to the neighbouring 5d orbital is via the f-d mixing energy due to the direct overlap, then again the d-electron experiences the strong d-f exchange interactions. The exchange integral has also the same sign as I_{df} and gives a ferromagnetic coupling which explains why the semiconducting compound EuO (T_c = 69.3 K) is a ferromagnet whereas EuTe (T_N = 9.6 K) is antiferromagnetic. This interaction mechanism based on the d-f mixing decreases very quickly with the RE-RE distance, nearly exponentially. The variations of the exchange integrals for Eu chalcogenides are given in figure 5.

All these exchange interaction mechanisms lead to an ordering temperature which is of the order of 10 K for semiconducting materials and only of a few Kelvin in more ionic compounds such as RE oxides or oxychalcogenides. When the RE-RE distance becomes large, as in RE gallium garnets for example, this contribution becomes negligible.

3.3. Dipolar interactions

In RE ionic compounds and especially for heavy RE as Tb, Dy, Ho, the dominant interaction is indeed the classical interaction between the dipolar magnetic moments.

$$H_{mn} = \frac{\vec{\mu}_m \cdot \vec{\mu}_n}{r_{mn}^3} - 3 \frac{(\vec{\mu}_m \cdot \vec{r}_{mn})(\vec{\mu}_n \cdot \vec{r}_{mn})}{r_{mn}^5} \tag{14}$$

This dipolar interaction is very anisotropic and can vary between 0.1 K to 10 K depending on the RE magnetic moment and the RE-RE distance.

Moreover, due to the non-spherical shape of the 4f cloud large magneto-elastic phenomena and quadrupolar interactions can be present which may induce structural phase transitions in ionic as well as in intermetallic compounds.

4.   MAGNETIC ANISOTROPY

In this section we want to emphasize the importance of the relative strength of the crystal field (H_c) and the magnetic interactions (H_{ex}) in determining the anisotropy, i.e. the moment direction. This question is purely a one ion approach which will depend very much on the relative ratio of H_c and H_{ex}.

4.1. Case $H_{ex} > H_c$

We will consider first the case where $H_{ex} > H_c$; this situation occurs when the exchange interaction is very large as in RE-intermetallics with 3d elements as cobalt (RCo_3, RCo_5) or iron (RFe_2), or when the CEF splitting is very weak as in gadolinium (or europium) compounds in which the Gd^{3+} (or Eu^{2+}) ions are in an S-state. In this latter case the CEF splitting is given by a second order perturbation theory.

In that case H_c can be treated as a perturbation in comparison with H_{ex}, then we get only an anisotropy of the energy which can be expressed by the classical formulation as :

$$E_A = K_1 \sin^2\theta + K_2 \sin^4\theta + \ldots \tag{15}$$

There is no anisotropy of the magnetic moment at the first order of perturbation, this means that the magnetic moment reachs its maximum value ($g_J J$) for a magnetic field applied along any crystal direction.

We can show that eq. (15) is valid only in the limit $H_c < H_{ex}$, in that case if $\vec{u}$ corresponds to the moment direction, the anisotropy energy at $T = 0$ is given by :

$$E_A = \langle J_u = J | H_c^u | J_u = J \rangle \tag{16}$$

H_c^u is the CEF Hamiltonian written with the quantization axis taken along $\vec{u}$. If H_c^z is the CEF Hamiltonian written with the quantization axis along the principal axis of the CEF, we have

$$H_c^z = \sum_{\ell m} V_\ell^m \theta_\ell Y_\ell^m(\theta,\Phi) \tag{17}$$

and
$$H_c^u = \sum_{\ell,m} V_\ell^m \theta_\ell \sum_{m'} Y_\ell^{m'}(\theta,\Phi) \langle m' | D_\ell(\theta,\Phi) | m \rangle \tag{18}$$

where $D_\ell(\theta,\Phi)$ is the representation of the rotation group. Then :

$$E_A = \sum_{\ell,m} V_\ell^m \theta_\ell \langle J_u = J | Y_\ell^0 | J_u = J \rangle \langle 0 | D_\ell(\theta,\Phi) | m \rangle \tag{19}$$

with $\langle 0 | D_\ell(\theta,\Phi) | m \rangle = P_\ell^m(\theta) \ell^{im\Phi}$ (20)

where $P_\ell^m(\theta)$ is the Legendre polynomial.
For example, the second order term gives :

$$\langle J_u = J | Y_2^0 | J_u = J \rangle \langle J | 3J_u^2 - J(J+1) | J \rangle = 2J(J+1)$$

Then

$$E_A = 2\alpha V_2^0 \ J(J+1)(3\cos^2\theta - 1) \tag{21}$$

Eq. (21) can be rewritten with the classical form given in (15)

$$E_A = K_1 \sin^2\theta$$

At finite temperatures, the thermal variation of the anisotropy coefficients is given by the Callen and Callen theory [15].

$$K_n(T) = K_n(0) \left(\frac{M(T)}{M(0)}\right)^{\frac{\ell(\ell+1)}{2}} \tag{22}$$

So the easy magnetization axis depends on the crystal structure via the CEF parameter V_2^0 and the number of 4f electrons via the α-coefficient. In particular in a given structure, the easy direction of Tb^{3+}, Dy^{3+} and Ho^{3+} will be perpendicular to that for Er^{3+} and Tm^{3+}. It corresponds to the case of RE metals in which Tb, Dy and Ho have their moments in the basal plane whereas for Er and Tm they are along the c-axis [16]. This simple picture allows us to explain quite well the moment direction in RE inter-metallics such as $RECo_3$ [17] (T_c = 506 K for $TbCo_3$) and $RENi_3$ [18] (T_c = 116 K for $GdNi_3$) in which there are two uniaxial RE-sites but with an opposite anisotropy. However on the hexagonal site the anisotropy is much larger than on the pseudo-cubic site. Therefore in the $RECo_3$ ferromagnets, due to the large RE-Co interaction, moments are aligned along the easy axis of the hexa-gonal site (the c-axis for Er, and the basal plane for Tb, Dy and Ho). In $RENi_3$ the exchange interactions are reduced to the RE-RE interactions because Ni is not magnetic, then the ordering is non-collinear. The moments on each site remain along their own easy axis with nevertheless a small canting of the moments on the cubic site towards those on the hexagonal site.

4.2. Case $H_{ex} < H_c$

This situation is the usual case for RE-ionic compounds, but it also occurs very often in RE-intermetallics which have a low ordering temperature (e.g., smaller than 20 K). We must define two parameters : Δ which is the total CEF-splitting and δ which is the energy of the first excited CEF level. A simple and more common situation corresponds to the case where the magnetic interactions are lower than δ ; then the Hamiltonian H_{ex} can be projected on the crystal field ground level.

In the case of Kramers ions when the ground state is a doublet we can use an effective spin S = 1/2 and define an anisotropic g-tensor such as $\vec{\mu} = \bar{\bar{g}} \vec{s}$ (in μ_B), then the most general form for H_{ex} can be written as

$$H_{ex} = -\sum_{m,n} \vec{s}_m \bar{\bar{K}}^{m,n} \vec{s}_n \qquad (23)$$

where the tensor $\bar{\bar{K}}$ contains both the CEF anisotropy via the g-tensor and the anisotropy of the interactions.

For non-Kramers ions the situation depends very much on the crystallographic structure by determining the symmetry of the RE site. In particular when the site symmetry is low, as in the intermetallic compounds RNi [19] or RSi [20] or in the ionic compounds as the RE-oxides (Tb_2O_3) [21], RE-perovskites [22] or RE-garnets [23], the CEF-levels are only singlets. If the ground state is a singlet, no magnetic ordering will take place ; indeed very often the ground state is composed of two singlet levels $|0_c>$ and $|1_c>$. In the sub-space spanned by $|0_c>$ and $|1_c>$ the magnetic moment has only one non vanishing matrix element, $\alpha = <0_c|J_u|1_c>$ along the u-direction. Then we get an Ising-like anisotropy and the magnetic ordering will develop only if the ratio between the interaction energy and the energy separation of the two singlets is larger than a critical value [24]. The results for a ferromagnetic ordering is given in figure 6 where is reported the dependence of T_c and the reduced moment $<J_u>/\alpha$ in function of the parameter $A = 4J(0)\alpha^2/\Delta$; in that case the critical ratio corresponds to $A = 1$. However the critical ratio disappears if the hyperfine interactions are taken into account because they introduce a mixing between the two singlets which always gives rise to a magnetic ordering. This ordering takes place in general at very low temperature (much smaller than one Kelvin) depending on the strength of the hyperfine coupling and how far we are from the critical ratio. Such a nuclear and electronic coupled ordering has been studied in ionic compounds as RE-garnets (TbGaG and HoGaG [25]) or in intermetallics such as $PrCu_5$ [26] or $PrNi_2$ [27].

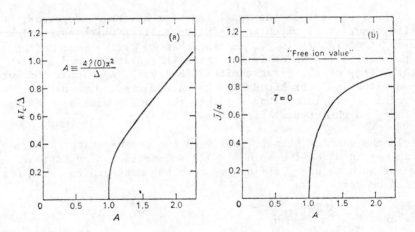

Fig. 6 (a) T_c/Δ vs. A for the two singlet level model. (b) Spontaneous magnetization (in dimensionless form) vs the parameter A (= $4J(0)\alpha^2/\Delta$) at T = 0.

The anisotropy of the magnetic moment is usually very large, except for the Γ_7 doublet in a cubic symmetry, and will depend on the value of matrix elements of the form $<\Gamma^{(\alpha)}|J_u|\Gamma^{(\beta)}>$ where $\Gamma^{(\alpha)}$ and $\Gamma^{(\beta)}$ are the irreductible representations according to which the ground state wave functions transform. These matrix elements will differ from zero only if the decomposition of the direct product $\Gamma^{(\alpha)} \times \Gamma$ contains the representation $\Gamma^{(u)}$

according to which the component J_u (u = x, y, z) is transforming. In table 2 are reported the irreducible representations and basis vectors for a few point groups. For a Kramers doublet we cannot predict the moment direction because it depends on the particular value of the CEF parameters via the g-tensor, except in a C_3 symmetry when the doublet has a wave function of the type $|^1\Gamma_5 + {}^1\Gamma_6\rangle$ which leads to an Ising anisotropy along the three fold axis.

Table 2. Irreducible representations and basis vectors for some point groups.

C_s, C_2		D_2, C_{2v}		C_{3v}		C_4, D_{2d}	
$^1\Gamma_1$	J_z	$^1\Gamma_1$		$^1\Gamma_1$		$^1\Gamma_1$	
$^1\Gamma_2$	J_x or J_y	$^1\Gamma_2$	J_y	$^1\Gamma_2$	J_z	$^1\Gamma_2$	J_z
		$^1\Gamma_3$	J_z	$^2\Gamma_3$	J_{xy}	$^1\Gamma_3$	
		$^1\Gamma_4$	J_x			$^1\Gamma_4$	
						$^2\Gamma_5$	J_{xy}
$^1\Gamma_3 + {}^1\Gamma_4$	$|\pm\frac{1}{2}\rangle$	$^2\Gamma_5$	$|\pm\frac{1}{2}\rangle$	$^2\Gamma_4$	$|\pm\frac{1}{2}\rangle$	$^2\Gamma_6$	$|\pm\frac{1}{2}\rangle$
				$^1\Gamma_5 + {}^1\Gamma_6$	$|\pm\frac{3}{2}\rangle$	$^2\Gamma_7$	

For non-Kramers ions a knowledge of the irreducible representations allows us to determine the moment direction. In particular in compounds with a low symmetry the moment is quenched along a well defined direction determined by the product $\Gamma^{(\alpha)} \times \Gamma^{(\beta)}$. For a site symmetry C_s or C_2 (RE-perovskites or RE-oxydes) the moments will be aligned along a direction within the mirror plane or pendicular to it, whereas for a C_{2v} or D_2 symmetry (RESi or RE-garnets) the moments will be always aligned along one of the three two-fold axes.

However, in compounds with a C_s or C_2 symmetry it may be possible to fix the moment direction within the mirror plane because for such a low symmetry the second order CEF Hamiltonian is usually predominant :

$$H_c = \alpha(V_2^0 O_2^0 + V_2^2 O_2^2 + V_2^{-2} O_2^{-2}) + 4^{\text{th}} \text{ order terms} + 6^{\text{th}} \text{ order terms} \tag{24}$$

By a rotation around the quantization axis (perpendicular to the mirror) it is possible to cancel the CEF coefficient V_2^{-2}, the rotation angle θ being determined by the relation

$tg m\theta = \dfrac{V_{\ell}^{-m}}{V_{\ell}^{m}}$ [28]. In these new (OXYZ) axes the Hamiltonian has a

pseudo C_{2v} symmetry and the moment must be along the X or Y-axis according to the nature of the RE ion. In RE-perovskites or RE-oxydes [28] and in intermetallic compounds with the FeB-type structure as RNi [19] or RSi [20] we got a very good agreement with the experimental angle value by computing the ratio $V_{\ell}^{-m}/V_{\ell}^{m}$ by a point charge model. In figure 7 we report the results for the RNi compounds. In the magnetic cell which is the same as the nuclear one there are four Bravais lattices (figure 7a) and the local axes are related by the symmetry elements (two fold axis along OY and inversion) as indicated in figure 7b. Then the moments lie along the X-axis (DyNi-type) when the α-parameter is positive (Tb, Dy, Ho) or along the Y-axis (ErNi type) when $\alpha < 0$ (Er, Tm) (figures 7c, 7d). Therefore, while the interactions are strongly ferromagnetic, the CEF anisotropy imposes a non collinear magnetic structure. In the case of RE-perovskites and RE-oxydes the agreement with experiment is also very good [28], the magnetic structure which looks very complex is indeed simply determined by the symmetry elements of the crystallographic structure and a strong CEF anisotropy.

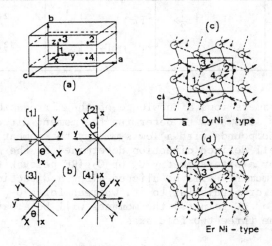

Fig. 7 Non-collinear magnetic structures in FeB-type compounds RNi. (a) and (b) local symmetry of the four Bravais lattices. (c) and (d) magnetic structures.

In fact not only in low symmetry compounds but in any RE-compounds in which the magnetic interactions are smaller than the energy δ of the first excited CEF-level, the magnetic behaviour is dominated by the CEF anisotropy.

Another interesting situation corresponds to the case where the ground state, instead to be a doublet or two singlets, is composed by ·a set of closely separated CEF levels which are well separated from the other excited levels. This situation can be described by the CEF level band model introduced by Trammell [29] to explain the unusual crossing of CEF levels in the Lea, Leask and Wolf [7] diagrams for the cubic symmetry. This model was also applied in RE compounds like DyVO$_4$ to explain the cooperative Jahn Teller transition [30]. In this paper we want to present the consequence of such a situation in RE-oxysulfides [31] as Tb$_2$O$_2$S [32]. In this compound it was shown [9] (see figure 2) that the ground multiplet $^6F_6$ is split by the CEF of C$_{3v}$ symmetry into 9 levels (5 singlets and 4 doublets) such as :

$$D_6 = 3^1\Gamma_1 + 2^1\Gamma_2 + 4^2\Gamma_3 \tag{25}$$

Optical experiments give effectively a set of levels close to the ground state with the energies : 0, 6, 15, 26, 106, 157, 188, 220 and 250 cm^{-1}. This CEF band, with a width of only 26 cm^{-1}, contains four levels, two singlets $^1\Gamma_1$ and $^1\Gamma_2$ and two doublets $^2\Gamma_3$ giving a six-fold degeneracy.

We can follow the idea proposed by Trammell in 1963 [33], which considers that a semi classical approach can be a good starting point for heavy lanthanides which have a large quantum number : J = 6 for Tb or Tm, J = 17/2 for Dy and Er, J = 8 for Ho. Then in the expression of the CEF Hamiltonian we can replace the operator $\vec{J}$ by a classical vector $\vec{J}(J, \theta, \phi)$. The CEF-parameters being known, we are able to compute the classical CEF energy in function of the angles θ and ϕ ($E_c(\theta, \phi)$). In Tb$_2$O$_2$S the energy E_c is minimum when $\vec{J}$ points along a direction u making an angle θ of about 50° with the C-axis and contained in a (b, c) plane (b being the orthohexagonal axis of the hexagonal unit cell). This easy axis direction is in fact determined by the main CEF parameter V_4^3 which leads to the classical energy of the form

$$E_4^3 = \beta\, V_4^3 J^4\, \cos\theta\, \sin^3\theta\, \cos3\phi \tag{26}$$

Taking into account the three-fold symmetry axis we get indeed six easy directions : $\pm\vec{u}_1$, $\pm\vec{u}_2$ and $\pm\vec{u}_3$ as shown in figure 8. Then the classical ground state is effectively 6-fold degenerate in agreement with optical experiments.

However, the classical description is not good enough because $\vec{J}$ is indeed an operator and these six states are not eigenstates of the system. As these levels are well separated from the excited levels we can, in a first approximation, write the CEF Hamiltonian within the basis spanned by the six states $|J, \vec{u}_1\rangle$, $|J,\vec{u}_2\rangle$, ..., $|J,\vec{u}_6\rangle$ such as $\vec{J}.\vec{u}_i |J,\vec{u}_i\rangle = J|J, \vec{u}_i\rangle$. These states are indeed not orthogonal and the overlap integral $\langle J, \vec{u}_i|J, \vec{u}_j\rangle$

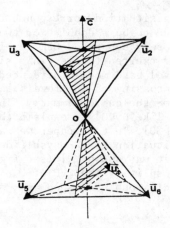

Fig. 8 Equivalent directions defining the six basic states of
the CEF band of Tb^{3+} in Tb_2O_2S.

gives the probability for a state $|J, \vec{u}_i>$ to tunnel to a state
$|J, \vec{u}_j>$. This tunneling effect splits the six fold degenerated
ground state to give a crystal field band. The number of levels
and their symmetry is determined by the reduction, upon the irreducible representations, of the representation $^{(6)}\Gamma$ induced by
the transformation of these six states with the symmetry elements.
We get

$$^{(6)}\Gamma = {^{(1)}}\Gamma_1 + {^{(1)}}\Gamma_2 + 2{^{(2)}}\Gamma_3 \qquad (27)$$

i.e. exactly the same levels as the direct quantum calculation.
The tunneling probability $<J, \vec{u}_i | J, \vec{u}_j>$, being proportional to
$(\cos \frac{\beta}{2})^{2J}$ (with $\beta = (\vec{u}_i, \vec{u}_j)$), is zero when $J \to \infty$, i.e. for the
classical limit, and for a finite J value it increases when the
β angle becomes smaller.

      The eigenstates and the eigenvectors are given by the dia-
gonalisation of the CEF matrix $H'_c = S^{-1/2} H_c S^{-1/2}$ where S is the
overlapping matrix. A very good agreement between experimental
(0, 6, 15, 26 cm^{-1}) and calculated (0, 6.3, 11.5, 31.8 cm^{-1})
values is obtained by taking an angle $\theta = 45°$ which is very close to the value $47 \pm 2°$ obtained by neutron experiments [32].
This good agreement confirms that in Tb_2O_2S the CEF anisotropy
leads to an easy axis which makes an angle with the principal
axis, so the Tb^{3+} states are characteristic of a six-state Potts
model.

      At high temperature the angular moment $\vec{J}$ precesses around
the $\vec{u}_i$-direction and jumps from $\vec{u}_i$-direction to another by tunneling effect. Whereas at low temperature (below T_N) the moments are

quenched along a well defined $\vec{u}_i$-direction, then the crystal undergoes a triclinic distortion. If the ordering temperature is low enough a cooperative Jahn Teller transition can occur as in RE-compounds of zircon type.

Therefore when the ground level is composed by several CEF levels it indicates that the CEF easy axis is not along the symmetry axes. These considerations allow us to understand why, in RE compounds with a low symmetry, the heavy RE-ions (Tb, Dy, Ho, Er, Tm) have very often a doublet or a pseudo-doublet as the ground state. The angular momentum being large and the semi-classical theory being a good approximation, the ground state must be doubly degenerate when the moment is along the two-fold symmetry axes or within the symmetry planes. These two states (e.g., $|J, \vec{u}_1\rangle$ and $|J, -\vec{u}_1\rangle$), being opposite, do not overlap and then the splitting of the ground state in two singlets will be very small if the excited levels are located at much higher energies.

4.3. Case $H_{ex} \sim H_c$

When $H_{ex} \cong H_c$, no approximation can be done and the complete Hamiltonian $H_{ex} + H_c$ must be diagonalized. The problem is then much more complex, but qualitative results can be obtained by using also a semiclassical description. By decreasing the temperature the population of excited levels decreases and a moment rotation may occur due to the competition between the CEF anisotropy and the entropy ; such a moment rotation has been observed for example,in HoAl$_2$ [34].

In fact to treat properly the problem we can proceed as follows. The crystal Hamiltonian will be taken as :

$$H = \sum_n H_{c_n} - \sum_{nm} J_{nm} \vec{J}_n \vec{J}_m \tag{28}$$

and we can start with a ground state $|\phi_0\rangle = \prod_n |\phi_{0n}\rangle$ where the product is over the various RE-ions and $|\phi_{0i}\rangle$ is the ground state eigenvector for the nth RE-ion which will be given by

$$\left[H_{cn} - 2 \sum_n J_{nm} \langle \vec{J}_m \rangle_0 \cdot \vec{J}_n \right] |\phi_{0n}\rangle = e_{0n} |\phi_{0n}\rangle \tag{29}$$

with $\langle \vec{J}_m \rangle_0 = \langle \phi_{0m} | \vec{J}_m | \phi_{0m} \rangle$

Eq. (29) corresponds to diagonalize the one ion Hamiltonian by using a molecular field approximation. Then the ground state energy is :

$$\langle\phi_0|H|\phi_0\rangle = \sum_n e_{0n} + \sum_{nm} J_{nm} \langle\vec{J}_m\rangle_0 \langle\vec{J}_n\rangle_0 \tag{30}$$

We can rewrite the total Hamiltonian H as :

$$H = H_0 + H_1 \tag{31}$$

where

$$H_0 = \sum_n H_{cn} - 2\sum_m J_{nm} \langle\vec{J}_m\rangle_0 \cdot \vec{J}_n + \frac{1}{2} \sum_{nm} J_{nm} \langle\vec{J}_m\rangle_0 \cdot \langle\vec{J}_n\rangle_0 \tag{32}$$

and

$$H_1 = - \sum_{nm} J_{nm} (\vec{J}_n - \langle\vec{J}_n\rangle_0) \cdot (\vec{J}_n - \langle\vec{J}_n\rangle_0) \tag{33}$$

H_0 is the static part which is diagonal in the states

$$|\phi_{p_1,p_2\ldots}\rangle = \prod_n |\Phi_{p,n}\rangle$$

where $|\Phi_{p,n}\rangle$ are the eigen states given by (27).
H_1 is the dynamical part which describes the dynamics of the system, i.e. a generalization of spin wave excitations. A more extensive analysis will be done in section 9.

In that case it is not possible to get a general trend, the magnetic properties will depend very much on the strengh of both the CEF and exchange terms. The symmetry of the RE site remains always an important parameter, but a large variety of anomalous magnetic behaviours can occur.

5.   MAGNETIC ORDERING

In this section we want to give a general view of the different kinds of magnetic structures and how to characterize them. The crystallographic structure is very important by the repartition of the RE ions in the nuclear cell which determines the dimensionality of the magnetic ordering (1D, 2D or 3D) and by the symmetry elements of the nuclear cell which limit the number of possible magnetic structures.

Before starting a magnetic structure determination it is important to identify the number of distinct magnetic sites n_s and for each site the number of equivalent positions which defines the number of Bravais lattices n_b associated to this particular site.

5.1. Neutron scattering and magnetic structure

The main tool to determine the arrangement of magnetic moments is the neutron diffraction technique. We must keep in mind
that a neutron experiment allows to determine only the value and
the direction of the Fourier components $\vec{m}_k^i$ and not the moment
distribution $\vec{m}_n^i$ which indeed must be deduced from the following
relation

$$\vec{m}_n^i = \sum_k \vec{m}_k^i \, e^{2\pi i \vec{k} \cdot \vec{R}_n} \tag{34}$$

where $\vec{R}_n$ is a lattice translation and $i = 1, \ldots, n_b$ labels the
Bravais lattices. In practice, a determination of a magnetic
structure needs first the identification of the wave vector of
each Fourier component by interpreting the Bragg angle of the
magnetic peaks and in a second step the magnetic coupling of the
Bravais lattices is obtained from Bragg peak intensities. It is
an easy task when there is only one Bravais lattice, but it becomes more tedious with several ones and when several sites exist.
Therefore considerations on phase transitions and group theory
must be taken into account.

According to Landau a transition is characterized by a
breaking of the symmetry, and the order parameter associated with
a second order phase transition transforms as an irreducible representation of the symmetry group G_0 of the paramagnetic phase.
For a space group G_0, each irreducible representation is classified according to a wave vector $\vec{k}$ of the Brillouin zone of the
Bravais lattice and to an index v. This index labels the irreducible representation Γ_k^v of the group G_k which leaves invariant
the wave vector $\vec{k}$. Indeed the representation $\Gamma_{\{k\}}^v$ must be considered as the direct sum of the representations Γ_k^v over the q-mem-
bers of the star of $\vec{k}$. Then the order parameter has $d_v \cdot q$
components, where d_v is the dimension of Γ_k^v and each component
is a linear combination of $\vec{m}_k^i$; for one Bravais lattice it reduces to $\vec{m}_k$. At the ordering temperature T_0 only one $\Gamma_{\{k\}}^v$ becomes
critical, so the magnetic structure will be characterized by the
parameters $(\vec{k}, v)$ and will be defined by the d_v components of the
order parameter $\Psi(\vec{k}_i, v)$. If this is not the case there exist
fourth order terms which induce a first order transition. So the
magnetic coupling of the Fourier components $\vec{m}_k^i$ is obtained, according to Landau theory, in such a way that all components $\Psi_1'(\vec{k}, v')$
are equal to zero except those transforming as the critical
representation Γ_k^v. The only problem which remains is the determination of the component $\Psi_1(\vec{k}, v)$. This can be solved by using a
method developed by Bertaut [35] based on group theory. It consists of introducing a representation Γ of G_k by the $3n_b$ Fourier
components $m_{k,\alpha}^i$, to reduce it on the irreducible representations

Γ_k^v and to determine the basis vectors of these representations using the projection operator method.

5.2. Collinear magnetic structures

Simple collinear magnetic structures are characterized by a wave vector $\vec{k}$ which corresponds to a symmetry point of the Brillouin zone, i.e. $\vec{k} = \vec{H}/2$ where $\vec{H}$ is a reciprocal lattice vector, and a magnetic site of high symmetry. However the long range nature of the RKKY interactions can lead to more complex ordering as we shall see in section 5. In figure 9 are reported the Brillouin zones of lattices which occur the most frequently in RE-compounds.

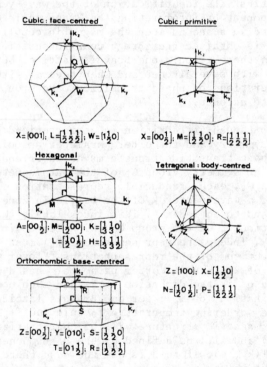

Fig. 9 Brillouin zones.

5.2.1. Cubic lattices

In a face centred cubic lattice there are two symmetry points corresponding to wave vectors $k = [001]$ (type I), $k = [\frac{1}{2} \frac{1}{2} \frac{1}{2}]$ (type II). Two large series crystallise with a f.c.c. lattice: the Laves-phase compounds RM_2 (M = Al, Fe, Co, Ni, Os,

Ir, Pt, Ru, Rh) and the monopnictides and monochalcogenides
RX (X = N, P, As, Sb, Bi, S, Se, Te). The compounds of the first
group are ferromagnets [13], except $CeAl_2$ [36], which exibits an
incommensurate structure (see section 7). The compounds of the
second group order mainly with a type II structure [1] except
cerium and neodymium monopnictides which have a type I structure.
In both cases the two irreducible representations of the group
G_k (D_{3d} or D_{4h}) define two sets of order parameters. For
$\vec{k} = [\frac{1}{2} \frac{1}{2} \frac{1}{2}]$, as an example, they are either $m_k^{\parallel}$, i.e. a moment
direction along <111> (TbX) or $m_k^{\perp}$, i.e. a moment direction per-
pendicular to <111> (ErX). But a collinear structure with a moment
direction along <100> (HoX, DyX) implies that a 4th order term
exists in the energy which couples the two irreducible representations.

In a primitive cubic lattice the Brillouin zone has three
symmetry points :

$$\vec{k}_1 = [00\frac{1}{2}], \quad \vec{k}_2 = [\frac{1}{2} \frac{1}{2} 0] \text{ and } \vec{k}_3 = [\frac{1}{2} \frac{1}{2} \frac{1}{2}]$$

The last one has a cubic symmetry while for the first two points
the symmetry is quadratic (D_{4h}). This implies that there are also
two kinds of order parameters : $m_k^{\parallel}$ and $\vec{m}_k^{\perp}$ (with respect to the
tetragonal axis). Two large series belong to this system : the
equiatomic rare earth intermetallic compounds RM (M = Cu, Ag, Au,
Zn, Al, Rh, etc...) with the CsCl structure and the compounds
RM_3 (M = In, Pb, Sn, Pd, Tl) with the $AuCu_3$ structure [1]. In the
RM group, RZn compounds are ferromagnets whereas RCu and RAg
compounds order antiferromagnetically with $\vec{k} = [\frac{1}{2} \frac{1}{2} 0]$ and $\vec{m}_k$ parallel to the tetragonal axis. But in HoCu $\vec{m}_k$ is along <111> [37]
RRh [38] and RMg [39] compounds order with $\vec{k} = [00\frac{1}{2}]$ and $\vec{m}_k$ is
parallel to $\vec{k}$, but in RMg complex magnetic structures have been
observed involving two Fourier components ($\vec{k} = [00\frac{1}{2}]$ and $\vec{k} = 0$).
The different values of the wave vector can be explained qualita-
tively by a RKKY model which leads to the sequence $\vec{k} = [00\frac{1}{2}]$,
$\vec{k} = [\frac{1}{2} \frac{1}{2} 0]$ and $\vec{k} = 0$ when the number of conduction electrons
increases. However some exceptions exist : Ce, Pr, NdZn are anti-
ferromagnets ($\vec{k} = [00\frac{1}{2}]$), Gd, Tb, DyRh are ferromagnets and RMg
compounds have a peculiar behaviour.

5.2.2. Hexagonal lattice

In the hexagonal symmetry there exist three symmetry points
defining two kinds of magnetic structures : hexagonal ($\vec{k} = 0$,
$\vec{k} = [00\frac{1}{2}]$), orthohexagonal ($\vec{k} = [\frac{1}{2}00]$, $\vec{k} = [\frac{1}{2}0\frac{1}{2}]$).

A large number of RE intermetallic compounds with Co, Ni or
Fe have a hexagonal structure, they order usually with a ferro
or ferrimagnetic structure [1]. All kinds of antiferromagnetic
structures have been observed in the heavy RE oxysulfides [31]:
a hexagonal structure in Yb_2O_2S ($\vec{h}$ = [00 1/2]), an orthohexagonal
magnetic cell in Ho_2O_2S ($\vec{k}$ = [1/2 00]) which is doubled along
the c-axis in Gd, Tb, Dy_2O_2S ($\vec{k}$ = [1/2 0 1/2]).

5.2.3. Body centered tetragonal lattice

This lattice has three symmetry points : $\vec{k}_1$ = [001] or [100],
$\vec{k}_2$ = [1/2 1/2 0], $\vec{k}_3$ = [1/2 0 1/2].

The rare earth intermetallic series RM_2 (M = C, Ag, Au) [1]
with the CaC_2-type structure and RCu_4Al_8 [1] with the $ThMn_{12}$-type
structure are good examples. The compounds of the first group
order at low temperature with a wave vector $\vec{k}$ = [1/2 1/2 0], but
with increasing temperature a first order transition occurs
towards a transverse sine wave modulation with $\vec{k}$ = [0.42, 0.42, 0].
The compounds RCu_4Al_8 exhibit an antiferromagnetic order with
$\vec{k}$ = [100] except $TbCu_4Al_8$ which has an incommensurate structure
along the c-axis with $\vec{k}$ = [1, 0, 0.16].

5.2.4. Face centered orthorhombic lattice

In this lattice only four symmetry points exist :
$\vec{k}_1$ = [00 1/2], $\vec{k}_2$ = [010], $\vec{k}_3$ =[1/2 1/2 0] and $\vec{k}_4$ = [01 1/2]. A
large number of equiatomic compounds crystallize with an ortho-
rhombic structure (RGa, RSi, RGe, RNi and with light elements
RCu, RRh), which can be of CrB-type (Cmcm) or of FeB-type (Pnma).
However, collinear magnetic structures occur only in the CrB-type
structure because rare earth ions are located in a site of C_{2v}
symmetry and there are two Bravais lattices related by the inversion symmetry element. In RSi compounds [40], TmSi and DySi order
with a wave vector $\vec{k}$ = [00 1/2] whereas TbSi, HoSi and ErSi exhibit a structure with $\vec{k}$ = [1/2 0 1/2] ; thus there is no continui-
ty across the serie which is quite difficult to understand in the
frame of RKKY interactions.

5.3. Non-collinear magnetic structures

Even with simple antiferromagnetic or ferromagnetic interactions the magnetic structure can be complex and non-collinear
because, as we have seen in section 4, the CEF anisotropy plays
an important role to determine the moment direction.

We may have a non-collinear magnetic structure if the unit cell contains several RE sites because there are no reasons that they have the same easy direction. This situation occurs in many RE intermetallics. For example RE-Ni intermetallics such as RNi_3 (R = Nd, Tb, Dy, Er) are non-collinear ferromagnets because the two RE sites have orthogonal easy axes.

Even if the unit cell contains only one RE site, a non-collinear structure is highly probable when the symmetry of the RE-site is low enough. We have shown in section 4 that this situation occurs in RE-intermetallics with the FeB-type structure (C_5), in RE-oxydes (C_2), in RE-perovskites (C_5) and in RE-garnets (C_{2v}), etc...

Non-collinear structures can exist also in highly symmetric lattices, even with a single Bravais lattice because fourth or higher order terms in the Hamiltonian couple the components of the order parameter to give rise to a multi-$\vec{k}$ ordering.

In section 5.1. we have seen that the moment distribution is in fact determined by the relation (34) :

$$\vec{m}_n = \sum_{k_i \{k\}} \vec{m}_{k_i} e^{2\pi i \vec{k}_i . \vec{R}_n}$$

Thus the problem is how to perform the sum : over one or several members of the star of $\vec{k}$? This gives rise to an ambiguity about the exact nature of the magnetic order which can be removed only by applying a pertinent perturbation to the crystal (a magnetic field or an uniaxial stress) or if there exists a crystallographic distortion at the magnetic ordering which lowers the symmetry. Then to each collinear structure described in section 5.2. we must associate multi-$\vec{k}$ structures. Their number will depend on the number of members in the star and on the direction of $\vec{m}_k$ with respect to $\vec{k}$. The number may be limited by the additional condition $\vec{m}_n^2 = \vec{m}_0^2$. For a f.c.c. lattice and $\vec{k} = [1/2\ 1/2\ 1/2]$ the analysis of multi-$\vec{k}$ structures has been given in [43] and we report in figure 10a the most symmetrical structures associated to one, two, three or four $\vec{k}$-vectors. For $\vec{k} = [001]$ there exist three kinds of multi-$\vec{k}$ structures, those corresponding to $\vec{m}_k$ parallel to $\vec{k}$ are given in figure 10b. For example, the 3k-vector structure has a cubic symmetry and the moment direction is along <111> whereas in the collinear structure the moment lies along <001>. Thus additional information concerning the crystal field is needed. Such a situation occurs in CeP and CeAs. In CeAs the triple-k structure, with moments along <111>, has been shown to exist in a neutron experiment by applying a magnetic field along a <110> axis of a single crystal [41].

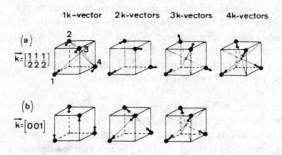

Fig. 10 Typical examples of the most symmetrical multi-$\vec{k}$ structures for a face-centered cubic lattice and a wave vector
$\vec{k}$ = [1/2 1/2 1/2] (a) or $\vec{k}$ = [001] (b). The ions (1), (2), (3)
and (4) are located respectively at (000), (0 1/2 1/2), (1/2 0 1/2)
and (1/2 1/2 0).

For the primitive cubic lattice multi-$\vec{k}$ structures give an
indetermination not only of the moment direction but also of the
size of the magnetic cell. For $\vec{k}$ = [00 1/2] and $\vec{k}$ = [1/2 1/2 0]
there are three members in the star giving rise to one, two or
three k-vector structure. For example for $\vec{k}$ = [00 1/2] the size
of the magnetic cell will be (a, a, 2a), (2a, 2a, a) and (2a, 2a,
2a) respectively. The 3k-vector structure has been effectively
evidenced in DyCu [42] and HoRh [43] in agreement with neutron
spectroscopy experiments [10] which indicate a <111> easy axis.

Non-collinear structures can be induced also by a fourth
order term which couples Fourier components with two distinct
wave vectors. The flip-flop structure of HoP [44] is a good
example, it is described by a Fourier component $m_k^\perp$ (for
$\vec{k}$ = [1/2 1/2 1/2] m_k is along [10$\bar{1}$]) and a ferromagnetic component m_0 along [101]. Thus the magnetic moments in successive
(111) ferromagnetic planes are aligned alternatively along [001]
or [100]. This kind of ordering results in strong quadrupolar
interactions.

Thus non-collinear magnetic structures result non only from
the 4f-electrons : one-ion anisotropy, coupling to a strain,
quadrupolar interactions, but also from the conduction electrons,
which in special cases gives rise to an anisotropic exchange
interaction with 4f electrons (see eq. (12)).

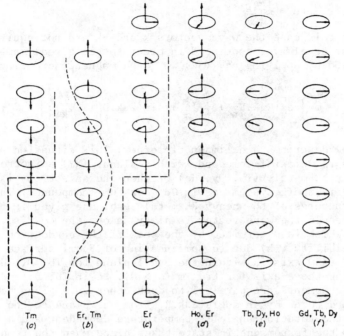

Fig. 11 The magnetic structures of heavy rare earth metals.

5.4. Incommensurate magnetic structures

The long-range and oscillatory nature of the RKKY exchange interactions may lead to magnetic structures which are incommensurate with the lattice. These structures are characterized by wave vectors corresponding to a point on a symmetry line of the Brillouin zone. The value of the wave vector, in the simple case of one Bravais lattice, is determined by the minimum of the Fourier transform of the exchange integral :

$$J(\vec{k}) = \sum_{n} e^{-2\pi i \vec{k}(\vec{R}_n - \vec{R}_m)} J_{nm} \tag{35}$$

Then to define the magnetic ordering, the symmetry of the group G_k is of first importance.

5.4.1. Helimagnetic structures

If $\vec{m}_k$ belongs to a two-dimensional representation we can write :

$$\vec{m}_k = \frac{m_0}{2}(\vec{u} - i\vec{v})e^{i\Phi_k} \tag{36}$$

In this case the wave vectors $\vec{k}$ and $-\vec{k}$ are not equivalent and must be always associated in order to get a real moment value. So by applying eq. (32) we get a magnetic ordering such as :

$$\vec{m}_n = m_0\left[\cos(2\pi\vec{k}.\vec{R}_n + \Phi_k)\vec{u} + \sin(2\pi\vec{k}.\vec{R}_n + \Phi_k)\vec{v}\right] \tag{37}$$

Eq. (35) describes a helimagnetic order. This situation occurs usually in uniaxial compounds with a planar (XY) anisotropy as in RE metals (Tb, Dy, Ho) or when the anisotropy is very weak (Heisenberg system) as in Gd^{3+} or Eu^{2+} compounds. The magnetic structures of the heavy RE metals [46] are given in figure 11. In RE-metals the wave vector is always along the c-axis, the value of which is determined by the peak of the conduction electron susceptibility $\chi(\vec{k})$ due to some nesting of Fermi surface pieces. The CEF easy axis is within the basal plane for Tb, Dy, Ho ($\alpha < 0$) and along the c-axis for Er, Tm ($\alpha > 0$). So in Tb, Dy and Ho the magnetic moments are parallel to each other in a given hexagonal layer and turn by an angle $2\pi\vec{k}.\vec{R}_n$ from one plane to the other. Indeed the rotation angle is temperature dependent and a transition to the ferromagnetic state takes place when the temperature is decreased.

5.4.2. Modulated structures

The Fourier components $\vec{m}_k$ are one dimensional if the irreducible representation is one dimensional or if the CEF anisotropy leads to an Ising-like behaviour. Then we can write

$$\vec{m}_k = \frac{A_k}{2}e^{i\Phi_k}\vec{u} \tag{38}$$

and $$\vec{m}_n = A_k\cos(2\pi\vec{k}.\vec{R}_n + \Phi_k)\vec{u} \tag{39}$$

The ordering corresponds to a sine wave modulation of the moment value propagating along the k-direction with a polarization along u. However a large entropy is associated with such a sine wave modulation, so in decreasing temperature it becomes unstable and a squaring up of the modulation must occur. A typical examples are given by Er and Tm metals [2] (see figure 11) where third, fifth and higher order harmonics progressively appear in decreasing temperature.

The entropy argument fails in the case of a two singlet system ; an example is provided by the intermetallic compound

TbNi$_{0.6}$Cu$_{0.4}$ [46] which exhibits a pure sine wave modulation down to T = 0. A similar behaviour has also been observed in CeAl$_2$ [36] but in that case the singlet ground state is expected to arise from the Kondo behaviour as we will see in section 7.

In some compounds there is a direct phase transition towards a commensurate phase instead of observing a squaring up of the modulation. This incommensurate-commensurate transition is observed to be either of first order as in TbAu$_2$ or TbZn$_2$ [1] or of second order as in TbSi [47]. The order could be related to the symmetry of the wave vector of the commensurate phase. Indeed the most common commensurate phases correspond to a wave vector $\vec{k} = \vec{H}/4$ which are induced by the fourth order terms in the Hamiltonian in presence of an Ising-like anisotropy. For example, in a f.c.c. lattice, we can get a wave vector $\vec{k}$ = <1/2 00> (type IA) or $\vec{k}$ = <1 1/2 0> (type III) (see figure 9). Moreover in some compounds, as CeSb [48], on decreasing the temperature many phase transitions between commensurate structures can be observed.

6.   DIPOLAR ORDERING

As we have seen in section 3, in ionic RE compounds the interaction between magnetic moments is mainly of dipolar type. The ordering temperature being of the order of one Kelvin or less we can describe the low temperature properties by projecting the Hamiltonian upon the doublet or pseudo-doublet ground state. In the first case we define a g-tensor by using an effective spin 1/2 whereas in the latter case the system will have an Ising-like anisotropy. The interaction Hamiltonian and the anisotropy being well defined, the ordering temperature and the nature of the magnetic structure can be predicted using a mean field theory. Such a prediction has been done by Capel in 1965 [49] in RE garnets as RGaG or RAlG. The main results will be briefly reported and compared with experimental data. In the garnet structure RE-ions occupy a site which has a D$_2$-symmetry and the cubic unit cell contains 12 Bravais lattices which are generate by the inversion, a three-fold axis and a two-fold axis. Therefore we must define six non equivalent local axes u,v,w, the quantization axis being along the $\vec{w}$-axis. As example for one site (let say C$_1$) the u-axis is along the [100] cubic axis and the v and w-axes are taken along the [01$\bar{1}$] and [011] axes respectively.

The magnetic interaction Hamiltonian can be written as :

$$H = \sum_{m \neq n} \vec{m}_n \vec{\vec{T}}_{nm} \vec{m}_m \tag{40}$$

$$\text{with } \vec{\vec{T}}_{nm} = \frac{1}{r_{nm}^5} [r_{nm}^2 \vec{u} - 3 \vec{r}_{nm} \vec{r}_{nm}] \tag{41}$$

and $\vec{m}_n = \mu_B \vec{g}_n \vec{s}$ $\qquad$ (s = 1/2)

In the molecular field approximation one get for a Kramers doublet :

$$\langle \vec{m}_n \rangle = \frac{\mu_B}{2} \frac{\vec{g}_n^2 \langle \vec{H}_n \rangle}{|\vec{g}_n \cdot \langle \vec{H}_n \rangle|} \, \text{th}\left(\frac{\mu_B}{2kT} |\vec{g}_n \cdot \langle \vec{H}_n \rangle|\right) \qquad (42)$$

where $\langle \vec{H}_n \rangle = \sum_{m \neq n} \vec{T}_{nm} \langle \vec{m}_m \rangle$

is the molecular field.

Then by linearizing eq. (40) near the ordering temperature, the Neel temperature is given by

$$\frac{4kT_N}{\mu_B^2} \vec{m}_i + \vec{g}_i^2 \sum_j \vec{T}_{ij} \vec{m}_j = 0 \qquad (43)$$

The index i = 1, ..., 12 labels the Bravais lattices.

This set of linear equations can be solved by using group theory and the largest eigenvalue will give the ordering temperature T_N.

When the anisotropy is Ising-like as for non-Kramers ion (TbGaG, HoGaG) [25] the ordering is antiferromagnetic with moments aligned along their own u-axis, i.e. a cubic axis. Whereas for DyGaG and YbGaG which have an XY and Heisenberg-type of anisotropy the theory predicts a ferrimagnetic structure schematized in figure 12 [50]. The ordering temperature for such a structure has been calculated to be about 0.7 K and 0.2 K for DyGaG and YbGaG respectively. These values are not in agreement with the experimental ones which have been found to be much lower : T_N = 0.375 K for DyGaG and T_N = 0.054 K for YbGaG. Indeed specific heat experiments have revealed a quite complex ordering process, in figure 13 we have reported the most typical result obtained on YbGaG [50]. In both DyGaG and YbGaG the ordering takes place in two steps : i/ at high temperature a kind of topological order develops at the temperature T_1 which is close to the value predicted by the mean field theory, ii/ at much lower temperature a long range order occurs. This behaviour is quite unusual and rises some fundamental questions upon the validity of the mean field theory to predict the order in dipolar systems.

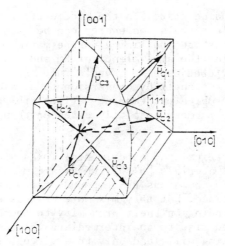

Fig. 12 Orientation of the magnetic moments in the ferrimagnetic structure predicted for the garnet RGaG.

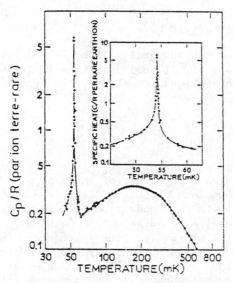

Fig. 13 Magnetic specific heat of the garnet compound YbGaG.

7.   ANOMALOUS LANTHANIDE COMPOUNDS

     In this section we are concerned by lanthanide compounds in which the 4f level is located close to the Fermi energy instead to be far from it ($\sim$ 5-10 eV) as for normal RE ions. In the gas phase most of RE are divalents, but in solid state they are triva-lent, due to the large cohesive energy gained by promoting a 4f electron into a bonding state. In the middle of the RE row, however, Hund's rule correlation energy can be gained by converting to a divalent state and lining up the maximum number of spins, which explains the ambivalency of Sm and Eu. At the end of the row (Tm and Yb) obedience to Hund's rule creates a similar wish to complete the shell by conversion to divalence. In the case of cerium the ambivalent tendency arises from the fact that the 4f orbital is more spatially extended than for the other rare earths.

A lot of theoretical and experimental works are presently undertaken on such materials [51,52] which are mainly Ce, Sm, Eu and Yb compounds. In anomalous RE-compounds the 4f electrons are generaly well localized, but they mix more or less strongly with band electrons depending on their proximity to the Fermi energy. A large mixing gives rise to an intermediate valence regime whereas a small mixing will lead only to a virtual valence fluctuation induced either by a mixing with the conduction s-d band or with the valence p-band.

7.1. The mixed-valent state [53]

A condition for non integral valence is that two bonding states $4f^n(5d6s)^m$ and $4f^{n-1}(5d6s)^{m+1}$ of the RE are nearly degenerate. As example for Sm ions in SmB_6 [55] the fundamental equation can be written at T = 0 as :

$$(Sm^{2+})_g \overset{\leftarrow}{\to} e^{5d}_{atE_F} + (Sm^{3+})_g - U^{gg}_{4f} \qquad (44)$$

where g means the ground state. If U^{gg}_{4f} is nearly zero, the crystal will contain a homogeneous mixture of Sm^{2+} and Sm^{3+} or generally of $4f^n$ and $4f^{n-1}$ ions. This mixture results from a strong hybridization (or mixing) with the conduction band. As indicated in figure 14 the f^6 level E_f of divalent Sm^{2+} in SmS, at atmospheric pressure, fall in the gap between the valence and the conduction bands. By applying a pressure the empty 5d6s conduction band goes down until it overlaps the f^6 level, at which point the latter empties electrons into the band. In the mixed-valent regime the f^6 level is thus pinned at the Fermi level E_F to get the proper mixture of Sm^{2+} and Sm^{3+}. Then 4f electrons tunnel through the crystal and at a given site the valence fluctuates. The 4f level takes a finite width, the inverse of this width can be identified as the valence fluctuation time. A similar sequence hold in cerium, except that the 4f level falls in the s-d band in the γ phase (nearly Ce^{3+}) and hence some d-f mixing is present, although it is sufficiently far below the Fermi level to ensure nearly complete occupency.

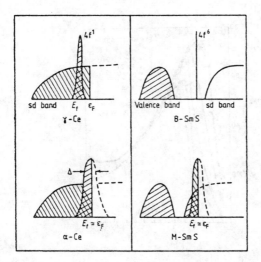

Fig. 14 Schematic density of state diagrams of cerium and SmS.

One of the key features of mixed-valent materials is that they exhibit non-magnetic ground states. It is believed that this non-magnetic behaviour comes from the fact that one valence state is non magnetic, indeed in TmSe where both valence states of Tm are magnetic, a magnetic ordering takes place at low temperature.

7.2. "Kondo"-like behaviour [54]

When the mixing between the conduction band and f-electron states (called d-f mixing) become weaker a Kondo-like behaviour is very often observed. This is particularly true for cerium intermetallics which exhibit resistance minima and regions of negative slope suggestive of a LnT contribution, as for the Kondo effect in dilute systems. In figure 15 is reported the resistivity of CeB$_6$ which is one of the most typical example [55].

While these compounds are trivalent, some have non magnetic ground states as CeAl$_3$ [54] or even a supraconducting transition has been observed in CeCu$_2$Si$_2$ [56]. Therefore it cannot be due to direct valence fluctuations but rather rises from virtual valence fluctuations indicating that the 4f level is very close to the Fermi level and then the compound is on the verge of a valence instability. Indeed most of the cerium intermetallics order magnetically but some of them exhibit a very unusual magnetic ordering, the most typical ones are CeAl$_2$ [36] and CeB$_6$ [57].

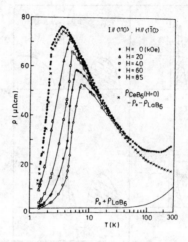

Fig. 15 Electrical resistivity of CeB_6 with and without an applied magnetic field.

CeAl$_2$, which has a cubic Laves phase structure, orders at T_N = 3.8 K with a transverse sine wave modulation of the moment length which propagates with a wave vector $\vec{k}$ = <1/2 + τ, 1/2 - τ, 1/2> and $\tau \cong$ 0.110 (figure 16). This incommensurate structure remains unchanged down to 0.4 K which is very unusual because Ce^{3+} has a Γ_7-Kramers doublet as the ground state. To explain this behaviour it was proposed that the ground state is a singlet Kondo-type state giving rise to a Kondo lattice. In CeB_6 two transition temperatures have been observed [55] at T_1 = 3.15 K and T_2 = 2.3 K (figure 17). Above T_1 (phase I) we have a dense Kondo behaviour in which incoherent Kondo scattering dominates. For $T_2 < T < T_1$ the most mysterious phase (phase II) exists which may correspond to the occurence of a coherent Kondo state. Below T_2 (phase III) an antiferromagnetic ordering takes place, but the structure is indeed a double-$\vec{k}\vec{k}'$ structure with $\vec{k}$ = <1/4 1/4 0> and $\vec{k}'$ = <1/4 1/4 1/2>. The understanding of these complex properties is far to be in good shape.

7.3. Mixing with the valence band [55]

A virtual valence fluctuation can be induced also by the mixing of the 4f electron with the valence p-band. Indeed in semi metallic compounds as CeSb and CeBi the p-f mixing is expected to be larger than the d-f because the anions are nearer than the n.n RE ions. The theory of this p-f mixing was recently done by Kasuya et al. [55] and this mechanism defines indeed a new class of   anomalous RE compounds.

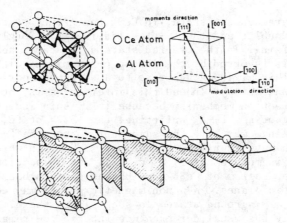

Fig. 16   Transverse sine wave magnetic structure of CeAl₂.

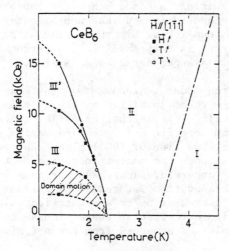

Fig. 17   (H, T) phase diagram of CeB₆.

Band calculations on LaX have shown that the gap disappears
between LaP and LaAs. LaSb and LaBi are semimetals with a very
small number of p-holes ($n_e = n_h \cong$ 1-2 %/La). Kasuya et al. [55]
have shown thus that a strong p-f mixing interaction exists only
between a given state of the Γ_8 quartet and the p-band of Γ_8 character. Then the 4f-Γ_8 states are pushed down to give a reduced
crystal field splitting and the valence band (Γ_8) moves up inducing an increase of the number of holes. Therefore this interaction is strongly non linear. On the other hand the Γ_7 states mix
with the lower part of the 5d(t_{2g}) conduction band which may
gives rise to a Kondo behaviour, such a behaviour has been indeed

observed in CeSb and CeBi by resistivity measurements [55].

In the paramagnetic state, in CeSb the ground state is pro-
bably the Γ_7 doublet and the Γ_7-Γ_8 splitting is strongly reduced
by this p-f mixing. In the ordered state a Γ_8 state, nearly a
$|5/2>$ state, is favoured by the strong p-fΓ_8 mixing within a given
(001) plane. This mixing gives rise to a strong planar coupling
and a strong anisotropy. Indeed this interaction is not an usual
exchange interaction mechanism because it is spin independent.
Then depending of the Γ_7-Γ_8 splitting, the $|5/2>$ state, favoured
by the p-f mixing, may have the same energy than the Γ_7 state.
This situation may occur in CeSb and not in CeBi in which the
CEF splitting is much smaller. As we will see in section 8, in the
magnetic phase diagram of CeSb some phases contain both magnetic
and non magnetic planes. These non magnetic planes may correspond
to Ce ions in a Γ_7 ground state. In such planes there is no p-f
mixing, so only the usual d-f exchange interaction remains, but
as it is much smaller, the cerium moments may not order. However,
as there is a d-fΓ_7 mixing, a Kondo behaviour may occur and
Kasuya et al. [55] proposed indeed that non magnetic planes result
from a strong Kondo coupling within these planes. As the number
of conduction electrons is very small (a few percent) Kasuya
suggests that CeSb is a super dense Kondo system.

Such a p-f mixing model, which explains the complex properties
of CeSb [58,59], seems to be justified by the experimental results
obtained by substituting Sb by either Bi or Te. In $CeSb_{0.95}Bi_{0.05}$
the magnetic order in zero field (figure 18) varies continously
with the temperature. Two kind of (001) planes can be distinguished. In the first ones the moment increases continously to the
full value when the temperature decreases ; whereas in the other
ones the moment increases from T_N, reaches a maximum, vanishes at
about $T_N/2$ and then increases up to a saturation value of $\sim 0.7 \mu_B$
This result could indicate that in these planes the ground
state is the Γ_7 doublet as expected from the p-f mixing model.

Indeed the most interesting effects have been observed in the
study of $CeSb_{1-x}Te_x$ solid solutions [59]. The substitution of Sb
by Te corresponds to add x conduction electrons and then to fill
the valence p-band. T_N drops abruptly from 16 K to less than 4 K
(figure 19) and the strong anisotropy disappears for x > 0.05.
Indeed a concentration x = 0.03 is large enough to destroy the
unusual magnetic behaviour of CeSb and to stabilize a type I
ordering. These results indicate that the number of holes in the
p-band is very small (a few %) but they are actually at the origin
of the unusual behaviour of CeSb and CeBi.

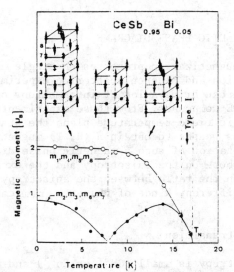

Fig. 18 Thermal variation of the magnetic moment and magnetic structures of CeSb$_{0.95}$Bi$_{0.05}$.

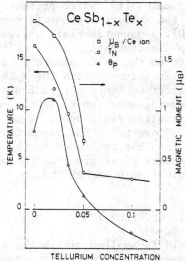

Fig. 19 Effect of the substitution of Sb by Te on the ordering and paramagnetic temperatures and on the ordered moment value.

These few examples of the magnetic properties of anomalous RE-compounds have shown that these compounds, especially those with Ce, exhibit a lot of unusual behaviours which attract many physicists. It is indeed a very active field in the RE research which may bring many new ideas to understand the more complex properties of actinide compounds.

8.   STUDIES OF MAGNETIC PHASE DIAGRAMS

The study of magnetization processes on single crystals by means of magnetization and neutron scattering experiments up to high magnetic fields can bring a lot of informations on the magnetic properties of RE-compounds. From these experiments we can drawn in the magnetic field-temperature plane the stability regions of the various phases to obtain a (H, T) magnetic phase diagram. The exploitation of such a magnetic phase diagram can bring information both on the anisotropy and the exchange parameters. Depending on the ratio between the anisotropy and the exchange energies, different kinds of phase diagrams can be expected.

8.1. The spin-flop transition

When the anisotropy is small (Gd^{3+} or Eu^{2+}) and the magnetic ordering consists of a simple antiferromagnetic structure, the magnetization processes are quite simple. A typical example is the perovskite compound $GdAlO_3$ [60] which presents a small uniaxial anisotropy with a two-sublattices antiferromagnetic structure [61] (see figure 20). A magnetic field applied along the AF direction induces a phase transition for a critical field H_{sf} which corresponds in fact to the value needed to suppress the spin wave gap. At this transition, called the spin-flop transition, the antiferromagnetic direction flops perpendicularly to the field direction. Then by increasing more the field the moments cant progressively along the applied field up to reach the ferromagnetic aligment of the moments. Close to T_N the transition is second order and occurs directly from the antiferro to the paramagnetic state. The transition becomes first order at the multicritical point below which the spin-flop phase exists.

8.2. Case of helimagnetic ordering [2]

When a magnetic field is applied within the plane which contains the magnetic moments, the transition from the helimagnetic to the ferromagnetic arrangement is complex and depends very much of the temperature and the magnetic field. If we take the case of the RE metals as Ho or Dy, for low fields there is a slight distortion of the structure with a small net moment along the field. For intermediate fields ($\sim$ 10 kOe) there may exist a fan structure (see figure 21) in which the moments oscillate sinusoidally about the field direction, while for fields greater than a critical value ($\sim$ 20-30 kOe) this structure collapses into

a ferromagnetic structure with moments parallel to the field
direction. If the basal plane anisotropy is sufficiently large,
the fan phase may not appear and only a heli-ferromagnetic
transition remains.

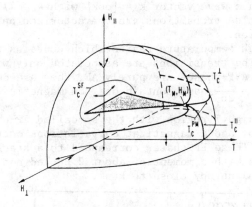

Fig. 20   Magnetic phase diagram of $GdAlO_3$.

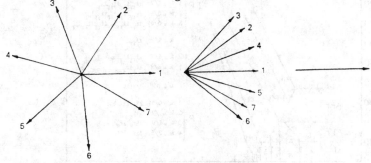

Fig. 21   Effect of a magnetic field applied on a helimagnetic
structure.

8.3. Metamagnetic transitions

When the anisotropy is large the magnetic field can produce
only a flip of the moments which are antiparallel to the field.
A direct transition to the ferromagnetic state may occur, but in
general there are several intermediate ferrimagnetic phases depending on the range and the value of the exchange integrals.

Then complex phase diagrams can be realized, the most typical
ones are certainly those obtained on CeSb and CeBi [48,59] which

have been extensively studied by neutron scattering experiments.
The most complex one is that of CeSb (figure 22) which contains
at least fourteen different magnetic structures [58,62]. Mainly
three regions can be distinguished :

i) A low temperature region containing phases, called AFF-phases,
which correspond to a stacking of ferromagnetic sheets with sequences associated with a wave vector $k = [ook]$ with $k = 4/7$, $2/3$
and 0 (see figure 22). The transitions are characterized only by
a change of magnetization.

ii) A low field and high temperature region which contains the
so-called AFP-phases. The transitions are associated only with an
entropy variation. The AFP-phases are purely AF, they are described by a wave vector $k = \dfrac{n}{2n-1}$ and contain one non magnetic layer
every $2n-1$ layers.

iii) A high field and high T region with the so-called FP-phases
in which both an entropy and a magnetization variations occur at
the phase transitions. These FP-phases correspond to a stacking
of ferromagnetic layers with a moment of about 2 μ_B and non
magnetic layers with an entropy close to kLn2.

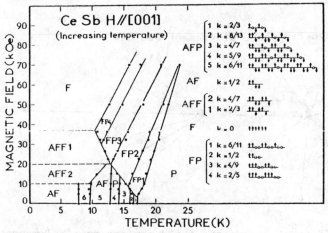

Fig. 22 Magnetic phase diagram of CeSb.

The main features of this phase diagram are the existence
of complicated magnetic structures in which coexist magnetic and
non magnetic (001) planes, the fact that all the transitions are
first order and the very strong anisotropy along a fourfold axis.

The (H,T) magnetic phase diagram of CeBi determined by magnetization and neutron experiments is given in figure 23 [48,59].
This phase diagram is also complex but the different phases contain only magnetic planes. Three main phases exist : a type I
($k = 1$, +−+−), a type IA ($k = 1/2$, ++−−) and a ferrimagnetic phase

(+++-). Instead of observing a direct transition between these phases, we discovered many structures with long range periods. As example between the type I (+-) and the ferri (+++-) phases the following sequences +++-:+++-:+- and +++-:+- have been observed in increasing temperature. A similar behaviour occurs at low temperatures. Moreover the investigation of the phase diagram by specific heat experiments has put in evidence a new phase close to the ordering temperature, the nature of which has not yet been determined (dotted line in figure 23).

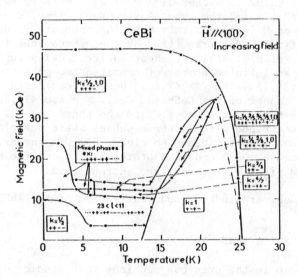

Fig. 23 Magnetic phase diagram of CeBi.

To understand the phase diagrams of these compounds we must remember that there is a strong coupling within (001) planes due to the p-f mixing, as explained in section 7.3, while the coupling between planes, probably of exchange type, is much smaller. Then these compounds can be considered as a quasi two-dimensional (layered) Ising system and the phase transition will be close to the Onsager transition. When an antiferromagnetic coupling between planes exists up to second neighbours a complex behaviour may occur with temperature or magnetic field. Many theoretical works have been done on such a system called the ANNNI (anisotropic next nearest neighbour Ising) model [63,64,65,66]. In such a model the Hamiltonian can be transformed into a one dimensional Ising Hamiltonian :

$$H = H_0 + H_1 \text{ with } H_0 \gg H_1$$

$$\text{and } H_1 = \sum_{m=1}^{N} \sum_{n=-\infty}^{+\infty} J(n) m_m \cdot m_n - H \sum_{m,n} m_m \qquad (45)$$

where H_0 is relative to N isolated (001) layers. $J(n)$ are effective inter-layer interactions, N is the number of layers, $|m_n| = 1$ and H is the field applied along <001>. Tacking into account only the first (J_1) and second (J_2) nearest neighbours, the theory gives at $T = 0$ and $H = 0$ a ferro or an antiferro type I structure according to wether $J_1 > 0$ or $J_1 < 0$ if $|J_2/J_1| < 1/2$, whereas if $|J_2/J_1| > 1/2$ the type IA structure (++--, $k = 1/2$) is stable. When T or H increases this type IA structure is destabilized through a first order transition by the spontaneous formation of interacting walls or solitons. Near the critical point $-J_2/|J_1| \cong 1/2$ an infinite number of commensurate phases can exist such as $<2^n,1>... <2^3,1> <2,1>$. These phases correspond indeed to that observed for CeSb and CeBi. A phase $<2^3,1>$ corresponds to a ++--++- sequence, i.e. to the phase $k = 4/7$. AFP phases can be described also by these phases where one plane is replaced by a non magnetic plane to give $<2^n, +0->$ phases. This is possible because $|5/2>$ and $|\Gamma_7>$ states have about the same energy and this gives rise to a large entropy gain.

CeSb and CeBi are indeed good realization of the ANNNI model.

9.   MAGNETIC EXCITATIONS

We have seen in the previous sections that studies of magnetic ordering in lanthanide compounds by neutron diffraction, together with bulk magnetization, susceptibility and specific heat measurements have yielded a great deal of information about the basic interactions in these compounds. Most of this information is obtained via a molecular field theory (see section 4.3) of the magnetic phenomena. A more stringent test of our understanding of the basic interactions (CEF, exchange, etc...) is to see wether we can successfully explain the magnetic excitations in these systems. Inelastic neutron scattering studies have been invaluable in probing such dynamics since they may be used to study the spectrum of the low lying magnetic excitations.

The differential neutron cross section is proportional to the dynamical structure functions [67] :

$$S^{\alpha\beta}(\vec{Q},\omega) = \frac{1}{2\pi N} \int dt \, e^{-i\omega t} \sum_{n,m} <J_n^\alpha(t) J_m^\beta(0)> e^{i\vec{Q}\cdot(\vec{R}_n-\vec{R}_m)} \qquad (46)$$

for an energy transfer $\hbar\omega$ and a scattering vector $\vec{Q}$. Inelastic neutron scattering experiments allow to study the pair correlation functions. There are two methods to perform such calculations either to use the Green's function technique or the conventional treatment of spin-waves by using the Hosltein-Primakoff transformation. However the latter is valid only when the exchange energy is much larger than the CEF energy splitting, this situation is almost never satisfied in RE-compounds. Therefore the former one gives a much better description of magnetic excitations in RE-compounds.

The general idea is to replace in the dynamical part of the Hamiltonian (eq. (31)) :

$$H_1 = \sum_{nm} J_{nm} (\vec{J}_n - \langle \vec{J}_n \rangle_0)(\vec{J}_m - \langle \vec{J}_m \rangle_0) \tag{33}$$

the components of the angular moment $\vec{J}$ by pseudo-fermion operators $C_p^+(n)$ and $C_p(n)$ that create and annihilate states $|p\rangle$ on the ion n. [68]. Then :

$$J_n^\alpha = \sum_{p,q} \langle p|J^\alpha|q\rangle \, C_p^+(n)C_q(n) \tag{47}$$

Using these operators the static part of the Hamiltonian (eq. (30)) is diagonal :

$$H_0 = \sum_n \sum_p \omega_p \, C_p^+(n)C_p(n) \tag{48}$$

where ω_p are the energies of the CEF and exchange levels.

Then eq. (31) can be written as :

$$H_1 = - \sum_{n,m} \sum_{\alpha,\beta=+,-,z} \sum_{pqrs} J_{nm} J_{pq}^\alpha J_{rs}^\beta \, C_p^+(n)C_q(n)C_r^+(m)C_s(m) \tag{49}$$

Indeed the product $C_p^+(n)C_q(n)$ represents an operator that induces at site n a transition from level q to level p. The dispersion relations $\omega(\vec{q})$ are obtained from the poles of the dynamical susceptibility. In the simple example of a magnetic excitation from the ground state to an excited state located at the energy $\Delta = \omega_1 - \omega_0$ and for a single Bravais lattice we can write the dispersion relation for a transverse magnetic excitation of wave vector q and energy ω as :

$$\omega^2(\vec{q}) = (\Delta - M^2 J(\vec{q}))(\Delta - M^2 J(\vec{q}+\vec{k})) \tag{50}$$

where $\vec{k}$ is the wave vector of the static magnetic structure and M is the matrix element $\langle 1|J^-|0\rangle$ between the ground state and the excited level.

In particular in the paramagnetic or the ferromagnetic state we get the simple relation :

$$\omega(q) = \Delta - M^2 J(q) \tag{51}$$

The dispersion of the excitations depends in fact on two factors : the value of the matrix element and the Fourier transform of the exchange integrals $J(\vec{q})$. If one of them is very small there will be no dispersion. In the case $J(q) \cong 0$, i.e. a very weak coupling between RE-ions, the energy of the levels will not depend on q and we will measure the energy of crystal field levels ; so to get them properly when the ordering temperature is large, it is suitable to dilute the compound with non magnetic ions as Y or Lu. In the case $M^2 \cong 0$ the excitations have also a small dispersion and moreover the intensity is very weak, hence it will be difficult to measure it.

Magnetic excitations can be observed, indeed, both in the paramagnetic and the ordered state. In figure 24 are reported the dispersion relations of magnetic excitations in Pr metal [69]. In the metal, Pr has a singlet ground state $|0\rangle$, the first excited doublet $|\pm 1\rangle$ is too far to allow a long range magnetic ordering. These excitations, observed along the various symmetry directions of the hexagonal lattice, are indeed magnetic excitons. As the unit cell contains two Bravais lattices we expect to observed two excitations, in fact for wave vectors within the basal plane each dispersion curve is split in two, indicating that the exchange interactions are anisotropic. Moreover the dispersion curve has a minimum along the <100> direction for a wave vector which would correspond to the magnetic ordering.

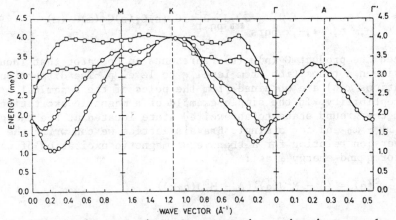

Fig. 24 Dispersion relations for magnetic excitations on the hexagonal sites of dhcp Pr at 6.4 K. The circles represent acoustic-type modes and the squares optic-type modes.

    In the ordered state we have chosen to present the magnetic
excitations measured in the compound CeBi [59,70]. In figure 25
are reported the dispersion curves measured along the [00q] and
[qq0] directions. They are highly unusual because when the excitations propagate perpendicularly to the ferromagnetic (001)
planes there is no dispersion, whereas within the (001) plane
the dispersion curves have a minimum at the zone boundary corresponding to the X-point ($\vec{q}$ = [100]). This spectrum cannot be
accounted for by a bilinear Hamiltonian as given in eq. (26).
Indeed an anisotropic form as :

$$H_{ex} = - \sum_{n,m} \sum_{\alpha\beta} J_{n,m}^{\alpha\beta} \, J_n^\alpha J_m^\beta$$

does not give a better agreement. However these dispersion curves
confirm the very weak coupling between (001) planes. From the
strong ferromagnetic coupling within the (001) plane we expect
that the dispersion curve has a minimum at the zone center when
the excitation propagates within the plane, but in fact we observe
only a minimum at the zone boundary. To explain this unusual
result, more experimental and theoretical works are needed, however we propose that these excitations may be the consequence of
the strong p-f mixing as was explained in section 7.3.

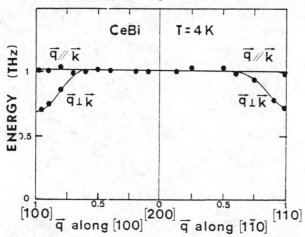

Fig. 25 Magnetic excitations in the type I structure of CeBi.

    Except in heavy RE metals where the magnetic excitations are
quite well explained by a RKKY interaction and a CEF contribution,
in light RE, in RE intermetallics or in Ce compounds the magnetic
excitations spectra indicate the presence of a much more complex
exchange interaction mechanism.

10. CONCLUSION

In this review, we have attempted to survey the main characteristic features which are involved in the magnetic properties
of lanthanide compounds. Due to the limited among of space it was
not possible to discuss all the properties involved in RE-compounds
such as magnetostriction effects, quadrupolar interactions, influence of magnetism on the supraconductivity, etc. Our understanding of the physics of lanthanide magnetism, at least at a pheno-
menological level, is in a reasonably satisfactory state if the
exchange and the CEF interactions are properly taken into account.
However in the case of anomalous RE-compounds and in particular
those with Ce a great deal of effort still is needed. At a more
microscopic level our understanding is far to be in good shape such
as calculations of the exchange integrals and CEF-parameters from
band calculations or the real coupling of RE-ions with conduction
electrons. Interesting problems remain to be solved such as the
magnitude of the anisotropic exchange interactions or the relative
importance of higher order exchange terms as biquadratic contribution. Finally we want to emphasize that many experimental results
must be put together and made consistent to get reliable values
for the microscopic parameters.

REFERENCES

1. K.A. Gschneider, L.R. Eyring (Editor), Handbook on the Physics and Chemistry of Rare Earths, North Holland, New-York 1978, Volume 1, 2, 3, 4.

2. B. Coqblin, The Electronic Structure of Rare Earth Metals and Alloys : the Magnetic Heavy Rare-Earths, Academic Press, London, 1977.

3. B.G. Wybourne, Spectroscopic Properties of Rare Earths, Willy, New-York, 1965.

4. D.J. Newman, Advances in Physics, 20, 197 (1971).

5. K.W.H. Stevens, Proc. Phys. Soc., 65A, 209 (1952).

6. A.J. Freeman, J.P. Desclaux
    J. Magn. Magn. Mat., 12, 11 (1979).

7. K.R. Lea, M.G.M. Leask, W.P. Wolf, J. Phys. Chem. Solids, 23, 1381 (1962).

8. G.F. Koster, J.O. Dimmock, R.G. Wheeler, H. Statz, Properties of the Thirty Two Point Groups, MIT Press, Cambridge, 1963.

9. J. Rossat-Mignod, J.C. Souillat, G. Quézel, Phys. Stat. Solidi (b), 62, 223 (1974).

10. J. Rossat-Mignod, R. Chamard-Bois, K. Knorr, W. Drexel, 11th Rare Earth Research Conference Proceeding, Traverse City, 1974, p. 317.

11. A. Yanase, T. Kasuya, Progr. Theoret. Phys., Suppl. n° 46, 46, 388 (1970).

12. T. Kasuya, Magnetism II B, (Rado and Suhl, Ed.), Academic Press, New-York, 1963.

13. P.W. Anderson, Magnetism I, (Rado and Suhl, Ed.) Academic Press, New-York, 1963.

14. T. Kasuya, IBM Res. Develop., 14, 214 (1970 b).

15. H.B. Callen, E. Callen, J. Phys. Chem. Solids, 27, 1271 (1966).

16. W.C. Koehler, Magnetic Properties of Rare Earth Metals, (R.J. Elliot, Ed.), Plenum, London, 1972.

17. J. Yakinthos, J. Rossat-Mignod, Phys. Stat. Sol. (b), 50,
     747 (1972).

18. J. Rossat-Mignod, J. Yakinthos, Phys. Stat. Sol. (b), 47,
     239 (1971).

19. D. Gignoux, J. Rossat-Mignod, F. Tchéou, Phys. Stat. Sol.(a),
     14, 483 (1972).

20. V.N. Nguyen, F. Tchéou, J. Rossat-Mignod, Solid State Commun.
     23, 821 (1977).

21. S. Quézel, E.F. Bertaut, G. Quézel, Acta Cryst., 25 A, S 252
     (1969).

22. E.F. Bertaut, Magnetism III, (Rado and Suhl, Ed.), Academic
     Press, New-York, 1963.

23. F. Tchéou, Thesis, Grenoble University, 1972.

24. B.R. Cooper, Magnetic Properties of Rare Earth Metals,
     (R.J. Elliot, Ed.), Plenum, London, 1972.

25. J. Hammann, M. Ocio, J. de Phys., 38, 463 (1977).

26. K. Andres, E. Bucher, J.P. Maita, A.S. Cooper, Phys. Rev.
     Lett., 28, 1652 (1972).

27. F.J.A.M. Greidanus, L.J. de Jongh, W.J. Huiskamp, K.H.J.
     Buschow, Physica (to be published).

28. J. Rossat-Mignod, F. Tchéou, J. de Phys., 33, 423 (1972)

29. G.T. Trammell, E.R. Seidel, Les Eléments des Terres Rares,
     Paris, Colloques Internationaux du CNRS, 2, 559 (1970).

30. J. Sivardière, Phys. Rev. (B), 8, 2004 (1973).

31. J. Rossat-Mignod, Thesis, Grenoble University, 1972.

32. Y. Abbas, J. Rossat-Mignod, F. Tchéou, G. Quézel, Physica,
     86-88 B, 115 (1977).

33. G.T. Trammell, Phys. Rev., 131, 932 (1963).

34. B. Barbara, J.X. Boucherle, B. Michelutti, M.F. Rossignol,
     Sol. Stat. Comm., 31, 477 (1979).

35. E.F. Bertaut, J. de phys. Colloq., 32, C1-462 (1971).

36. B. Barbara, J.X. Boucherle, J.L. Buevoz, M.F. Rossignol,
    J. Schweizer, Solid State Commun., 24 481 (1977).

37. E. Grüneisen, Ann. Phys., 39, 257 (1972).

38. R. Chamard-Bois, V.N. Nguyen, J. Yakinthos, M. Wintenberger,
    Solid State Commun., 10, 685 (1972).

39. A. Aléonard, P. Morin, J. Pierre, D. Schmidt, J. Phys. F.,
    6, 1361 (1876).

40. V.N. Nguyen, A. Barlet, J. Laforest, J. de Phys. Colloq. 32,
    C1-1133 (1971).

41. S. Quézel, P. Burlet, J. Rossat-Mignod, O. Vogt, (to be
    published).

42. M. Wintenberger, R. Chamard-Bois, M. Belakhovsky, J. Pierre,
    Sol. Stat. Commun., 48, 705 (1971).

43. R. Chamard-Bois, Thesis, Grenoble University, 1974.

44. J.W. Cable, W.C. Koehler, E.O. Wollan, Phys. Rev., 136, 240
    (1964).

45. J. Rossat-Mignod, J. de Phys., C5-40, 95 (1979).

46. D. Gignoux, J.C. Gomez Sal, J. Magn. Magn. Mat., 1, 203 (1976)

47. V.N. Nguyen, F. Tchéou, J. Rossat-Mignod, (to be published).

48. J. Rossat-Mignod, P. Burlet, S. Quézel, O. Vogt, Physica,
    102 B, 237 (1980)

49. H.W. Capel, Physica, 31, 1152 (1965).

50. J. Filippi, Thesis, Grenoble University, 1981
    J. Filippi, J.C. Lasjaunias, B. Hebral, J. Rossat-Mignod,
    F. Tchéou, J. Magn. Magn. Mat., 15-18, 527 (1980).

51. L.M. Falicov, W. Hanbe, M.B. Maples (Ed.), Valence fluctua-
    tions in Solids, North Holland, New-York, 1982.

52. P. Watcher (Ed.), Zurich Conference on Valence Instabilities,
    North-Holland, 1982, in press.

53. J.M. Lawrence, P.S. Riseborough, R.D. Parks, Valence fluctua-
    tion Phenomena, Rep. Progr. Physics, 44, 1 (1981).

54. J.X. Boucherle, J. Flouquet, P. Haen, J. Schweizer,
     C. Vettier, Proceedings of Rare Earths and Actinides
     Conference, Durbam, U.K. (1982).

55. T. Kasuya, K. Takegahara, Y. Aoki, T. Suzuki, S. Kunii,
     M. Sera, N. Sato, T. Fujita, T. Goto, A. Tamaki,
     T. Komatsubara, Zurich, Conference on Valence Instabilities,
     1982, (to be published).

56. F. Steglich, J. Aarts, C.D. Bredl, W. Lieke, D. Meschede,
     W. Franz, H. Schäfer, Phys. Rev. Lett., 43, 1892 (1979).

57. J.M. Effantin, P. Burlet, J. Rossat-Mignod, S. Kunii,
     T. Kasuya, Zurich Conference on Valence Instabilities, 1982,
     (to be published).

58. J. Rossat-Mignod, P. Burlet, J. Villain, H. Bartholin,
     T.S. Wang, D. Florence, O. Vogt, Phys. Rev., B 16, 440 (1977)

59. J. Rossat-Mignod, P. Burlet, S. Quézel, J.M. Effantin,
     O. Vogt, H. Bartholin, Conference on Solid Compounds of
     Transition Elements, Grenoble 1982. To be published in the
     Annales de Chimie by Masson.

60. S. Quézel, J. Rossat-Mignod, F. Tchéou, Sol. Stat. Commun.,
     42, 103 (1982).

61. H. Rohrer, C.H. Gerber, Phys. Rev. Lett., 38, 99 (1977).

62. J. Rossat-Mignod, P. Burlet, H. Bartholin, O. Vogt,
     R. Lagnier, J. Phys. C., 13 (1980).

63. P. Bak, J. Von Boehm, Phys. Rev., B 21, 5297 (1980).

64. W. Selke, M.E. Fischer, Phys. Rev., B 20, 257 (1979).

65. J. Villain, M.B. Gordon, J. Phys. C., 13, 3137 (1980).

66. V.L. Pokrovsky, G.V. Uimin, (to be published in Zh E T F).

67. W. Marshall, S.W. Lovesey, Theory of Thermal Neutron Scatte-
     ring, Oxford University Press, London, 1971.

68. W.J.L. Buyers, T.M. Holden, A. Perreault, Phys. Rev. B, 11,
     266 (1975).

69. J.G. Houmann, M. Chapellier, A.R. Mackintosh, P. Bak,
     O.D. Mc Masters, K.A. Gschneidner, Phys. Rev. Lett., 34, 587
     (1975).

70. D. Delacote, Thesis, Grenoble University (1981).

J. Rossat-Mignod, D. Delacote, J.M. Effantin, C. Vettier, O. Vogt, Yamada Conference VI on Neutron Scattering of Condensed Matters, Japan, Sept. 1982.

DISCUSSION

URLAND (question)

You still use the Stevens' approach for calculating the matrix
elements. It is much more consistent to use the tensor operator
method of Racah. If you include the higher states, calculations
of the off-diagonal terms by Stevens' method is very complicated.

ROSSAT-MIGNOD (answer)

For the ground state, it does not matter so much which method
you use.

DE LONG (comment)

It was said that there was no good experimental evidence for the
tensor terms in the exchange interaction at the beginning of the
talk. I believe that in the dilute alloys of lanthanides-Sn(III)
with Nd impurities are becoming much more well documented now.
There is evidence from superconductivity data as well as transport
and other measurements that such anisotropic terms do in fact
exist and have an effect on the magnetic properties and the depression of the superconductivity properties.
I would like to ask a question on the form factor measurements.
Do you think there is a possibility that the form factor in Ce
intermetallics be caused by relatively obscured crystal field
effect in some of the mixed valent systems rather than some sort
of a Mott transition.

ROSSAT-MIGNOD (answer)

You are thinking of Ce-Sn(III). In Ce-compounds where the crystal
field splitting is quite large, I think the situation is clear.
But it is not so clear when the splitting is small or when you
have a mixed valence system. In a mixed valence system the electron is not all the time in the f-state; it can be in the d-state
or in the "free state". In such case we are not sure how to handle
the data. If you consider that you have 50% 4f and 50% 5d and the
5d is expanded more, there will be a contribution close to the
origin for your form factor. In the case of Ce-Sn(III) we have
seen a very large contribution appearing when we go down in temperature. The contribution has mainly d-character because it has
expanded in real space.

NETZ   (question)

How do you explain the non-magnetic behaviour of Ce(III) ion in
your compound? Are you sure that Ce is in the 3+ state. I would
like to add that Pr and Nd show intermediate valence behaviour
too and these two ions should be included in your list.

ROSSAT-MIGNOD   (answer)

I became aware during this conference that Pr and Nd show intermediate valence behaviour.
Yes, Ce is in the 3+ state and non-magnetic. In a sense this is
unique. At this moment, there is no theory that explains this behaviour. I believe that this non-magnetic behaviour is due to
strong coupling between f-electron and conduction electrons,
which eliminates the degree of freedom of your f-electron. But
this is only a qualitative picture.

DE LONG   (comment)

You may be aware of some recent work in rare earth-Rh-borides,
where an explanation for the trend in the magnetic ordering temperature is given. The magnetic ordering temperature peaks at Dy.
This is supposedly due to crystal field anisotropy. What is your
feeling on the role of crystal field anisotropy in this particular
type of material?

ROSSAT-MIGNOD   (answer)

In rare earth intermetallics the interaction is via conduction
electrons. For this interaction between the spin of the 4f electron and the spin of the conduction electron, you get a maximum
value if you have a s-band. But if the exchange interaction is
more complex this does not occur.

CHOPRA   (question)

It has been reported that metallic Sm and Ce surfaces exhibit divalent and tetravalent ions respectively. Please comment.

BREWER   (comment)

If your Ce somehow gets oxidized you will have tetravalent Ce.
But another phenomenon may be important. In order to have metallic
bonding you have to have many neighbours. The structure of the surface may not be the same as the bulk.

DE LONG   (comment)

People looked into Sm compounds and claim has been made that divalent Sm is stable on the surface. Later on it was found that
the material responsible is actually SmO. However, recent papers
really address this problem. They claim now that with clean Sm
surface they see a non-oxidized divalent Sm. What Prof. Brewer
says makes sense because you do not have the availability of
bonding in the half-plane. This raises an interesting question,
what would happen to Nd surface! It is really a very important
problem in the field of catalysis how the surface behaves.

SINHA   (comment)

Of course the reaction occuring on the metal surface is of enormous industrial importance. My attention was recently drawn to an
interesting lecture by Prof. E. L. Muettertis of Berkeley(Royal
Soc. Chem. Centenary Lecture, Dalton Division, May 13, 1982)
reviewing the reactions of hydrocarbons at metal centers. In his
lecture, Prof. Muettertis pointed out how the coordination numbers
of the metal atoms present on the surface may change, depending
on the packing(structure). Thus for a fcc lattice the following
coordination numbers of the surface metal atoms are proposed:

    111 closed packed plane: CN = 9
    flat 100 plane   : CN = 8
    110 superstepped plane : CN = 7(in step sites)
    just below, in next plane: CN = 11

[This work has just appeared in print: see E. L. Muettertis, Chem.
Soc. Rev. $\underline{11}$, 283 (1982) -- Editor]

Spectroscopic Properties

OPTICAL SPECTROSCOPY OF LANTHANIDES IN CRYSTALLINE MATRIX

S. Hüfner

Universität des Saarlandes, Saarbrücken, Germany

ABSTRACT

The energy level structure of free trivalent rare earth (RE)
ions is discussed and its modification by the crystal field
described. Modification of the crystal field spectra in
imperfect crystals are discussed. Intensities and linewidth
are considered. The spectroscopy of solids employing lasers
is described followed by the consideration of hyperfine interactions, energy transfer and glasses contains RE's. Finally
divalent RE ions and magnetic interactions are mentioned
briefly.

1. INTRODUCTION

The most impressive feature about the spectra of rare earth
(RE) ions in ionic crystals is the sharpness of many lines
in their absorption and emission spectra. As early as 1908
[1] it was realized that in many cases these lines can be
as narrow as those commonly observed in the spectra of free
atoms of free molecules. If one thinks of spectra of solids
in terms of very broad absorption lines or bands, this feature must seem attractive. It means that in principle it is
possible to investigate interactions in a solid by optical
means with a degree of accuracy similar to that usually
possible for free atoms or ions.

The narrow optical lines suggest that the interaction or RE
ions with the crystalline environment is relatively weak.
One can thus describe RE energy levels to a good degree of
accuracy with a one-ion model - a very attractive feature

313

S. P. Sinha (ed.), Systematics and the Properties of the Lanthanides, 313–388.

in solid state physics.

We list in Table 1 some basic information about the RE:
atomic number, elements, outer electronic shells, free ion
ground terms and Landé g_J-factors for the trivalent state.

Table 1.   The Rare Earth Elements and Some of Their Basic
           Spectroscopic Properties

Atomic number	Element	Electron configuration RE^{3+}	Ground term RE^{3+}	Landé factor $g_J RE^{3+}$
57	Lanthanum	$4f^0 5s^2 5p^6$	$^1S_0$	0
58	Cerium	$4f^1 5s^2 5p^6$	$^2F_{5/2}$	$\frac{6}{7}$
59	Praseodym	$4f^2 5s^2 5p^6$	$^3H_4$	$\frac{4}{5}$
60	Neodymium	$4f^3 5s^2 5p^6$	$^4I_{9/2}$	$\frac{8}{11}$
61	Promethium	$4f^4 5s^2 5p^6$	$^5I_4$	$\frac{3}{5}$
62	Samarium	$4f^5 5s^2 5p^6$	$^6H_{5/2}$	$\frac{2}{7}$
63	Europium	$4f^6 5s^2 5p^6$	$^7F_0$	0
64	Gadolinium	$4f^7 5s^2 5p^6$	$^8S_{7/2}$	2
65	Terbium	$4f^8 5s^2 5p^6$	$^7F_6$	$\frac{3}{2}$
66	Dysprosium	$4f^9 5s^2 5p^6$	$^6H_{15/2}$	$\frac{4}{3}$
67	Holmium	$4f^{10} 5s^2 5p^6$	$^5I_8$	$\frac{5}{4}$
68	Erbium	$4f^{11} 5s^2 5p^6$	$^4I_{15/2}$	$\frac{6}{5}$
69	Thulium	$4f^{12} 5s^2 5p^6$	$^3H_6$	$\frac{6}{7}$
70	Ytterbium	$4f^{13} 5s^2 5p^6$	$^2F_{7/2}$	$\frac{8}{7}$
71	Lutecium	$4f^{14} 5s^2 5p^6$	$^1S_0$	0

It has become customary in spectrocsopy to give the energy
in wave numbers or cm^{-1} - a unit that, in order to be a
"true energy unit", has to be multiplied by hc (h is Planck's
constant and c the velocity of light). We note finally that

a number of reviews exist about the topics (or part of them) covered here [2-30], and we refer the reader to them for further details.

2. FREE IONS

The REs in solids are either divalent or trivalent. Their electronic configuration is $4f^N 5s^2 5p^6$ or $4f^{N-1} 5s^2 5p^6$, respectively. By far the most common valence state of the RE ions in solids is the trivalent one.

The 4f electrons are obviously not the outermost ones. They are "shielded" from external fields by two electronic shells with larger radial extension ($5s^2 5p^6$), which explains the "atomic" nature of their solid state spectra. Due to the shielding the 4f electrons are only weakly perturbed by the charges of the surrounding ligands.

This is why the RE ions are such a useful probe in a solid: the crystal environment constitutes only a small perturbation on the atomic energy levels, and many of their solid state, and hence spectroscopic, properties can be understood from a consideration of the free ions. In turn, the wavefunctions of the free ions constitute a good zero order approximation for a description of solid state properties.

The energy levels of the 4f configuration for trivalent free RE ions have been analyzed in arc spectra for Pr^{3+} [31], Gd^{3+} [32], and Er^{3+} [33]. Table 2 lists the experimentally observed free ion energy levels together with the centers of gravity for $PrCl_3$. The Russell-Saunders notation for the energy levels is used. The fact apparent from the table is that the positions of the terms in the free ion and in the ionic host are the same to within a few hundred wave numbers. This finding had been anticipated from the electronic structure but is nicely confirmed by the experimental data.

Having recognized that all the electronic shells except the 4f shell are spherically symmetric, and therefore do not contribute significantly to the relative positions of the 4f energy levels, we can write the Hamiltonian that determines the 4f energy levels as:

$$H = -\frac{\hbar^2}{2m} \sum_{i=1}^{N} \Delta_i - \sum_{i=1}^{N} \frac{Z^* e^2}{r_i} + \sum_{i<j}^{N} \frac{e^2}{r_{ij}} + \sum_{i=1}^{N} \zeta(r_i) \underline{s}_i \cdot \underline{l}_i \quad (1.1)$$

where $N = 1,\ldots,14$ is the number of the 4f electrons, $Z^* e$ the screened charge of the nucleus because we have neglected the closed electronic shells, and $\zeta(r_i)$ the spin orbit coup-

Table 2: Energy Levels for Pr^{3+} as Obtained from Arc Spectra
Yielding Those of the Pr^{3+} Ion, and for PrCl$_3$.After[31]

	E(Vapor state) (cm^{-1})	E(Crystal) (cm^{-1})		E(Vapor state) (cm^{-1})	E(Crystal) (cm^{-1})
$^3H_4$	0	0	$^1G_4$	9921.4	9697.6
$^3H_5$	2152.2	2117.4	$^1D_2$	17334.5	16639.3
$^3H_6$	4389.1	4306.3	$^3P_0$	21390.1	20383.4
$^3F_2$	4996.7	4846.6	$^3P_1$	22007.6	20984.9
$^3F_3$	6415.4	6232.3	$^1I_6$	22211.6	21324.5
$^3F_4$	6854.9	6681.7	$^3P_2$	23160.9	22139.1
			$^1S_0$		47200.0

ling function

$$\zeta(r_i) = \frac{\hbar^2}{2m^2c^2r_i} \frac{dU(r_i)}{dr_i} \tag{1.2}$$

where $U(r_i)$ is the potential in which the electron i is moving. The first term in the Hamiltonian (1.1) is the kinetic
energy of the 4f electrons and the second term represents
their Coulomb interaction with the nucleus.

The first two terms of the Hamiltonian are spherically
symmetric and therefore do not remove any of the degenera-
cies with the configuration of the 4f electrons. We neglect
them in the following discussion.

The next two terms, which represent the mutual Coulomb interaction of the 4f electrons (H$_c$) and their spin-orbit interaction (H$_{so}$) are responsible for the energy level structure
of the 4f electrons. In REs, the two last terms in eq.(1.1)
are of about equal magnitude and the energy level calculations are therefore mathematically more involved; this situation is called intermediate coupling.

The Hamiltonian in eq.(1.1) is solved in a one-electron
approximation; that is, electron correlations are neglected.
To calculate the energy levels in the intermediate coupling
approximation we have to calculate the matrix elements for
the Hamiltonian

$$H_1 = H_c + H_{SO} \tag{1.3}$$

in a set of basis functions and then diagonalize the matrix
for the specific $4f^N$ configuration that interests us. It is
now common practice to use a basis set of Russell-Saunders
eigenfunctions. The Hamiltonian H_1 is diagonal in J, and
therefore the total matrix of energies for the $4f^N$ configuration can be split up into submatrices for states with the
same J. These states are then still degenerate in M_J and are
a linear combination of states with different L and S but
the same J. There are also configurations with the same L
and S that occur more than once; to distinguish them, new
quantum numbers must be introduced. Details of the matrix
element calculations of H_c and H_{SO} will not be given here,
since there are many excellent reviews in which the deriva-
tions can be found [16,30].

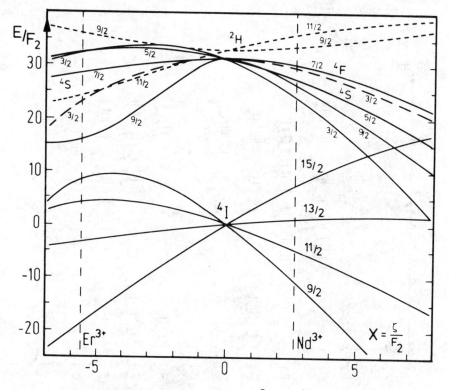

Fig. 1. The lowest terms of the $4f^3 = 4f^{11}$ configuration as a
function of the normalized spin-orbit coupling constant
$X = \zeta_{4f}/F_2$. After [8]
The value of this constant where the best agreement with experimental data is achieved is also indicated for Er^{3+} and Nd^{3+}.
The numbers on the various curves are the total angular momenta J of the terms. Energies at the ordinate are in units of F_2.

A scheme that illustrates well the deviations from pure
Russell-Saunders coupling is one in which the energies of the
terms obtained by diagonalizing the Coulomb and the spin-
orbit interaction matrices of a $4f^N$ configuration for every
value of J are plotted as a function of ζ/F_2. As an example
in Fig. 1 these results are shown for the $4f^3$ and $4f^{11} =$
$4f^{-3}$ configurations representing Nd^{3+} and Er^{3+}[8]. The electrostatic matrix elements are the same for a configuration of N
electrons and N holes. On the other hand, ·the spin-orbit
interaction changes sign, being negative for the $4f^{-N}$ configuration. The energies of the $4f^N$ to $4f^{-N}$ configurations
are thus obtained from the same matrices where only the sign
of the spin-orbit matrix elements changes in going from
$4f^N$ to $4f^{-N}$. The same plot of the energies as a function of
ζ/F_2 therefore contains for $(\zeta/F_2)< 0$, the energies for the
$4f^{-N}$ configuration, and for $(\zeta/F_2)> 0$, the energies for the
$4f^N$ configuration.

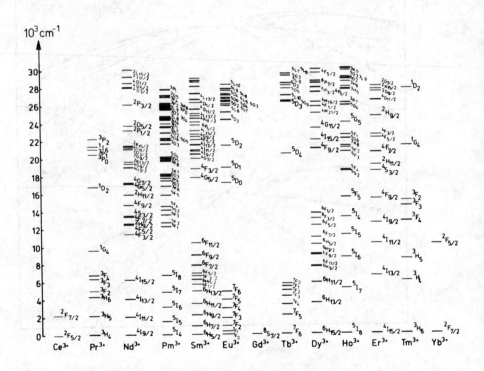

Fig. 2 Energies of the RE^{3+} levels. From [2]

The Russell-Saunders approximation holds as long as the energies are a linear function of the spin-orbit interaction. One sees from Fig. 1 that already for modest values of ζ strong nonlinearities occur indicating the deviation from the Russell-Saunders coupling.

In the Fig. 1 also those values of ζ/F_2 for Er^{3+} and Nd^{3+} are indicated at which the best agreement with the experimental energy levels is observed. In this way the experimental energy levels can be explained in the intermediate coupling scheme.

Fig. 2 shows free ion energy levels of all RE^{3+} determined in this way. We can see that up to the near ultraviolett region for all the REs the energy level schemes are established.

The wavefunctions are now no longer pure Russell-Saunders wavefunctions, but a linear combination of them. For example, the ground term wavefunction of Er^{3+} is in the intermediate coupling approximation [34]:

$$|^4I'_{15/2}) = 0.984\,|^4I_{15/2}) + 0.176\,|^2K_{15/2}) + 0.019\,|^2L_{15/2})\quad (1.4)$$

(instead of pure $^4I_{15/2}$) where the primed symbol denotes the intermediate coupling state. In this case the admixture is small and can almost be neglected. This is not so for some of the higher terms for example [34]:

$$|^4I'_{9/2}) = -0.399\,|^4F_{9/2}) - 0.303\,|^2G^1_{9/2}) + 0.239\,|^2G^2_{9/2}$$
$$-0.017\,|^4G_{9/2}) - 0.200\,|^2H^1_{9/2}) + 0.419\,|^2H^2_{9/2})\quad (1.5)$$
$$+0.690\,|^4I_{9/2})$$

where the superscripts 1 and 2 distinguish the two ^2G and ^2H states for $4f^{-3}$. These wavefunctions can be used to calculate various properties. For example the g_J-factor of a state is obtained by

$$g_J = \sum_i a_i^2 g_{J_i}\,(RS)\quad\quad\quad (1.6)$$

yielding for the $^4I'_{15/2}$ term in Er^{3+}

$$g_J(^4I'_{15/2}) = (0.984)^2 g_J(^4I_{15/2}) + (0.176)^2 g_J(^2K_{15/2})$$
$$+ (0.019)^2 g_J(^2L_{15/2}) = 1.196\quad (1.7)$$

instead of the pure Russell-Saunders value 1.20.

The situation described so far, gives the basic physics of
the problem. However there are higher order terms to the
Hamiltonian in eq.(1.1) that have to be taken into account
if a truly good agreement between theory and experiment is
to be achieved. We refer the reader to the appropriate literature [2,30,35,36].

3.   TRIVALENT IONS IN THE CRYSTAL FIELD

The 4f shell of RE ions is a nonclosed shell and therefore
has an aspherical charge distribution. If the ion is introduced into a crystal, the ion experiences an inhomogeneous

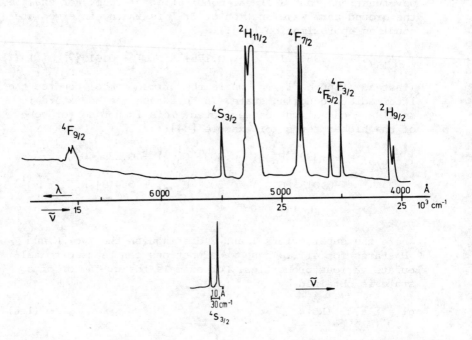

Fig. 3   Absorption spectrum of Er-ethylsulfate (Er(C$_2$H$_5$SO$_4$)$_3$ ·
9H$_2$O). Scan of the total visible range with low resolution,
showing the wellseparated groups of lines. The transition
$^4I_{15/2} \rightarrow {}^4S_{3/2}$ (at 4.2 K) is enlarged, in order to show the
sharpness of the absorption lines. After [2]

electrostatic field, the so-called crystal field, which is produced by the charge distribution in the crystal. It removes to a certain degree (which depends on the crystal symmetry), the M_J degeneracy of the free ion 4f levels. Experiments have shown that for trivalent RE ions the crystal field interaction is smaller than the energy separation of the various free ion terms. This means that an understanding of this crystal field splitting can be obtained by applying the potential produced by the crystal on the free ion 4f wavefunctions.

Fig. 3 shows a typical absorption spectrum of a RE, where one can easily see the effect of the crystal field (Er-ethylsulfate). At 77 K and low resolution well resolved groups of lines which are separate from each other, are observed. At 4.2 K and at high resolution the M_J splitting of the J levels is resolved: the $^4S_{3/2}$ state splits into two crystal field levels, one with $M_J = \pm 1/2$ and another one with $M_J = \pm 3/2$.

Crystal field splitting has two aspects: one concerns the symmetry of the problem, namely, the number of levels into which an ion's free ion J terms are split in a crystal field of a given symmetry; the second concerns the actual size of the crystal field splitting. Whereas the first topic is now completely understood, the second is being actively researched, and the various mechanisms that determine the magnitude of the crystal field splitting are by no means completely understood (see, e.g. [20]).

Therefore a parametrization scheme has been developed that makes it possible to adsorb a measured crystal field spectrum into a few parameters [37].

We calculate the crystal potential $V(r_i, \varphi_i, \theta_i)$ at the site of the 4f electrons and the potential energy of the 4f electrons in this potential. If the crystal has charge density $\rho(R)$ and the 4f electrons have radius r_i, we obtain

$$V = - \sum_i \int \frac{e_i \rho(R)}{|R - r_i|} \, d\tau = - \sum_{k,i} e_i \int \rho(R) P_k (\cos(R, r_i)) \frac{r_<^k}{r_>^{k+1}} \, d\tau \quad (3.1)$$

where $r_<$ and $r_>$ are, respectively, the smaller and larger value of R and r_i.

$$V = - \int \sum_{k,q,i} e_i \int \rho(R) \frac{4\pi}{2k+1} (-1)^q Y_{kq}(\theta_i, \varphi_i) Y_{k-q}(\theta, \varphi) \frac{r_<^k}{r_>^{k+1}} \, d\tau \quad (3.2)$$

$$V = - \sum_{k,q,i} e_i \int (-1)^q \rho(\underline{R}) C_{kq}(\theta_i,\varphi_i) C_{k-q}(\theta,\varphi) \frac{r_<^k}{r_>^{k+1}} d\tau \qquad (3.3)$$

$$V \equiv \sum_{k,q,i} B_{kq} C_{kq}(\theta_i,\varphi_i) \qquad (3.4)$$

where the crystal field parameter B_{kq} is defined as

$$B_{kq} = -e \int (-1)^q \varsigma(\underline{R}) C_{k-q}(\theta,\varphi) \frac{r_<^k}{r_>^{k-1}} d\tau \qquad (3.5)$$

To calculate the matrix elements of V we apply the rules for calculating the matrix elements of tensor operators [16,30]. The radial integrals that are hard to obtain exactly are absorbed (together with the effect of the crystal potential) into the crystal field parameters B_{kq}.

In early treatments of the crystal field interaction it was assumed that in RE compounds the point charge contribution would be the dominant part of the crystal field interaction. Then in eq.(3.2) the integral over the lattice can be replaced by a sum over all lattice points and $r_<^k$ can be replaced by r_i^k. The latter replacement can be performed as long as the charge distribution of the crystal does not enter that of the 4f electrons (as long as $r_i < \underline{R}$), which implies that the potential acting on the 4f electrons obeys the Laplace equation ($\Delta\phi(r_i,\theta_i,\varphi_i) = 0$) at the position of the 4f electrons. This has led to slightly different notation, which we mention, because it is used so frequently. Since the symbols originate from a Hamiltonian that was derived from a point charge model, the radial $<r^k>$ integrals for the 4f electrons show up explicitly. Abragam and Bleaney [4] use the Hamiltonian

$$V = \sum_{k,q,i} \bar{B}_{kq} r_i^k Y_{kq}(\theta_i,\varphi_i) \qquad (3.6)$$

which yields crystal field parameters $\bar{B}_{kq} <r^k>$

$$\bar{B}_{kq} = -e \int \frac{\rho(\underline{R})}{R^{k+1}} (-1)^q C_{k-q}(\theta,\varphi) d\tau \qquad (3.7)$$

Then Stevens [37] in his original work on crystal field parametrization used the potential energy form.

$$V = \sum_{k,q,i} A_{kq} P_{kq}(x_i,y_i,z_i), \quad q \geq 0 \qquad (3.8)$$

where the P_{kq} are special polynomials. Here the calculation
of the matrix elements again results in a factor $<r^k>$ and
crystal field parameters $A_{kq}<r^k>$ are obtained. These $A_{kq}<r^k>$
parameters are the ones most often used in the literature.
We emphasize that experiments always give the products $B_{kq}<r^k>$
and $A_{kq}<r^k>$ and there is no way to separate them. The situation with respect to the crystal field parameters is further
complicated because the Johns Hopkins group has frequently
used the definition $B_{kq} = A_{kq}<r^k>$ (where the $A_{kq}<r^k>$ are those
of eq.(3.8)), and recently Crosswhite has used the notation
$C_{kq} = B_{kq}$ (where the B_{kq} are those of eq.(3.4)). Tables that
relate the crystal field parameters derived from the different
definitions can be found eg. in [2,4].

In actual calculations a problem can occur. As mentioned, the
free ion energy levels are not perfectly described by the
Hamiltonian in eq.(1.1). Discrepancies of the order of
100 cm^{-1} are possible. If the procedure just described is
applied, the crystal field matrix elements nondiagonal in J
will immediately try to compensate for this discrepancy, producing an artificial fit. Two approaches have been used to
correct this. The better method is to incorporate all the
additional parameters, that determine the free ion energy
levels, into the fitting routine. This has been done for RE^{3+}
in $LaCl_3$ [35,38,39]. However, a more practical and quite successful way can also be used: first, the centers of gravity
of the terms are fitted to the free ion Hamiltonian containing only the contributions with ζ, F_2, F_4 and F_6, and the
remaining discrepancies are adjusted by small additional energies. With these "perfect" free ion energies the total crystal
field Hamiltonian (including J mixing) is then applied, and
the small free ion correction energies just mentioned are also
varied iteratively [40].

Inclusion of the off-diagonal crystal field matrix elements
(in J) is essential in order to obtain a reasonable fit, especially in systems with a large crystal field interaction.
Another problem lies in the fact that in many instances the
description of a measured crystal field splitting is markedly
improved if different sets of crystal field parameters are
used for different multiplets. This indicates that the charge
distribution of the 4f electrons depends on the nature of the
state [40].

This means that one can no longer regard all 4f electrons as
equal but has to introduce two and more electron operators
into the crystal field Hamiltonian. This introduces a great
number of additional parameters [41,42,43,44,45]. Judd [46]
has, however, found an easy way to handle the leading contribution to the two electron Hamiltonian. He points out that

the strong attractive exchange forces between 4f electrons
with parallel spin leads to a less extended wavefunction and
therefore also to a smaller crystal field parameter. The
easiest way to take this into account is to replace the operator C_{kq} $(\varphi_i, \theta_i,)$ in eq. (3.4) by

$$C_{kq} \ (\varphi_i, \theta_i) + c_k \ (\underline{S} \cdot \underline{s}_i) \ C_{kq} \ (\varphi_i, \theta_i) \tag{3.9}$$

The parameters c_k must be negative if the contraction of the
radial function corresponds to parallel spins.

This replacement introduces three new parameters c_k (k = 2,4,6)
but it seems to improve matters considerably in cases where
the simple crystal field approach leads to not completely
satisfactory results.

The crystal field splitting of the $^7F_1$ and the $^5D_1$ term in
$Eu(C_2H_5SO_4)_3 \cdot 9H_2O$ may serve as an example. A J = 1 state
in the hexagonal ethylsulfate lattice is split into two levels
$M_J = 0$ and $M_J = \pm 1$ and the splitting is proportional to the
operator equivalent factor α_J [37]. Numerically one finds
$\alpha(^5D_1)/\alpha(^7F_1) = 0.24$ [46], whereas the ratio of the crystal
field splittings is 0.12 [47]. Using the substitution in eq.
(3.9) one finds

$$\alpha(^5D_1)/\alpha(^7F_1) = \frac{11}{45} + \frac{79}{90} c_2 \tag{3.10}$$

A value of $c_2 = -0.14$ brings the measured and calculated ratio
of the crystal field splittings of the $^5D_1$ and $^7F_1$ state into
agreement. Judd [46] points out that similar small corrections
also improve substantially the crystal field analysis in other
cases.

With respect to the results for Nd^{3+}:$LaCl_3$, the system for
which perhaps the most detailed crystal field analysis has
been performed, the data of Crosswhite et al. [35] were obtained with a 23-parameter fit (although some parameters were
constrained) to 101 crystal field energy levels, and the mean
deviation between calculated and experimental energy levels
is 8 cm^{-1}. On the other hand, the original calculation by Judd
[48] was performed with much less experimental material. In
view of this, his results are also quite remarkable: the deviation between his parameters and those of [38] is less than
10 % except for the $A_{20}\langle r^2 \rangle$ parameter, where it amounts to
25 %. Thus, such seemingly naive calculations [48] reproduce
the physics of the problem very well.

Now we consider the theoretical consequences of experimental

results on the crystal field splitting of RE ions presented.
We have seen that the Stevens [37] parametrization scheme is
very successful in describing the experimental data. This
leads to two questions.

(1)   Why is the parametrization scheme used for the analysis
of the crystal field interaction so successful?
(2)   Is it possible to determine the magnitude of the crystal
field parameters from an a priori calculation?

The first question was basically answered earlier when we
introduced the crystal field parameters. The parametrization
scheme makes no specific assumptions about the nature of the
crystal field and therefore holds equally well for more or
less ionic crystals. The two main restrictions entering the
scheme are that we assume that the crystal field acts on all
4f electrons independently and equally, and that the radial
distribution function of the 4f electrons is the same for
all terms considered. Both these approximations seem justifiable for 4f electrons and therefore make the parametrization
scheme's success on a qualitative basis understandable.

Newman [20,43] has tried to put the parametrization scheme on
a more mathematical basis in what he calls the superposition
model. The reasoning underlying this model can be found in
earlier work [50,51,52], where the model of Jorgensen has
been called the angular overlap model and is also frequently
used. The model of Newman assumes that each ion in the crystal
contributes separately and independently to the crystal field
of a reference ion. Therefore, all like ions contribute the
same amount if their distance is the same and if the geometrical factors are taken into account. We can therefore define
intrinsic parameters $\bar{A}_k(R)$, which depend only on the distance
R. The experimental parameters can then be written in the
form

$$A_{kq}\langle r^k\rangle = \sum_j g_{kq}(j)\bar{A}_k(R_j) \tag{3.11}$$

where the g_{kq} are the geometrical factors, which can be determined from the x-ray structure data, and the $\bar{A}_k(R_j)$ are
called intrinsic parameters; j is the running index for the
various ions. To make the superposition model applicable one
has to introduce the further assumption that only the ligands
contribute significantly to the crystal field. This is valid
for the k = 4 and k = 6 terms, whose contributions fall off
very rapidly with distance; it is, however, a bad approximation for the k = 2 terms, which are therefore not considered
in this model.

Analysis of a set of crystal field parameters in terms of
the superposition model has the advantage, especially in the
case of low symmetry, of reducing the number of parameters;
it also shows, at least qualitatively, the effect of two- and
more-body interactions on the size of the crystal field splitting. It also has value in the a priori calculation of crystal
field parameters, which can now be performed in terms of interactions between two ions.

Superposition analysis seems a very useful way to check the
consistency of a set of experimentally determined crystal
field parameters; it is also a good starting point for an
a priori calculation of crystal field parameters. First prin-
ziple calculations of crystal field parameters that give
satisfying results are still rare, having been performed so
far only by Newman and his collaborators [20] (for earlier
attempts see, e.g. [53,54]).

The superposition model shows that the $k = 4$ and $k = 6$ terms
of the crystal field are dominated by the interaction of the
central ion with its ligands; in addition, two- and three-
body interactions seem small. These findings provide the
starting point for the ab initio calculation of the crystal
field parameters as performed by Newman and his collaborators
for some systems; we shall deal here with the results for
$PrCl_3$, because they seem to be the most detailed [20]. In
these calculations the crystal is split into two parts: a
$PrCl_9^{-6}$ cluster, which is the central ion with all its ligands,
and the rest of the crystal; the cluster is then treated
"exactly", whereas only the point charge contributions from
the rest of the crystal are used.

Altogether so far ten different contributions to the crystal
field parameters in $PrCl_3$ have been calculated. Table 3 lists
the results of the crystal field parameters calculations and
compares them with the experimental parameters. The contributions (1) - (4) are self explanatory. The polarization
terms (3), (4) were calculated by using free ion polarizabi-
lities. The charge penetration contribution (5) is a correction to (1) where the effect of the ligands is modified due
to the finite size of their charge distribution. The overlap
and covalency term have just been explained. The ligand lanthanide exchange charge term (8) comes from the penetration
of the ligand charges into the $5s^2 5p^6$ shell; this charge distribution occurs by the mutual interpenetration of the ligand
charges (9). Finally, there is a three-body interaction between two ligands (10). Contributions (9) and (10) are not
encompassed by the superposition model but they are small.

The comparison between theory and experiment is quite grati-

Table 3. Results of a First Principle Calculation of the Crystal Field Parameters in $PrCl_3$ in cm^{-1}. After [20], see also [54a]

	Mechanism	$A_{20}<r^2>$	$A_{40}<r^4>$	$A_{60}<r^6>$	$A_{66}<r^6>$
(1)	Ligand point charges	158	-17	-2.7	33
(2)	Remaining point charges	638	8	-0.9	24
(3)	Dipolar polarization	469	3	-0.3	5
(4)	Quadrupolar polarization	-705	-14	1.0	-5
(5)	Charge penetration	-39	17	6.8	-82
(6)	Overlap	13	-27	-27	326
(7)	Covalency	7	-14	-13.1	160
(8)	Ligand-lanthanide exchange charge	-105	8	0.6	-7
(9)	Ligand-ligand exchange charge	-57	-3	-0.5	6
(10)	Triangular path contributions	2	-5	-5.1	76
	Total theory	(381)	-44	-41.2	536
	Experiment	47	-44.6	-39.6	405

fying for the $A_{40}<r^4>$, $A_{60}<r^6>$ and $A_{66}<r^6>$ terms. We also note that covalency and overlap contribute significantly to the size of the parameters. As expected, the calculated $A_{20}<r^2>$ parameter is considerably larger than the experimental one. In contrast to the fourth and sixth order parameters the second order parameter is given mostly by the point charge and polarization contributions. This fact introduces a number of problems. The polarization contributions were calculated by using free ion polarizations; it is not known how much these values are changed in going to the solid. Moreover, it is known that the electrostatic second order contributions are considerably shielded. This has often been expressed as

$$A_{kq}<r^k> = (1 - \sigma_k)A_{kq}<r^k> \quad \text{(point charge)} \qquad (3.12)$$

In the simplest interpretation the point charge contribution is understood to comprise only the first two terms in Table 3; the next step is to include contributions (3) and (4) as well.

At this point, it is not clear how the contributions in Table 3 are affected by shielding. Calculations give values for σ_2, of the order of 0.8 and for σ_4 and σ_6 of the order of 0.1 [55]. This shows that a reliable theoretical estimate of

$A_{20}<r^2>$ requires a careful shielding calculation.

4.   SATELLITES

In the simplest view the rare earth (RE) crystal spectra are
composed of pure electronic transitions between states which
have their origin from the crystal field split free ion
states. There are, however, in these spectra a great number
of additional lines which fall into two classes. One series
of lines comes from the simultaneous excitation of a phonon
or a magnon with the electronic transition. Another group of
lines are the so-called satellite lines which are generally
quite close to the main electronic lines and mostly show
properties (polarization and Zeeman effect) very similar to
those of the pure electronic transition. These satellite
lines have already been observed when the first spectra of RE
ions were analysed.[2]. For decades these satellite lines
where a puzzle and many different suggestions as to their
origin were put forward. But none of them could be proven.
Faulhaber [56] and Yamamoto et al.[57] could for the first
time trace down exactly the origin of at least a few of these
satellites. Faulhaber showed that if the neighbouring ligand
ions to the RE ion were replaced by similar ions (as e.g. a
replacement of Al by Ga in $3Dy_2O_3 \cdot 5Al_2O_3$) satellite lines
were produced; and he suggested therefore that satellite
lines were produced by impurities on regular lattice sites
although of different ionicity and volume than the original
ions. Burns [58] analysed the data of Faulhaber and showed
that the substitution of impurities can also produce considerable changes in crystal field and not only covalency shift
as suggested by Faulhaber. Yamamoto et al. [57] showed that
if one replaced a RE ion in a crystal by another RE ion,
satellites could occur because, due to different ionic radii,
the lattice was slightly distorted which in turn was responsible for a different crystal field producing additional
(satellite) lines.

In all the work except that of Faulhaber [56] the dominant
distortion produced by the impurity was a change of the
symmetry of the crystal field and therefore the separation
of the satellite lines from the main lines was relatively
small. The work of Faulhaber had indicated that in addition
to the crystal field distortion sizeable shifts due to the
changes of the ionicity of the impurity as compared to the
original ion could be produced. It therefore seemed interesting to persue this effect in order to possibly learn more
about the contributions to the crystal field interaction in
RE compounds. This was appealing especially in view of the
fact that theoretical efforts [20] have produced considerable
insight into this problem. A system which seemed very suited

for this kind of study are the rare earth trihalides because
here ligands with very different ionicity from F to I are
available and on the other hand the crystal structure of
these compounds is relatively simple.

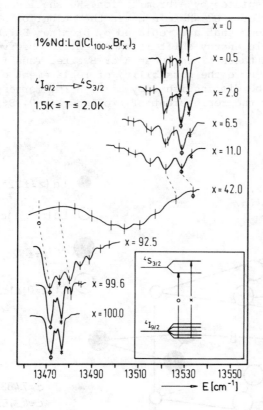

Fig. 4   Absorption spectra of the $^4I_{9/2} \rightarrow {}^4S_{3/2}$ transition
for Nd^{3+} in $La(Cl_{100-x} Br_x)_3$ ($0<x<100$) at 1.6 K.
After [59]

The system $Nd: LaCl_{3-x}Br_x$ ($0<x<3$) was investigated over the
whole concentration range. A typical set of spectra from the
transition $^4I_{9/2} \rightarrow S_{3/2}$ is shown in Fig. 4. Starting from
pure $LaCl_3$ where essentially the two crystal field levels
of the $^4S_{3/2}$ state are observable one notices with increasing Br concentration the apparence of satellite lines [59].

We see that the doping produces extra lines, generally called
satellites. A full interpretation of these satellites can
be given with the crystal structure of Fig. 5 [60]. It shows
that each RE ion in $LaCl_3$ is surrounded by 9 chlorines, three
in the plane of the RE ion with distance B = 2.96 Å and three

each above and below the RE ion, with distance A = 2.91 Å.
The general interpretation of the satellite now assumes, that
they correspond to ions in which relative to the unperturbed
lattice one or more of the nine nearest neighbour chlorine
atoms are replaced by a bromine ion. We shall call the satellites henceforth S^n where n gives the number of near neighbour
chlorine atoms that are replaced by bromine. Furthermore a
difference in energy shift is expected and observed if the
replacement takes place on an A or B site. Thus for one ion
replaced we have the possibility of replacement on two sites,
which will be distinguished by $S^{1,A}$, $S^{1,B}$. Correspondingly
we have for the replacement of two ions $S^{2,2A}$, $S^{2,2B}$ and
$S^{2,A,B}$.

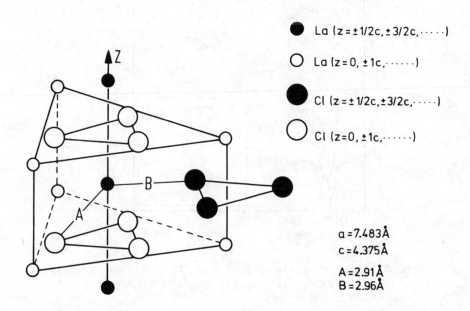

Fig. 5   Crystal structure of $LaCl_3$. After [60]

The analysis can be started with spectra taken from samples
with very small doping (Fig. 6, x = 0, 0.02, 0.5, 2.8). The
x = 0 spectrum shows two absorption lines as expected for a
transition to an excited state with J = 3/2 at low temperatures. In Fig. 6 all transitions that belong to $I_{9/2}(1) \rightarrow$
$^4S_{3/2}(1)$ have been marked by a circle and these belonging to

$I_{9/2}(1) \rightarrow {}^4S_{3/2}(2)$ by a cross. In the x = 0.5 spectrum two
groups of satellites at lower transition energy can be clearly
seen. Their energy separation corresponds roughly to the
${}^4S_{3/2}$ crystal field splitting, and therefore we attribute them
to the S^{1A}, S^{1B} satellites. We see that the intensity is
roughly 2:1 in agreement with the occupation of the A and B
sites respectively. One also notices that the shift by substitution on a A site (2.91 Å) is larger than that on a B
site (2.96 Å). For a higher concentration the S^2 satellites
can be made visible and the identification of $S^2,2A$, S^2,AB,
and $S^2,2B$ has been made according to the seize of the separation from the main line.

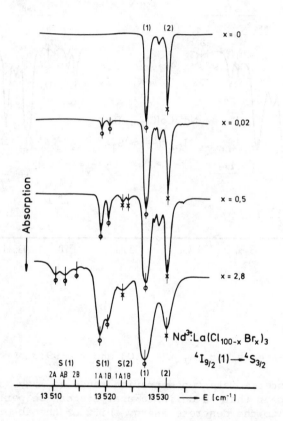

Fig. 6   Absorption spectra of the ${}^4I_{9/2} \rightarrow {}^4S_{3/2}$ transition
for Nd in La$(Cl_{100-x} Br)_3$ $(0 \leq x \leq 2.8)$ at 1.6 K.
After [59]

The situation can be seen most clearly by comparing the
spectra of LaCl$_3$ and LaBr$_3$ with a very weak doping of Br and
Cl respectively (see Fig. 7). The spectra are mirror images

of each other, the transitions to the $^4S_{3/2}(2)$ state always being somewhat weaker than that to the $^4S_{3/2}$ (1) state.

From Fig. 5 we see that the doping of $LaCl_3$ with Br(or of $LaBr_3$ with Cl) has two effects. First satellite lines appear, which are transitions to ions that have a discretely different ionic environment from that of the undoped crystal.

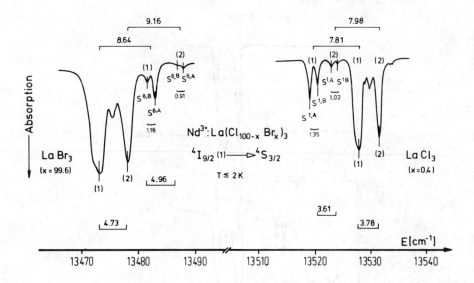

Fig. 7 Same as Fig. 4, x = 0.04 and x = 99.6. After [59]

In the intermediate concentration range so many satellite lines appear that no definite analysis can be performed. In addition to the descrete energy shift of the absorption lines as a consequence of the replacement of one halogen ion by another also a shift of the center of gravity of the two main lines $^4I_{9/2}(1) \rightarrow {}^4S_{3/2}(1)$, $^4S_{3/2}(2)$ is observed. This center of gravity shift is to higher transition energies for $LaCl_3$ upon Br substitution and to lower energies for $LaBr_3$ upon Cl doping. Fig. 8 shows the experimental data. This center of gravity shift can only be measured very accurately for small dopant concentrations and is than extrapolated over

the whole concentration range. We see, however, from Fig. 8
that in both cases, as expected, the same shift of $\varepsilon = 18$ cm^{-1}
is observed. This shift is interpreted as one which is pro-
duced by the internal pressure of the crystal. If LaCl$_3$ is
doped with Br the pressure decreases and the extrapolated
value (13547 cm^{-1}) of the transition energy is the center of
gravity of the $^4I_{9/2}(1) \to {}^4S_{3/2}$ transition in a fictions
LaCl$_3$ lattice with the LaBr$_3$ lattice constant. Correspondingly the extrapolated value (13458 cm^{-1}) of the transition
energy starting at the LaBr$_3$ side is the position of this
center of gravity in fictious LaBr$_3$ lattice with a LaCl$_3$
lattice constant.

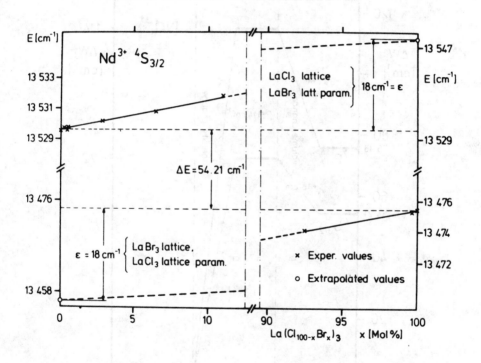

Fig. 8   Center of gravity shift of the $^4I_{9/2}(1) \to {}^4S_{3/2}$
         transition in LaCl$_3$ upon doping with Br and in LaBr$_3$
         upon doping with Cl extrapolated to full doping.
         After [59]

An analysis of these data is given in Fig. 9. It shows that
the center of gravities of the $^4I_{9/2}(1) \to {}^4S_{3/2}$ transition

in LaBr$_3$ and in LaCl$_3$ with a LaBr$_3$ lattice constant differ
by 72 cm^{-1}. Now the average satellite separation for this
transition for the replacement of one bromine ion by a
chlorine ion is δ = 8 cm^{-1}. Thus if all the 9 Br ions are
replaced by Cl, keeping all other parameters as eg. the
lattice constant fixed is 9 · 8 = 72 cm^{-1}. This is however
exactly the value obtained above by comparing directly the
center of gravity in LaBr$_3$ and LaCl$_3$ with LaBr$_3$ lattice constant. Thus we arrive at the conclusion that the replacement
of one ligand by another one has two effects: one which is
related to the chemical bonding between the RE ion and the
ligand and another one which is caused by the internal pressure in the crystal.

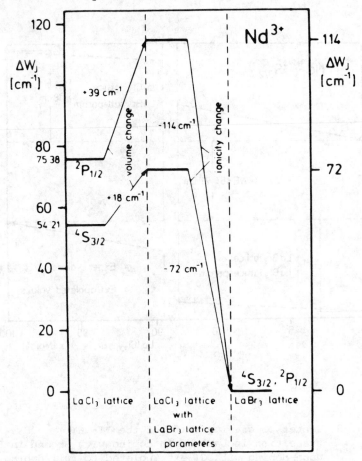

Fig. 9   Analysis of the LaCl$_3$/LaBr$_3$ data. The position of
         the $^4I_{9/2} \rightarrow {}^4S_{3/2}$, $^2P_{1/2}$ in LaCl$_3$ are set to zero.
         After[59]

The results for the $^4I_{9/2}(1) \rightarrow {}^4S_{3/2}$ transition have been checked by measuring also the $^4I_{9/2}(1) \rightarrow {}^2P_{1/2}$ transition. Here the average satellite separation is $\delta = 12.4$ cm^{-1}, and the volume shift from LaCl$_3$ to LaCl$_3$ with the LaBr$_3$ lattice constant is 39 cm^{-1} giving a total energy separation of this transition from LaCl$_3$ to LaBr$_3$ of 114 cm^{-1}. Indeed we find again that $9\delta(^2P_{1/2}) = 111.5$ cm^{-1} in very good agreement with the extrapolated value.

We find other interesting correlation between the data of the two transitions (data from Fig. 9):

$$\frac{E(^4S_{3/2})_{LaCl_3} - E(^4S_{3/2})_{LaBr_3}}{E(^2P_{1/2})_{LaCl_3} - E(^2P_{1/2})_{LaBr_3}} = \frac{54}{75} = 0.72$$

$$\frac{\delta(^4S_{3/2})}{\delta(^2P_{1/2})} = \frac{8}{12.4} = 0.64$$

$$\frac{\varepsilon(^4S_{3/2})}{\varepsilon(^2P_{1/2})} = \frac{18}{39} = 0.46$$

This indicates that these different effects have a common cause. Experiments have also been performed by doping LaCl$_3$ crystals with I. For this impurity the descrete shifts due to replacement of Cl ions are much larger than for Br doping as can be seen from the data in Fig. 10. The most interesting aspect for this doping is however that now the splitting of the satellite lines, which we attributed to population of the A and B halogene sites is no longer visible.

Finally the hypothetical pressure shift which we required above has been obtained qualitatively by a measurement with an externally applied pressure. Fig. 11 shows the energy dependence of the $^4I_{9/2}$ (1) $\rightarrow {}^2P_{1/2}$ transition as a function of the applied pressure parallel and perpendicular to the crystallographic c-axex. As expected the energy decreases with increasing pressure where:

$$\Delta E_{\|}(^2P_{1/2}) = -0.85 \text{ cm}^{-1}/\text{kbar}$$

$$\Delta E_{\perp}(^2P_{1/2}) = -0.3 \text{ cm}^{-1}/\text{kbar}$$

In a similar experiment one finds

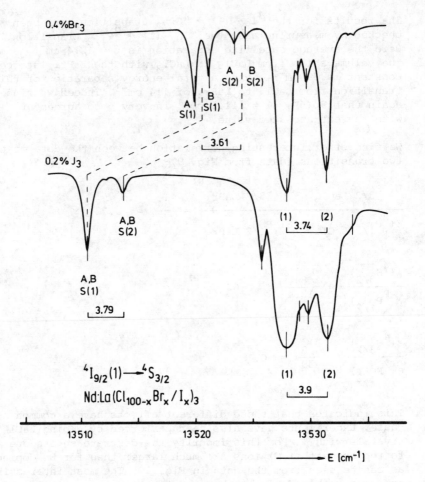

Fig. 10   Absorption spectra of the $^4I_{9/12} \to {}^4S_{3/2}$ transition
for Nd^{3+} in $La(Cl_{100-x} I_x)_3$ and Nd^{3+} in $La(Cl_{100-x}Br_x)$.
After [59]

$$\Delta E_{\parallel}(^4S_{3/2}) = -0.50 \text{ cm}^{-1}/\text{kbar}$$

where we find again for the ratio:

$$\frac{\Delta E_{\parallel}(^4S_{3/2})}{\Delta E_{\parallel}(^2P_{1/2})} = 0.60$$

In agreement with the above ratios.

Doping experiments have also been performed for Pr^{3+} in $LaCl_3$.
Pr^{3+} has an even number of 4f electrons and in a crystal

field of high symmetry such as in $LaCl_3$ doubly degenerate crystal field energy levels are possible. By doping the local symmetry is reduced and therefore the symmetry degeneracy is lifted.

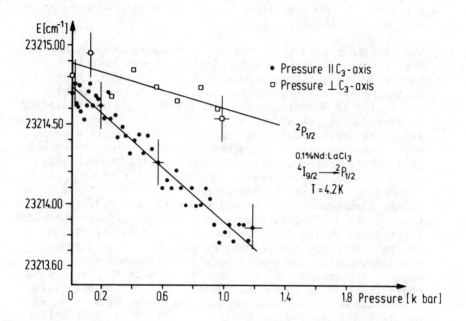

Fig. 11   Pressure shift of the $^4I_{9/2} \rightarrow {}^2P_{1/2}$ transition in $LaCl_3$ for pressure parallel and perpendicular to the crystallographic c-axis. After [59]

In summary the investigation of the satellites seems to be a very promising tool for studying the crystal field interaction in RE compounds.

5.   INTENSITIES

The question what intensity a particular transition between two crystal field levels has, is one that can be broken up into two aspects. First, we have to work out the selection

rules, which are imposed by the symmetry of the crystal;
they can be obtained by the application of group theory.
Second, we have to ascertain which selection rules of the
free RE ion "survive" the transformation of the RE ion from
the gas into the crystal, and how they may be modified by the
properties of the crystal. This quite important second aspect
was dealt with systematically for the first time by van Vleck
[61]. All the important points set down in this work are
still valid, and have since been brought into a more mathematical form [62,63].

The experimental data on the solid state spectra of RE show
that the radiation is mostly electric dipole in nature, though
some cases of magnetic dipole radiation are also observed.
Since the optical transitions take place between levels of a
particular $4f^N$ configuration, the electric dipole radiation
is forbidden in first order, because the electric dipole operator has uneven parity and the transition matrix element
must have even parity (Laportes selection rule). Van Vleck
[61] pointed out that electric dipole radiation can only
occur because the $4f^N$ states have admixtures of $4f^{N-1}nl$ configurations (nl will be mostly 5d), where $4f^{N-1}nl$ has to be
chosen such that it has opposite parity from $4f^N$. The admixture of the $4f^{N-1}nl$ wavefunctions into the $4f^N$ wavefunctions
is produced by interactions that have odd parity. In the
crystal there are two mechanisms that can produce these admixtures, namely odd parity crystal field components and
crystal vibrations of odd symmetry. We note that in crystals
where there is an inversion center, the odd crystal field
components are zero, meaning that there can be no electric
dipole radiation induced by the crystal field. Of course,
the possibility of inducing it by odd symmetry vibrations
remains. The four dominant sources of optical radiation in RE
crystal spectra are

(I)   forced electric dipole radiation induced by odd terms
        of the crystal field;
(II)   forced electric dipole radiation induced by lattice
        vibrations;
(III)   allowed magnetic dipole radiation;
(IV)   allowed electric quadrupole radiation.

In free atoms, magnetic dipole radiation is about six orders
of magnitude weaker than electric dipole radiation; thus since
the latter occurs only as a consequence of a perturbation,
both kinds of radiation show up in the RE spectra with about
the same intensity; so far, quadrupole radiation has not
been observed in RE crystal spectra. With the advent of
tunable lasers (dye lasers), however it may be possible to
induce quadrupole transitions.

The rate for spontaneous emission by electric dipole radiation is then given by

$$A_{AA'}^{ed} = \frac{64\pi^4 e^2 \nu^3}{3hc^3 (2J+1)} \quad \chi \quad \sum_{\lambda=2,4,6} \Omega_\lambda (f^N J \| U_\lambda \| f^N J')^2 \qquad (5.1)$$

$$\chi = \left(\frac{n^2+2}{3}\right)^2 \cdot n$$

In applying the formula to the transition between two particular J terms, we have to assume that all crystal field states are equally populated and that the transitions between two terms all have the same energy. These assumptions are questionable but they simplify the actual calculations considerably, and if we keep in mind that the Judd-Ofelt theory involves other drastic approximations, those just mentioned are not inordinate.

The Ω_λ parameters have so far been assumed to arise solely from the crystal field; however, they also contain contributions from admixtures by the lattice vibrations. If for small elongations ΔR, the vibrational term is written as $(\partial B_{kq}/\partial R_j)\Delta R_j$, the contributions from this term are incorporated in the Ω_λ provided the matrix elements of the vibrational coordinates are evaluated properly [62].

The transition rate for magnetic dipole radiation is given by:

$$A_{AA}^{md} = \frac{64\pi^4 \nu^3}{3hc^3 (2J+1)} \chi' |\mu_B|^2 |(f^N J \| L + 2S \| f^N J')|^2 ; \chi' = n^3 \qquad (5.2)$$

The Judd-Ofelt formalism has been applied to the analysis of a number of systems. In most of these analyses the crystal field splitting of the terms was neglected; therefore, the total absorption intensities between the ground term and the excited terms were analyzed with only three impirical parameters Ω_λ ($\lambda = 2,4,6$) and the success may be taken as a justification for the procedure.

6.   SPIN-PHONON-INTERACTION, LINEWIDTH

The interaction of the magnetic moment of an ion with the phonon field in an insulator has been dealt with by many authors attempting to explain paramagnetic relaxation phenomena [64,65]. This approach was also applied to the interpretation of optical linewidths and line shifts for similar systems and we shall follow these authors [66]. The relevant Hamiltonian contais the terms

$$H = H_I + H_p + H_{IP} \tag{6.1}$$

where H_I stands for the energy of the paramagnetic ion (free ion plus static crystal field energy), H_p stands for the phonon field energy, which we assume to be an ensemble of harmonic oscillators; and H_{IP} is the term that represents the interaction between the two systems. This last term is the important one because it is responsible for the transitions between various crystal field levels by emission and absorption of phonons. Orbach [64] suggested developing the crystal potential in a power series of the strain. The static crystal field Hamiltonian is

$$H_{CF} = V = \sum_{kq} A_{kq} P_{kq} \tag{6.2}$$

We now assume that the Hamiltonian for a position $R + \Delta R$, where R is the equilibrium position, can be obtained by a series expansion

$$V(R + \Delta R) = V(R) + \frac{\partial V}{\partial R} \Delta R + \frac{1}{2} \frac{\partial^2 V}{\partial R^2} (\Delta R)^2 + \dots \tag{6.3}$$

If we introduce a homogeneous lattice strain $\varepsilon = \Delta R/R$, this expansion can be written as

$$V(R + \Delta R) = V_o + V_1 \varepsilon + \frac{1}{2} V_2 \varepsilon^2 + \dots \tag{6.4}$$

Using the Hamiltonian for the crystal field interaction we obtain

$$V = V_o + \frac{\partial}{\partial R} \left(\sum_{k,q} A_{kq} \right) \Delta R \, P_{kq} + \frac{1}{2} \frac{\partial^2}{\partial R \partial R'} \left(\sum_{kq} A_{kq} \right) \Delta R \Delta R' P_{kq} + \dots \tag{6.5}$$

and by using the strain ε we get

$$V = V_o + \varepsilon R \frac{\partial}{\partial R} \left(\sum_{kq} A_{kq} \right) P_{kq} + \frac{1}{2} (\varepsilon \varepsilon' RR') \frac{\partial^2}{\partial R \partial R'} \left(\sum_{kq} A_{kq} \right) P_{kq} + \dots \tag{6.6}$$

If we assume that $A_{kq} \sim 1/R^n$, we obtain as an order of magnitude estimate

$$V_{IP} = \varepsilon \sum_{k,q} A_{kq} P_{kq} + \frac{1}{2} \varepsilon^2 \sum_{kq} A_{kq} P_{kq} + \dots \tag{6.7}$$

which means that in the first approximation the dynamical crystal field is just the strain times the static crystal field.

So we obtain for the transition rate

$$w_{i \to j}^{d} = \frac{3}{2\pi\rho^5 \hbar} \left(\frac{W_{ij}}{\hbar}\right)^3 |<i|\Sigma A_{kq}P_{kq}|j>|^2 [p(W_{ij}) + 1] \qquad (6.8)$$

where $\rho = M/V$ is the density of the crystal, v the sound velocity and $p(W_{ij}) = [\exp(W_{ij}/k_B \cdot T) - 1]^{-1}$.

Another process that can product transitions from crystal
field level i to another crystal field level j is the Raman
process, or the inelastic scattering of phonons. This process
is calculated by taking in eq.(6.7) the term linear in ε to
second order or by taking the term in ε^2 to first order,
which (after setting $x = W/k_B \cdot T$)yields

$$w_i^R = H_i \left(\frac{T}{\Theta}\right)^7 \int_0^{\Theta/T} dx \frac{x^6 e^x}{(e^x - 1)^2}$$

$$(6.9)$$

$$= H_i \left(\frac{T}{\Theta}\right)^7 I_6 \left(\frac{\Theta}{T}\right) \; ; \quad I_n = \int_0^{\Theta/T} \frac{x^n dx}{(e^x - 1)(1 - e^x)}$$

where the $I_6(\Theta/T)$ integrals are tabulated by Ziman [67]and $H_i$is
a combination of crystal field matrix elements (including
some constants) whose exact form does not interest us, because we are interested only in the temperature dependence
of the transition rate.

In some cases the energy separation between two crystal field
levels is larger than the Debye energy $k\Theta$. Then the crystal
field level can decay only via the spontaneous emission of
several phonons. This decay rate is temperature independent
and relatively small

$$w_i^s = \sum_{j<i} \alpha_{ij} \qquad (6.10)$$

The total linewidth $\Delta\tilde{\nu}$ (in cm^{-1}) of a crystal field level i
is then given as the sum of the contributions mentioned
earlier:

$$\Delta\tilde{\nu} = \frac{1}{2\pi c} \sum_{j<i} \alpha_{ij} + \frac{1}{2\pi c} \sum_{j<i} K_{ij} [p(W_{ij}) + 1]$$

$$(6.11)$$

$$+ \frac{1}{2\pi c} \sum_{j<i} K_{ij} p(W_{ij}) + \frac{1}{2\pi c} H_i \left(\frac{T}{\Theta}\right)^7 I_6 \left(\frac{\Theta}{T}\right)$$

The width of an optical absorption or emission line is always
the sum of the linewidth of the two levels A and B between
which the actual transition takes place.

$$\Delta \tilde{\nu}_{A\ B} = \Delta \tilde{\nu}_A + \Delta \tilde{\nu}_B \qquad\qquad (6.12)$$

The mechanisms described so far yield a homogeneous broadened optical line, meaning a Lorentzian line shape.

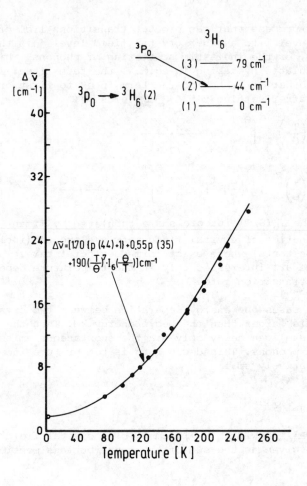

Fig. 12a Temperature dependence of the linewidth $\Delta \tilde{\nu}$ of the $^3P_o \rightarrow {}^3H_6(2)$ transition in LaF$_3$: Pr^{3+}. The solid line is calculated with the equation shown in the figure. After [66]

There is a great number of investigations of linewidth in RE crystals and we shall only discuss briefly two examples. First we deal with the work of Yen et al. [66] and discuss the data

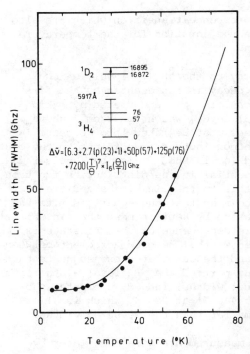

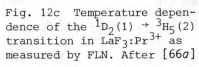

Fig. 12b Temperature depen-
dence of the $^3H_4(1) \rightarrow \,^1D_2(1)$
transition in $LaF_3:Pr^{3+}$ as
measured by FLN. After [66a]

Fig. 12c Temperature depen-
dence of the $^1D_2(1) \rightarrow \,^3H_5(2)$
transition in $LaF_3:Pr^{3+}$ as
measured by FLN. After [66a]

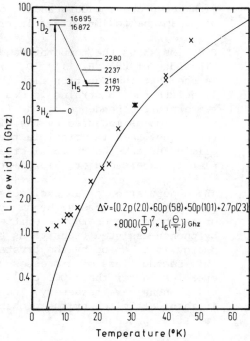

on the $^3P_0 \rightarrow \, ^3H_6(2)$ fluorescence transition which are also
given in Fig. 12a, where the equation for the temperature
dependance is also shown.

The $^3H_6(2)$ level can decay to the $^3H_6(1)$ level, which is
44 cm^{-1} lower, and this decay is represented by the first
term. The $^3H_6(2)$ level can also be depopulated by making
transitions to the $^3H_6(3)$ level, which is 35 cm^{-1} above it.
Finally, the Raman process must be taken into account. In
principle, of course, phonon-induced transitions to the other
(i = 4-13) crystal field levels should also be considered,
but because of the exponential in the distribution function
they contribute very little to the linewidths and are therefore neglected. It can be seen that the equations fits the
data quite nicely. It has to be kept in mind that in this
procedure the temperature dependence of the $^3P_0$ state is
contained in the measured width. The good agreement between
the measured linewidth and the calculated one suggests that
the temperature-dependent part of the $^3P_0$ level width contributes little to the actual widths; indeed, estimates show
that it can give at most 4 cm^{-1} at 300 K, which is almost
an order of magnitude smaller than the measured widths.

More detailed linewidth measurements were made later by
Erickson [66a] employing lasers. The results are in agreement
with the present understanding of electron phonon interaction
of paramagnetic impurities in crystals.

Fig. 12b shows the temperature dependence of the absorption
linewidth of the $^3H_4(1) \rightarrow \, ^1D_2(2)$ transition in Pr^{3+}:LaF$_3$.
We see that the equation fits the data nicely. The residual
linewidth due to strains is 6.3 GHz. Fig. 12c shows the
temperature dependence of the linewidth of the $^1D_2(1) \rightarrow \, ^3H_5(1)$
transition as measured by the fluorescence line narrowing
technique. Here in principle the strain broadening should be
eliminated and the residual broadening of $\sim$ 1 GHz is not completely understood at this point.

The depopulation of an excited state of an RE ion can take
place by two processes: by direct transition to a lower
state where the energy is emitted as radiation (light); and
by phonon-assisted transitions to a lower state, where the
energy is taken up by the crystal in the form of phonons. Since
this second class of processes usually involves transition
energies larger than 1000 cm^{-1}, high energy phonons (optical phonons) are necessary. The total lifetime of a state
is then determined by the probability for radiative and non-
radiative processes

$$1/\tau = w^R + w^{NR}$$

$$(6.13)$$

It was shown that transitions within a crystal field multiplet are mediated by acoustical phonons, and that the lifetimes governing these processes are 10^{-10} sec or shorter. By contrast, the measured optical lifetimes are orders of magnitude longer (10^{-6} sec and longer). Therefore, the emission of light is predominantly observed to originate from the lowest energy level of a particular crystal field multiplet because its lifetime is determined by the modes of deexcitation to the other crystal field multiplets.

The most straightforward approach to obtaining the radiationless decay rate is to calculate the radiative probability of a particular transition, subtract it from the measured transition probability, and thus determine the radiationless contribution. To obtain the total radiation probability of a state, we have to calculate A^{ed} and A^{md} to all the states energetically below the one in question; the radiative lifetime is then given by

$$\tau_j = 1 \Bigg/ \left(\sum_{i<j} A^{ed}_{ji} + \sum_{i<j} A^{md}_{ji} \right) \tag{6.14}$$

The nonradiative lifetime is hard to calculate. The theory of multiquantum emission shows, however, that a simple relationsship should hold here; if w_{no} is the rate for the spontaneous emission of n phonons at $T = 0$ K, the temperature dependence of the transition probability for the simultaneous emission of n phonons becomes [68,69]

$$w^{NR}_{nT} = w^{NR}_{nO} \left(1 - \exp(-\hbar\omega/k_B T) \right)^{-n} \tag{6.15}$$

where $\hbar\omega$ is the energy of the phonon under consideration. The constant w_{no} is, of course, highly dependent on the order n of the process as well. This means that with increasing magnitude of the energy gap between which a transition takes place, the order of the process gets higher and the radiative decay becomes more important than the nonradiative decay. Therefore, strong emission of light is always observed from terms that are separated from the next lower term by a large energy gap. We see that by this reasoning terms with strong light emission should be $^3P_0$ (Pr^{3+}), $^4F_{3/2}$ (Nd^{3+}), $^5D_0$ (Eu^{3+}), $^5D_4$ (Tb^{3+}). These predictions agree with experimental observations.

The temperature and energy gap dependence of multiphonon transitions has been investigated in a number of hosts [23, 68,69]. The data confirm the previous reasoning. The rates for spontaneous multiphonon decay seem to depend exponentially on the power of the order of the process because they show a straight line (Fig. 13) if plotted semilogarithmically against the energy gap, which the decay has to bridge ($w_{no}^{NR} \sim e^{b\Delta W}$, where ΔW is the gap energy).

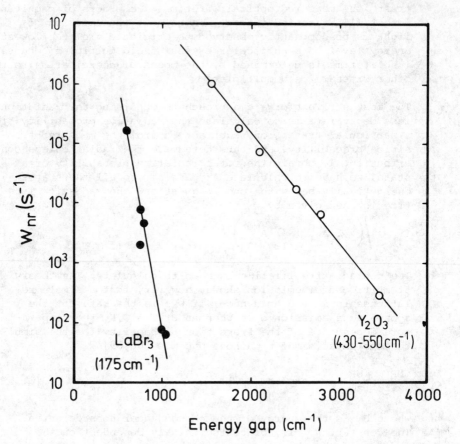

Fig. 13   Zero temperature multiphonon decay rate as a function of the energy gap in $LaBr_3$ and Y_2O_3. After [68][69] [22]

An example for this is shown in Fig. 14 [23,68,69,70,71],where the nonradiative decay rates in two host of different hardness namely $LaBr_3$ and Y_2O_3 is plotted. One can see that the exponential behaviour is nicely fullfilled.

The electron-phonon coupling that drives the nonradiative decay is proportional to the strain ε. An n phonon process can be obtained by taking the second term in eq.(6.4) to nth order; thus the nonradiative transition probability at T = O K, is written as

$$w_{no}^{NR} = C\varepsilon^n \tag{6.16}$$

Now we assume that the gap energy ΔW can be divided into n phonons $\hbar\omega$ such that $n \cdot \hbar\omega = \Delta W$. This yields

$$w_{no}^{NR} = C\varepsilon^{\Delta W/\hbar\omega} \tag{6.17}$$

$$w_{no}^{NR} = C \exp\left(-(\Delta W/\hbar\omega)\,|\ln\varepsilon|\right) \tag{6.18}$$

which gives just the behaviour seen in the data of Fig. 14.

7.   OPTICAL SPECTROSCOPY EMPLOYING LASERS

In ordinary absorption and emission experiments the sample is irridiated with a classical light source (eg. a Xenon high pressure lamp) and after interacting with the ions in the crystal, the light is analysed with a grating spectrograph. Such a spectrograph, if equipped with good gratings, can have a resolution of 10^5 of slightly more. This is sufficient to measure in RE absorption and emission spectra the actual linewidth. These linewidth are in the most favourable cases ~ 0.1 cm^{-1} (3 GHz). If we compare these measured linewidth however with measured radiation decay times which are often in the range of μsec (or even longer) one realizes that the radiation decay cannot be the dominating source of the linewidth. In favorable case like the ^3P_o and the $^1D_2$ state in Pr^{3+} or the ^5D_o state in Eu^{3+} also the nonradiation decay does not contribute sizably to the observed linewidth. The most likely source of the large linewidth detected with conventional means is therefore an inhomogeneous broadening caused by inhomogenities of the crystal field.

It was not before the advant of the tunable dye laser that "true" homogeneous linewidth could be measured. They proved indeed to be of the order of the measured decay times. At this point the narrowst line measured directly is that of the $^1D_2 \to {}^3H_4$ transition in Pr^{3+} : LaF_3 which yielded 200 kHz (6.6 x 10^{-6} cm^{-1})[72] and the largest "lifetime" is the dephasing time of the $^1D_2 \to {}^3H_4$ transition in Pr^{3+} : $LaAlO_3$ which was measured as 78 μsec (2.0 kHz)[73].

In order to make this chapter reasonably selfcontained we shall now give a brief description of the methods employing lasers that allow to study the radiative properties of these narrow levels [74].

Fluorescence line narrowing (FLN)

This is the technique most widely used. It can be employed
for linewidth measurements and, if a pulsed laser is used,
for the study of excitation dynamics.

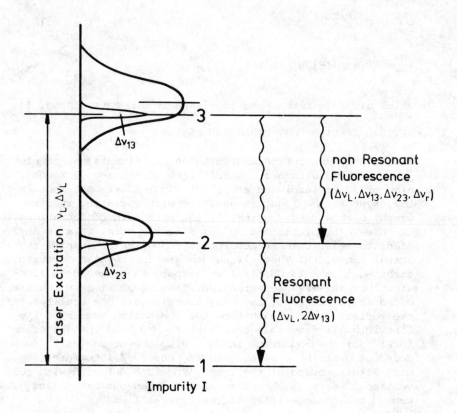

Fig. 14 Principles of FLN experiments. The narrow laser line
         $\Delta\nu_L$ excites only a small package of ions within the
         inhomogeneous linewidth, from which they decay,
         assuming no energy migration or diffusion.

The principles of the method are shown in Fig. 14. An absorption line observed in a conventional experiment with broadband white excitation is a convolution of the ensemble of
states with narrow lines which experience different crystal
fields (strains). These "broad" lines are shown for two
excited states 2 and 3. $\Delta\nu_{13}$ is the true linewidth for a
transition between the groundstate 1 and the excited state
3 for a particular ion I, $\Delta\nu_{23}$ is defined correspondingly. Now

laser excitation takes place with ν_L and a laser linewidth $\Delta\nu_L$. Thereby only that packet of ions is excited in the level 3, which falls into $\Delta\nu_L$.

If now resonance fluorescence takes place the linewidth of ν_{13} is given by the convolution of $2\Delta\nu_{13}$ and ν_L. The condition for such an experiment to work in the way just described is, that during the lifetime of the excited state negligeable energy migration takes place. This condition is fullfilled in systems, where the RE ions are incorporated as dilute impurities.

Emission from the excited state 3 can also take place to an intermediate level 2. Than the linewidth is given by a convolution of $\Delta\nu_L$, $\Delta\nu_{1,3}$, $\Delta\nu_{2,3}$ and $\Delta\nu_r$ where $\Delta\nu_r$ is some residual broadening in level 2.

We note that only resonance fluorescence allows the measurement of the "true" linewidth of an excited state. However resonant experiments are difficult because one has to detect a small fluorescence signal under a high background of light from the exciting laser, where the wavelength of the two are very similar. In nonresonant experiments the discrimination between the fluorescence light and the laser light is much easier.

Fig. 15 shows a schematic experimental arrangement for FLN experiments. A dye laser ($\Delta\nu_L \sim 1$ MHz) irridiates a sample and the fluorescence light is analysed with a Fabry-Pérot interferometer (resolving power 10^6). A grating spectrograph acts as a broadband filter in order to discriminate against other fluorescence lines. A chopper, which acts both in the exciting beam and the fluorescent beam is arranged such that each of the two beams is closed if the other one is opened.

Hole burning

The hole burning technique can be though of as the reverse to the FLN technique. A powerful narrow line laser is used to deplete the groundstate of a particular package of ions sufficiently. If now a second weak laser beam is scanned in wavelength over the total line its absorption will be weaker for the ions that are excited by the pumping laser. Thus a "hole" will be found in the absorption crossection as monitored by the second laser. A typical experimental arrangement is shown in Fig. 16. A laser at fixed frequency produces the depletion of the ions in a particular area of a line. The tunable laser interrogates the hole position of the line. The advantage of this technique relative to the FLN techniques lies obviously in the fact, that one does not need a spectro-

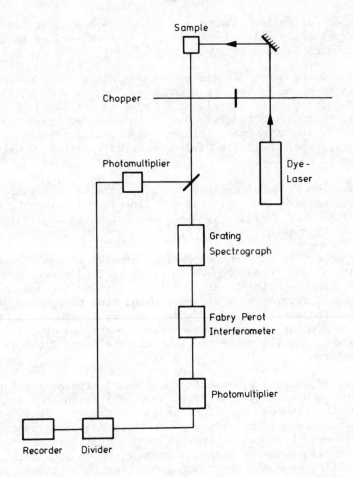

Fig. 15 Schematic experimental arrangement for FLN
          experiment.

meter.

A technique very similar to the hole burning method uses the
change in optical birefringens of the sample produced by the
pump laser rather than the change of population.

Coherent Transient Spectroscopy

These techniques are borrowed from NMR spectroscopy, where
they have been first and successfully introduced [75]. In
principle they consist of producing a coherent set of ions
and to monitor the decay (or dephasing) of this coherent
state. In terms of NMR language one measures by this technique
the relaxation time T_2. In this technique the coherence of the

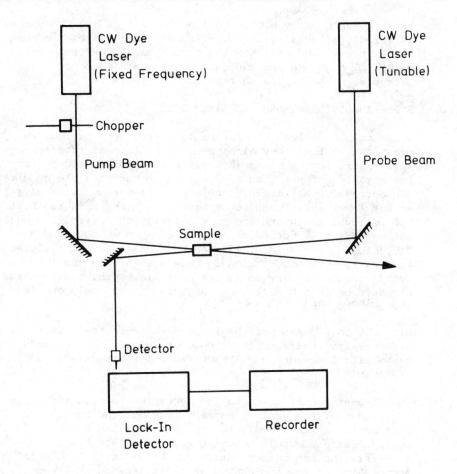

Fig. 16   Schematic experimental arrangement for hole burning
          experiments.

laser light is essential, while in the ones mentioned before,
only the narrowness and intensity of the laser light was used.
There is a number of different ways to detect the decay of
the coherence which will be treated in turn.
a) Photon echos. This technique is borrowed directly from
the NMR work of Hahn [75]. One needs two pulses of proper
length ($\pi/2$) at a time distance τ to observe an echo at time
2τ after the first pulse. The intensity of this echo pulse
relative to the original pulses as a function of the time τ
between the two $\pi/2$ pulses is than a measure of the dephasing
time of the coherent excited state.
b) Free induction decay (FID). In this method a package of
ions is produced with a laser of frequency ν_L for a time Δt.
Than the laser frequency is shifted suddenly slightly within

the inhomogeneously broadened line to $\nu_L + \Delta\nu$ and the sample
is irridiated equally for a period Δt, after which time the
laser is switched off. Now the sample units the frequencies
ν_L and $\Delta\nu + \Delta\nu$ and one can detect the beat frequency, which
decays with the dephasing time of the system.

8. HYPERFINE INTERACTIONS AND LINEWIDTHS

Since the RE optical absorption lines are very narrow, we
might ask whether they also permit detection of interactions
with the nuclear moments (hyperfine structure). Except for
one notable exception such detection is not possible by direct
conventional spectroscopy: the hyperfine interactions give
energy level splittings of at most 1 cm^{-1} and this is about
the residual linewidth due to strains in the crystal. However,
the hyperfine structure of holmium in $Ho^{3+} : LaCl_3$ and $Ho^{3+} :$
$Y(C_2H_5SO_4)_3 \cdot 9H_2O$ (Fig. 17) has been observed and analyzed
in detail. This observation is made possible by the fortunate
coincidence of a very small residual linewidth, a very large
magnetic nuclear moment, and the existence of only one isotope, ^{165}Ho [76].

With the development of the laser techniques the investigation
of hyperfine interactions is however a "trivial problem", because with linewidth of $\sim 1 MHz$ these interactions are easily
resolvable.

The largest contribution to the hyperfine interaction is generally that of the nuclear magnetic moment; the quadrupole
interaction is much smaller. The latter is the product of
the nuclear quadrupole moment Q and the electrical field
gradient (the second derivative of the electrical potential
at the site of the nucleus) seen by it. For an RE ion sitting
in a crystal, this electric field gradient has two contributions: one from an aspherical charge distribution of the 4f
electrons and one from the aspherical charge distribution of
the surrounding ions. Both contributions however are modified
from their pure values by the fact that the closed shells of
the RE ions are polarized by these electrical field gradients
and can thus enhance or diminish them; this effect is called
the Sternheimer [77] effect. The quadrupole interaction is
therefore written as

$$W_Q \sim Q\{(1 - R_Q)q_e + (1 - \gamma_\infty)q_{lat}\} \tag{8.1}$$

where q_e is the field gradient produced by 4f electrons at
the nucleus, q_{lat} is the field gradient produced by the
lattice at the nucleus, and $(1 - R_Q)$ and $(1 - \gamma_\infty)$ are the
Sternheimer factors, which correct for the polarizability of

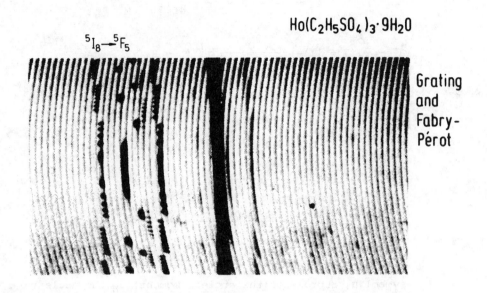

$^5I_8 \rightarrow {}^5F_5$

$Ho(C_2H_5SO_4)_3 \cdot 9H_2O$

Grating and Fabry-Pérot

Fig. 17 Hyperfine structure of Ho ethylsulfate. The transition $^5I_8 \rightarrow {}^5F_5$ in $Ho^{3+}_{0.05} : (C_2H_5SO_4)_3 \cdot 9H_2O$ recorded with a grating and a Fabry-Pérot interferometer. After [2,76]

the closed shells of the RE ion. These factors are of interest because they can be compared with those obtained from theoretical atomic wavefunctions.

We now discuss the case of Ho^{3+} ethylsulfate and chloride. For axial symmetry the hyperfine interaction Hamiltonian can be written as [4]

$$H_{hfs} = A_J \underline{I} \cdot \underline{J} + \{P^J_{4f}(J^2_z - \tfrac{1}{3}J(J + 1)) + P_{lat}\}\{I^2_z - \tfrac{1}{3}I(I + 1)\}$$

$$-g_N \mu_B \underline{I} \cdot \underline{H}_{ext} \qquad (8.2)$$

where

$$A_J = \frac{2\mu_I \mu_N \mu_B}{I} \, N_J \langle r^{-3}\rangle_{4f} (1 - R_M)$$

$$P_{4f}^J (J_z^2 - \frac{1}{3}J(J+1)) = \frac{3e^2 Q}{4I(2I-1)} \, q_{zz}^{4f} (1 - R_Q)$$

(8.3)

$$q_{zz}^{4f} = -3\alpha_J \langle r^{-3}\rangle_{4f} (J_z^2 - \frac{1}{3}J(J+1))$$

$$Q = \langle I \, M_I = I | 3z_n^2 - r_n^2 | I \, M_I = I\rangle$$

$$P_{lat} = \frac{3e^2 Q}{4I(2I-1)} \, q_{zz}^{lat} (1 - \gamma_\infty)$$

$$q_{zz}^{lat} = -\frac{4A_{20}\langle r^2\rangle}{e^2 \langle r^2\rangle_{4f} (1 - \sigma_2)}$$

The symbol μ_I represents the nuclear moment, μ_N the nuclear magneton, N_J (often written $(J\|N\|J)$) the equivalent operator for the hyperfine interaction. R_M the Sternheimer [77] factor for the magnetic hyperfine interaction, r_n the radius vector of the nuclear coordinate, and z_n its z components; q_{zz} equals q_e if the ion is in a lattice of axial symmetry (as is the case for the ethylsulfates).

The optical experiments were performed by placing a Fabry-Pérot interferometer in front of the grating spectrograph in order to increase the resolution. A typical spectrum taken with such an arrangement was shown in Fig. 17 where the hyperfine splitting of the electronic lines is clearly seen.

Analysis of the measurements yielded the shielding constant values.

We shall now describe the investigation of a similar system, namely Pr^{3+} : $LaCl_3$ by the FLN technique [78]. Fig. 18 shows the principles of the experiment. The groundstate of Pr^{3+} : $LaCl_3$ is a doublet, meaning it has a magnetic moment, which in turn produces a magnetic field at the nucleus, which produces a magnetic hyperfine splitting. This is the dominant contribution to the hyperfine interaction, the quadrupole interaction being much smaller. The excited state, the lowest crystal field level of the $^1D_2$ term, is a singulet. Thus in first order it has no hyperfine interaction and the level is degenerate in M_I. However there is an inhomogeneous crystal

field giving rise to the inhomogeneous linewidth of this transition (6011 Å, 16630.5 cm^{-1}). The experiment is now performed
such, that the laser line is tuned to the center of the 6011 Å
emission line.

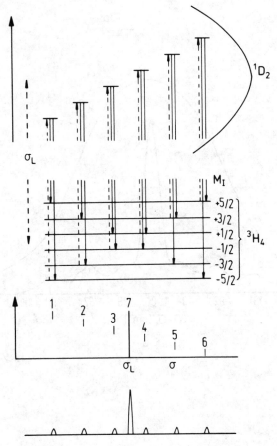

Fig. 18   Energy level diagram showing the FLN detection of
          hfs in Pr^{3+}:LaCl$_3$. The transition is inhomogeneously
          broadened and in the excited state the ±M$_I$ components
          are degenerate such that they can be excited simul-
          taneously. After [78]

Than the transitions shown in Fig. 18 are possible. The laser
transition from the M$_I$ = -5/2 state takes the ion to the
M$_I$ = ±5/2 excited state (line 1), which are degenerate. From
these an emission can take place to the M$_I$ = -5/2 and to the
M$_I$ = +5/2 state one of them being the laser frequency ν_L and
the other one the frequency ν_L - ν_{hfs} where ν_{hfs} is the total
hyperfine splitting of the groundstate. Similar transitions
take place from all the hyperfine levels of the groundstate.

Thus one observes seven fluorescence transitions, where the
middle one is six times as intense as the other ones, which
should all be of similar intensity as is roughly observed
(Fig. 19).

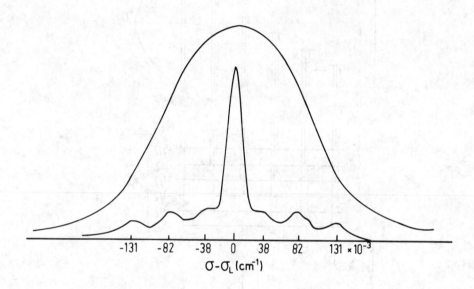

Fig. 19   FLN signal for the 6011 Å line of Pr^{3+}:$LaCl_3$. The
          inhomogeneous line is shown in addition to the FLN
          signal which shows the seven lines anticipated from
          Fig. 18. After [78]

The hyperfine parameter of the groundstate calculated from
the data is

$$A_I = (25.6 \pm 0.2) \; 10^{-3} cm^{-1}$$

which is in good agreement with the one measured by EPR namely
$A_I = (25.1 \pm 0.3) \; 10^{-3} cm^{-1}$ [79].

The interpretation of the observed spectrum makes use of two facts. One has to assume that the inhomogeneous broading of the transition is larger than the hyperfine interaction and that no crossrelacion between different M_I levels takes place in the excited state between absorption and emission. This is obviously the case.

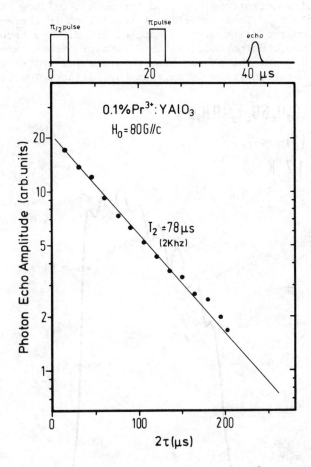

Fig. 20   Pulse echo investigation of the $^3H_4(1) \rightarrow {}^1D_2$ transition in Pr^{3+}:LaAlO$_3$. On top the principles of the pulse sequence are shown. The figure gives the decay of the echo signal as a function of delay time between the two $\pi/2$ pulses. Measurements were performed in an external field of 80 G. After [73],[73a]

Pr^{3+} has also been the favorite ion for the investigation of the linewidth. The narrowest line obtained by direct observation was the 800 kHz found by Ericksen [72] from a hole burning experiment on the $^3H_4(1) \rightarrow {}^1D_2(1)$ transition of LaF$_3$:

Pr^{3+}. From this he convoluted an internal homogeneous width
of 200 kHz. The narrowest lines with any technique are the
2 kHz found for the $^3H_4(1) \rightarrow {}^1D_2$ transition for $Pr^{3+}:LaAlO_3$
observed by Chen et al. [73] with a pulse echo method (Fig.20).

Optical spectroscopy was also the method to show, that it is
possible to detect hyperfine interactions in concentrated
paramagnetic salts with no applied field. This technique has
later had considerable use in Mössbauer effect experiments.
Fig. 21 shows the photometric recording of a line in the
$^5I_8 \rightarrow {}^5F_5$ transition in holmiumethylsulfate. The eight

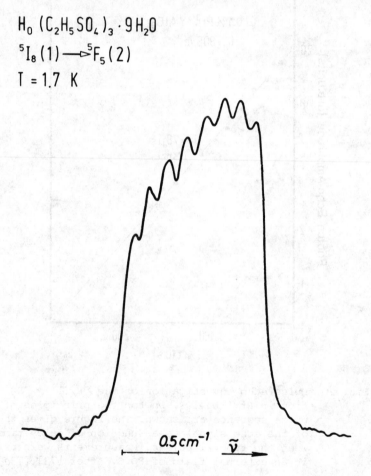

$$H_0 (C_2H_5SO_4)_3 \cdot 9H_2O$$
$$^5I_8(1) \longrightarrow {}^5F_5(2)$$
$$T = 1.7 \ K$$

$$0.5 cm^{-1} \quad \tilde{\nu}$$

Fig. 21   Photometric recording of the transition $^5I_8(1) \rightarrow$
$^5F_5(2)$ in holmiumethylsulfate at 1.7°K.

hyperfine lines are clearly seen [79a]. This observation is

not trivial. In paramagnetic salts even at low temperatures
the spin spin interaction broadenes the energy levels to such
a degree that observation of hyperfine structures is impossible. Therefore EPR experiments are generally performed in highly
deluted samples. The delution removes the ions from each other
and thereby makes the spin spin relaxation time considerably
longer. The same holds of course also in general for optical
experiments.

The spin spin interaction is of the type S_+ or S_-, and can act
only between states differing in spin by one unit; in other
terms this means that besides g_z a state must have finite g_x
and g_y such that spin spin transitions are possible. It can
however easily be shown [2,4] that in even electron systems
(such as holmiumethylsulfate) g_x and g_y by definition have to
be zero. Thus in the case shown in Fig. 21 the observation of
paramagnetic hyperfine structure is possible in a concentrated
RE salt, because since g_x, g_y are zero the spin spin relaxation
time is very long. This reasoning was later confirmed by Mössbauer effect experiments [79b].

9.   ENERGY TRANSFER AND ITS MEASUREMENT

In the basic interpretation of the RE spectra the emission and
absorption is considered a one ion process. There are however
also a number of processes in which two or more ions partici-
pate.The most important ones are those that transfer energy from
one ion called the donor to another ion called the acceptor.
The simplest possibilities of energy transfer processes are
shown in Fig. 22, where the so-called resonant and phonon
assisted energy transfer are indicated schematically.

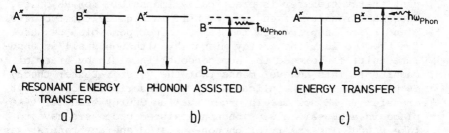

Fig. 22   Principles of resonant and phonon assisted energy
          transfer. After [2]

The original theory of energy transfer has been worked out by
Förster [80,81] and Dexter [82,83]. There exist a number of
reviews on this subject [see eg.84,85]. The rate of transfer
of energy from state A to state B is given by

$$w_{AB} = \frac{2\pi}{\hbar}|<A^*B|H_{int}|AB^*>|^2 \int g_A(W)g_B(W)\,dW \qquad (9.1)$$

Here we denote the quantum numbers of the states by A and B, and H_{int} is the interaction Hamiltonian for the transition between the two ions. $g_A(W)$ and $g_B(W)$ are the two line shape functions of the transitions $A* \to A$ and $B \to B*$. We see that this formula is valid only for resonant energy transfer because the integral gets small very rapidly as a mismatch between the donor and acceptor energies occurs. The transition matrix elements between the ions can be of three types: electric multipole interaction, exchange interaction, and virtual phonon exchange interaction. So far experiments on REs have been interpreted mostly in terms of electric multipole interactions between the ions (exchange interactions are important for the 3d ions). When discussing experiments, only the range dependence of the interaction, given by

Dipolar-dipolar $1/(R_{AB})^6$

Dipolar-quadrupolar $1/(R_{AB})^8$

Quadrupolar-quadrupolar $1/(R_{AB})^{10}$ (9.2)

Magnetic dipole $1/(R_{AB})^6$

Exchange $\exp[-2z|\ln S|]$

(z is the number of ligands and S the overlap integral) is important. The lineshape measurements on RE compound performed in recent year with lasers have shown, that the apparent linewidths, generally observed in absorption or emission experiments are due to random strains. The homogeneous widths at low temperatures are of the order of MHz, whereas the inhomogeneous width are of the order of GHz. This however means that resonant energy transfer is practically impossible for RE salts. Thus the transfer of energy, even if in Fig. 22 the two levels A and B are in principle the same, is not possible, because for the two ions the energy mismatch, although small in absolute units can exceed the homogeneous linewidth by order of magnitude. This however means that the Förster-Dexter theory is practically not applicable for the description of energy transfer in RE solids. In order that an energy transfer take place in a RE salt always phonon assisted processes have to take place. The theory of phonon-assisted energy transfer was first developed by Orbach [86] and then worked out in detail by Miyakawa and Dexter [86a], Orbach [86b] and Holstein et al. [84]. The transition rate is

$$w_{AB} = \frac{2\pi}{\hbar}|<A*B|H_{int}|AB*>|^2 K[p(\omega) + 1] \int g_A(W)g_B^+(W-\hbar\omega)dW \quad (9.3)$$

where K is a combination of electron-phonon matrix elements and $p(\omega)$ is the Bose function for the phonon occupation; $p(\omega)$ and $(W + \hbar\omega)$ stand for process (c), and $p(\omega) + 1$ and

$(W - \hbar\omega)$ for process (b).

We note that this treatment of the problem gives a simple result only for cases with large energy mismatch ΔE_{AB}, large meaning $\Delta E_{AB} \gg k\Theta$ where Θ is the Debye temperature. We can than write the energy transfer rate as a function of the energy mismatch ΔE_{AB} (see chapter 6, eqs.(6.15),(6.18)

$$w_{AB} = B \exp(-b\Delta E_{AB}) \tag{9.4}$$

and as a function of temperature

$$w_{AB}(T) = w_{AB}(T = 0)(1 - \exp(-\hbar\omega/k_B T))^{-n} \tag{9.5}$$

where $\hbar\omega$ is the average phonon energy needed to bridge the energy gap in an n-phonon process ($\Delta E_{AB} = n\hbar\omega$).

The formulation given in eq.(9.3) is difficult to evaluate further. Therefore Holstein et al. [84] have employed the t matrix approach in order to uncover the nature of phonon assisted energy transfer. The mathematics of this approach is involved and we shall not give any details here.

The simple process depicted in Fig. 22 involves one phonon. We shall however realize that this process gets unlikely under certain condition, which will make it necessary to invoke higher order processes, that involve two (and more) phonons. For that reason we shall distinguish here explicitly these different cases.

a) One phonon process
The initial state of the system is described as

$$|i\rangle = |A^*, B, n_q\rangle \tag{9.6}$$

and the final state as

$$|f\rangle = |A, B^*, n_q \pm 1\rangle \tag{9.7}$$

Here A and B are the two states of the crystal and n_q is the phonon occupation number (where we have dropped for convenience the polarisator index). The coupling of the two sites A and B is given as

$$J = \langle A, B^*|H_{A,B}|A^*, B\rangle \tag{9.8}$$

The matrix elements of the phonon Hamiltonian can be written for ion j in its groundstate:

$$\langle j, n_q \pm 1 \,|\, H_{ph}(j) \,|\, j, n_q \rangle$$

$$= f(j) \langle n_q \pm 1 \,|\, \varepsilon(j) \,|\, n_q \rangle \tag{9.9}$$

Here $f(j)$ is the coupling strength between the electronic and phonon states of ion j in the groundstate and $\varepsilon(j)$ is the strain operator at site j.

A similar equation can be defined for the excited state of that ion:

$$\langle j^*, nq \pm 1 \,|\, H_{ph}(j) \,|\, j^*, n_q \rangle$$

$$= g(j) \langle n_q \pm 1 \,|\, \varepsilon(j) \,|\, n_q \rangle \tag{9.10}$$

The evaluation of the spatial part of the matrixelement of the strain operator yields:

$$\langle n_q \pm 1 \,|\, \varepsilon(j) \,|\, n_q \rangle =$$

$$\langle n_q \pm 1 \,|\, \varepsilon \,|\, n_q \rangle \exp [\mp iqr_j] \tag{9.11}$$

The energy transfer probability is evaluated with the "golden rule" to yield:

$$w_{AB} = \left(\frac{2\pi}{\hbar}\right) \sum_q |t_{i \to f}|^2 \delta \left[\sum (\Delta n_q \hbar \omega_q) \pm \Delta E_{AB}\right] \tag{9.12}$$

Here $t_{i \to f}$ is the t-matrix for the transition between the initial and the final state; Δn_q is the change in phonon occupation number in going from the initial to the final state.

The evaluation of transition rate for the one phonon process yields for the case of like ions (A = B, and the energy mismatch is produced by inhomogeneous linewidth):

$$w_{AB} = \frac{2\pi J^2 (f-g)^2}{\hbar \, \Delta E_{AB}}$$

$$\times \sum_q |n_q \pm 1 \,|\, \varepsilon \,|\, n_q|^2 |\exp [(iq \cdot r)] - 1|^2 \tag{9.13}$$

$$\times \delta(\hbar \omega_q \mp \Delta E_{AB})$$

The interesting factor in this transition probability is the coherence factor $\{\exp[iqr]-1\}$, which for $q \cdot r \ll 1$ gets very small and makes the energy transfer vanish.

One therefore has to distinguish two regimes:

α) Large energy mismatch. This case applies especially to glasses where the inhomogeneous linewidth can amount to ~ 100 cm^{-1} or more. Under this condition q is large and the coherence factor averages out yielding $|\exp[iqr]-1|^2 \approx 2$ so that the transition probability reduces to

$$w_{AB} = \frac{J^2(f-g)^2 \Delta E_{AB}}{\pi \hbar^4 \rho \cdot v^5} \left\{ \frac{p(\Delta E_{AB}) + 1}{p(\Delta E_{AB})} \right\} \tag{9.14}$$

Where the upper part stands for the emission of a phonon of energy E_{AB} and the lower part for the absorption; p is the Bose factor, ρ the density of the crystal and v the sound velocity.

At high temperatures, if $\Delta E_{AB}/k_B \cdot T \gg 1$ the Bose factor can be developed and the transition probability reduces to

$$w_{AB} = \frac{J^2(f-g)^2}{\pi \hbar^4 \rho \cdot v^5} k_B T \tag{9.15}$$

β) Small energy mismatch. In this case the phonon wavelength is large, and correspondingly q small. Thus the argument of the coherence factor can be developed yielding:

$$\left| \exp[iq \cdot r] - 1 \right|^2 \approx (qr)^2 / 3 \tag{9.16}$$

which gives for the transition probability in the high temperature approximation:

$$w_{AB} = \frac{J^2(f-g)^2}{\pi \hbar^4 \rho \cdot v^5} \left(\frac{qr}{3}\right)^2 k_B T \tag{9.17}$$

Now

$$\lambda = \frac{v}{h \cdot \nu} = \frac{2\pi \hbar \cdot v}{\Delta E_{AB}} \tag{9.18}$$

and

$$\lambda = \frac{2\pi}{q} \tag{9.19}$$

yielding

$$q = \frac{\Delta E_{AB}}{\hbar v} \qquad (9.20)$$

which leads to

$$w_{AB} = \frac{J^2 (f-g)^2 |\Delta E_{AB}|^2 r^2}{3\pi \hbar^6 \rho v^7} k_B T \qquad (9.21$$

which is a small rate and is responsible for the fact that the one phonon assisted energy transfer is rarely observed.

b) Two phonon process.
The two phonon processes are usually dominant because they can involve higher energy phonons and thus avoid the destructive interference which produces the cancellation for the one phonon process.

The energy conservation is than given as:

$$\hbar (\omega_q - \omega_{q'}) = \Delta E_{AB} \qquad (9.22)$$

and the transfer probability reads:

$$w_{A\ B} = \frac{2\pi}{\hbar} \sum_{q,q'} |t_{i \to f}|^2 \delta(\hbar\omega_q - \hbar\omega_{q'} - \Delta E_{AB}) \qquad (9.23)$$

There are a number of ways in which the two phonon processes can be active and we do not want to give the details but rather only quote the final formula [84].

I) Two site nonresonant process.
Here the energy transfer is assisted by a phonon absorption on site A and a phonon emission on site B.

$$w_{AB} = \frac{J^2 (f-g)^4}{2\pi^3 \hbar^7 \rho^2} \frac{1}{v} k_B T^3 I_2 (x); \qquad (9.24)$$

$$x = \hbar\omega_q / k_B T \qquad (9.25)$$

for $I_n (x)$ see eq. (6.9).

The energy mismatch ΔE_{AB} drops out at high temperatures.

II) One site Raman process.
Here the energy transfer is assisted by a phonon scattering process on site A or B.

$$w_{AB} = \frac{J^2 (f_B - g_B)^2}{4\pi^3 \hbar^7 \rho^2 \Delta E_{AB}^2} \left(\frac{1}{v^5}\right) I_6 (x) \qquad (9.26)$$

III) One site nonresonant process.
Here the energy transfer is assisted by a one phonon transition that takes ion A (or B) to another electronic level (A**)
while the second phonon acting on the same ion makes up for
the energy mismatch.

$$w_{AB} = \frac{16J^2 (k_B T)^7 \Gamma_2^2 I_6 (x)}{\pi \hbar \Delta^6 (\Delta E_{AB})^2} \qquad (9.27)$$

where

$$\Gamma_2 = \frac{|F^2| \Delta^3}{4\pi h^3 \rho} \frac{1}{v^5} (p(\Delta) + 1) \qquad (9.28)$$

and

$$<A^{**}|H_{ph}(A)|A^*> = F \cdot \varepsilon(A) \qquad (9.29)$$

and $\Delta = E(A^*) - E(A^{**})$ $\qquad$ (9.30)

where A** is an intermediate level.

The question now arises how to obtain w_{AB} from an experiment.
For the donor ions the possibility to transfer energy to the
acceptor ions constitutes an additional possibility to decay
from an excited state (in addition to radiative and nonradiative decay). This modifies the donor excited state lifetime.
A good way to investigate energy transfer is therefore to
measure the donor fluorescence lifetime. This means, we need
equations that relate w_{AB} to, e.g., the time dependence of
that fluorescence lifetime. To obtain them we assume, for a
crystal of volume V, that there is a donor concentration c_A
and an acceptor concentration c_B; that the intrinsic donor
lifetime is τ_A (measured for small donor concentrations and
in the absence of any acceptor); and that the donor-acceptor
transfer rate has the form $w_{AB} = a_s/R^s$. The critical transfer
radius is introduced in the form $1/\tau_A = a_s/R_o^s$, which means that
it is the distance at which $w_{AB} = 1/\tau_A$. The donor-acceptor
transfer rate can then be expressed in the form $w_{AB} = (R_o/R)^s$
$\cdot 1/\tau_A$. The critical transfer radius R_o can be related to a
critical concentration by $c_o^{-1} = 4\pi R_o^3/3o$. Finally, we have
to take into account that a donor can decay not only radiati-

vely and nonradiatively, or by transferring energy to the
acceptor system, but also by exchanging energy with another
donor. The last process, which depends on the donor concentration, can be thought of as energy migration or diffusion
in the donor system and is therefore described by a diffusion
constant D.

Now the rate equation for the population in the donor system,
taking into account the various decay channels, is set up;
from it, the time dependence of the donor fluorescence intensity can be calculated.[85].

We call $P_n(t)$ the probability that an atom n is at a time t
in an excited state. The time evolution of this probability
is than given by:

$$\frac{dP_n(t)}{dt} = \left(-\gamma_A + X_n + \sum_{n \neq n'} w_{nn'}\right) P(t)$$

$$+ \sum_{n' \neq n} w_{n'n} P_{n'}(t)$$

(9.31)

The meaning of the symbols is as follows

$$\gamma_A = \frac{1}{\tau_A} ,$$

X_n = transfer rate from donor atom n to acceptor atoms,

$\sum_{n \neq n'} w_{nn'}$ = donor-donor transfer rate,

$\sum_{n'n} w_{n'n} P_{n'}(t)$ = acceptor-donor transfer rate.

The general equation (9.31) will now be discussed with respect to some situations, which have been investigated by
experiment.

The simplest experiment is the monitoring of the time dependence of a FLN signal as a function of time. This means we
measure the quantity

$$R(t) = \frac{I_n(t)}{I_t(t)} = \frac{\text{Intensity in the FLN signal at time t}}{\text{Total intensity at time t}}$$

(9.32)

To evaluate this quantity we assume:

a) that no acceptors are present, $c_B = 0$,

b) $k_B T$ >> inhomogeneous linewidth,

c) $w_{nn'}$ independent of the energy mismatch.

The evaluation R(t) from equ.(9.31) for the case of a lattice
with a random distribution of donor ions (which is the case
normally encountered in experiments) is a tremendously difficult problem. We therefore present here only an approximation
which reads as [85], [87], [88]

$$R(t) = \prod_e \left\{ 1-c_A + c_A \left[\exp(-w_{oe}t) \right] f(w_{oe}t) \right\}$$ (9.33)

here w_{oe} is the energy transfer rate of a donor at site 0 to
a donor at site e. The form of the function $f(w_{oe}t)$ depends
on the donor concentration.

For $c_A \geq 0.5$ one has:

$$f(w_{oe}t) = 1 + 1/2 (w_{oe}t)^2$$ (9.34)

and for $c_A \leq 0.2$ one has:

$$f(w_{oe}t) = \cosh(w_{oe}t)$$ (9.35)

An example for an analysis along these lines is shown in
Fig. 23. It shows R(t) as measured by a FLN experiment in the
transition $^3P_0 \rightarrow {}^3H_6(1)$ of $Pr_{0.2}La_{0.8}F_3$. The full courve is
a fit to the data with $f(w_{oe}t) = \cosh(w_{oe}t)$ and one can realize quite good agreement between theory and experiment [87].

The foregoing considerations it was assumed that there were
no acceptors (or traps). We now turn to the situation where
this restriction is dropped. The occupation number of the
donor under these circumstances is given by

$$N_A(t) = N_A(0) \left\{ \exp\left[-\gamma_A t \right] \right\} \cdot f(t)$$ (9.36)

An exact solution for f(t) is possible for two limiting cases.

α) no donor-donor transfer

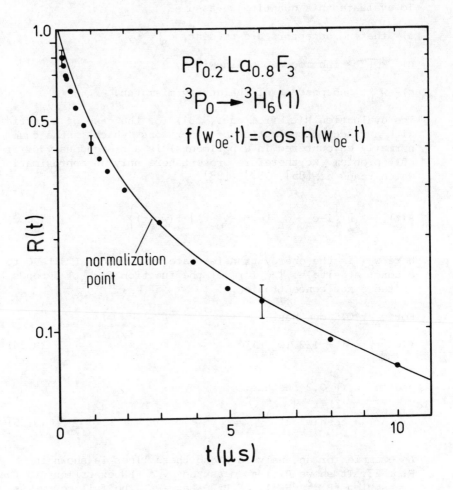

Fig. 23 Comparison of the measured and calculated values
for $R(t)$ in the $^3P_0 \rightarrow {}^3H_6(1)$ of Pr^{3+} in $Pr_{0.2}La_{0.8}$
F_3, assuming a r^{-6} donor-donor transfer rate and
$f(w_{oe}t) = \cosh(w_{oe}t)$. After [85]

Then we have [85],[89]

$$f(t) = \prod_{e}\left[1 - c_B + c_B \exp(-X_{oe} t)\right] \tag{9.37}$$

where X_{oe} is the transfer rate from a donor at site o to an acceptor at site e.

β) infinitely rapid donor-donor transfer

In this limiting case it is assumed that all donors have the same probability of beeing excited. We than have [90,91]

$$f(t) = \exp[-c_B \sum_{e} X_{oe}] \tag{9.38}$$

In the intermediate regime meaning finite donor-donor transfer the behaviour of f(t) is very complicated and various approximations have been used. The most commonly used models are the diffusion model [92] and the hopping model [93]. Both approximations give nonexponential decay curves for small t, and an simple exponential for $t \to \infty$. For the diffusion model the result is particularly simple yielding for $t \to \infty$:

$$f(t) = \exp\left[-\left\{4\pi \, Dc_B \, R_s\right\}t\right] \tag{9.39}$$

where D is the diffusion constant associated with the donor-donor transfer.

In the evaluation of experiments one usually assumes that the interactions are of the dipol-dipol type $X_{ee'} = \alpha/r_{ee'}$.

Than one can rewrite

$$R_S = 0.68 \, (\alpha/D)^{1/4} \tag{9.40}$$

yielding [94]

$$f(t) = \exp\left[-\{8.55 \, c_B \, \alpha^{1/4} \, D^{3/4}\}t\right] \tag{9.41}$$

for the longterm limit of the diffusion model.

In practice however also the intermediate time regime is of importance.

In order to analyze experimental data some approximations

which were calculated from the diffusion model are used. If we assume that the donor-donor transfer is small $D \approx 0$ only the donor decay and the donor $\rightarrow$ acceptor transfer are operative. In this case and for an interaction of the type α/r^6 one obtains:

$$f(t) = \exp\left[-\left(4\pi^{3/2}/3\right) \cdot c_B \left(\alpha \cdot t\right)^{1/2}\right] \tag{9.42}$$

which in essence is also obtained by equ. (9.33) for small times.

For finite donor-donor transfer the diffusion approach also yield an approximate solution for intermediate times namely [95]

$$f(t) = \exp\left\{-\left(4\pi^{3/2}/3\right) \cdot c_B \left(\alpha \cdot t\right)^{1/2} \cdot \frac{1+10.87x+15.50x^2}{1 + 8.74 \, x}\right\} \tag{9.43}$$

where $x = D \, \alpha^{-1/3} t^{2/3}$.

The change from the behaviour with no donor-donor interaction to that with strong donor-donor interaction can be seen, e.g. in the system of YF_3 doped with Yb and Ho [96]. Here Yb acts as a donor (via the $^2F_{5/2} \rightarrow \, ^2F_{7/2}$ transition) and Ho is the acceptor. The acceptor can take up energy by a two-step process where the first is $^5I_8 \rightarrow \, ^5I_6$, a phonon-assisted transition, and the second is $^5I_6 \rightarrow \, ^5S_2$, a resonant energy transfer process. Since the first step is phonon assisted, we can expect a relatively small donor-acceptor transfer rate. An example of donor decay curves for small donor concentrations ($x = 0.003$) and correspondingly small donor-donor transfer is shown in Fig. 24a. We note the anticipated nonexponential decay at the beginning of the decay curve and the exponential decay for large times, determined approximately by the donor decay time τ_A. These curves were fitted with eq. (9.47). Figure 24b shows an example with large donor concentrations ($x = 0.1$) the full curve beeing given by eq. (9.43). Analysis of the curves for a number of donor/acceptor combinations yields $\tau_A = 1.7 \times 10^{-3}$ sec, $R_0 = 6$ Å, and $a_6 = 1.8 \times 10^{-41}$ cm^6/sec. a_6 is approximately independent of donor and acceptor concentrations. It has a relatively small value (due to the nonresonant nature of the energy transfer) Krasutsky and Moos [97] find for energy transfer in $LaCl_3$ between Pr and Nd ions in the infrared spectral region, a coupling constant $a_6 = 6 \times 10^{-38}$ cm^6/sec, considerably larger than in the present case showing the large spread possible for the strength of the coupling. The numbers for YF_3 enable us to calculate $w_{AB} \sim 10^4$ sec^{-1} - again a relatively small number.

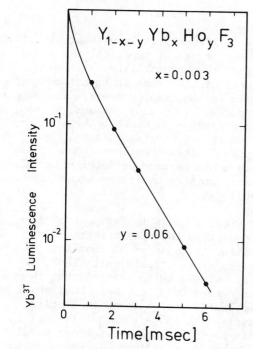

Fig. 24a

Fluorescence decay of Yb^{3+} in $Y_{1-x-y} Yb_x Ho_y F_3$. The donor concentration is low (x = 0.003). After [96]

Fig. 24b

Fluorescence decay of Yb^{3+} in $Y_{1-x-y} Yb_x Ho_y F_3$. The donor concentration is high (x = 0.1). After [96]

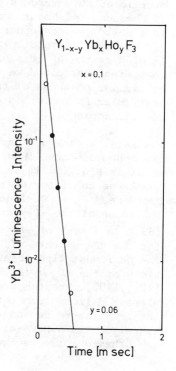

10.   RARE EARTH GLASSES

The preceding chapters have dealt exclusively with crystalline
solids. However, many solids that are of practical use today
are amorphous or glassy rather than crystalline. This means,
that the immediate environment of like ions in such substances
is similar, but that there is no long range order in the sample.
RE ions can be easily incorporated into many glasses. It was
noticed quite early that in glasses - as might be expected -
the most prominent feature of the RE crystal spectra, the
extreme sharpness of the optical lines, vanishes. When it was
realized that REs in glasses could be used as laser materials,
many investigations into their spectra and fluorescence properties were undertaken [98].

From a simplified point of view, a glass is a supercooled
liquid. It can therefore be assumed that the spectra of RE
ions in glasses will be similar to those of RE ions in liquids.
The spectra in liquids show a "crystal field splitting", although with very wide lines. This is an indication that the
RE ions in a liquid are surrounded by a near neighbour shell
of ligands - similar to the configuration found in a solid
and the same for every dissolved RE ion - and that the un-
correlated structure is only beyond the near neighbour shell.
If the near neighbour coordination in a liquid is the same as
in a solid, we can understand the similarity in the magnitude
of the crystal field splitting of the crystal and the solution.
In glasses the RE ions are incorporated as oxides. From the
reasoning just cited we can expect RE spectra in glasses to
be similar to those of the stable oxide modification of the
particular RE ion; this expectation is verified by the experimental findings.

The laser techniques have allowed to investigate in detail
the inhomogeneous linewidth in glasses. A simple example is
shown in Fig. 25, it shows the energies of the $^7F_1$ and $^7F_2$
levels as the laser energy is tuned through the inhomogeneous
broadened $^7F_0 \rightarrow {}^5D_0$ linewidth. In this process the total
splitting of the $^7F_1$ level increases from 150 cm^{-1} to 600 cm^{-1}
[99]. In a similar way the total decay rates vary as is shown
for the $^4F_{3/2}$ state of Nd^{3+} in silicate glass (Fig. 26)[100].
The narrowest line observed to data in a glass is the 20 MHz
width of the $^5D_0 \rightarrow {}^7F_0$ transition of Eu^{3+} measured at 1.7 K
[101],[102]. The temperature dependence of the linewidth in
glasses differs from that found in crystals. Fig. 27 shows
the temperature dependence of the linewidth of the $^3P_0 \rightarrow {}^3H_4(1)$
transition in a BeF$_2$-glass, where over a wide temperature range
a T^2 behaviour is observed. At his point it is not clear what

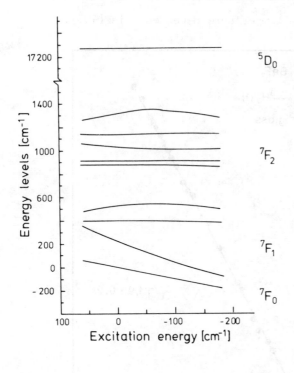

Fig. 25
Relative position of
the crystal field energy
levels in a silicate glas
as a function of excita-
tion energy for the $^7F_1$
and $^7F_2$ excited states
as measured by FLN.
After [99]

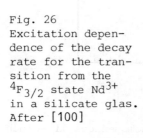

Fig. 26
Excitation depen-
dence of the decay
rate for the tran-
sition from the
$^4F_{3/2}$ state Nd^{3+}
in a silicate glas.
After [100]

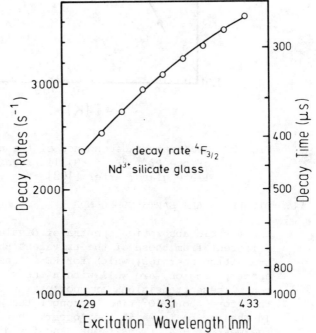

decay rate $^4F_{3/2}$
Nd^{3+} silicate glass

the origin of this temperature dependence is [103].

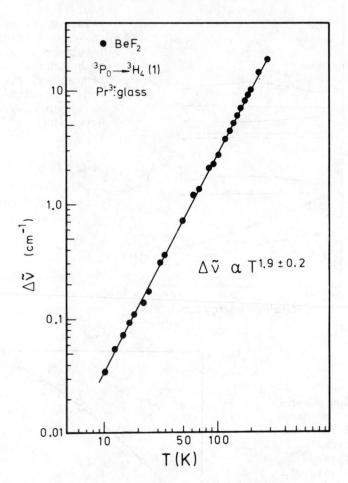

Fig. 27   Temperature dependence of the homogeneous linewidth
          of the Pr^{3+} $^3P_o \rightarrow ^3H_4$ (1) transition in a fluorobery-
          llate glass. After [103]

11.   DIVALENT RARE EARTH COMPOUNDS

The visible absorption spectra of divalent REs are noticeably
different from those of the trivalent REs. The $4f^N \rightarrow 4f^{N-1}5d$
transition energies, which for RE^{3+} ions are in the ultraviolet
spectral region, are well within the visible region for RE^{2+}
ions. These transitions are not parity forbidden, and are
therefore about 10^6 times stronger than the $4f \rightarrow 4f$ transitions.
In the region in which they occur, they can completely swamp
the weak $4f \rightarrow 4f$ transitions, and for divalent REs we are

limited to observing them mostly in the red and infrared
spectral regions. The visible spectra of a divalent RE compound like EuF_2 are therefore dominated by the $4f^7 \rightarrow 4f^6 5d$
transitions.

The crystal field theory can also be used to describe the
crystal field interaction of a 5d electron. The 5d electrons
extend spatially farther out than the 4f electrons, which
means that the crystal field interaction for a 5d electron is
larger than for a 4f electron; in fact, in comparison to it
the spin-orbit interaction can be neglected in first order[4].

For an octahedral (cubic) coordination, the crystal field
Hamiltonian in spin Hamiltonian notation has the form

$$H_{CF} = A_{40} <r^4> \beta (O_{40} + 5 O_{44}) \tag{11.1}$$

where one often writes

$$A_{40}<r^4> = Dq/12; \quad \beta = 2/63 \tag{11.2}$$

The states of a 5d electron in a cubic crystal field are calculated by applying the Hamiltonian (11.1) to its fivefold
orbitally degenerate state (^2D). This means the basic functions
are those characterized by L and M_L, as in the case of the
4f electrons. The matrix elements of (11.1) can be calculated
and solving the resulting secular equation yields energies at
-4 Dq $= (-96/63)A_{40}<r^4>$, which is an orbital triplet state,
and at $+5$ Dq $= (144/63)A_{40}<r^4>$, which is an orbital doublet
state. The total splitting between the triplet and doublet
state is 10 Dq $= (80/21)A_{40}<r^4>$. The triplet state is a Γ_5
state and the doublet state is a Γ_3 state. More conventional,
however, is the molecular orbital terminology in which the Γ_5
state is called at t_{2g} state and the Γ_3 state an e_g state.

In most compounds REs are trivalent and in some they are divalent. Yet in some substances called mixed valence compounds,
both divalent and trivalent RE ions are present (for a review
of this topic see[103a,103b]. The most common mixed valence
RE compounds are metallic. Some, however, are semiconductors
(or even insulators), and optical investigation of them has
produced insight into the energy level diagrams of these substances

In Eu_3O_4 ($Eu^{2+}Eu_2^{3+}O_4^{2-}$) the divalent and trivalent Eu ions
have a ratio of 1:2. Since Eu_3O_4 cannot be grown in thin films
its opaqueness in the visible and ultraviolet regions prevents
taking other than reflection spectra. From them via a Kramers-
Kronig analysis the imaginary part of the dielectric constant

(the absorption) is obtained; the results are shown in Fig. 28, which also gives the analysis of the spectral structure.

Since we again observe very strong absorption, the transition must be parity allowed, meaning that for the 4f electrons we see $4f^N \rightarrow 4f^{N-1}5d$ transitions, specifically, $4f^7 \rightarrow 4f^65d$ (for divalent Eu) and $4f^6 \rightarrow 4f^55d$ (for trivalent Eu). As in the cubic case, the 5d state is split into two levels. In Eu_3O_4 neither the Eu^{2+} nor the Eu^{3+} ions occupy cubic sites, but the crystal field can be approximated by a cubic one, such that the two hump structure is preserved. In a $4f^N \rightarrow 4f^{N-1}5d$ transition the $4f^{N-1}$ structure of the RE ions after the transition (final state) must not be in the ground state, but can also be in some excited state. For the $4f^7 \rightarrow 4f^65d$ transition the $4f^65d$ final state is $4f^6(^7F_J)5d$ (the next terms are too high energetically to be excited) and for the $4f^6 \rightarrow 4f^55d$ transition the $4f^55d$ final state is $4f^5(^6H_J, {}^6F_J, {}^6P_J)5d$.

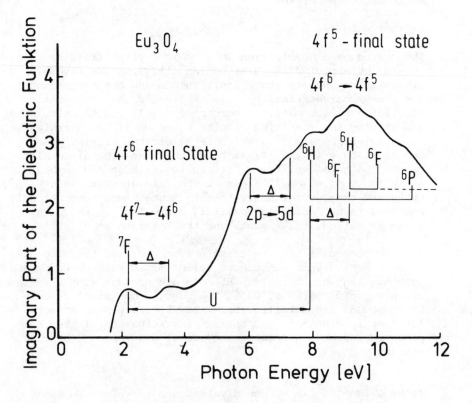

Fig. 28 Imaginary part of the dielectric function of Eu_3O_4.
         Δ = crystal field splitting of the 5d band, U = Coulomb correlation energy. After [103c]

The internal ^7F_J structure cannot be resolved in the spectra. One can, however, see the $4f^5(^6H, \, ^6F, \, ^6P)5d$ final state structure in Fig. 28.

The production of excited states in the photoabsorption process can easily be seen in the conceptually simpler experiments of photoemission spectroscopy because here the final state of the electron is a plane wave. Therefore the spectra reflect directly the states produced from the groundstate configuration after the excitation of an electron. As an example we show in Fig. 29 the photoemission spectrum of SmTe, Sm beeing divalent in this compound and therefore electronically equal to Eu^{3+}.

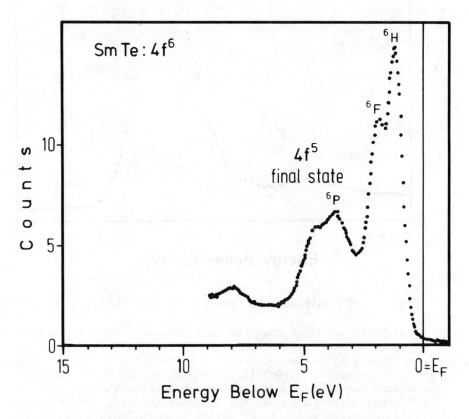

Fig. 29 XPS valence band spectrum of SmTe (Sm^{2+}) showing the final state structure. After [104]

The spectrum clearly shows three peaks, attributable to the ^6H, ^6F and ^6P final state multiplets of the $4f^5$ final state configuration, which is produced from the $4f^6$ initial state

configuration by the photoemission process. Fig. 30 shows the
photoemission spectrum of EuO, which has a $Eu^{2+}(4f^7)$ initial
state and a $4f^6(^7F)$ final state configuration exhibiting only
one peak in the spectrum as is actually observed.

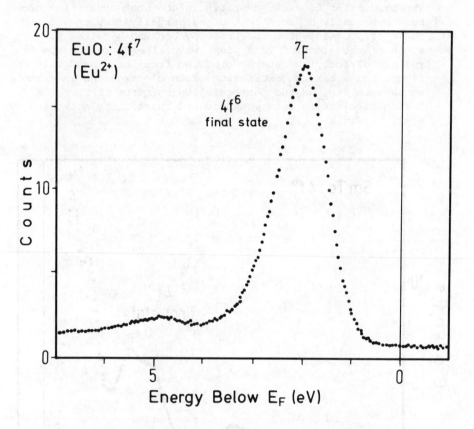

Fig. 30 XPS valence band spectrum of EuO (Eu^{2+}). After [104]

A property uniquely observed in the spectra of mixed valence
compounds is the so-called Coulomb correlation energy U. This
is the energy difference between the $4f^7$ and the $4f^6$ configuration in solids (generally, the $4f^N$ and the $4f^{N-1}$ configuration); in other words, the energy necessary to produce the
$4f^7$ configuration out of the $4f^6$ configuration by putting an
extra electron into its 4f shell. In Eu_3O_4 this energy difference is measured between the position of the 4f ground state
configurations for Eu^{2+} ions ($4f^7$) and Eu^{3+} ions ($4f^6$) and it
comes out to be 5.7 eV. This value is much smaller than we
would estimate from free ion spectra, because in the crystal
there is screening by the charge of the ligands. Fig. 31 shows
the energy level diagram consistent with the interpretation

of the spectrum in Fig. 28. The band states (which are one
electron states) and 4f states (which are many electron states)
are separated.

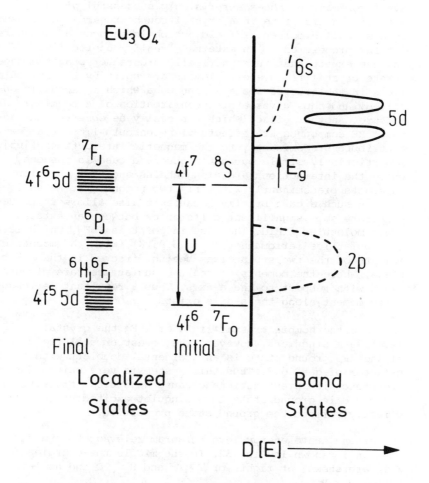

Fig. 31   Energy level diagram for Eu_3O_4. At the left of the
diagram are the localized states, at the right the
band states. After [103a]

12.   MAGNETIC INTERACTIONS

Since the optical lines in concentrated RE crystals can be
as narrow as 1 cm^{-1} one is in principle able to measure magnetic interactions with them [2]. Magnetic interactions in RE
compounds are however relatively small. The dipole-dipole
contribution is of the order of a few cm^{-1} and exchange interactions cannot be larger because of the localized nature of

the 4f electrons which precludes large 4f - 4f overlap. There
is yet an obvious   incentive for the investigation of magnetic interactions by optical means. In the most naive, however, for many purposes surprisingly successful picture,
magnetic interactions in a crystal can be regarded as caused
by a so-called molecular field, which is thought of as having
similar properties as an external "real" magnetic field. Thus
one can expect that in magnetically ordered crystals the degenerate crystal field energy levels are split by the molecular
field in the ordered state a phenomena which is actually observed, and thus gives a nice confirmation of a seemingly naive
picture. Another effect which can easily be observed in magnetic RE compounds are effects of neighbours if only a few
contribute predominantly to the magnetic interactions [105].
A particularly simple but often observed case is the one,
where the interactions are along chains and thus each ion
interacts predominantly with only two neighbours. We shall
discuss such a case briefly because it also allows us to demonstrate some significant differences between an external
and a molecular field. The case in point is TbF_3, which is a
two sublattice ferromagnet (T_c = 3.95 K), with the magnetic
moments of the two sublattices beeing directed in the a - c
plane, where the moments of the two sublattices are directed
$\pm$ 26° with respect to the a-axis. Thus a resultant ferromagnetic moment along the a-axis exists.

In the orthorhombic crystal field of TbF_3 the crystal field
levels are singulets. However, the lowest crystal field level
of the $^7F_6$ ground state is an accidental doublett with a
g-factor of g = 16.3. Thus this system is very suited for
spectroscopic investigations because an optical transition from
the magnetic ground state to a singulet excited state will
directly reflect the ground state properties.

The ground state energy level diagram relevant to the discussion is shown in Fig. 32. In the middle the energies for
$T > T_C$ are shown, at right for $T > T_C$ and $H_{ext} > 0$ and at left for
$T < T_c$, $H = H_{mol}$.

For $T > T_C$, H_{ext} = 0 one has three possible energies because
the neighbour configuration of each ion ($\uparrow$) can have a particular arrangement, from pure ferromagnetic $\uparrow$ ($\uparrow$) $\uparrow$ to pure
antiferromagnetic $\downarrow$ ($\uparrow$) $\downarrow$, where the occupation of the different configurations is given by the Boltzmann distribution.
We also see that each of the configurations is degenerate.
This degeneracy can be lifted by an external applied field
and instead of three energy levels in the case with an external magnetic field one has six.

These considerations are borne out by experiment as can be

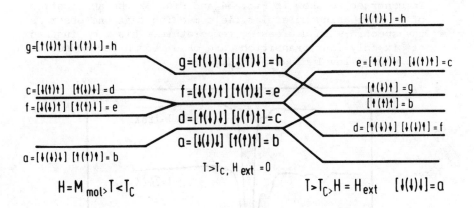

Fig. 32   Schematic energy level diagram for interaction of an
          ion (↑) with two neighbours only. After [106]

seen from Fig. 33 [106], which shows the magnetic field dependence of the optical transition $^7F_6(1) \rightarrow {}^5D_4(4)$ and one
clearly see the six transitions.

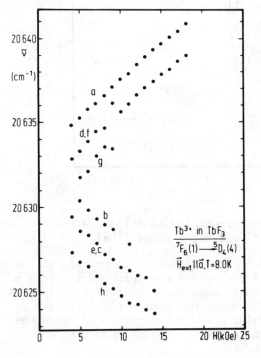

Fig. 33
Magnetic field dependence
of the transition $^7F_6(1) \rightarrow$
$^5D_4(4)$ in TbF$_3$; magnetic
field parallel to the a-axis.
After [106]

In contrast we show in Fig. 34 and Fig. 35 the application
of an increasing molecular field meaning that one observes
the spectra with decreasing temperature. This time instead
of six only four transitions are observed in accordance with
the energy level diagram in Fig. 32.

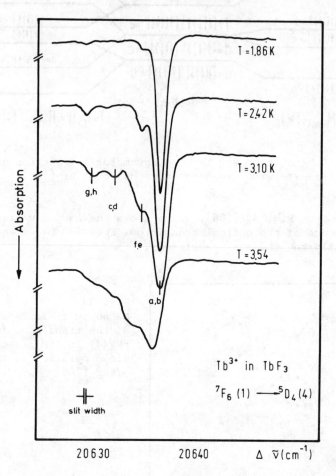

Fig. 34 $^7F_6(1) \rightarrow {}^5D_4(4)$ transition in TbF$_3$ as a function of
temperature. T_c = 3.95 K. After [106]

The molecular field is a local field, which we can imagine
to be produced by the direction of the neighbouring ions of
an ion (↑) or (↓). In this sence the configurations ↑ (↑) ↑
and ↓ (↓) ↓ which can be split by an external field are still
degenerate in an molecular field because in both cases the
molecular field is parallel to the direction of the ion.

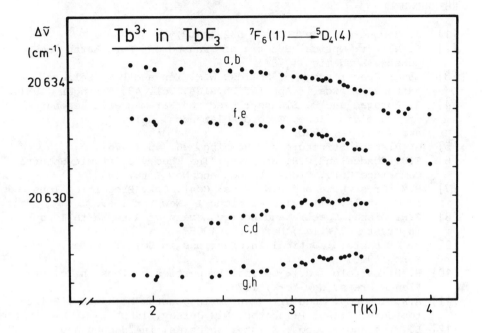

Fig. 35 Temperature dependence of the splitting of the
$^7F_6(1) \rightarrow {}^5D_4(4)$ transition in TbF$_3$. After [106]

The same reasoning holds of course for the configurations
↑ (↓) ↑ and ↓ (↑) ↓, for both of which the molecular field
is antiparallel to the ion however in both cases of the same
relative direction.

The situation is different for the intermediate spin configuration. If we start with ↑ (↑) ↑, where the molecular field
is assumed to be parallel to (↑) the two first excited configurations ↑ (↑) ↓ and ↑ (↓) ↓, which are degenerate for fi-
H$_{mol}$ = 0 are now clearly energetically different by 2μ · H,
which results on the observed splitting of the middle line
of the triplet (see Fig. 32). Spectra demonstrating this behaviour are shown in Fig. 34, where clearly the four line
pattern in the molecular field can be seen.

Acknowledgement:
Thanks are due to R. Baumert, J. Brinkmann and R. Courths
for allowing me tu use material from their dissertation.
I am very much indebted to the Deutsche Forschungsgemeinschaft
for generous support over many years.

REFERENCES

[1]   J. Becquerel, Phys. Z. 8, 632 (1908).
[2]   S. Hüfner: Optical Spectra of Transparent Rare Earth Com-
      pounds, Academic (1978).
[3]   Laser Spectroscopy of Solids, W.M. Yen and P.M. Selzers
      editors, Topics in Applied Physics, Vol. 49, Springer (1981.
[4]   A. Abragam and B. Bleaney: Electron Paramagnetic Resonance
      of Transition Ions, Oxford (Clarendon) Univ. Press, London
      and New York, 1970.
[5]   G. Blasse: Structure and Bonding, 26, 43 (1976).
[6]   E.U. Condon and G.H. Shortley: The Theory of Atomic Spectra,
      Cambridge Univ. Press, London and New York, 1953.
[7]   H.M. Crosswhite and H.W. Moos: Conf. Opt. Properties Ions
      in Crystals, Wiley (Interscience), New York, 1967.
[8]   G.H. Dieke: Spectra and Energy Levels of Rare Earth Ions
      in Crystals, Wiley, New York, 1968.
[9]   B. Di Bartolo: Optical Interactions in Solids, Wiley, New
      York, 1968.
[10] B. Di Bartolo (ed.): Optical Properties of Ions in Solids,
      Plenum Press, New York, 1975.
[11] A.R. Edmonds: Angular Momentum in Quantum Mechanics, Prince-
      ton Univ. Press, Princeton, New Jersey, 1960.
[12] E. Fick and G. Joos, Kristallspektren, in "Handbuch der
      Physik" (S. Flügge, ed.), Vol. 28, p. 205, Springer-Verlag,
      Berlin and New York, 1957.
[13] J.S. Griffith: The Theory of Transition Metal Ions, Cam-
      bridge Univ. Press, London and New York, 1964.
[14] M.T. Hutchings, Solid State Phys. 16, 227 (1964).
[15] S. Hüfner in "Magnetism Current Topics" (S. Foner, ed.),
      p. 641, Gordon and Breach, New York, 1976.
[16] B.R. Judd: "Operator Techniques in Atomic Spectroscopy",
      McGraw-Hill, New York, 1963.
[17] A.A. Manenkov and R. Orbach (eds.): Spin Lattice Relaxation
      in Ionic Solids, Harper and Row, New York, 1966.
[18] D.S. McClure, Solid State Phys. 9, 399 (1959).
[19] S. Methfessel and D. Matthis, Springer Tracts Phys. 18,
      389 (1968).
[20] D.J. Newman, Adv.Phys. 20, 197 (1971).
[21] R.D. Peacock, Structure and Bonding, 22, 83 (1975).
[22] R. Reisfeld and C.K. Jorgensen, Inorg.Chem.Concepts 1 (1977).
[23] L.A. Riseberg and M.J. Weber: Relaxation Phenomena in
      Rare Earth Luminescence, in "Progress in Optics" (E.Wolf,
      ed.), Vol XIV, North-Holland Publ., Amsterdam, 1976.
[24] M.E. Rose: Elementary Theory of Angular Momentum, Wiley,
      New York, 1957.
[25] S. Sugano, Y. Tanabe, and H. Kamimura: Multiplets of Tran-
      sition Metal Ions in Crystals, Academic Press, New York,
      1970.
[26] M.Tinkham:Group Theory and Quantum Mechanics, McGraw-Hill,

New York, 1964.
[27]   P. Wachter, CRC Critical Rev. Solid State Sci. 3, 189 (1973).
[28]   P. Wachter, in "Handbook of the Physics and Chemistry of
       the Rare Earths" (K.A. Gschneidner and L. Eyring, eds.),
       Vol. 1, Chapter 19, North-Holland Publ. Amsterdam, 1977.
[29]   J.C. Wright, in "Topics in Applied Physics" (F.K. Fong, ed.),
       p. 239, Springer-Verlag, Berlin and New York, 1976.
[30]   B.G. Wyborne: Spectroscopic Properties of Rare Earths,
       Wiley, New York, 1965.
[31]   J. Sugar, Phys.Rev.Lett. 14, 731 (1965).
[32]   J.F. Kielkopf and H.M. Crosswhite, J. Opt. Soc. Am. 60,
       347 (1970).
[33]   H.M. Crosswhite and H.W. Moos: ref.[7], p.3.
[34]   H.G. Kahle, Z.Phys. 161, 486 (1961).
[35]   H.M. Crosswhite, H. Crosswhite, F.W. Kaseta, and R. Sarup,
       J.Chem.Phys. 64, 1981 (1976).
[36]   K. Rajnak and B.G. Wybourne, Phys. Rev. 132, 280 (1963).
[37]   K.W.H. Stevens, Proc. Phys. Soc. A65, 209 (1952).
[38]   W.T. Carnall, H. Crosswhite, H.M. Crosswhite, and J.G.
       Conway, J.Chem. Phys. 64, 3584 (1976).
[39]   W.T. Carnall, H. Crosswhite, H.M. Crosswhite: Energy Level
       Structure and Transition Probabilities of Trivalent Lanthan-
       ides in LaF3, Argonne Nat. Lab. Rep. (1977).
[40]   P. Grünberg, S. Hüfner, E. Orlich, and J. Schmitt, Phys.
       Rev. 184, 285 (1969).
[41]   S.S. Bishton and D.J. Newman: J.Phys. C, Solid State Phys.
       3, 1753 (1970).
[42]   B.R. Judd, J. Chem. Phys. 66, 3163 (1977).
[43]   D.J. Newman, J. Phys. C, Solid State Phys. 10, 4753 (1977).
[44]   D.J. Newman, J. Phys. C, Solid State Phys. 10, L617 (1977).
[45]   G.M. Copland, G. Balasubramanian, and D.J. Newman, J.Phys.C,
       Solid State Phys. 11, 2029 (1978).
[46]   B.R. Judd, Phys. Rev. Lett. 39, 242 (1977).
[47]   K.H. Hellwege, U. Johnson, H.G. Kahle, and G. Schaack,
       Z.Phys. 148, 112 (1957).
[48]   B.R. Judd, Mol. Phys. 2, 407 (1959).
[49]   M.K. Bradbury and D.J. Newman, Chem. Phys. Lett. 1, 44 (1967).
[50]   M.T. Hutchings and W.P. Wolf, J. Chem. Phys. 41, 617 (1964).
[51]   C.K. Jørgensen, R. Pappalardo, and H.H. Schmidtke, J. Chem.
       Phys. 39, 1422 (1963).
[52]   C.E. Schäffer and C.K. Jørgensen, Mol. Phys. 9, 401 (1965).
[53]   J.D. Axe and G. Burns, Phys. Rev. 152, 331 (1966).
[54]   R.E. Watson and A.J. Freeman, Phys. Rev. 156, 251 (1967).
[54a] W. Urland, Chem. Phys. Lett. 53, 296 (1978).
[55]   D. Sengupta and J.O. Artman, Phys. Rev. B1, 2986 (1970).
[56]   R. Faulhaber,Thesis Darmstadt (1968), see also R. Faulhaber
       and S. Hüfner, Solid State Commun. 7, 389 (1969); Z. Phys.
       228, 235 (1969).
[57]   H. Yamamoto, Y. Otomo, and T. Kano, J. Phys. Soc. Jpn. 26,
       137 (1969).

[58] G. Burns, Solid State Commun. 7, 1073 (1969).
[59] R. Baumert, J. Pelzl, and S. Hüfner, Solid State Commun. 16, 345 (1975), and R. Baumert, Dissertation, Berlin (1981).
[60] W.H. Zachariasen, J. Chem. Phys. 16, 254 (1954).
[61] J.H. van Vleck, J. Phys. Chem. 41, 67 (1937).
[62] B.R. Judd, Phys. Rev. 127, 750 (1962).
[63] G.S. Ofelt, J. Chem. Phys. 37, 511 (1962).
[64] R. Orbach, Proc. Roy. Soc. A264, 458, 485 (1961).
[65] P.L. Scott and C.D. Jeffries, Phys. Rev. 127, 32 (1962).
[66] W.M. Yen, W.C. Scott and A.L. Schawlow, Phys. Rev. 136, A271 (1964).
[66a] L.E. Erickson, Phys. Rev. B 11, 77 (1975).
[67] J.M. Ziman, Proc. Roy. Soc. (London) A226, 436 (1954).
[68] M.J. Weber, Phys. Rev. B 8, 54 (1973).
[69] H.W. Moos, J. Lum. 1,2, 106 (1970).
[70] G.F. Imbusch and R. Kopelman: in ref. [3].
[71] L.A. Riseberg and H.W. Moos, Phys. Rev. 174, 429 (1968).
[72] L.E. Erickson, Phys. Rev. B 16, 4731 (1977).
[73] Y.C. Chen, K.P. Chiang, S.R. Hartmann, Opt. Commun. 28, 181 (1979).
[73a] W.M.Yen and P.M. Selzer, in ref. [3].
[74] D.S. Selzer in ref. [3].
[75] E.L. Hahn, Phys. Rev. 77, 297 (1950).
[76] J. Pelzl, S. Hüfner and S. Scheller, Z. Phys. 231, 377 (1970).
[77] R.M. Sternheimer, Phys. Rev. 84, 244 (1951).
[78] C. Delsart, N. Pelletier-Allard and R. Pelletier, J. Phys. B, Atom. Molec. Phys., 8, 2771 (1975).
[79] C.A. Hutchison and E. Wong, J. Chem. Phys. 29, 754 (1958).
[79a] S. Hüfner, Z. für Naturforschung, 179, 825 (1962).
[79b] A.J. Dekker in "Hyperfine Interactions", A.J. Freeman and R.B. Frankel eds., Academic Press, New York, 1967, p. 679.
[80] T. Förster, Ann. Phys. 2, 55 (1948).
[81] T. Förster in "Modern Quantum Chemistry" (O. Sinanoglu, ed.), Part III, Chapter B.1., Academic Press, New York, 1965.
[82] D.L. Dexter, J. Chem. Phys. 21, 836 (1953).
[83] D.L. Dexter, Phys. Rev. 126, 1962 (1962).
[84] T. Holstein, S.K. Lyo and R. Orbach, in ref. [3].
[85] D.L. Huber, in ref. [3].
[86] R. Orbach, in "Conf. Properties of Ions in Crystals", p. 445, Wiley (Interscience), New York, 1967.
[86a] T. Miyakawa and D.L. Dexter, Phys. Rev. B 1, 2961 (1970).
[86b] R. Orbach in "Optical Properties of Ions in Crystals", p.445, Wiley (Interscience), New York, 1967.
[87] D.L. Huber, D.S. Hamilton, B. Barnett, Phys. Rev. B 16, 4642 (1977).
[88] W.Y. Ching, D.L. Huber, B. Barnett, Phys. Rev. B 17, 5025 (1978).
[89] M. Inokuti, F. Hirayama, J. Chem. Phys. 43, 1978 (1965).
[90] D.L. Huber, Phys. Rev. B 20, 2307 (1979).
[91] D. Fay, G.Huber, W.Lenth, Opt.Commun. 28, 117 (1979).

[92]   P.G. de Gennes, J. Phys. Chem. Sol. 7, 345 (1958).
[93]   A.L. Burshtein, Zh. Eksp. Teor. Fiz. 62, 1695 (1972).[Sov.
       Phys. - JETP 35, 882 (1972)]. See also L.D. Zusman, Zh. Eksp.
       Teor. Fiz. 73, 662 (1977). [Sov. Phys. - JETP 46, 347 (1977)]
[94]   P.M. Richards, Phys. Rev. B 20, 2965 (1979).
[95]   M. Yokota, I. Tanimoto, J. Phys. Soc.(Jpn.) 22, 779 (1967).
[96]   R.K. Watts and H.J. Richter, Phys. Rev. B6, 1584 (1972).
[97]   N. Krasutsky and H.W. Moos, Phys. Rev. B8, 1010 (1973).
[98]   M.J. Weber, in ref. [3]
[99]   C. Brecher, L.A. Riseberg, J. Non-Cryst. Solids 40, 469
       (1980).
[100] C. Brecher, L.A. Riseberg, M.J. Weber, Phys. Rev. B18, 5799
       (1978).
[101] P.M. Selzer, D.L. Huber, D.S. Hamilton, W.M. Yen, M.J. Weber,
       Phys. Rev. Lett. 36, 813 (1976).
[102] P.M. Selzer, D.L. Huber, D.S. Hamilton, W.M. Yen, M.J. Weber,
       in Structure and Excitations in Amorphous Solids, AIP Con-
       ference Proc. No. 31 (1976), p. 328.
[103] J. Hegarty, W.M. Yen, Phys. Rev. Lett. 43, 1126 (1979).
[103a] D.K. Wohlleben and B.R. Coles "Magnetism" (H. Suhl, ed.)
       Vol. V, p. 3, Academic Press (1973).
[103b] C.M. Varma, Rev. Mod. Phys. 48, 219 (1976).
[103c] B. Batlogg, E. Kaldis, A. Schlegel, and P. Wachter, Phys.
       Rev. B14, 5503 (1976).
[104] M. Campagna, G.K. Wertheim and Y. Baer, Topic in Applied
       Physics, Vol. 27, L. Ley and M. Cardona eds., p. 215 (1979),
       Springer Verlag, Heidelberg.
[105] G.A. Prinz, Phys. Lett. 20, 323 (1966); Phys. Rev. 152,
       474 (1966).
[106] J. Brinkmann, Dissertation, Berlin (1980).

DISCUSSION

DE LONG   (question)

How general is the effect of pressure on the crystal field parameters? In praseodymium antimonide, for example, the crystal field
level scheme appears to decrease under pressure, which is a surprise. There are other examples as well. If you could comment on
this.

HÜFNER   (answer)

In our case also, the energy levels get compressed so the cyrstal
field decreases by applying pressure, which surprised us also.
There are very few experiments in this line and we need to have
many more before we could generalize.

GROUP I   (question)

Could you comment briefly on the spectra of the divalent rare
earth ions?

HÜFNER   (answer)

First of all the spectrum profiles of the divalent lanthanides are
much different from those we have seen for the trivalent ones. Let
us take the case of EuF_2 and Eu(III) in KBr. Here we see two broad
bands. The shifts and splittings are different. The origin of these
broad bands may be traced to the excitation of $4f^7$ to $4f^6$ 5d and
the splitting of the 5d electron under the crystal field, which
we assume to be cubic, into e_g and t_{2g} levels. Notice, however,
superimposed on the spectrum are the multiplets of the $4f^6$ configuration.

CARNALL   (comment)

In lanthanide sulfide you do see the f→f transitions. In actinides,
however, the f→d configurations are shifted up a little bit higher
in energy. There are a few cases where you see even more multiplets of the divalent salt-like compounds. From this you can get
an idea that the crystal field apparently is considerably less
than the spin-orbit or electrostatic interactions.

SPECTROSCOPIC PROPERTIES OF THE f-ELEMENTS IN COMPOUNDS AND SOLUTIONS

W. T. Carnall, J. V. Beitz, H. Crosswhite

Chemistry Division, Argonne National Laboratory,
9700 South Cass Avenue, Argonne, Illinois 60439

K. Rajnak

Physics Department, Kalamazoo College,
Kalamazoo, Michigan 49007

J. B. Mann

Los Alamos National Laboratory, Los Alamos,
New Mexico 87545

ABSTRACT

In this systematic examination of some of the spectroscopic
properties of the f-elements we deal with both the trivalent lanthanides and actinides. We summarize the present status of our
energy level calculations in single crystal matrices and in
aqueous solution, and compare the predicted crystal-field structure in certain low-symmetry sites with that observed. Some interesting new structural insights are thereby gained. The state
eigenvectors from these calculations are then used in part in
reassessing and interpreting the intensities of transitions in
aqueous solution via the Judd-Ofelt theory. The parameters of
this theory derived from fitting experimental data are compared
with those computed from model considerations. Finally, we discuss some recent contributions to the interpretation of excited
state relaxation processes in aqueous solution.

389

S. P. Sinha (ed.), Systematics and the Properties of the Lanthanides, 389–450.
Copyright © 1983 by D. Reidel Publishing Company.

I. SYSTEMATIC ANALYSIS OF THE ENERGIES OF LANTHANIDE TRANSITIONS
   IN SOLIDS AND IN AQUEOUS SOLUTION

In this section we begin with a brief summary of the theoretical models used to compute the energy level structure within the
f^N-configurations of the trivalent lanthanides (Ln) and actinides
(An). The discussion focuses on relating observed absorption
band structure, interpreted in terms of transitions within the
f^N-configuration, to the complete set of energy levels for the
configuration. The interpretations in subsequent sections all
depend directly on the ability to model the interactions that
give rise to the electronic structure. The purpose of the summary is not only to give a status report on an area that is still
under development but to try to indicate the extent to which the
models are sensitive to apparent changes in the observed spectra.

I(A). Model Interactions for f^N-Configurations

The process of developing a complete Hamiltonian for f^N-
configurations is approached in stages. The first deals with the
energy level structure of the gaseous free-ion, and the second
with the additional (crystal-field) interactions which arise when
the ion is in a condensed phase. The free-ion Hamiltonian is
assumed to be the same in both cases, and the centers of gravity
of groups of crystal field levels are interpreted on the same
basis as the degenerate levels of the gaseous free ion. Because
of the abundance of data in condensed media, and the paucity of
true gaseous free ion data, the Hamiltonian for the ion in condensed phases has been much more extensively studied. Thus, unless explicitly noted, subsequent *data* on the "free ion" Hamiltonian will refer to studies of ions in condensed media.

The fundamental interactions that give rise to the free-ion
structure in trivalent lanthanides and actinides are the electrostatic repulsion between electrons in the f^N-configuration and
the coupling of their spin and orbital angular momenta. For details of the development see [1-4]. There are two different
approaches to modeling these types of interaction--the Hartree-
Fock (HF) approach and what we will call the Parametric approach.
Both begin with Schrödinger equation for the steady state of a
many electron system, Fig. I-1.

The actual form of the Hamiltonian assumes that the nucleus
can be treated as a point charge with infinite mass. Since exact
solutions are only known in the one-electron case, some method of
approximation must be used. In both the HF and parametric approaches, the first step is to obtain approximate total wavefunctions based on the central field approximation. Each electron
is assumed to move independently in the field of the nucleus and
a central field composed of the spherically averaged potential

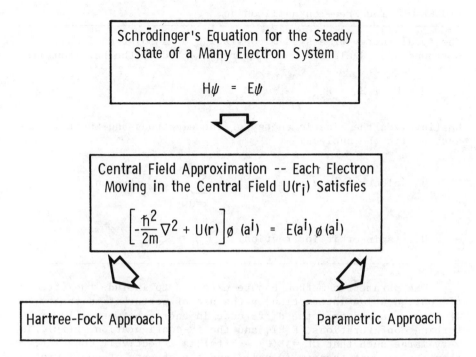

FIGURE I-1.

fields of each of the other electrons in the system. In other
words, each electron is treated as if it moved independently in a
spherically symmetric potential.

The HF-approach seeks the evaluation of this potential using
the variational principle [5]. Computed values of the desired
integrals can be obtained to varying degrees of approximation
depending upon the sophistication of the computer codes used.
The effects of configuration interaction can in principle be
introduced, but in practice this is not normally done.

In the parametric approach, each electron is assumed to move
in a central field satisfying an equation similar to the Schrödinger
equation for the hydrogen atom, except that the Coulomb potential
$-e^2/r$ is replaced by an undefined central field potential $U(r)$.
Variables are separated as with the hydrogen atom, and the angular
parts of the interaction are evaluated explicitly. Since the
radial equation contains the undefined function, $U(r)$, it cannot
be solved. The radial integrals are treated as parameters to be
evaluated from experimental data via an appropriate fitting
procedure. The expression for the energy, Fig. I-2, has the same
form as that of the HF-approach, but there is no radial function
from which to evaluate the F^k and ζ.

FIGURE I-2. The Free-Ion Hamiltonian.

The total energy of a system consisting of a point nucleus
surrounded by N electrons can be represented by the Hamiltonian:

$$H = H_0 + H_E + H_{SO}$$

H_0 (involves the kinetic energy of the electrons and their interaction with the nucleus)

H_E (electrostatic term) $E_e = \sum_{k=0}^{6} f_k F^k$ (k even)

H_{SO} (spin-orbit interaction) $E_{SO} = A_{SO} \zeta_f$

The parametric method can be extended to include the effects
of configuration interaction by the use of perturbation theory.
If it is assumed that the difference in energy between all perturbing configurations, $E(P)$, and the f^N-configuration, $E(f^N)$, is
very large such that $\Delta E = E(P) - E(f^N)$ is effectively constant,
then the closure theorem is valid and the effects of configuration
interaction can be represented by certain operators acting *within*
the f^N configuration. These result in

1. changes in the original F^k

2. additional 2- and 3-body (effective) operators operating
   within the f^N-configuration

Within the above context, the new F^k integrals should not be
identified as the integrals of the HF model but as parameters
that absorb some of the effects of configuration interaction.
For further discussion and references see [6,7].

Model calculations which include only the electrostatic
interaction in terms of the F^k-integrals, and the spin-orbit
interaction, ζ, result in correlations between calculated and observed gaseous free-ion states that are only marginally useful.
It was pointed out some 40 years ago, for example, that in the
relatively simple cases of Pr^{3+} and Tm^{3+} ($4f^2$ and $4f^{12}$) differences
between calculated and observed energy levels were in some cases
>500 cm^{-1} [8]. A poor correlation of this magnitude severely
limits the usefulness of the calculations for analyzing data [9].

The two-body effective operator correction terms incorporated into systematic parameter evaluations for the lanthanides to account for the effects of configuration interaction are usually expressed in the form given by Rajnak and Wybourne [10]. The principal terms of the Hamiltonian including the two-body (scalar) operators for configuration interaction can be written:

$$H = \sum_{k=0}^{6} F^k(nf,nf)f_K + \zeta_f A_{SO} + \alpha L(L+1) + \beta G(G_2) +$$

$$\gamma G(R_7) \quad (k \text{ even}) \qquad\qquad (I-1)$$

where f_k and A_{SO} represent the angular parts of the electrostatic and spin-orbit interactions, respectively. Similarly α, β, and γ are the parameters of the two-body correction terms while $G(G_2)$ and $G(R_7)$ are Casimir's operators for the groups G_2 and R_7. The effects of configuration interaction that can be expressed in the same form as the f_k are of course automatically absorbed in the F^k radial integrals when they are treated as parameters. The additional terms, α, β, and γ thus represent effects that do not transform as the f_k.

The values of α, β, and γ arising from electrostatic configuration interaction have been calculated for Pr^{3+} by Morrison and Rajnak [11], using *ab initio* methods, Table I-1. A particularly useful insight gained from this work was that higher energy processes such as excitation of one or two particles to the continuum made large contributions to the parameter values. The fact that the energies of the continuum states relative to the f^N-configurations did not change significantly with atomic number could be correlated with the near constancy of the fitted parameter values across the lanthanides series [12,13]. A subsequent perturbed-function approach to the calculation of the same continuum interactions addressed in [11] confirmed the results for Pr^{3+} and extended the calculation to other 3+ lanthanides as well as to Pu^{3+} [14].

For configurations of three or more equivalent f electrons, three-particle configuration interaction terms have been added to the model in the form given by Judd [15] and Crosswhite *et al.* [16], Table I-1. Such terms arise from the perturbing effects of those configurations that differ from f^N in the quantum numbers of a single electron, and are expressed as $T^i t_i$ (i = 2, 3, 4, 6, 7, 8) where T^i are the parameters and t_i are matrix elements of three-particle operators within the f^N-configuration. As in the case of the two-body terms, values of the three-particle correction parameters have been calculated by *ab initio* methods and found to agree with those defined by fitting experimental data,

TABLE I-1. Elements of the Parametric Hamiltonian. The Fitted
Values are for Nd^{3+}:$LaCl_3$ [19]. [a]Values computed using a relativistic HF-code, [b]Ref. 11, [c]Ref. 17.

		Fitted Value (cm^{-1})	Computed Value (cm^{-1})
H_E (Electrostatic Term)	F^2	71866	102720[a]
	F^4	52132	64462
	F^6	35473	46386
$E_e = \sum\limits_{k=0}^{6} f_k F^k$ (k-even)			
H_{SO} (Spin-Orbit Interaction)	ζ_f	880	950.5
$E_{SO} = A_{SO}\zeta_f$			
$H_{CF(2)}$ (Two-body Configuration Interaction)	α	22.1	28 ⎫
	β	-650	-615 ⎬ Pr IV[b]
	γ	1586	1611 ⎭
$E_{CF(2)} = \alpha L(L+1) + \beta G(G_2) + \gamma G(R_7)$			
$H_{CF(3)}$ (Three Particle Configuration Interaction Operators)	T^2	377	394 ⎫
	T^3	40	-34	
	T^4	63	89 ⎬ Pr III[c]
	T^6	-292	-214	
	T^7	358	314	
	T^8	354	274 ⎭
$E_{CF(3)} = \sum\limits_{i} t_i T^i$			
Electrostatically Correlated Spin-Orbit Interaction (Two-Body Pseudo Magnetic Operators)	P^2	225	
	P^4	R	
	P^6	R	
Spin-other-orbit and spin-spin effects: Marvin Integrals	M^0		
	M^2 ⎬ HF values used directly		
	M^4		
Crystal Field Interaction	$\sum\limits_{k,q,i} B_q^k \left(C_q^{(k)}\right)_i$ (terms appropriate to the crystal symmetry)		

Balasubramanian *et al.* [17]. The values of these parameters have
also been shown to be nearly constant over the lanthanide series
[12].

Magnetically correlated corrections to the interactions included in Eq. I-1 have been introduced in the form suggested by
Judd *et al.* [18]. Values of the Marvin integrals, M^h (h = 0, 2,

4), which represent spin-spin and spin-other-orbit relativistic corrections, were initially determined from parametric fits to experimental data. However, the values obtained were essentially identical to those computed using HF-methods, so that more recently the latter values either have been used directly and not optimized, or M^0 has been varied while M^2 and M^4 were fixed in their HF ratios to M^0. The two-body magnetic corrections, $a_i Z_i$, appear to be dominated by the electrostatically correlated spin-orbit perturbation which involves the excitation of an f electron into a higher-lying f-shell. The corresponding parameters P^f (f = 2, 4, 6) show a regular increase across the lanthanide series, but have exclusively been evaluated by parametric fitting [12].

Although extensive corrections to the free-ion Hamiltonian have been developed, practically all crystal-field calculations are carried out using a single-particle crystal-field theory in which the parameters are appropriate to a given site symmetry, $H_{CF} = \sum_{k,q,i} \left[B_q^{(k)} (C_q^{(k)})_i \right]$. For details and references to the original literature see [1-3]. Thus to complete the interactions given in Eq. I-1, the following terms are included in the Hamiltonian currently used in the parametric fitting of the experimental data:

$$\sum_{i=2,3,4,6,7,8} T^i t_i + \sum_{k=0,2,4} M^h m_h + \sum_{f=2,4,6} P^f p_f +$$

$$\sum_{k,q,i} \left[B_q^k C_q^{(k)} \right]_i$$

Typical values of the atomic parameters appropriate to $Nd^{3+}:LaCl_3$ [19], are included in Table I-1.

I(B). Interpretation of the Model Parameter Values

Hartree-fock values of the F^k's and ζ_f are always larger than those obtained by allowing them to vary as parameters. There are several reasons for this:

1) The usual HF-calculation is non-relativistic; inclusion of relativistic effects can improve the agreement with experiment, but discrepancies remain. The HFR code of Cowan and Griffin [20], has the advantage of nearly reproducing the relativistic results, via a pseudorelativistic correction to the potential, while maintaining the simpler non-relativistic formalism. It gives remarkably good agreement with empirical values of ζ but the F^k's, while smaller than HF values, remain considerably larger than the empirical ones, (Table I-1). This suggests that

differences between HF and empirical ζ values arise largely from neglect of relativistic effects. There is also a relativistic effect on the F^k's but that is not the major consideration.

2) Even a relativistic HF calculation is usually based on interactions in a pure f^N-configuration whereas the f-electrons spend some time in higher-lying configurations where they move in larger orbits and interact less than the *ab initio* model assumes. In addition, the experimental results are frequently for an ion in a condensed phase, not for a gaseous free-ion. However, this effect is not very large. The value of F^2 for Pr IV (free-ion) is only $\sim6\%$ larger than that for $Pr^{3+}:LaCl_3$, Table I-2 [21].

TABLE I-2. Comparison of Values of Free-Ion Parameters (cm^{-1}).

	HFR[a]	Pr IV[c] Free-Ion	$Pr^{3+}:LaCl_3$[d]
F^2	98723	72553	68368
F^4	61937	53681	50008
F^6	44564	36072	32743
ζ	820.22	769.91	744
α	$(28)^b$	23.786	22.9
β	$(-615)^b$	-613.24	-674
γ	$(1611)^b$	745.73	1520
M^0	1.991	1.588	1.76
P^2	-	-	275

[a]Reference 20.   [c]Reference 21.
[b]Reference 11.   [d]Reference 12.

While the HFR values of F^k are too large, the differences between the HFR and the experimental (parametrized) values, F^k (HFR) - F^k (EXP) = ΔF^k, have been shown to be nearly constant over the lanthanide series, as illustrated in Figs. I-3 and I-4 [12]. Both ζ_{4f} and F^2 $vs.$ Z change slope at Z = 64 (Gd^{3+}), Figs. I-3 and I-5. While fitting of data near the center of the series poses special problems, the parametric fit results appear to be consistent with HFR calculations. Similar results have been obtained with the actinides [13].

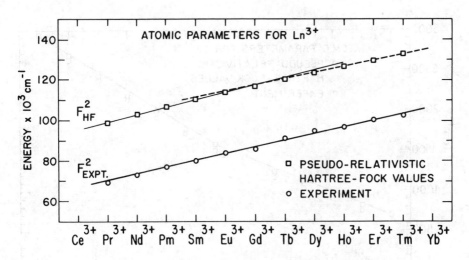

FIGURE I-3. Variation of Slater Integral F^2 with lanthanide atomic number.

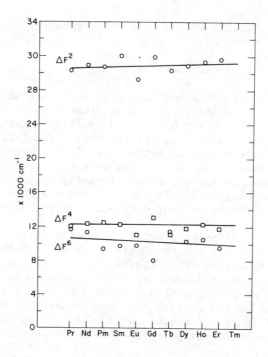

FIGURE I-4. Variation of the difference between the pseudo relativistic Hartree-Fock values of the Slater integrals F^k and those determined experimentally as a function of lanthanide atomic number.

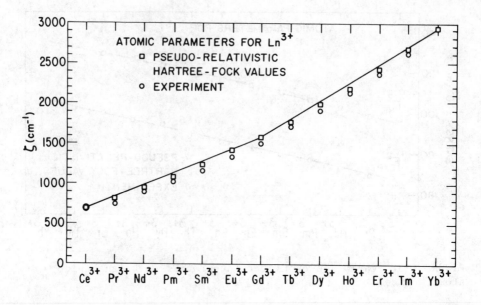

FIGURE I-5. Variation of the spin-orbit integral (ζ) with lanthanide atomic number.

I(C). Crystal-Field Calculations

Having defined the atomic (or free-ion) portion of the parametric model, and indicated that consistent results are obtained
using it over both the lanthanide and actinide series, we will
briefly address the status of crystal-field calculations and some
problems that have arisen as better correlation with experiment
has been achieved. Judd [22,23] has recently drawn attention to
some of the inadequacies of the single-particle model currently
in use, and has suggested refinements. However, the following discussion deals with the general conception of the model.

As Wybourne [1] pointed out, the crystal field was originally
thought of as a purely electrostatic interaction between the central ion and the surrounding ligands with the latter replaced by
point charges. A more general approach considers the potential
energy $V(r)$ of an electron of a central ion where r is the radius
associated with an f-electron. The environment is then represented
by a classical charge distribution $\rho(R)$ where R is the radius
associated with a general point in that environment:

$$V(r) = -\int \frac{e\,\rho(R)\,dr}{|R-r|} \qquad (I-2)$$

When this potential is expanded in terms of Legendre polymonials and the spherical harmonic addition theorem applied, the result can be written

$$V(r) = \sum_{k,q} B_q^k \, C_q^{(k)} \tag{I-3}$$

$$B_q^k = -e \int \rho(R) \, \frac{r_<^k}{r_>^{k+1}} \, (-1)^q \, C_{-q}^{(k)} \, (\Theta, \, \Phi) \, d\tau \tag{I-4}$$

$$= <r^k> \left[-e \int \frac{\rho(R)}{R^{k+1}} \, (-1)^q \, C_{-q}^{(k)} \, (\Theta, \, \Phi) \, d\tau \right] \tag{I-5}$$

The expression given in Eq. (I-3) is the usual form of the crystal-field potential where B_q^k are the parameters and $C_q^{(k)}$ are tensor operators which represent the angular part of the crystal field interaction. The values of k and q for which the B_q^k are non-zero are determined by the symmetry of the crystal field. Hüfner [3] pointed out in his recent book that, whereas the aspect of crystal field splitting that is symmetry related is well understood, the mechanisms that determine the magnitude of the splitting are "by no means completely understood".

In obtaining a fit to experimental crystal field levels we use Eq. (I-3), treating the B_q^k as parameters. Since complete atomic and crystal field matrices can now be diagonalized simultaneously, we allow for J-mixing, but we have not introduced any corrections to the single particle crystal-field model. Parametrization of the crystal field has been extremely successful in correlating a large amount of the data, particularly with the $Ln^{3+}:LaCl_3$ system where the data base includes polarization and Zeeman effect measurements [12,13]

The free-ion and crystal-field parameters for a well characterized system, $Nd^{3+}:LaCl_3$ are shown in Table I-3. The complete model parameters are indicated in column C, and they reproduce 101 experimentally verified levels (i.e., polarization and Zeeman spectra were also taken) of the total set of 182 with a root mean square deviation of 8.1 cm^{-1} [19]. In column B the free-ion parameter set was reduced to include only the two-body configuration interaction operators. As a result the parameter values are distorted compared to those in Column C as they rather unsuccessfully attempt to fit the same 101 crystal-field components. The crystal-field interaction in lanthanide spectra can usually be treated as a perturbation and consequently even early attempts to define the parameters were successful if the energies of the

TABLE I-3.   Crystal Field Parameter Fits for $Nd^{3+}:LaCl_3$ (in cm^{-1}).

	A^a	B^b	C^c
F^2	.73686	72959(76)^d	71866(42)
F^4	52996	52318(317)	52132(77)
F^6	39429	34384(156)	35473(41)
α		21.4(0.6)	22.08(0.1)
β		−650(33)	−650(5)
γ		1770(59)	1586(12)
T^2			377(15)
T^3			40(1)
T^4			63(3)
T^6			−292(5)
T^7			358(8)
T^8			354(11)
ζ	884.58	878(4)	880(1)
M^k			(HF)
P^2			255(23)
B^2_0	195	68(65)	163(8)
B^4_0	−309	−309(178)	−336(22)
B^6_0	−711	−730(174)	−713(22)
B^6_6	466	463(138)	462(17)
σ	>100	>65	8.1
No. of Levels	22	101	101

[a]J. C. Eisenstein, J. Chem. Phys. <u>39</u>, 2134 (1963).

[b]Fit same data as (c).

[c]Reference 19.

[d]Numbers in parentheses are the rms errors on the parameter values.

crystal field components for a given state were arbitrarily adjusted to fit the centers of gravity of the observed free-ion
groups (Column A). For the experimentalist, the significance of
the refined model is clear. It is a working tool. It provides
the basis for predicting the energies of crystal-field components
in unanalyzed regions of the spectrum, sometimes calling attention
to very weak features of the spectrum. The error in the predicted energy is expected to be small in comparison to the usual
energy separation of crystal-field components.

Attempts to calculate the crystal-field parameters from
first principles are still in progress. The early work of
Hutchings and Ray [24], which explored and indicated the limitations of the point charge model, and more recent work by Faucher
et al. [25] can be contrasted with attempts of other groups, particularly Newman and coworkers, who have expressed their results
in terms of the angular overlap and covalency contributions to
the crystal-field parameters [26-28]. For D_{3h}-symmetry, only the
B_0^2 term is not in agreement with experiment, Fig. I-6.

In the following section we focus on some new insights into
both the symmetry and the magnitude of the crystal field. The
expression given in Eq. (I-4) is general. If $r_<^k$ is associated
with an f-orbital and $r_>^{k+1}$ with R, then integration over the f-
electron wave function gives the expectation value, $<r^k>$, and its
coefficient is the potential due to the ligands. Energy level
analyses of U^{3+}:$LaCl_3$ [29] and Np^{3+}:$LaCl_3(LaBr_3)$ [7] were recently
published. The values of B_q^k were approximately twice the magnitude
of corresponding values for the lanthanides. One question that
arose was whether this increased *magnitude* is consistent with,
larger than, or smaller than expectations.

The literature contains a number of examples of HF-calculations
which indicate that the 5f-orbitals are less well shielded by
filled s and p shells than is the case for lanthanides. In
Fig. I-7 the results of calculations using an HFR program are
shown with the 4f and 5f vertical scales increased relative to
those of the cores [29]. For Nd^{3+}, the radius corresponding to
the maximum in the probability function $(r^2\psi^2)$ for the 4f electrons
is well inside that for the 5s and 5p electrons. In contrast the
probability function maxima for the 5f, 6s, and 6p electrons all
occur at essentially the same radius.

We were interested in determining whether the factor of two
difference in magnitude of the crystal field parameters for
An^{3+}:$LaCl_3$ compared to Ln^{3+}:$LaCl_3$ could be correlated with the
apparent increased potential for overlap between the 5f and
ligand wavefunctions. A purely electrostatic point of view, the
point charge model, was adopted [7]. Since values of $<r^k>$,

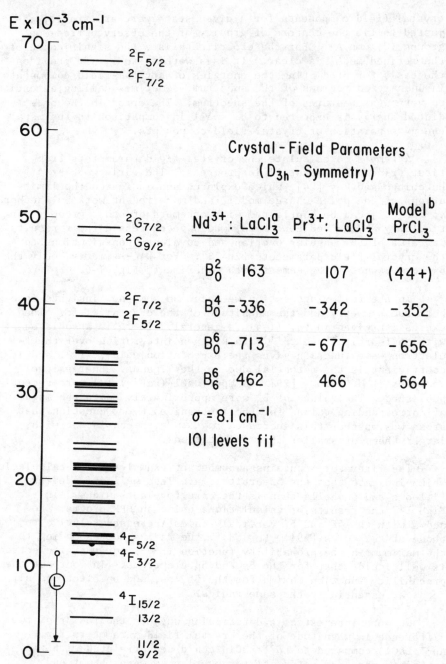

FIGURE I-6. Crystal-field parameters and energy level structure for $Nd^{3+}(4f^3)$ which has a total of 41 free-ion states and 182 crystal-field states (D_{3h}-symmetry). [a]Ref. 12, [b]Ref. 26.

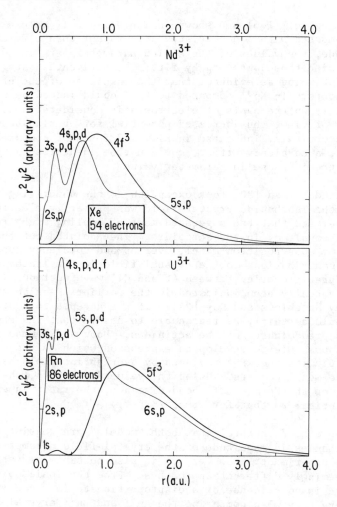

FIGURE I-7.   Hartree-Fock calculations for the outer orbitals in
Nd^{3+} and U^{3+}.   The 4f and 5f vertical scales are increased relative to those of the cores [29].

Eq. (I-5), have been computed (both with relativistic and non-
relativistic codes), and since the crystal-field parameters of
both Np^{3+} and the lanthanide analog have been determined from
spectra in the $LaCl_3$ host, we argued that to a first approximation
the terms from Eq. (I-5) which involve the crystal host would
cancel, giving the expression:

$$B_q^k(An^{3+} \text{ predicted}) = \frac{\langle r^k \rangle_{An^{3+}}}{\langle r^k \rangle_{Ln^{3+}}} B_q^k(Ln^{3+}) \qquad (I-6)$$

Results using Eq. (I-6) are shown in Table I-4. If we use values
of $<r^k>$ computed with a non-relativistic rather than a relativistic HF code, the values of the scaled parameters are in better
agreement with those obtained by fitting experimental data. We
took the foregoing as evidence that the magnitude of the crystal
field parameters in Np^{3+} compared to Pm^{3+} could indeed be rationalized solely on the basis of electrostatic considerations. From
this point of view, the increased potential for overlap between f
and ligand orbitals, indicated in Fig. I-7 for a 5f compared to a
4f species, a covalency effect, could not be identified as the
source of the increase in parameter values.

Poon and Newman [30] took exception to the underlying assumption that the lanthanide crystal-field was primarily electrostatic
in character and that the suggested mode of scaling might reveal
pronounced covalency in the actinides. They pointed out that
based on the superposition model, overlap and covalency represent
the major contributions to the crystal field in the lanthanides.
The considerable overlap between 4f and Cl^- wavefunctions in
$PrCl_3$ had actually been indicated in the earlier HF-calculations
reported by Hutchings and Ray [24]. If one accepts this argument,
then it would certainly be reasonable to assume that the same interactions predominate for the actinides. Poon and Newman showed
that it is possible to develop an expression which is the analog
of Eq. (I-6) with the $<r^k>$ replaced by overlap integrals between
Cl^- and the metal centers. Using the value $d = 2.94$ Å, characteristic of the $Np^{3+} - 6Cl^-$ bond distance, the ratio is essentially identical to that for $<r^k>_{5f}/<r^k>_{4f}$.

It appears, therefore, that both "models" are consistent
with the changes in magnitude of the crystal-field parameters
defined by experiment. However, it is also possible to rationalize
the two seemingly different approaches. From the ionic point of
view, there is no evidence of a disproportionate increase in the
role of covalency when comparing the An^{3+} and Ln^{3+} crystal-field
parameters. From the viewpoint advanced by Poon and Newman, the
increased overlap between metal and Cl^- centers, and the increased
magnitude of the crystal field parameters are mutually consistent.
From the latter standpoint, Fig. I-7 may tend to suggest a greater
potential for radial overlap in An^{3+} compared to Ln^{3+} than is
actually borne out by the present experiment. Nevertheless, the
increase in covalency is not to be considered disproportionate.
It is perfectly consistent with expectations.

I(D). Approximate Symmetries

We have carried out calculations of the crystal-field parameters with the usual assumption that the symmetry-related dependence of the field, the $C_q^{(k)}$ operators in Eq. (I-3), is well
understood. However there are aspects of this part of the treat-

TABLE I-4. Comparison of Crystal Field Parameters for the $4f^4$ and $5f^4$ ions Pm^{3+} and Np^{3+}, diluted in $LaCl_3$, with those computed for Np^{3+} based on point charge considerations.[a]

	Np^{3+}	Pm^{3+}	Pr^{3+}	Predicted B^k_q (Np^{3+})[b]	
				Scaled on Pm^{3+}	Scaled on Pr^{3+}
B^2_0 (cm^{-1})	165(26)	143(18)	107	290	186
B^4_0 (cm^{-1})	-623(44)	-395(29)	-342	-1317	-853
B^6_0 (cm^{-1})	-1615(48)	-666(30)	-677	-3229	-2186
B^6_6 (cm^{-1})	1041(33)	448(21)	466	2172	1505
$\langle r^2 \rangle$ (a.u.)	2.36	1.16	1.356		
$\langle r^4 \rangle$ (a.u.)	11.65	3.49	4.673		
$\langle r^6 \rangle$ (a.u.)	107.6	22.19	33.32		

[a] Reference 7.

[b] Derived from the expression $(\langle r^k \rangle\ Np^{3+}/\langle r^k \rangle\ Ln^{3+})\ B^k_q(Ln^{3+})$, where Ln = Pm or Pr.

ment that are difficult to define. For example, there are many crystal lattices in which the central ion (Ln^{3+} or An^{3+}) resides at a low symmetry site. We are interested in knowing more about the circumstances under which it is appropriate to adopt a higher, more mathematically tractable, approximate symmetry for such cases that would allow computation of crystal-field levels in agreement with those observed experimentally.

Some useful insights into the problem have been derived from recent analyses of the spectra of $Ln^{3+}:LaF_3$. The actual site symmetry in LaF_3 is C_2 [31,32] but the approximate symmetry approaches that of $LaCl_3$ (D_{3h}) [33]. As a point of reference, the crystal-field analysis of $Nd^{3+}:LaCl_3$, Table I-3, is a good example of the degree to which the present model can reproduce experimental data taken in a well characterized lattice over a broad range of the optical spectrum, i.e. $\sigma = \sim 8 \text{ cm}^{-1}$.

It was pointed out by Onopko [34] that the energies of the crystal-field components of several of the lowest-lying free-ion states in $Nd^{3+}:LaF_3$ and $Er^{3+}:LaF_3$ could be computed in reasonable agreement with experiment by assuming that the site symmetry approached D_{3h}. A point charge calculation confirmed the signs of the parameters shown in Table I-5, as did a molecular orbital treatment carried out by Newman and Curtis [35]. A subsequent analysis based on Onopko's crystal-field parameters and reasonable sets of free-ion parameters, showed that the energies of crystal-field transitions in Nd^{3+}, Sm^{3+}, Gd^{3+}, Dy^{3+}, and Er^{3+}, each doped into single crystal LaF_3, could be calculated in good agreement with experiment over the whole of the optical range to 50000 cm^{-1} [36]. As indicated in Table I-5, the deviations between calculated and assigned energy levels for the $Ln^{3+}:LaF_3$ spectra compared well with those for the $LaCl_3$ host considering the approximation to the actual symmetry.

In the 3+ lanthanides with an odd number of f-electrons, the crystal-field will induce a splitting of each free-ion state into $J+1/2$-components in all site symmetries except cubic or O_h. Thus the number of components is the same for C_2 as for D_{3h}. On the contrary, in even f-electron systems such as the oft studied $Pr^{3+}:LaF_3$, ($4f^2$), crystal-field calculations in D_{3h} symmetry do not remove the degeneracy of the $\mu = \pm 1$ or $\mu = \pm 2$ states, yet the number of lines observed in the spectrum of $Pr^{3+}:LaF_3$ does imply a lower site symmetry consistent with the complete removal of symmetry-related degeneracy. Thus analysis of the crystal-field in even f-electron systems was not attempted when using the D_{3h} approximation [36].

Subsequently Morrison and Leavitt [37] published an analysis of the spectra of $Ln^{3+}:LaF_3$ using the actual C_2 symmetry crystal-field. Initial values of the 14 crystal-field parameters for

TABLE I-5. Comparison of Crystal Field Parameters.

	$Nd^{3+}:LaF_3$		$Nd^{3+}:LaCl_3$ [c]
B_0^2	276^a cm^{-1}	210^b cm^{-1}	163 cm^{-1}
B_0^4	1408	1239	-336
B_0^6	1600	1500	-713
B_6^6	679	773	462
		$\sigma = 16$ cm^{-1}	$\sigma = 8.1$ cm^{-1}
		145 levels fitted	101 levels fitted

	$Er^{3+}:LaF_3$		$Er^{3+}:LaCl_3$ [c]
B_0^2	282^a cm^{-1}	229^b cm^{-1}	216 cm^{-1}
B_0^4	1160	965	-271
B_0^6	773	909	-411
B_6^6	463	484	272
		$\sigma = 12.1$ cm^{-1}	$\sigma = 5.0$ cm^{-1}
		117 levels fitted	80 levels fitted

[a] Reference 34.

[b] Reference 36.

[c] References 19,21.

ions with an odd number of f electrons were obtained from lattice sum calculations referred to a coordinate system in which the crystal axis was parallel to the C_2-axis and perpendicular to that in D_{3h}-symmetry. For practical computational reasons, they adopted a modified free-ion Hamiltonian which allowed adjustment of the centroids of the crystal-field levels associated with a given free-ion state so as to maximize the fitting of experiment-ally established sets of crystal-field components. Only the 9-10 lowest energy multiplets of ions with odd numbers of f-electrons were involved in the fitting process. These results provided the

initial parameters used in the present fitting of even f-electron
cases. The magnitudes of crystal-field parameters relevant to
this study are indicated in Table I-6.

The calculation in C_2-symmetry requires determination of 14
independent crystal-field parameters, and this is a major computational problem, particularly when coupled with the extensive
free-ion treatment. We therefore sought a middle ground by exploring the possibility of using an approximate C_{2v}-symmetry,
which is low enough to completely remove the symmetry-related
degeneracy of crystal-field states, yet allowed us to retain the
extensive free-ion model and to continue to simultaneously diagonalize the atomic and crystal field portions of the Hamiltonians.

There are at least two approaches to use the C_{2v} approximation. One is to maintain the D_{3h} symmetry axis and add the
additional parameters introduced in the C_{2v} case to simulate the
distortion from D_{3h} symmetry. This was the original course we
chose, since by fitting an odd-f electron case in this manner,
the extra parameters could be determined and subsequently used as
a trial set to interpret the spectra of even f-electron systems.
One of the problems encountered is that the D_{3h} approximation
provides such a good correlation between experiment and theory
that it is difficult to adequately determine the extra parameters
arising in the C_{2v} symmetry so that their magnitude and signs may
be arbitrary, depending upon the data being fit. This approach
has been discussed by Caro and coworkers [38].

The other approach is to fit the crystal-field states of an
odd f-electron system using as an initial set of parameters
either a set related to those computed for LaF_3 in C_2-symmetry
(C_2-axis) [37] or a set derived from the D_{3h}-approximation of the
LaF_3 structure (D_{3h}-axis) transformed by a suitable rotation of
axes. The experimental data for $Er^{3+}:LaF_3$ are particularly useful for testing such parametrization methods, because practically
all of the crystal-field states to $\sim$40000 cm^{-1} have been assigned
[36]. A comparison of different sets of crystal-field parameters
fit to $Er^{3+}:LaF_3$ data is shown in Table I-7. There is little
change in the values of B_0^2, B_0^4, B_0^6 and B_6^6 between the D_{3h} case
and the C_{2v} set (D_{3h}-axis). However starting from the C_2-
parameters (C_2-axis) obtained by Morrison and Leavitt in a limited
fit to $Er^{3+}:LaF_3$ data, it is apparent that a final set can be
derived (C_2-axis) in which the parameters themselves are better
determined.

A reasonable test of the C_{2v} approximation would involve its
application to even f-electron systems. Thus we have combined
the crystal field parameters shown in Table I-7 (C_{2v}-C_2 axis) for
$Er^{3+}:LaF_3$ with a set of free-ion parameters for neighboring Ho^{3+}

TABLE I-6. Crystal-field Parameter Values (in terms of A_{nm} in cm^{-1}) obtained from lattice sum calculations (LaF$_3$). Only the values of the even (nm) components which determine the level energy are given [37].

kq	$B_q^{(k)}$ (D$_{3h}$)[a] Real[a]	$B_q^{(k)}$ (C$_2$)[b] Real	Imag.	$B_q^{(k)}$ (C$_2$)[c] Real	Imag.	Nd^{3+}:LaF$_3$ $B_q^{(k)}$ (C$_2$)[d] Real	Imag.
20	465	66	0	-145	0	-216	0
22		-46	79	5	0	-36	0
40	1849	994	0	652	0	700	0
42		-103	178	422	118	197	71
44		-56	-96	397	241	229	181
60	949	844	0	523	0	490	0
62		17	-30	-793	66	-928	-23
64		14	24	-113	-342	-131	-449
66	862	784	0	-442	-442	-427	-653

[a] Crystal structure data of K. Schylter, Arkiv Kemi 5, 73 (1953) similar to results of Onopko [34]. The c-axis is parallel to the D$_{3h}$ axis.

[b] Crystal structure of Cheetham et al. [31], but with the z-axis of the A$_{nm}$ parallel to the crystal axis.

[c] Crystal structure of Cheetham et al. [31] with the z-axis of the A$_{nm}$ perpendicular to the crystal axis.

[d] Limited fit of experimental data for Nd^{3+}:LaF$_3$ [37].

TABLE I-7. Crystal Field Parameters (cm^{-1}) for $Er^{3+}:LaF_3$.

	D_{3h}[a]	D_{3h}-axis[a] C_{2v}-symmetry	C_2-axis[b] C_2-symmetry (real part only)	C_2-axis[a] C_{2v}-symmetry
B_0^2	226(18)	220(17)	-228	-226(1)
B_0^4	965(39)	953(36)	545	552(3)
B_0^6	899(34)	897(41)	275	261(10)
B_6^6	477(26)	478(30)	-307	-460(22)
B_2^2		69(22)	-119	-87(12)
B_2^4		67(57)	301	276(9)
B_4^4		-40(70)	358	415(10)
B_2^6		-36(67)	-520	-620(11)
B_4^6		144(54)	56	1(23)
σ	16	14	17	16

[a]Present work, 117 of 183 states assigned.
[b]Reference 37.

derived from a fit to the approximate centers of gravity of observed crystal-field components, but also constrained to be consistent with the trends observed in similar parameters for other
$Ln^{3+}:LaF_3$ [36]. Diagonalization of the combined set provided a
model calculation with which observed structure in $Ho^{3+}:LaF_3$
could be compared. Much of the multiplet structure was very satisfactorily fit; however, some useful insights were immediately
gained. The experimental results for the $^5I_7$-state [39], are
compared with the computed structure of that state in Fig. I-8.

There are several features of the figure that make it a good
example of the importance of the interaction between theory and
experiment. Considering the left side of Fig. I-8 (larger range
of energies) it is apparent that the calculated crystal-field
components are grouped from $\sim$5200-5300 cm^{-1} and there is no computed analog for levels $Y_{11}-Y_{15}$. These were levels that were

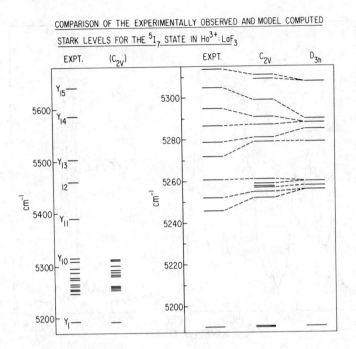

FIGURE I-8. Comparison of the experimentally observed and model computed Stark levels for the $^5I_7$ state in Ho^{3+}:LaF_3 at 4 K, [39].

relatively poorly resolved and only observed in fluorescence; however, there was no experimental basis for excluding them. It was noted that some of the levels might be vibronic in origin. The present computed set would have been extremely helpful in encouraging the experimentalist to look for a possible assignment to another group. On the right side of Fig. I-8 the crystal-field structures computed assuming C_{2v} (C_2-axis) and D_{3h} symmetries are compared using an expanded energy scale. The fundamental grouping of levels is reproduced by the approximate D_{3h}-symmetry. Recourse to C_{2v}-symmetry removes the degeneracy of several levels consistent with experiment, but it also indicates the energy range in which additional structure should be observed. While assignments can be made based on the model calculation, significant improvement of the fit would not be expected because of the very small adjustments required. The spectrum of the $^5I_7$ group at $\sim4°K$ is shown in Fig. I-9. Some of the indicated structural features may be vibronic in character, so that in general the model predictions will also be of value in avoiding assignment to some relatively prominent structure of that possible origin. The advantages of the interaction of the type of model calculation discussed here with experimental data is apparent. However one of the deficiencies of the present analysis is the lack of indedent methods (other than corresponding energies) for assigning

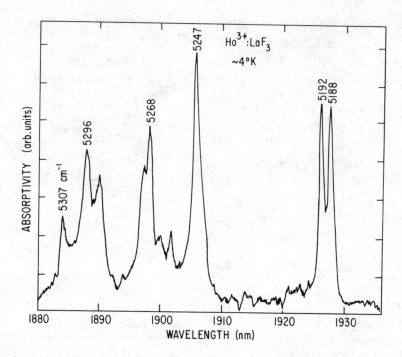

FIGURE I-9. Spectrum of the $^5I_7$ state of $Ho^{3+}:LaF_3$ at 4 K.

crystal-field components in $La^{3+}:LaF_3$. The attempts to compute
crystal-field component intensities based on model calculations
is one important new direction [40]; another stringent test is the
correlation between observed magnetic properties and values
calculated from crystal-field eigenvectors [38].

II.   THE INTENSITIES OF f→f TRANSITIONS IN ABSORPTION AND IN
        FLUORESCENCE

The use of absorption spectra to monitor changes in the environment of transition-metal ions is a widely applied technique.
The normally sharp well-defined absorption bands characteristic
of the trivalent lanthanides and actinides, are particularly useful in this respect. Moreover, while shifts in band energies and
intensities can be a qualitative indicator of environmental
changes, it was realized in the 1930's and 40's that spectra of
the f-elements could in principle be subjected to a quantitative
intensity analysis if the knowledge of their energy level structures in various media were sufficiently detailed. Such an
analysis would require some modeling of the ionic environment.

In the early 1940's a group of Dutch scientists [41] carried out the first comprehensive quantitative study of the intensities of rare earth absorption bands in solution. They measured the area under the band envelopes for solutions of known concentration and thereby established values for the oscillator strengths, P, of various transitions:

$$P = \frac{2303 \ mc^2}{N\pi e^2} \int \epsilon_i(\sigma)d\sigma = 4.32 \times 10^{-9} \int \epsilon_i(\sigma)d\sigma \qquad (II-1)$$

where $\epsilon = \frac{1}{C\ell} \log I_o/I$; C is the molar concentration of the f-element, ℓ is the light path in the solution (cm), and $\log I_o/I$ is the absorptivity at a given energy σ (cm^{-1}) within the band envelope.

The general form of the theoretical model for the oscillator strength of a transition corresponding to Eq. II-1 was written [41],

$$P = \frac{8 \ \pi^2 \ mc\sigma}{3he^2 \ (2J+1)} \left[\chi \ \overline{F}^2 + n\overline{M}^2 \right] \qquad (II-2)$$

where $\chi = \frac{(n^2+2)^2}{9n}$ is the refractive index (n) correction, and $\overline{F}^2$ and $\overline{M}^2$ are respectively the matrix elements of the electric dipole and magnetic dipole operators between the ground state and a particular excited state. The problem was to be able to calculate the required matrix elements for transitions of interest.

Practically all of the lanthanide absorption bands usually observed in the near-infrared to near-ultraviolet range of the spectrum are attributed to electric dipole transitions, although in the strict sense such transitions are parity forbidden since they occur between states within the same configuration. The intensities of lanthanide transitions, which result in bands with $P \sim 10^{-6}$, indeed reflect a highly forbidden character compared to allowed transitions (such as f→d) where $P = \sim 1$. However, it was pointed out by Broer et al. [41], based on earlier work of van Vleck [42] that the observed intensities could be accounted for by assuming a small mixing of the higher-lying opposite parity configurations, i.e. $f^N \rightarrow f^{N-1}d$, $f^{N-1}g$, and others into the f^N-states via the odd terms in the potential due to the ligand field.

The work of Broer and coworkers was based on the analysis of aqueous solution spectra. While of considerable interest in its

own right, this medium has obvious experimental advantages, including the assurance of homogeneity of lanthanide distribution.
A number of analyses of the spectral intensities of lanthanides
either neat or doped into crystal hosts have by now also been published.

While the character of the f^N-transitions had been established prior to 1962, the problem of computing the matrix elements
of F (Eq. II-2) for lanthanide ions in an arbitrary crystal or
ligand-field environment had not been fully addressed. On the
other hand, computation of the matrix elements of M, the magnetic
dipole operator between the ground state and any given excited
state was well understood. For a summary of the method see [43].
Very few of the observed transitions exhibit any appreciable
magnetic dipole character.

Thus in 1962 when Judd [44] and independently Ofelt [45]
derived closed expressions for $\overline{F}^2$ they opened the way for a new
dimension in the analysis of rare earth absorption spectra. At
about the same time, the early 1960's, there developed a very
active interest in the mechanisms of excited state relaxation of
rare earths both in solutions and in crystals. As the analysis
proceeded, it was shown that the Judd-Ofelt theory could be used to
compute the total *radiative* relaxation rates of excited states of
interest. This made it possible to predict pathways of excited
state relaxation; although, most states were found to relax
primarily via non-radiative mechanisms. In recent years, the
field of rare earth laser engineering has emerged to identify
potential lasing transitions for rare earths in various host
crystals. Concepts evolved in use of the Judd-Ofelt theory are
also applicable to the screening of rare earth doped glasses to
maximize the efficiency of high power lasers for use in both
fusion and fission energy applications [46,47].

At this point, 20 years after publication of the Judd-Ofelt
theory, we recall particularly its successful use in extending
our knowledge of the energy level schemes of the lanthanides, and
its contribution to the study of excited state relaxation [48].
In this section we reexamine and extend our previous efforts to
understand the intensity patterns exhibited by lanthanide transitions in solution, and project the discussion to the trivalent
actinides where new analyses have been carried out. A knowledge
of the energy level structure which allows identification of
transitions observed in solution in terms of a useful coupling
scheme is the basis for the intensity analysis. Consequently,
the results of the energy level analysis of systems such as
$Ln^{3+}:LaF_3$ are directly applicable to the developments discussed
here.

II(A). Judd-Ofelt Theory

In effect, the efforts to interpret rare earth solution spectra prior to 1962 had established that:

$$P_{EXPT} = P_{ED} + P_{MD}$$

that is, an experimentally measured quantity, the oscillator strength (or probability for absorption of radiant energy) could be expressed to a good approximation in terms of absorption of light by electric and magnetic dipole mechanisms without recourse to higher multipoles. P_{MD} was known to be important in only a few transitions, so principal interest focused on P_{ED}.

Judd derived the expression [44]:

$$P_{ED} = \sum_{\lambda = 2,4,6} T_\lambda \, \nu \, (\psi J||U^{(\lambda)}||\psi'J')^2 \qquad (II\text{-}3)$$

where $\nu(\sec^{-1})$ is the frequency of the transition $\psi J \to \psi'J'$, $U^{(\lambda)}$ is a tensor operator of rank λ, and the T_λ are quantities which contain the description of the immediate environment of the rare earth ion as well as overlap integrals and energy differences. The beauty of this result was that Judd was able to substitute three parameters, T_λ, for those interactions that constitute the model of the ion in its environment. Since the matrix elements of $U^{(\lambda)}$ could be calculated from a knowledge of the free-ion structure of the ion of interest, the parameters could be determined empirically from experimental data.

$$T_\lambda = \frac{8\pi^2 m}{3h} \frac{\chi}{2J+1} (2\lambda + 1) \sum_t (2t + 1) \, B_t \, \Xi^2(t,\lambda) \qquad (II\text{-}4)$$

- $B_t = \sum_p |A_{tp}|^2/(2t + 1)^2$ expresses the influence of the environment on the central ion. The A_{tp} are the odd components of the crystal field.

- $\Xi^2(t,\lambda) = f\,[(nl|r|n'l')\,(nl|r^t|n'l')/\Delta(n'l')]$ involves radial integrals coupling the f^N to perturbing configurations and energy differences. See II(C).

Since T_λ is composed of several parts, it is difficult to extract explicit information about the environment. The problem is
similar to that met in crystal-field calculations. The parametrization in terms of $U^{(\lambda)}$ does not imply a unique model.

A useful alternate parametrization of Judd's expression [48],
is adopted here and written:

$$P_{ED} = \frac{8\pi^2 mc}{3h} \frac{\sigma}{(2J+1)} \chi \sum_{\lambda=2,4,6} \Omega_\lambda \, (\psi J||U^{(\lambda)}||\psi'J')^2 \qquad (II-5)$$

The values of Judd's T_λ and Ω_λ defined above are related by
$\Omega_\lambda = (2J+1) \, [3.618\chi]^{-1} T_\lambda$ (for transition frequency in sec^{-1}).
Most experimental results are now quoted in terms of the energy
in cm^{-1} of the transitions. When this framework is used the
appropriate conversion factor is $\Omega_\lambda = (2J+1) \, [1.085 \times 10^{11}\chi]^{-1} T_\lambda$.
The rationale for modifying Judd's original notation and adopting
that of Eq. II-5 is that the latter is more directly related to
the subsequent calculation of intensities in emission.

In our original use of the Judd-Ofelt theory [49], we showed
that a single set of three parameters could reproduce the observed
intensities of all the absorption bands for a given Ln^{3+}(aquo)
ion, within a reasonable degree of accuracy. However, in the
early stages of this work our understanding of the energy level
schemes was fragmentary, particularly the structure at higher
energies. Thus as soon as the Judd-Ofelt theory had been tested
and shown to be capable of reproducing the intensities of well
characterized lower-energy absorption bands, it was realized that
the theory could also be utilized as a basis for identifying the
transitions involved in isolated, more intense absorption bands
at higher energies. With the resulting new assignments it was
possible to explore the parametrization of configuration interaction in f^N-configurations in a much broader and more systematic
manner than had previously been attempted.

There were of course limitations on the extent to which even
the improved parameterization scheme could represent the data.
Subsequent developments have culminated in the extensive analysis
of energy level structure in lanthanide spectra discussed in
section I. At this point it is of some interest to return to the
analysis of intensities for Ln^{3+}(aquo) and reexamine the intensity parameters which can now be derived based on an independent and consistent analysis of the energy level structure. Inspection shows that the free-ion states computed for Ln^{3+}:LaF_3
[36] correlate well with the energies of bands observed in the
spectra of Ln^{3+}(aquo). It comes as no surprise that in a few
cases the original assignments must be modified. Moreover, in

addition to the attempt to derive a more self consistent set of intensity parameters for Ln^{3+}(aquo) based on an improved understanding of the energy level scheme, it was of interest to use such a set as the basis for comparison with recently determined values of Ω_λ for An^{3+}(aquo) in the heavy half of the 5f-series. Finally, we are presently in a better position than Judd was in 1962 with respect to the evaluation of overlap integrals and energy differences included in the interactions which contribute to the magnitude of Ω_λ, both for the lanthanides and the actinides; so the results of new model calculations are reported.

II(B).   Intensity Analysis of Lanthanide Solution Spectra

As suggested in the previous discussion, the free-ion energy level calculations for Ln^{3+}:LaF_3 provide both a basis for comparison with the original assignments made to absorption bands observed for Ln^{3+}(aquo) ions and in some cases require slightly modified values of $[U^{(\lambda)}]^2$. However no large changes in the values of Ω_λ previously computed were to be expected, and none have been observed. What does emerge is a similar, somewhat more self-consistent, but on the basis of independent confirmation more firmly-based set of Ω_λ, Table (II-1). Details of the calculations are readily available from other sources [43].

In the new evaluation, we also examined changes in the values of Ω_λ that could result from changing the nature of the fitting algorithm. Given an equation of the form

$$P/\sigma = \xi = \Omega_2 [U^{(2)}]^2 + \Omega_4 [U^{(4)}]^2 + \Omega_6 [U^{(6)}]^2$$

we originally chose to directly minimize the differences in ξ(observed) − ξ(calculated), using a least-squares fitting procedure to obtain the optimum values of Ω_λ. This method automatically weights the fit in favor of the transitions with the largest values of ξ. We have now also examined the values of Ω_λ arising when the expression minimized was $\{1 - (\xi(\text{calculated})/\xi(\text{observed}))\}^2$ and it is these values that are given in Table II-1. While there are small differences between the results of the two methods of fitting, no major discrepancies emerge. In several cases individual levels which tend to distort the fit when included in the parametrization can be readily identified. These are almost exclusively cases in which the $U^{(2)}$ matrix elements are very large. An important aspect of the parametrization as a whole is an apparent small average decrease in the magnitude of Ω_λ over the series. The consistency of the new parameter values tends to emphasize the disproportionately large values of Ω_2 and Ω_6 for Pr^{3+}, and of Ω_6 for Nd^{3+}. Since the values of the matrix elements of $U^{(\lambda)}$ for Pr^{3+} and Nd^{3+} are consistent with those calculated for other members of the series, the larger

TABLE II-1.   Energy Level Parameters for Ln^{3+}(aquo).

	$\Omega_2 \times 10^{20}$ cm^2	$\Omega_4 \times 10^{20}$ cm^2	$\Omega_6 \times 10^{20}$ cm^2
Pr^{3+}	$28.0+72$	$5.89+2.50$	$32.2+3.0$
Nd^{3+}	$2.25+1.7$	$4.08+.80$	$9.47+1.3$
Pm^{3+}	$1.30+.26$	$4.36+.48$	$3.94+.34$
Sm^{3+}	$1.08+.42$	$3.67+.70$	$2.87+.56$
Eu^{3+a}	(1.46)	(6.66)	(5.40)
Gd^{3+}	$1.94+.43$	$5.27+1.7$	$4.46+1.1$
Tb^{3+}	$2.76+5.3$	$7.95+6.2$	$2.87+1.0$
Dy^{3+}	$.584+6.3$	$3.54+.74$	$3.90+.62$
Ho^{3+}	$.791+.79$	$3.13+.40$	$2.86+.26$
Er^{3+}	$1.34+.37$	$2.19+.25$	$1.88+.11$
Tm^{3+}	$.646+1.0$	$2.31+.60$	$1.47+.20$

[a]For comparison with other members of the series, the parameters
for Eu^{3+} were adjusted for the effects of a low-lying excited
state [49].

values of Ω_λ reflect disproportionately intense transitions. A
qualitative difference in the intensities of several bands in
Pr^{3+} and Nd^{3+} compared to other members of the series can also be
seen in Figs. II-1, II-2, which are adapted from [43].

Several general characteristics of the fit values of Ω_λ in
Table II-1 that were discussed in our first publications on the
subject are still evident. The value of Ω_2 in aqueous solution
is generally small and poorly determined. It does not enter in
the calculation of P for many of the levels since for many transitions the value of $[U^{(2)}]^2$ is zero or very small. As a consequence, the intensities of transitions for the Ln^{3+} aquo ion
spectra are almost entirely reproduced by a two parameter model.

Considerable interest in the Judd-Ofelt theory has been generated by the fact that it readily accomodates and indeed it
predicted those transitions which were subsequently designated as
hypersensitive [50-52]. This refers to characteristic band intensity patterns of Ln^{3+} ions in crystals, solutions, and vapor
complexes, where one or two lanthanide transitions gain significant
intensity relative to all of the other transitions depending upon

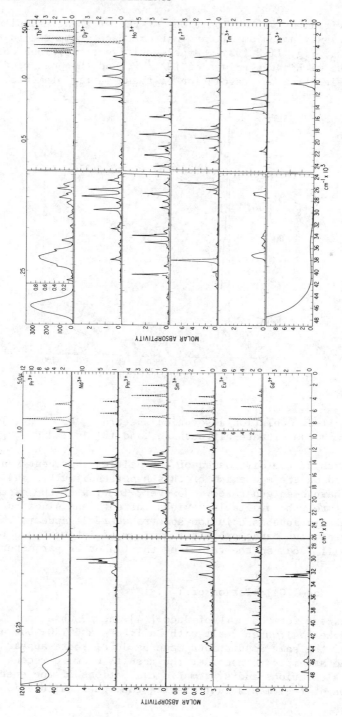

FIGURE II-1,2. Absorption Spectra of the light and heavy trivalent lanthanide aquo ions.

the environment of the ion. One of the striking examples is Ho^{3+} in tetrabutylammonium nitrate-nitromethane, compared to Ho^{3+} in 1 M HNO_3, Fig. II-3 [53]. Hypersensitive transitions correspond to those with large values of $[U^{(2)}]^2$, and thus can readily be identified from any tabulation of these matrix elements.

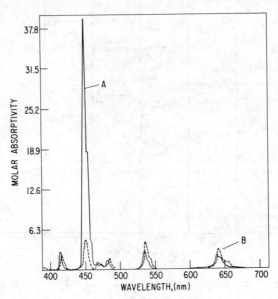

FIGURE II-3. Solution absorption spectra of Ho^{3+} in (A) tetra-butylammonium nitrate-nitroethane, and (B) in 1 M HNO_3 [53].

Much of the discussion of intensities in recent years has centered on the mechanism of this hypersensitivity, and useful ideas have been generated. However, in view of the extensive literature, the available reviews thereof, and since our interests here are in aqueous solution spectra where it appears that the hypersensitive bands exhibit a minimum of intensity, our atten-tion will focus on the values of the two other parameters, Ω_4 and Ω_6.

II(C). Model Calculation of T_λ

Except for the work of Judd [44] and a subsequent publication by Krupke [54] which dealt with $Ln^{3+}:LaF_3$ and Y_2O_3 but utilized most of the same assumptions made by Judd, there appear to have been no attempts to consider the results of more recent Hartree-Fock calculations and systematic evaluations of the energy level schemes of f-elements in reviewing the model calculation of Ω_λ.

In particular no effort has been made to carry out a model calculation of Ω_λ for the 5f-series.

Although a recent review of the intensity calculations for An^{3+}(aquo) had to characterize them as poorly determined and incomplete [13], new experimental data have been obtained and on the basis of extensive analyses of Bk^{3+}(aquo) and Cf^{3+}(aquo) spectra, it is now possible to discuss intensity relationships over the whole of the series. This type of approach, the comparison of actinide and lanthanide spectra based on model calculations, as well as the comparison of intensities in the lighter and heavier members of each series, is in any case probably more relevant and illuminating, than the question of how accurately the model reproduces the values of Ω_λ obtained experimentally for a particular ion.

If in Judd's expression, Eq. II-3, we replace $\nu(\sec^{-1})$ by $c\sigma(cm^{-1})$ and correspondingly in Eq. (II-4) make the substitution which leads to Eq. (II-5), we have

$$\Omega_\lambda = (2\lambda + 1) \sum_t (2t + 1) B_t \, \Xi^2(t, \lambda) \qquad (II-6)$$

where t is an index which takes values permitted by a 3-j symbol in the expression for $\Xi(t,\lambda)$. There are two quantities to be evaluated, B_t and $\Xi(t,\lambda)$.

As noted earlier, B_t expresses the interaction of the environment with the central ion. In his evaluation of this quantity for Ln^{3+}(aquo), Judd [44] chose $GdCl_3 \cdot 6H_2O$ [55] as a structural model and obtained an expression for B_t as a function of an appropriate water dipole-metal ion separation R, and a quantity related to the dipole moment of a typical water molecule in the inner coordination sphere, μ.

$$B_t = \left[\frac{\mu e(t+1)}{(2t+1) \, R^{t+2}} \right]^2 \sum_{i,j} P_t(\cos \omega_{ij}) \qquad (II-7)$$

With this expression, assuming a similar geometric arrangement of the dipoles for all Ln^{3+}, $\sum_{i,j} P_t(\cos \omega_{ij})$, Judd computed values of B_t for a typical light lanthanide, Nd^{3+}, and a typical heavy lanthanide, Er^{3+}, using appropriate values of μ and R.

We could go to the trouble of constructing a different model that would take into account the fact that in the light half of the 4f series the coordination is probably 9-fold, possibly similar to that in $Nd(Et\ SO_4)_3 \cdot 9H_2O$ or $LaCl_3$, while for the heavy

members of the series there is evidence for 8-fold coordination
[56] with a possible square antiprismatic structure. However, in
attempting to develop a new structural model we would be left
with many of the same approximations that Judd found necessary to
invoke such as neglect of all but a first coordination sphere and
the treatment of the water molecules as point charges. Judd made
allowance for some difference in the value of B_t as a function of
different ionic radius and effective charge, and it is unlikely
that this term would represent anything more than a gradual
variation over the series even if a "superior" ionic model could
be constructed. We have consequently adopted the original mode
of calculation. A new set of effective radii, R, and dipole
moments, μ, were computed after averaging the nearest neighbor
metal-H_2O distances from available data [57].

$$\Xi(t,\lambda) = 2\sum_t (2\ell+1)(2\ell'+1)(-1)^{\ell+\ell'} \begin{Bmatrix} 1 & \lambda & t \\ \ell & \ell' & \ell \end{Bmatrix} \begin{pmatrix} \ell & 1 & \ell' \\ 0 & 0 & 0 \end{pmatrix} \begin{pmatrix} \ell' & t & \ell \\ 0 & 0 & 0 \end{pmatrix}$$

$$\frac{<n\ell|r|n'\ell'><n\ell|r^t|n'\ell'>}{\Delta(n'\ell')} \qquad\qquad (II-8)$$

Since we are only interested in f^N-configurations and we assume
that the $f^{N-1}d$-configuration is the excited configuration that
most strongly interacts with f^N, we have:

$$\Xi(t,\lambda) = -70 \sum_t \begin{Bmatrix} 1 & \lambda & t \\ 3 & 2 & 3 \end{Bmatrix} \begin{pmatrix} 3 & 1 & 2 \\ 0 & 0 & 0 \end{pmatrix} \begin{pmatrix} 2 & t & 3 \\ 0 & 0 & 0 \end{pmatrix}$$

$$\frac{(nf|r|n'd)(nf|r^t|n'd)}{\Delta(n'd)} \qquad\qquad (II-9)$$

We have to evaluate one 6-j and two 3-j symbols expressing the
coupling of angular momenta. From the second 3-j symbol in (II-9) we have $t \leqslant 5$ and odd, so for Ω_4 and Ω_6 only t = 3 or 5 can be
involved. Variation in $\Xi(t,\lambda)$ along the series is clearly a
function of the radial integrals,

$$(nf|r^t|n'd) = \int_0^\infty R(nf)\ r^t\ R(n'd)dr$$

and of the energy differences between f^N and $f^{N-1}d$ electronic
states.

Judd [44] used values of radial integrals available at the
time. Since then, relativistic Hartree-Fock codes have been used
to compute the requisite integrals for both the 4f and 5f-elements.

Typical results for Ln^{3+} are shown in Table II-2. It is apparent that the relativistic values are somewhat smaller than those used originally by Judd. They are also smaller than the expectation values $\langle r^n \rangle$ computed earlier by Lewis [58]. This further emphasizes the fact that meaningful comparisons between experiment and theory are primarily to be made in systematic trends and in this case in comparisons between two series, 4f and 5f, using a single consistent method of calculation.

The other quantity to be evaluated in Eq. II-9 is $\Delta(n'd)$. In 1971, Brewer [59] published a very important compilation of spectroscopic data that for the most part were not available to Judd in 1962. Some pertinent results are illustrated in Fig. II-4 which shows the energies of the ground states of several excited configurations relative to the f^N. We could take the results for $f^{N-1}d$ directly, but they refer to gaseous free-ion spectra and we are concerned with solutions. The appropriate lowering of energy due to the effects of the condensed phase is 15-20000 cm^{-1} [60]. It can be seen from Table II-3 that while this does not appreciably change the value of $\Delta(n'd)$ originally used by Judd for Nd^{3+}, the new value for Er^{3+} is lower. In considering the role of the lowest-lying $f^{N-1}d$-configuration in Eq. II-9, Judd showed that contributions from related configurations of the type

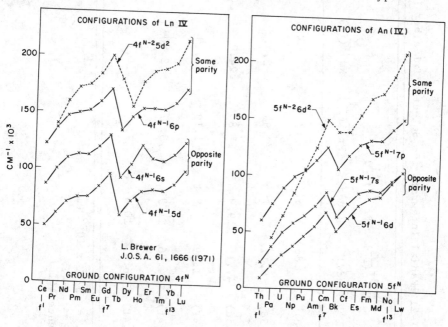

FIGURE II-4. Variation of the lowest energy state in the lower-lying configurations of Ln^{3+} and An^{3+} with atomic number. The reference energy is that of f^N [59].

TABLE II-2. Radial Integrals for Selected Trivalent Lanthanides. Radial Integrals (atomic units).[a]

Integral	$Nd^{3+}(f^3)$ (Judd)[b]	$Er^{3+}(f^{11})$ (Judd)[b]	Relativistic Hartree-Fock Calculation					
			$Pr^{3+}(f^2)$	$Nd^{3+}(f^3)$	$Pm^{3+}(f^4)$	$Ho^{3+}(f^{10})$	$Er^{3+}(f^{11})$	$Tm^{3+}(f^{12})$
$(4f\|r\|5d)$	.869	.615	.778	.735	.699	.562	.547	.534
$(4f\|r^3\|5d)$	5.17	2.75	3.85	3.47	3.16	2.08	1.97	1.88
$(4f\|r^5\|5d)$	47.1	19.9	27.8	24.0	20.9	11.7	10.9	10.2
$(4f\|r^2\|4f)$	1.394	.831	1.19	1.10	1.01	.726	.693	.664
$(4f\|r^4\|4f)$	4.96	1.95	3.27	2.80	2.44	1.30	1.20	1.11
$(4f\|r^6\|4f)$	36.4	10.5	17.6	14.1	11.6	4.85	4.36	3.94

[a] 1 a.u. = .5292 x 10^{-8} cm.

[b] Reference 44.

TABLE II-3.   Energy Differences $\Delta(f^N \to f^{N-1})$ for Ln^{3+} corrected for Ln^{3+}(aquo) Stabilization Energy.

	Judd[a] (cm^{-1})	This Work[b] (cm^{-1})
Pr^{3+}		45000
Nd^{3+}	58000	56000
Pm^{3+}		59000
Gd^{3+}		78000
Tb^{3+}		38000
Ho^{3+}		63000
Er^{3+}	92000	65000
Tm^{3+}		64000

[a]Reference 44.

[b]From Figs. 1,2 corrected in part using Reference 59.

$f^{N-1}n'd$ and $n'd^9f^{N+1}$ could be neglected while those for the $f^{N-1}n'g$ could be approximated and thus included in the sum. Overlap with continuum functions was not considered. In the present development we limit consideration to the perturbing effects of lowest excited $n'd$-configuration.

The new results for Ω_λ, Table II-4, do not differ importantly from those computed by Judd, whose values corresponding to Ω_4 and Ω_6 were a close enough approximation to experiment to make it possible to argue that small adjustments in the magnitude of the interactions considered would reasonably account for the difference and no additional mechanisms for enhancement of the model-computed intensity needed be invoked. We have already commented on the fact that the radial integrals quoted here are considerably smaller than those used by Judd. What does become apparent from the form of Eq. II-8 and the results shown in Table II-2 and Fig. II-4, is that the values of Ω_4 and Ω_6 are predicted to follow a pattern in which there is a decrease in magnitude from f^2 to f^7, an increase from f^7 to f^8 and a second pattern of decrease from f^8 through f^{12}. The values of the fit parameters given in Table II-4 are consistent with the prediction for the heavy lanthanides, but the pattern is less well established for the light members of the series. However, it is apparent that the experimentally established parameter values for Gd^{3+}(aquo) do not follow the expected pattern. The disproportionately large values for Ω_2 and Ω_6 in Pr^{3+}, and for Ω_6 in Nd^{3+} reflect more intense transitions than revealed in other comparable members of the series.

TABLE II-4. Comparison of Calculated Values of Ω_4 and Ω_6 with Those Fit to the Experimental Results for $Ln^{3+}(aquo)$.

	Ω_λ $(x10^{20}$ $cm^2)$			
	Fit	Calc'd	Fit	Calc'd
	Ω_4	Ω_4	Ω_6	Ω_6
Pr^{3+}	5.89	1.13	32.2	1.73
Nd^{3+}	4.08	.553	9.47	.791
Pm^{3+}	4.36	.405	3.94	.523
Gd^{3+}	5.27	.127	4.46	.146
Tb^{3+}	7.95	.462	2.87	.515
Ho^{3+}	3.13	.131	2.86	.141
Er^{3+}	2.19	.110	1.88	.116
Tm^{3+}	2.31	.104	1.47	.106

II(D). Intensity Analysis of Actinide Solution Spectra

In developing the basis for comparing intensities of transitions characteristic of $An^{3+}(aquo)$ and $Ln^{3+}(aquo)$, it is important to be aware of the relative density of states for the $4f^N$ and $5f^N$-configurations as indicated for the light half of the two series in Fig. II-5. It is apparent that the f^N-states occur within a smaller energy range in the actinides. As a consequence, particularly in U^{3+}, Np^{3+}, and Pu^{3+}, the population density over the optical range of prime interest here is large. It is difficult to make very many meaningful assignments to $An^{3+}(aquo)$ even though the corresponding states have been well characterized in crystals [13]. In addition to the higher density of states in the light actinides, the intensities of individual $An^{3+}(aquo)$ transitions can be as much as a factor of 10-100 greater than for corresponding lanthanides. As indicated in Fig. II-6, the average intensity of a transition decreases significantly with increasing Z. It reaches a minimum near Cm^{3+} and remains essentially constant over the heavy half of the series, Fig. II-7.

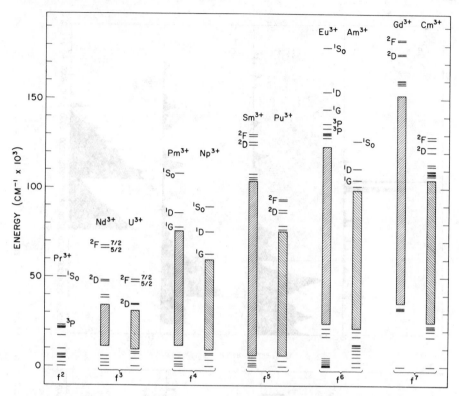

FIGURE II-5. Computed energy span of the f^N-configurations for the light trivalent lanthanides and actinides.

While an intensity analysis of Cm^{3+}(aquo), the analog of Gd^{3+}(aquo), was published earlier [61], and intensity-related arguments were used to aid the interpretation of the energy level structure in Es^{3+}(aquo) [62], it is only very recently that new experimental work has made possible a detailed interpretation of the spectra of Bk^{3+}(aquo) and Cf^{3+}(aquo).

The Judd intensity parameters for the heavier actinides are set out in Table II-5. The individually determined parameters for Cm^{3+}, Bk^{3+}, and Cf^{3+} were extrapolated to give a set for Es^{3+} which was not only found to be consistent with earlier work but actually gave an improved correlation with the observed spectrum. The absorption spectrum for Es^{3+}(aquo) shown in Fig. II-7 is a composite of a number of measurements made using micro absorption cells. The increasing background upon which the spectrum is superimposed is due to light scattering and radiolysis products.

The intensity parameters derived from fitting the experimental data for Bk^{3+}(aquo) and Cf^{3+}(aquo) could be interpreted in detail

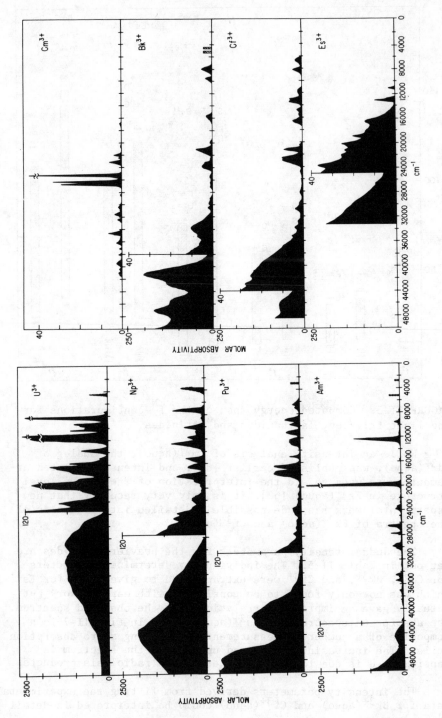

FIGURE II-6,7. Absorption spectra of the light and heavy trivalent actinide aquo ions.

TABLE II-5. Intensity Parameters for the Heavy An^{3+}(aquo) Ions.

	$\Omega_2 \times 10^{20}$ cm^2	$\Omega_4 \times 10^{20}$ cm^2	$\Omega_6 \times 10^{20}$ cm^2
Cm^{3+}	15.2	16.8	38.1
Bk^{3+}	6.96	12.2	18.7
Cf^{3+}	3.39	15.4	16.6
Es^{3+}	1.32	15.8	18.5

by the Judd theory, as indicated in Table II-6 for Bk^{3+}. The highly forbidden character of the spectrum in the lanthanide analog, Tb^{3+}(aquo), which is manifest in weak absorption bands and smaller than average values for $[U^{(\lambda)}]^2$, is not evidenced in Bk^{3+}(aquo) where intensities and matrix element magnitudes are average for the heavy actinides. This difference in character can be traced to the increased spin-orbit coupling which results in a greater mixing of states in Bk^{3+} as indicated in the ground term eigenvector which is 95.6% ^7F in Tb^{3+} and only 72.2% ^7F in Bk^{3+}. The first f→d transitions in Tb^{3+} and Bk^{3+} both occur in the near ultraviolet range with the spin-forbidden ^9D ($f^8 \to f^7d$) state in Tb^{3+} centered near 38000 cm^{-1} and that in Bk^{3+} near 34000 cm^{-1}, Fig. II-8. As was the case for this band in Tb^{3+} [48], the superimposed structure in Bk^{3+} can be analyzed in terms of f→f transitions.

TABLE II-6. Observed and Calculated Band Intensities for Bk^{3+}(aquo).

Band Center (cm^{-1})	$P \times 10^6$ Expt.	Theory	Band Center (cm^{-1})	$P \times 10^6$ Expt.	Theory
8000	17.4	14.5	26450	3.46	3.35
9560	7.08	9.24	27100	4.32	3.01
15700	11.7	7.49	28000	0.4	0.2
19840	3.24	2.14	28820	2.76	2.21
21190	32.4	27.7	29675	4.28	1.86
23590	28.5	33.6	30500	8.20	8.68
25250	5.18	4.98			

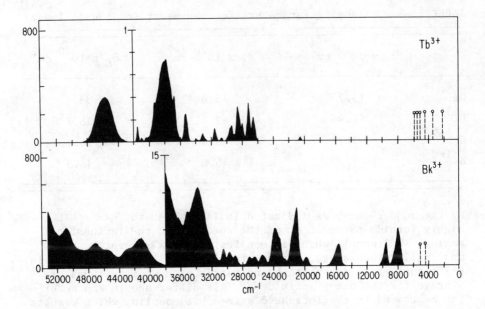

FIGURE II-8. Absorption spectra of the aquo ions of $Tb^{3+}(4f^8)$ and $Bk^{3+}(5f^8)$.

Comparison of the intensity parameters for analog members of the lanthanide and actinide series in Tables II-1 and II-5 shows a factor of 5-10 increase in intensity in the heavy actinides compared to the lanthanides. At this point it is evident that Judd's formalism is fully applicable to the experimentally observed intensity patterns observed for 5f as well as 4f-elements so that it remains to explore the correlation with the model calculations of Ω_λ--again restricted to Ω_4 and Ω_6.

In approaching the model calculation of Ω_4 and Ω_6 for An^{3+}(aquo), we consider first the analog of Eq. II-7. The problem is simplified because of the close relationship between the ionic radii characteristic of the trivalent ions in the two series. As an approximation the radii of $Ln^{3+}(f^N)$ and $An^{3+}(f^{N+2})$ are usually quite comparable. Thus for example for the $MCl_3 \cdot 6H_2O$ structure in the lanthanides M $\leftrightarrow$ (2)H_2O distances range from 2.449 Å in M=Nd to 2.312 Å in M=Lu [57] while for M=Am the distance is 2.440 Å [63]. We consequently do not expect any significant change in the contribution of B_t, Eq. II-7, computed for a characteristic light (or for a heavy) lanthanide ion or actinide ion, and the method of approximation used by Judd appears to be equally applicable to the two series.

One of the interesting questions that arises is whether there is a change in inner sphere hydration number in the

An^{3+}(aquo) series corresponding to that deduced for the lanthanides. Recent electromigration studies suggest that such a
change could be occurring in the region of Cf^{3+} - Es^{3+} [56].
However neither the absorption spectra of Ln^{3+}(aquo), nor that of
An^{3+}(aquo) appear to provide an indication of this change. The
hypersensitive transitions in both the 4f and 5f-series, normally
the most sensitive spectroscopic monitors of a changing ionic
environment, uniformly show a minimum of intensity in the aquo
ions. Thus aquo ion spectra serve as a standard for judging
increased intensity in other environments, without providing any
internal evidence for structural modification. However, in the
role of providing a standard for judging increased intensity,
comparisons to spectra in crystal matrices can indicate where in
a series structural changes appear to be occurring.

If we compare the spectrum of Cf^{3+}(aquo), Fig. (II-7), and
that of $CfCl_3$ (hexagonal $LaCl_3$-type structure) [64], there is a
similar energy and band intensity pattern with the exception of
the increased intensity in the hypersensitive transition in $CfCl_3$
near 11400 cm^{-1}. Thus, if the structure is correctly identified,
we have a case of what would appear to be a small distortion
giving rise to an enhancement in intensity, a unique case for the
$LaCl_3$-type structure. On the other hand, if there is a distortion
toward the $AlCl_3$(YCl_3)-phase which is characteristic of the heavy
lanthanide chlorides [65] then having examined other cases we
would expect that there would be increased intensity in the indicated hypersensitive transition. In our absorption spectra
study of thin films of $CfCl_3$ [66] we identified the sample by X-
ray analysis as exhibiting the hexagonal $LaCl_3$-type structure.
While there was no apparent increased intensity near 11400 cm^{-1}
relative to Cf^{3+}(aquo), the 298 K spectra of the thin film were
admittedly broad and diffuse and thus not definitive in this
respect. What can be said is that there is some evidence of
distortion suggesting a tendency toward structural change at this
point in the An^{3+} series.

For a related case, Es^{3+}(aquo) compared to Es^{3+}:$LaCl_3$ [67],
the correlation between the two spectra is very good both in
terms of energy and band intensity including the hypersensitive
transition. This conforms to similar behavior for the lanthanides
in this matrix. On the contrary there is evidence of increased
intensity in the hypersensitive band in EsF_3 near 20000 cm^{-1}
[68], a sample for which an X-ray diffraction pattern could not
be obtained. Alternative structures [65] are the LaF_3-type (not
conducive to increased intensity in hypersensitive transitions)
and the YF_3-type which is known to induce increased intensity in
the hypersensitive transitions. Since the results are consistent
with a change in crystal structure from the LaF_3-type typical of
preceding members of the series, a change in hydration number
occurring at this point in the series would not be unexpected.

Returning to the calculation of Ω_λ, in additon to the B_t term, Eq. II-7, the overlap integrals and energy differences indicated in Eq. II-9 must be computed. We limit consideration, as was done for the lanthanides, to overlap with the n'd configuration. Consequently the 6-j and 3-j symbols of Eq. II-8 are identical for the 4f- and 5f-series. The overlap integrals computed using a relativistic HF code and analogous to those for the lanthanides (Table II-2) are set out in Table II-7.

The $f^N \to f^{N-1}d$ transitions in the actinides occur at somewhat lower energies than for the corresponding lanthanides, as noted earlier, and the hydration energy correction relative to the gaseous free-ion data is apparently less in the light than in the heavy portion of the 5f-series. The estimated values of $\Delta(n'd)$ are given in Table II-8. The appearance of intense absorption bands in the visible to ultraviolet range of the spectra of many of the An^{3+}(aquo) ions Fig. II-3, II-4, provides a much better basis for a consistent assignment of an energy for the first $f \to d$ transition than is the case for Ln^{3+}(aquo) where few such transitions lie low enough in energy to be observed.

In comparing the intensity parameters for An^{3+}(aquo) determined by fitting the experimental data and those based on a model calculation, Table II-9, it is at once apparent that the agreement is similar to that found for the lanthanides. Thus the very intense transitions in the light actinides are well correlated with the close proximity of, and the increased overlap with the n'd configurations. The model computed intensity decreases markedly with Z except for the break at f^7 corresponding to a sudden change in $\Delta(n'd)$, while this trend is less pronounced in the heavy members of both the 4f and 5f series. Since we have considered only the effects of the (n'd) configurations, inclusion of the effects of other opposite parity interactions should be investigated. Judd's estimates of the effects of the n'g-configurations did show a greater contribution to the heavy members of the Ln^{3+}(aquo) series [44] and this was also emphasized by Krupke [54].

A comparison of the model-computed intensity parameter values for Ln^{3+}(aquo) and An^{3+}(aquo), Tables II-4 and II-9, further shows that the magnitude of the increased intensity between the two series is correctly predicted, and thus can reasonably be correlated with the larger values computed for the radial overlap integrals. The similarity in the values for other Ln^{3+}(aquo) and An^{3+}(aquo) dependent quantities included in the model makes identification of the role of the overlap integrals particularly clear in this case.

TABLE II-7. Radial Integrals for Selected Trivalent Actinides. Radial Integrals (atomic units)

| | | | Relativistic Hartree - Fock Calculation | | | |
Integral	$U^{3+}(f^3)$	$Np^{3+}(f^4)$	$Pu^{3+}(f^5)$	$Bk^{3+}(f^8)$	$Cf^{3+}(f^9)$	$Es^{3+}(f^{10})$
$(5f|r|5d)$	1.22	1.14	1.07	.911	.869	.831
$(5f|r^3|5d)$	7.91	6.95	6.18	4.57	4.20	3.87
$(5f|r^5|5d)$	74.4	61.4	51.6	33.3	29.5	26.3
$(5f|r^2|5f)$	2.28	2.07	1.90	1.53	1.43	1.35
$(5f|r^4|5f)$	10.2	8.42	7.06	4.54	4.00	3.55
$(5f|r^6|5f)$	81.8	61.1	47.0	24.4	20.3	17.0

TABLE II-8.   Energy Differences $\Delta(f^N \rightarrow f^{N-1}d)$ for An^{3+} Corrected for An^{3+}(aquo) Stabilization Energy.[a]

	cm^{-1}		cm^{-1}
U^{3+}	25000	Cm^{3+}	(52000)[b]
Np^{3+}	32000	Bk^{3+}	34000
Pu^{3+}	38000	Cf^{3+}	46000
Am^{3+}	44000	Es^{3+}	(57000)[b]

[a]From Figs. II-4, II-6, and II-7.   [b]Estimated from Fig. II-7.

TABLE II-9.   Comparison of Calculated Values of Ω_4 and Ω_6 with Those Fit to Experimental Results for An^{3+}(aquo).

	Ω_λ $(x10^{20}$ $cm^2)$			
	Fit Ω_4	Calc'd Ω_4	Fit Ω_6	Calc'd Ω_6
U^{3+}	55^a	33.1	186^a	80.8
Np^{3+}	77^a	14.3	155^a	31.3
Pu^{3+}	26^a	7.41	66^a	14.8
Cm^{3+}	16.8	2.28	38.1	3.89
Bk^{3+}	12.2	4.21	18.7	6.79
Cf^{3+}	15.4	1.86	16.6	2.83
Es^{3+}	15.8	.994	18.5	1.44

[a]Estimated values.

Not all of the observed intensity patterns are satisfactorily accounted for in the model. The fact that the fit parameters for f^7 are larger than those for f^8 in both the 4f- and 5f-series is not explained. The role of J-mixing as a possible explanation for the observation of nominally forbidden transitions, particularly $^7F_0 \rightarrow$ odd J transitions in the f^6-configurations needs to be explored, as do similar perturbing effects in Bk^{3+}. However it is clear that Judd's development of the intensity analysis continues to give us very important insights into the physics of these unique f-elements.

III. FLUORESCENCE STUDIES OF THE TRIVALENT LANTHANIDE AND ACTINIDE AQUO IONS

One of the important features of a successful analysis of the intensities of the f-elements in solution is that it provides the basis for calculating the radiative lifetime characteristic of any given excited state [43]. Most excited states are primarily relaxed by non-radiative mechanisms, but for those that exhibit a measurable fluorescence it is useful to point out the relationship:

$$(\tau_T)^{-1} = A_T(\psi J) + W_T(\psi J) \tag{III-1}$$

where τ_T is the observed lifetime of a particular excited state, $A_T(\psi J)$ is the total radiative relaxation rate and $W_T(\psi J)$ is the total non-radiative relaxation rate. The rate of relaxation computed from the intensity analysis corresponds to the case in which $W_T=0$. Thus there is a limited range of relative rates of A_T and W_T in which both contribute importantly, and where consequently, the difference $(\tau_T)^{-1} - A_T(\psi J) = W_T(\psi J)$ can provide an independent method of assessing W_T. This can be very useful information when exploring the mechanisms of non-radiative relaxation.

Another instructive use of $A_T(\psi J)$ is in the prediction of pathways of radiative relaxation. A given excited state is coupled to each state that is lower in energy. Knowledge of which states are most strongly coupled to the excited state leads to the prediction of the energies of strong fluorescing lines. Of course if the predominant mechanism of relaxation involves non-radiative processes, no fluorescence will be observed.

The general form of the probability for spontaneous relaxation of an excited state [69], is

$$A(\psi J, \psi'J') = \frac{64\pi^4\sigma^3}{3h} (\psi J|\underset{\sim}{D}|\psi'J')^2 \tag{III-2}$$

where ψJ and $\psi'J'$ are the initial and final states, A is the spontaneous transition probability per unit time, $\sigma(cm^{-1})$ the energy gap between the states, and $\underset{\sim}{D}$ is the dipole operator which we can replace by the electric and magnetic dipole line strengths. Following Axe [48], Eq. III-2 can be written

$$A(\psi J, \psi' J') = \frac{64\pi^4 \sigma^3}{3h\ (2J+1)} \left[\chi'\ \overline{F}^2 + n^3\ \overline{M}^2 \right] \tag{III-3}$$

where J in this case refers to the initial excited state, *not* the ground state, and $\chi' = {}_2 n\ (n^2 + 2)^2/9$. The electric dipole $(\overline{F}^2)$ and magnetic dipole $(\overline{M}^2)$ operators are those defined previously in terms of static absorption intensities, so the required matrix elements for transitions of interest can be readily computed.

$$\overline{F}^2 = e^2 \sum_{\lambda=2,4,6} \Omega_\lambda\ (\psi J || U^{(\lambda)} || \psi' J')^2 \tag{III-4}$$

$$\overline{M}^2 = e^2 (\psi J || L + 2S || \psi' J')^2 / 4m^2 c^2 \tag{III-5}$$

Further details of the computation are given in [43].

Calculation of spontaneous emission rates and the lifetimes are summarized by Eq. III-1, and by the expression:

$$A_T(\psi J) = \sum_i (\psi J,\ \psi' J')_i \tag{III-6}$$

where the sum runs over all states with energy less than that of ψJ. Thus in absorption a particular state can be populated. In emission, relaxation to all lower-lying states is summed over to give a total rate. The branching ratio, β_R, from the relaxing state to a particular final state is given by

$$\beta_R\ (\psi J,\ \psi' J') = \frac{A(\psi J, \psi' J')}{A_T(\psi J)} \tag{III-7}$$

The principal fluorescing states of the lanthanide aquo ions were known long before their lifetimes could be accurately measured. Thus, based on work completed in the 1930's, Fig. III-1, the excited states of interest to the type of analysis developed here are well characterized. Selected values of the measured radiative lifetimes of some of the excited states of Ln^{3+} (aquo) are compared with those computed from spontaneous radiative relaxation rates in Table III-1. It is apparent that even for Gd^{3+} (aquo), non-radiative relaxation is of major importance since based on Eq. III-1, assuming $\tau H_2O = 2$ msec, and computing $A_T(^6P_{7/2})$ = 92 sec^{-1}, we see that $W_T(^6P_{7/2})$ = 408 sec^{-1}.

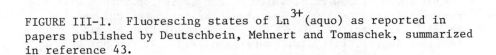

FIGURE III-1. Fluorescing states of Ln^{3+}(aquo) as reported in papers published by Deutschbein, Mehnert and Tomaschek, summarized in reference 43.

TABLE III-1. Radiative Lifetimes of Excited States of Ln^{3+} (aquo).

	Excited State	Energy Gap $\Delta E (cm^{-1})$	Experimental (msec)		Theory (msec)
			τ_{H_2O}	τ_{D_2O}	
Gd^{3+}	$^6P_{7/2}$	32,200	2.3[a]		10.9
Tb^{3+}	$^5D_4$	14,800	0.39[b]	3.3[b]	9.02
Eu^{3+}	$^5D_0$	12,300	0.10[b]	1.9[b]	9.67
Dy^{3+}	$^4F_{9/2}$	7,400	0.0023[a]	0.038[a]	1.85
Sm^{3+}	$^4G_{5/2}$	7,400	0.0023[a]	0.053[a]	6.26
Pm^{3+}	$^5F_1$	5,800			0.65
Nd^{3+}	$^4F_{3/2}$	5,380	0.00003[c]	0.00017[c]	0.42
Er^{3+}	$^4S_{3/2}$	3,100			0.66
Ho^{3+}	$^5S_2$	3,000			0.37

[a] Reference 77.   [b] Reference 75.   [c] Reference 79.

III(A). Branching Ratio Calculations

     While the experimental observation of fluorescence of lan-
thanide ions in aqueous solution predated our ability to predict
the patterns of excited state relaxation, it is still useful to
point out some of the features of the calculation of branching
ratios. Fig. III-2 illustrates the fact that a fluorescing excited state, in this case $^5D_4$ of Tb^{3+}, may be particularly strongly
coupled to one or two of the lower-lying states, and not necessarily to the ground state. Of course the σ^3 dependence in Eq. III-3
will strongly select a large energy gap. For Tb^{3+}(aquo), the $^5D_4$
state is most strongly coupled to the (excited) $^7F_5$ state. Consequently the prediction would have been that the strongest band
in fluorescence originating from the $^5D_4$-state would be found at
$20450-2100 = \sim 18350$ cm^{-1}; which of course is consistent with experimental results [70]. The branching ratios calculated for
$Tb(NO_3)_3$ in a molten $LiNO_3$-KNO_3 eutectic solution at $\sim 150°$ are
included in Fig. III-2 for comparison [71]. It was pointed out
earlier that of the three parameters, Ω_λ, Ω_2 is most sensitive to
changes in the ionic environment. In the molten nitrate salt

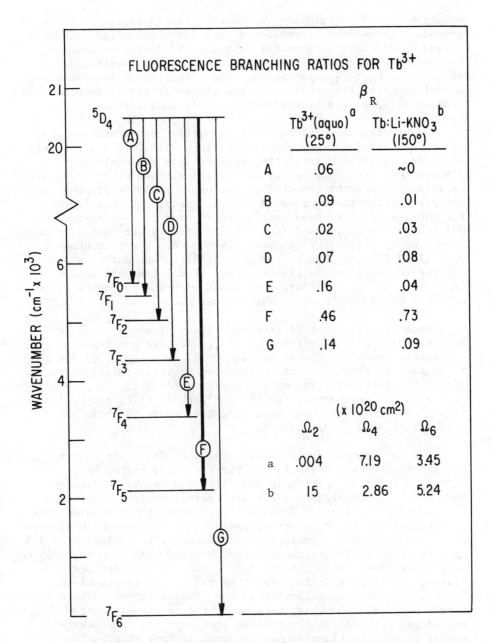

FIGURE III-2. Computed fluorescence branching ratios for Tb^{3+}(aquo) and for Tb(NO$_3$)$_3$ in a molten LiNO$_3$–KNO$_3$ eutectic at 150°, [43,71].

spectrum of Tb^{3+} a relative increase in the intensity of $U^{(2)}$-dependent transtions is reflected in a large value for Ω_2. Both Ω_4 and Ω_6 also differ somewhat from the Tb^{3+}(aquo) values. However, the calculated branching ratios even more strongly select the $^5D_4 \rightarrow {}^5F_5$ transition. The principal lasing frequency associated with fluorescing excited states of the lanthanides can usually be predicted from branching ratio calculations.

Absorption and excited state relaxation are processes that are in equilibrium; consequently, it is possible to use properly corrected oscillator strengths obtained from the analysis of fluorescence spectra as input to the calculation of Ω_λ along with oscillator strengths computed in the usual analysis of absorption spectra. Although recorded in the gas phase, not in solution, the $TbCl_3$-$AlCl_3$ gas-phase complex is an example of a system in which both absorption and fluorescence spectra were utilized in computing Ω_λ [72]. The measured fluorescence spectrum shown in Fig. III-3 includes bands from several different parent states, but the relative intensities of those transitions attributed to $^5D_4 \rightarrow {}^7F_5$ and $^5D_4 \rightarrow {}^7F_6$ are consistent with the intensity patterns expected based on the branching ratios.

Recently the first fluorescence study of an An^{3+} ion in aqueous solution was carried out [73]. We have now begun a systematic investigation of An^{3+}(aquo) fluorescence [74] and it is as part of this study that branching ratio calculations are aiding the interpretation of new experimental data. The results of the intensity analyses for the Cm^{3+}, Bk^{3+}, Cf^{3+}, and Es^{3+} aquo ions cited previously have been used to compute the branching ratios indicated in Fig. III-4.

For Cm^{3+}(aquo), the actinide analog of Gd^{3+}, the most important fluorescing state is the first excited state, $^6P_{7/2}$. Selective laser excitation in D_2O yielded a measured lifetime of 940 μsec [73] compared to a computed purely radiative lifetime of 1.3 msec. Excitation at 25445 cm^{-1} was found to be relaxed non-radiatively to the $^6P_{7/2}$ state, the only state from which fluorescence was observed. Similarly, excitation of Bk^{3+}(aquo) at 25575 cm^{-1} resulted in fluorescence being observed only from the J=6 state near 16000 cm^{-1}. In the latter case the results are consistent with the indicated branching ratio. No intermediate level is more strongly coupled to the fluorescing state than the ground state is. However, because of the relatively narrow energy gap between the fluorescing state and the next lower-energy state, the fluorescence lifetime is very short. The dependence of lifetime on energy gap is discussed in a subsequent section.

Branching ratio calculations for the most probable fluorescing transitions of Cf^{3+}(aquo) are shown in Fig. III-4. Fluores-

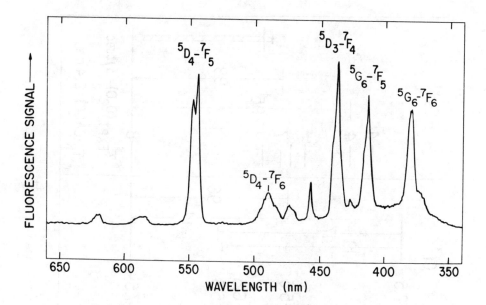

FIGURE III-3. Fluorescence spectrum of the gas phase $TbCl_3(AlCl_3)_x$ complex at 535 K showing the bands arising from the $^5D_4 \rightarrow {}^7F_0 - {}^7F_6$ transitions [72].

cence from both of these Cf^{3+}(aquo) transitions is predicted to be quite weak with lifetimes much less than 0.1 microseconds expected since the energy gaps are even narrower than in the fluorescing transition of Bk^{3+}(aquo). We exclude the 11/2 state at about 6500 cm^{-1} from consideration since nanosecond response photodetectors for this spectral region are quite insensitive compared to photomultipliers. Selective laser excitation of Es^{3+}(aquo) at 20070 cm^{-1} gives rise to detectable fluorescence from the J=5 state at about 9500 cm^{-1} (Fig. III-4). In detecting fluorescence characteristic of Es^{3+}(aquo), it should be noted that we are dealing with an intensely α-active isotope ($t_{1/2}$ = 21.5 days). Use of current laser and photodetection technology has enabled us to measure the fluorescence lifetime of Es^{3+}(aquo) in D_2O solution at concentrations as low as 5×10^{-6} molar.

We have not addressed the calculation of branching ratios for the light members of the 3+ aquo actinide ion series. The high density of 5f states in the energy level structures of these ions makes it improbable, with the possible exception of Am^{3+}, that measurable fluorescence will be found even in D_2O solution using the techniques at hand [73,74].

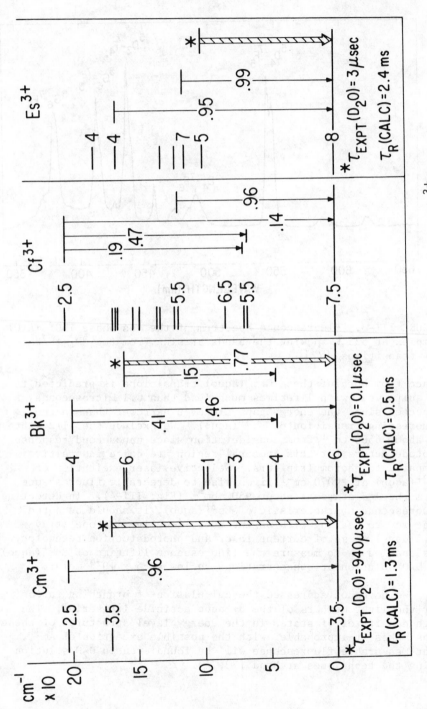

FIGURE III-4.   Computed radiative branching ratios for the heavier An³⁺ aquo ions.

III(B). Non-Radiative Relaxation

In crystalline hosts, it has been shown that the quantitative treatment of non-radiative relaxation mechanisms is extremely complex. Multiphonon orbit-lattice relaxation has been recognized as an important mode of decay; experimentally, the dependence of fluorescence quenching on the energy gap (ΔE) between the fluorescing state and the next lower-energy state has been well established.

Investigations of the temperature dependence of the multi-phonon relaxation rates have shown that the decay usually involves emission of high-energy optical phonons. In $LaCl_3$, for example, the phonon density of states cuts off at ~ 260 cm^{-1}. It was found that for crystals the process was adequately represented by,

$$W = \beta e^{\alpha \Delta E}$$

(III-8)

where β and α are parameters characteristic of the particular lattice. The energy gap, ΔE, thus determines the number of high energy phonons that must be emitted simultaneously to absorb the requisite energy. The multiphonon process becomes less probable as the number of phonons that must be simultaneously excited to conserve energy increases. Recalling the rule of ~ 1000 cm^{-1}, which corresponds to a minimum ΔE required before fluorescence would be expected in $LaCl_3$ host, it is clear that the emission of less than 4-phonons is a process that efficiently competes with the radiative relaxation modes.

Experiments exploring non-radiative modes of relaxation in solution actually paralleled work in the solid state in the early 1960's. In a particularly enlightening series of papers Kropp and Windsor [75] examined the effects that the substitution of D_2O for H_2O had on the fluorescent lifetimes of a number of different states in rare earth ion spectra. These results illustrate the sensitivity of the lifetimes to changes in the ionic environment when for all practical purposes no change is observed in the energy or intensity of the absorption bands, i.e., in D_2O and H_2O.

In their pioneering work Kropp and Windsor pointed out that the ratio of the intensity of fluorescence of a given Ln^{3+} state in D_2O to that of the same state in H_2O was inversely proportional to the energy gap, ΔE. Subsequently, they concluded that the quenching of fluorescence in aqueous solution occurred via OH coupled modes and the rate was proportional to the number of such modes associated with the lanthanide ions. Gallagher [76] reached the same conclusion showing that the introduction of a single OH-group into the inner coordination sphere in Eu^{3+} was sufficient

to reduce the fluorescence lifetime of the $^5D_0$ state from 3.9
(pure D_2O) to 0.12 msec.

This led to the interpretation of the quenching of fluorescence in H_2O (D_2O) in terms of a multiphonon mechanism involving
the transfer of energy to a single vibration mode (OH) excited to
high vibrational states. This parallels the interpretation in
crystals. The number of phonons required to bridge the gaps
characteristic of fluorescing states in Ln^{3+}(aquo) using ν_1(OH) =
3405 cm^{-1} and ν(OD) = 2520 cm^{-1}, are:

	$\Delta E/\nu_1$	
	OH	OD
Gd^{3+}	10	13
Tb^{3+}	5	6
Eu^{3+}	4	5
Dy^{3+}	3	4
Sm^{3+}	3	4
Nd^{3+}	1	2

The lifetimes observed in aqueous solution follow a somewhat
similar pattern to that observed in crystals: the higher the order
of the quenching process (i.e. the larger the number of phonons
needed to bridge the energy gap), the lower the non-radiative
decay rate W_T. In Fig. III-5, the logarithm of the non-radiative
decay rate constant has been plotted versus the energy gap for
the emitting Ln^{3+}(aquo) state in H_2O and D_2O solution, using Eq.
III-1 and the data in Table III-1. By analogy with doped single
crystal studies of lanthanide ion non-radiative decay, a linear
relationship might have been expected (see Eq. III-8), but, as is
evident from Fig. III-5, only rough linearity is found. As noted
by Stein and Wurzberg [77], non-radiative decay of lanthanide
ions in solution is most appropriately considered in the "large
molecule" limit where the lanthanide ion and its solvation sphere
are treated as a single quantum mechanical system. However, argue-
ments against use of the latter approach have been given [80].

Non-radiative relaxation of aquo lanthanide and actinide
ions in solution is even more complex than is the case in doped
single crystal studies of these same ions. The greater complexity
arises from the wide energy range spanned by the normal vibrational
mode frequencies of water and the lack of long range order which
is characteristic of liquids. Even inner coordination sphere
water molecules exchange on a time scale short compared to the

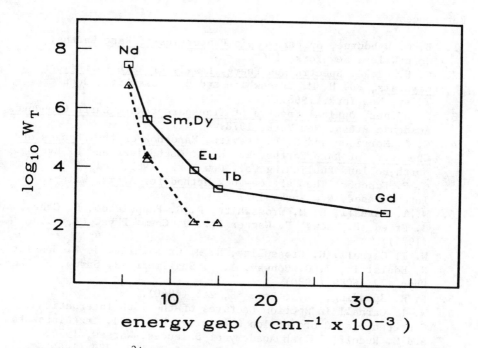

FIGURE III-5. Ln^{3+}(aquo) non-radiative decay rate versus energy gap in H_2O and D_2O solutions. —— H_2O, ----- D_2O.

lifetimes of large energy gap fluorescing f states [78]. Initial efforts have been made to understand f electron non-radiative decay in solution [79], but this is an area in which there is significant potential for development. The Judd-Ofelt theory provides a good approximation to the radiative portion of the experimentally observed lifetimes of 3+ aquo lanthanide and actinide ion fluorescing f electron states, but it is evident that several mechanisms in addition to multiphonon-like processes must be explored to adequately represent the observed non-radiative relaxation rates [80].

Acknowledgment

Work performed under the auspices of the Office of Basic Energy Sciences, Division of Nuclear Sciences, U. S. Department of Energy under contract number W-31-109-ENG-38.

References

1. B. G. Wybourne, Spectroscopic Properties of Rare Earths,
    John Wiley, New York, 1965.
2. G. H. Dieke, Spectra and Energy Levels of Rare Earth Ions in
    Crystals, Ed. H. M. Crosswhite and H. Crosswhite, John
    Wiley, New York, 1968.
3. S. Hüfner, Optical Spectra of Transparent Rare Earth Compounds,
    Academic Press, New York, 1978.
4. C. A. Morrison and R. P. Leavitt, Handbook of the Physics and
    Chemistry of Rare Earths, Ed. K. A. Gschneider and L. Eyring,
    North-Holland Publishing Co., New York, 1982, Vol. 5.
5. C. F. Fischer, The Hartree-Fock Method for Atoms, Wiley-
    Interscience, New York, 1977.
6. W. T. Carnall, H. M. Crosswhite, R. G. Pappalardo, D. Cohen,
    S. Fried, P. Lucas, F. Wagner Jr., J. Chem. Phys. $\underline{61}$, 4993
    (1974).
7. W. T. Carnall, H. Crosswhite, H. M. Crosswhite, J. P. Hessler,
    N. Edelstein, J. G. Conway, G. V. Shalimoff, R. Sarup, J. Chem.
    Phys. $\underline{72}$, 5089 (1980).
8. F. H. Spedding, Phys. Rev. $\underline{58}$, 255 (1940).
9. W. T. Carnall in Section Lectures of the 13th International
    Conference on Coordination Chemistry, Ed. B. J. Trzebiatowska
    and M. Rudolf, Polish Academy of Sciences, Warsaw, 1974.
10. K. Rajnak and B. G. Wybourne, Phys. Rev. $\underline{132}$, 280 (1965).
11. J. C. Morrison and K. Rajnak, Phys. Rev. $\underline{A4}$, 536 (1971).
12. H. M. Crosswhite, Colloques Internationaux du C. N. R. S.,
    Spectroscopie des Elements de Transition et des Elements
    Lourds dans les Solides, 28 Juin - 3 Juillet 1976, Editions
    due C. N. R. S., Paris, 1977, p. 65.
13. Jan P. Hessler and W. T. Carnall, ACS Symposium Series No.
    131, 349 (1980).
14. J. C. Morrison, Phys. Rev. $\underline{A6}$, 643 (1972).
15. B. R. Judd, Phys. Rev. $\underline{141}$, 4 (1966).
16. H. Crosswhite, H. M. Crosswhite, and B. R. Judd, Phys. Rev.
    $\underline{174}$, 89 (1968).
17. G. Balasubramanian, M. M. Islam, and D. J. Newman, J. Phys.
    $\underline{B8}$, 2601 (1975).
18. B. R. Judd, H. M. Crosswhite, and H. Crosswhite, Phys. Rev.
    $\underline{169}$, 130 (1968).
19. H. M. Crosswhite, H. Crosswhite, F. W. Kaseta, and R. Sarup,
    J. Chem. Phys. $\underline{64}$, 1981 (1976).
20. R. D. Cowan and D. C. Griffin, J. Opt. Soc. Am. $\underline{66}$, 1010
    (1976).
21. H. M. Crosswhite and H. Crosswhite, Private Communication.
22. B. R. Judd, J. Lumin, $\underline{18/19}$, 604 (1979).
23. B. R. Judd, Phys. Rev. $\underline{C13}$, 2695 (1980).
24. M. T. Hutchings and D. K. Ray, Proc. Phys. Soc. $\underline{81}$, 663
    (1963).

25. M. Faucher, J. Dexpert-Ghys, and P. Caro, Phys. Rev. B21, 3689 (1980).

26. D. J. Newman, Adv. Phys. 20, 197 (1971).

27. D. J. Newman, J. Phys. C10, 4753 (1977).

28. D. J. Newman, Aust. J. Phys. 31, 489 (1978).

29. H. M. Crosswhite, H. Crosswhite, W. T. Carnall, A. P. Paszek, J. Chem. Phys. 72, 5103 (1980).

30. Y. M. Poon and D. J. Newman, J. Chem. Phys. 75, 3646 (1981).

31. A. K. Cheetham, B. E. F. Fender, H. Fuess, A. F. Wright, Acta Cryst. B32, 94 (1976).

32. A. Zalkin, D. H. Templeton, and T. E. Hopkins, Inorg. Chem. 5, 1466 (1966).

33. W. H. Zachariasen, J. Chem. Phys. 16, 254 (1948).

34. D. E. Onopko, Optics and Spectro. Suppl. 4, USSR Academy of Sciences, 1968; Optics and Spectro. 24, 301 (1968).

35. D. J. Newman and M. M. Curtis, J. Phys. Chem. Solids 30, 2731 (1969).

36. W. T. Carnall, H. Crosswhite, H. M. Crosswhite, Energy Level Structure and Transition Probabilities of the Trivalent Lanthanides in LaF$_3$, Argonne National Laboratory Report (1977).

37. C. A. Morrison and R. P. Leavitt, J. Chem. Phys. 71, 2366 (1979).

38. P. Caro, J. Derouet, L. Beaury, G. Teste de Sagey, J. P. Chaminade, J. Aride, M. Pouchard, J. Chem. Phys. 74, 2698 (1981).

39. H. H. Caspers, H. E. Rast, and J. L. Fry, J. Chem. Phys. 53, 3208 (1970).

40. R. P. Leavitt and C. A. Morrison, J. Chem. Phys. 73, 749 (1980).

41. L. J. F. Broer, C. J. Gorter, and J. Hoogschagen, Physica 11, 231 (1945).

42. J. H. Van Vleck, J. Phys. Chem. 41, 67 (1937).

43. W. T. Carnall, Handbook on the Physics and Chemistry of Rare Earths, Ed. K. A. Gschneidner Jr. and L. Eyring, North-Holland, New York, 1979, Vol. 3, p. 171.

44. B. R. Judd, Phys. Rev. 127, 750 (1962).

45. G. S. Ofelt, J. Chem. Phys. 37, 511 (1962).

46. R. Reisfeld and C. J. Jørgensen, Lasers and Excited States of Rare Earths, Springer, New York, 1977.

47. M. J. Weber, Handbook on the Physics and Chemistry of Rare Earths, Ed. K. A. Gschneidner Jr. and L. Eyring, North-Holland, New York, 1979, Vol. 4, p. 275.

48. J. D. Axe, J. Chem. Phys. 39, 1154 (1963).

49. W. T. Carnall, P. R. Fields, and K. Rajnak, J. Chem. Phys. 49, 4224, 4443, 4447, 4450 (1968).

50. R. D. Peacock, Struct. Bonding 22, 83 (1975).

51. B. R. Judd, J. Chem. Phys. 70, 4830 (1979).

52. S. F. Mason, Struct. Bonding 39, 43 (1980).

53. W. J. Maeck, M. E. Kussy, and J. E. Rein, Anal. Chem. 37, 103 (1965).

54. W. F. Krupke, Phys. Rev. 145, 325 (1966).

55. M. Marezio, H. A. Plettinger, and W. H. Zachariasen, Acta Cryst. 14, 234 (1961).

56. R. Lundquist, E. K. Hulet, and P. A. Baisden, Acta Chem. Scand. A35, 653 (1981).

57. Gmelin Handbuch der Anorganischen Chemie, 8th Ed., Rare Earth Elements, Part C4a, Springer Verlag, Heidelberg, 1982, p. 185-194.

58. W. B. Lewis, Los Alamos Scientific Laboratory Report LA-DC-11574 (1970).

59. L. Brewer, J. Opt. Soc. Am. 61, 1101, 1666 (1971).

60. E. Loh, Phys. Rev. 147, 332 (1966).

61. W. T. Carnall and K. Rajnak, J. Chem. Phys. 63, 3510 (1975).

62. W. T. Carnall, D. Cohen, P. R. Fields, R. K. Sjoblom, and R. F. Barnes, J. Chem. Phys. 59, 1785 (1973).

63. J. H. Burns and J. P. Peterson, Inorg. Chem. 10, 147 (1971).

64. J. R. Peterson, R. L. Fellows, J. P. Young, and R. G. Haire, Radiochem. Radioanal. Ltrs. 31, 277 (1977).

65. D. Brown, Halides of the Lanthanides and Actinides, J. Wiley, New York, 1968.

66. W. T. Carnall, S. Fried and F. Wagner, Jr., J. Chem. Phys. 58, 1938 (1973).

67. R. L. Fellows, J. R. Peterson, J. P. Young and R. G. Haire, The Rare Earths in Modern Science and Technology (G. J. McCarthy and J. J. Rhyne, Ed.) Plenum Press, New York, 1978, p. 493.

68. D. D. Ensor, J. R. Peterson, R. G. Haire, and J. P. Young, J. Inorg. Nucl. Chem. 43, 2425 (1981).

69. E. U. Condon and G. H. Shortley, The Theory of Atomic Spectra, Cambridge University Press, London, 1957, pp. 91-109.

70. W. R. Dawson, J. L. Kropp, and M. W. Windsor, J. Chem. Phys. 45, 2410 (1966).

71. W. T. Carnall, J. P. Hessler, and F. Wagner, Jr., J. Phys. Chem. 82, 252 (1978).

72. J. A. Caird, W. T. Carnall, J. P. Hessler, J. Chem. Phys. 74, 3225 (1981).

73. J. V. Beitz and J. P. Hessler, Nucl. Tech. 51, 169 (1980).

74. J. V. Beitz, W. T. Carnall, D. W. Wester and C. W. Williams, Lawrence Berkeley Laboratory Report LBL-12441, 1981.

75. For a review of some of the relevant work of Kropp and Windsor see Reference 43.

76. P. K. Gallagher, J. Chem. Phys. 43, 1742 (1965).

77. G. Stein and E. Würzberg, J. Chem. Phys. 62, 208 (1975).

78. J. Burgess, Metal Ions in Solution, John Wiley, Sussex, 1978, p. 317.

79. J. V. Beitz, Unpublished Results, Argonne National Laboratory, 1982.

80. M. Stavola, L. Isganitis, and M. G. Sceats, J. Chem. Phys. 74, 4228 (1981).

DISCUSSION

GROUP I   (question)

What is hypersensitivity and what is the present status of your understanding of the mechanism by which it occurs?

CARNALL   (answer)

The intensities of one or two transitions for each lanthanide are extremely sensitive to the environment! These transitions exhibit a normal intensity for example in aquo ions, but may increase markedly in intensity and increase relative to all other observed transition intensities, in certain environment. These are the hypersensitive transitions.

A number of mechanisms has been proposed and rejected over the last few years, but Mason, Peacock and Stewart [Chem.Phys.Lett. 29, 149 (1974)] appeared to be successful in proposing that the f-electrons on the Ln(III) ion can polarize the ligand. This has the effect of greatly extending the sphere of influence of the f-electrons. As long as there is no center of inversion, the induced dipoles can continue and yield a non-vanishing dipole moment. This moment interacts with the radiation field, and the result is that a normally forbidden electric dipole transition becomes allowed.

Computations for a number of known cases of intense hypersensitive transitions appeared to be in line with experiment. Judd [J.Chem. Phys. 70, 4830 (1979)] consequently points out that this dynamic coupling was indeed equivalent to a process he had proposed much earlier. Unfortunately this earlier proposal was criticised based on what later turned out to be incorrect data in the literature. However, Judd also pointed out the necessity of introducing a screening factor in the computation that greatly reduced the ability of dynamic coupling mechanism to account for the results. So at this point let us say that there have been important developments that need further consideration. It would not be accurate to say that there is an existing generally accepted solution to the problem.

AMBERGER   (question)

Did you ever observe hypersensitivity in emission or only in absorption spectra?

CARNALL   (answer)

You probably do observe hypersensitivity in emission. If you look
at the right band you will certainly see it.

SINHA   (comment)

I guess Bill is right. We have observed hypersensitivity in the
fluorescence spectra of complexes of Eu(III), Sm(III) and Dy(III)
with various types of organic ligands both in solution and in the
solid state when compared to the intensities of the aquo ions.
There is a general increase in the fluorescence intensity with
complexation, but hypersensitivity, as the name suggests, is a
very special case. As you have seen in my talk that carbonate produced an overall increase of Eu(III) fluorescence intensity by a
factor of about 10 for most peaks but over 100 fold for $^5D_0 \rightarrow {}^7F_2$
transition (Chapter 10, Table 4). This transition is indeed a hypersensitive transition ($\Delta J = 2$) in Eu(III) complexes. While most
ligands are expected to produce hypersensitivity effects with respect to the aquo ions, there may be some ligands for which the
effect may not be observed. It is also very important to compare
the area under the curves (oscillator strength) rather than the
linear increase in the peak heights when evaluating the hypersensitive transitions.

10

FLUORESCENCE SPECTRA AND LIFETIMES OF THE LANTHANIDE AQUO IONS AND THEIR COMPLEXES

Shyama P. Sinha

Hahn-Meitner-Institut, Postfach 390128,
D-1000 Berlin 39, Fed. Rep. Germany

ABSTRACT

High resolution fluorescence spectra of the aquo Sm(III), Eu(III), Gd(III), Tb(III) and Dy(III) ions have been investigated in order to establish the energy levels of the ground state multiplets. The energy levels of the aquo Eu(III) ion have been established using both fluorescence and excitation spectra. The effect of coordination of simple inorganic and organic ligands have been investigated using changes in the fluorescence intensities and lifetimes of the aquo ions. The lanthanides-carbonate systems have received special attention due to the important role of carbonates in the geochemical process. We have observed strong hypersensitive fluorescence transitions for the ions mentioned above and including Tm(III) in the anionic tetracarbonato complexes. Extreme hypersensitivity was observed for the $^5D_0 \rightarrow {}^7F_2$ transition of the Eu allowing us to detect an Eu concentration $\sim 10^{-7}$M in solution. The lifetime of the excited states in the tetracarbonato complexes are Sm($^4G_{5/2}$) 12, Eu($^5D_0$)465, Gd($^6P_{7/2}$)1460, Tb($^5D_4$)1100, Dy($^4F_{9/2}$)11 and Tm($^1G_4$) 7 μsec. Energy transfer from the $^5D_4$ level of Tb to the $^5D_0$ level of Eu in the mixed tetracarbonato system was also observed. It was monitored by measuring the change of lifetimes of Eu and Tb ions in this mixed system. The exchange time is ~ 25 μsec. Mixed fluorocarbonato complex of Eu(III) was selected as a model system for the geologically occuring bastnasite. Indication of the presence of a mono-fluoro-Eu(III)-carbonato complex was obtained.

S. P. Sinha (ed.), Systematics and the Properties of the Lanthanides, 451–500.
Copyright © 1983 by D. Reidel Publishing Company.

I. INTRODUCTION

Trivalent lanthanide ions in their salts and doped crystals usually show sharp line fluorescence originating within the $4f^q$ configurations, when optically excited[1]. In aqueous solution, however, fluorescence only from the middle of the lanthanide series (Sm,Eu,Gd,Tb,Dy) is observed. Contrary to the claims of Stein and Würzberg[2], we are unable to observe any fluorescence from Pr(III) aquo ion even using the sensitive photon counting technique. Tm(III) aquo ion, on the other hand, produced a broad shoulder at ~450nm (22220cm^{-1}), but this is intensified to a nice peak at 21880cm^{-1} in anionic carbonato complex, which we shall discuss later on.

We have recently undertaken a thorough study of the luminescence spectra of these aquo ions in order to investigate the energy levels of the ground state multiplets in particular and the effect of complexation on the intensities and lifetimes of the fluorescent transitions.

Most of the ground state multiplets of the trivalent Sm,Eu, Tb,Dy, and Tm ions occur within zero and 12000 wavenumbers, and this region is not usually available for absorption spectral measurement of the aqueous solutions. Thus fluorescence measurements remain as the only means of obtaining information regarding the ground state multiplets of the lanthanides in solution and in crystalline matrix.

II. FLUORESCENT LANTHANIDE AQUO IONS

We shall first take the case of Eu(III) aquo ion having $4f^6$ electronic configuration. The ground state of Eu(III) is ^7F and this is split into seven components($^7F_0$-$^7F_6$) due to first order spin orbit coupling with J=0-6. The first excited level of Eu(III) is $^5D_0$ situated at ~17300cm^{-1} above the ground level $^7F_0$. In the absorption spectrum of Eu(III) aquo ion only weak lines in the visible and in the ultraviolet regions are observed[3].

Excitation of Eu(III) aquo ion with near UV light results in fluorescence originating not only from the $^5D_0$ level but also from the excited $^5D_1$ level(~19050cm^{-1}). However, the intensities of the $^5D_1 \rightarrow {}^7F_J$ transitions are much lower than those of the $^5D_0 \rightarrow {}^7F_J$ transitions. The excitation spectrum of a 9.28×10^{-2}M Eu(III) aquo ion is shown in Fig.1 together with the assignment of the energy levels. Such assignments are possible as the excitation spectrum is a replica of the absorption spectrum of the lanthanide(III) ions where virtually no Stokes shift occurs due to relatively shielded nature of the f-electrons from the lattice or medium interaction. We have selected the 396nm(25250cm^{-1}) band($^5L_6$) for exciting the Eu(III) aquo ion and using the sensitive

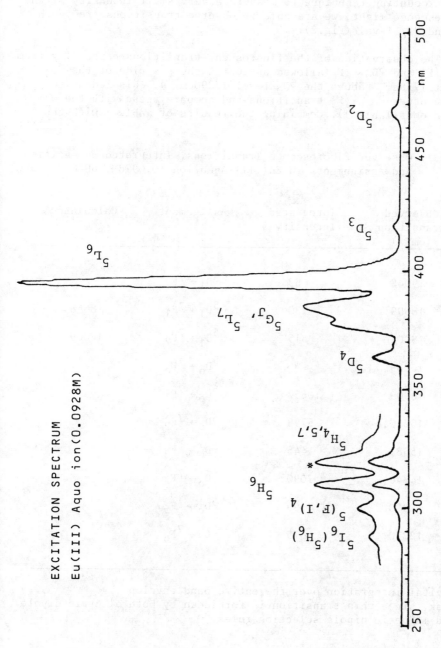

Fig. 1 Ecitation spectrum of 0.0928M Eu(III) aquo ion(pH 0.8) (Em. 616 nm). Peak marked with * is the second order contribution.

photon counting technique to measure a very small intensity of the fluorescence light, we are able to observe transitions from both $^5D_1$ and $^5D_0$ levels(Fig.2).

The observation of the fluorescent transitions especially from $^5D_1 \rightarrow {}^7F_0$ and $^5D_0 \rightarrow {}^7F_0$ allowed us to fix the position of the $^5D_1$ level at $1758cm^{-1}$ above the $^5D_0$ level($17290cm^{-1}$). This led us to calculate other $^5D_1 \rightarrow {}^7F_J$ transitions and compare these with the observed ones(Table 1). Combining the results of Table 1 with the

Table 1. Observed Fluorescence Transitions, Integrated Intensities and Assignments in Eu(III) Aquo Ion (0.0928M, pH 0.8, 19°C)

Observed Transition (cm^{-1})	Integrated Intensity §	Assignment	Calculated (cm^{-1})
19048	83	$^5D_1 \rightarrow {}^7F_0$	
18699	230	$^5D_1 \rightarrow {}^7F_1$	18668
18002	515	$^5D_1 \rightarrow {}^7F_2$	18009
17289	*	$^5D_0 \rightarrow {}^7F_0$	
16909	10652	$^5D_0 \rightarrow {}^7F_1$	
16250	6798	$^5D_0 \rightarrow {}^7F_2$	
15385	445	$^5D_0 \rightarrow {}^7F_3$	
14340	7090	$^5D_0 \rightarrow {}^7F_4$	
13300	99	$^5D_0 \rightarrow {}^7F_5$	
12270	75	$^5D_0 \rightarrow {}^7F_6$	

§ Digital integration over the entire band envelop
* Weak, since this transition is forbidden by both electric dipole and magnetic dipole selection rules.

observed levels in the excitation spectrum a fairly good energy level scheme for Eu(III) aquo ion is constructed (Table 2).

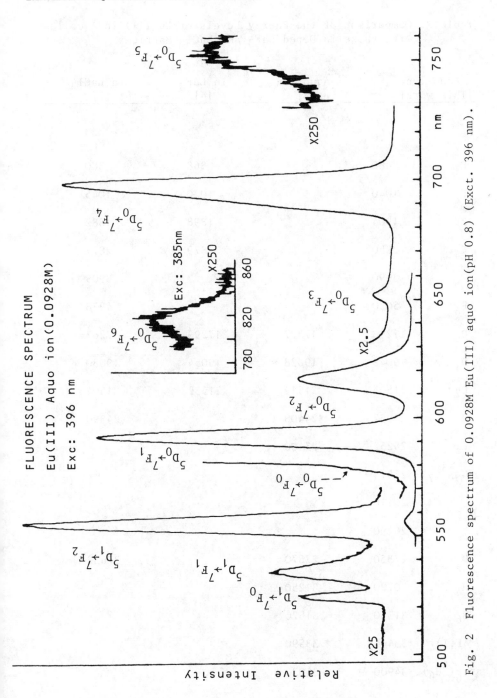

Fig. 2 Fluorescence spectrum of 0.0928M Eu(III) aquo ion(pH 0.8) (Exct. 396 nm).

Table 2. Comparison of the Energy Levels of Eu(III) in Aquo Ion
with those in Doped LaF_3 and $LaCl_3$ Matrices

Aquo ion [This work]		Aquo ion [3]	In LaF_3 [4]	In $LaCl_3$ [5]
^7F_O	0.0			
F_1	380		368	380
F_2	1040		1020	1045
F_3	1904		1898	1882
F_4	2950		2879	2877
F_5	3990			3909
F_6	5020			4978
^5D_O	17290	17277	17293	17267
D_1	19048	19028	19054	19030
D_2	21505	21519	21532	21504
D_3	23952	24408		24390
$^5L_6$	25252	25400		
^5G_J, $^5L_7$	25974			
	26110			
	26490			
$^5D_4$	27550	27670		27632
$^5H_{4,5,7}$	30720	31250		
$^5H_6$	31350	31520		
$^5(F,I)_4$	33445	33590		
$^5I_6(^5H_6)$	34900	35030		

Our results agree well with the literature data on the aquo ion[3] especially for the higher energy levels and those found for Eu(III) in LaF$_3$[4] and in LaCl$_3$[5]. We shall remark here that the nephelauxetic effect[1] is very small in going from the aquo ion to Eu(III) doped LaF$_3$ and LaCl$_3$. It has been emphasized by the present author several times in the past that except in the case of Pr(III) and Nd(III)[6-10] nephelauxetic shift may not have any useful meaning.

The levels used for excitation of the lanthanide aquo ions and the observed fluorescent transitions are shown in Fig.3 and the assignments for Sm(III), Gd(III), Tb(III), and Dy(III) ions are given in Table 3. These assignments agree with those that would be expected from the absorption spectral measurements[3].

Table 3. Observed Fluorescent Transitions in Sm(III), Gd(III), Tb(III), and Dy(III) Aquo Ions

Transition	Energy(cm^{-1})	Transition	Energy(cm^{-1})
Sm(III)[Exc. 406nm]		Gd(III)[Exc. 272nm]	
$^6G_{5/2} \rightarrow {}^6H_{5/2}$	17840	$^6P_{5/2} \rightarrow {}^8S_{7/2}$	32895
$\rightarrow {}^6H_{7/2}$	16792	$^6P_{7/2} \rightarrow {}^8S_{7/2}$	32150
	16666sh		
$\rightarrow {}^6H_{9/2}$	15504		
Tb(III)[Exc. 380nm]		Dy(III)[Exc. 354nm]	
$^5D_4 \rightarrow {}^7F_6$	20336	$^4F_{9/2} \rightarrow {}^6H_{15/2}$	20877
$\rightarrow {}^7F_5$	18350		20534sh
$\rightarrow {}^7F_4$	17035	$\rightarrow {}^6H_{13/2}$	17422
$\rightarrow {}^7F_3$	16050		

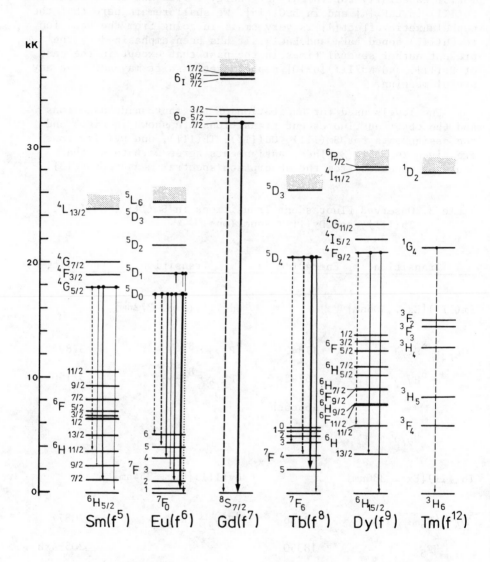

Fig. 3 Energy levels and the observed fluorescence transitions
in several trivalent lanthanide aquo ions. The levels used for
excitation are shown as broad ▓▓▓▓ bands.

III. EFFECT OF COMPLEXATION

   Like the absorption bands, the fluorescent transitions of the lanthanide aquo ions also exhibit (a) very small red shifts and (b) change in intensity due to complex formation. However, one or two of the transitions of the aquo ions show remarkable increase in intensity (hypersensitive transitions), like the case of the anionic carbonato $[M(CO_3)_4^{5-}]$ complex of Eu(III), where the $^5D_0 \rightarrow {}^7F_2$ transition shows (Fig.4) an increase of ~ 100 fold over the aquo ion (see later). Beside these, a drastic change in lifetime of the fluorescent transitions is observed due to complexation, which we shall discuss in section IV.

   The carbonato complexes of the lanthanides are not only chemically interesting, but carbonatites form an important class of ores beside the silicates. The distribution of the lanthanides in carbonatites is different from that in the silicates. Carbonatites show preference for the light lanthanides[11] and the percentage of the lanthanides in carbonatite is relatively high[12].

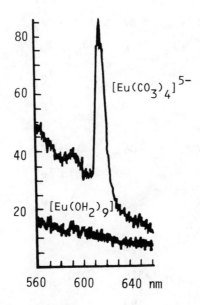

Fig. 4   Fluorescence spectrum of the $[Eu(CO_3)_4]^{5-}$ complex (Eu = 1.29 x 10^{-6}M) compared to the Eu(III) aquo ion of same concentration and exhibiting extreme hypersensitivity of the 620 nm transition (Exct. 396 nm). Notice a very small intensity of the 590 nm transition still observable at this low concentration of the aquo ion.

We have recently undertaken a systematic investigation of the spectroscopic behaviour and complex formation characteristics of the lanthanides-carbonate systems[13]. A comparison of the integrated intensities of the aquo ions and their anionic $[M(CO_3)_4^{5-}]$ complexes is made in Table 4. Beside the remarkable increase in

Table 4. A Comparison of the Integrated Intensities of Aquo
Sm(III), Eu(III), Gd(III), and Dy(III) Ions and their
Anionic Carbonato Complexes

Transition	Integrated Intensity Aquo ion	Complex	Transition	Integrated Intensity Aquo ion	Complex
Sm(III)			Eu(III)		
$^4G_{5/2} \rightarrow ^6H_{5/2}$	1	18	$^5D_1 \rightarrow ^7F_0$	2	–
$\rightarrow ^6H_{7/2}$	11	63	$\rightarrow ^7F_1$	3	42
$\rightarrow ^6H_{9/2}$	2	87	$\rightarrow ^7F_2$	17	13
$\rightarrow ^6H_{11/2}$	–	6	$^5D_0 \rightarrow ^7F_0$	–	205
			$\rightarrow ^7F_1$	337	2347
Gd(III)			$\rightarrow ^7F_2$	201	21101
$^6P_{7/2} \rightarrow ^8S_{7/2}$	83	804	$\rightarrow ^7F_3$	7	417
			$\rightarrow ^7F_4$	269	1868
Dy(III)			$\rightarrow ^7F_5$	2	48
$^4F_{9/2} \rightarrow ^6H_{15/2}$	344	1480			
$\rightarrow ^6H_{13/2}$	327	4556			
$\rightarrow ^6H_{11/2}$	–	190			
$\rightarrow ^6H_{9/2}$	–	60			

[Sm] = 1.0×10^{-3}M, [Eu] = 1.29×10^{-3}M, [Gd] = 1.08×10^{-3}M, [Dy] = 1.01×10^{-3}M. The [CO_3] for all lanthanides was 2.7M, except for Eu where it was 3.01M.
The anionic complex is most probably $[M(CO_3)_4]^{5-}$ type.

intensity (Fig.4) of the $^5D_0 \rightarrow {}^7F_2$ hypersensitive transition of
Eu(III), we have observed an increase in the intensities of
$^4G_{5/2} \rightarrow {}^6H_{7/2}$ of Sm(III)(6 fold), $^6P_{7/2} \rightarrow {}^8S_{7/2}$ of Gd(III)(10 fold),
$^5D_4 \rightarrow {}^7F_5$ of Tb(III)(7 fold), and $^4F_{9/2} \rightarrow {}^6H_{13/2}$ of Dy(III)(14 fold)
in anionic carbonato complexes (Fig.5). It is difficult to give
an estimate of the intensity increase for Tm(III) anionic carbonato complex, as we are unable to observe any well-formed fluorescent peak for Tm(III) aquo ion (Fig.6). We notice here that all
transitions which exhibit an increase in intensity do not rigorously follow the hypersensitive selection rule $\Delta J = 2$[14].

As mentioned earlier carbonato complexes are chemically interesting. Sherry and Marinsky[15] described an ion exchange
method of separating lanthanides using carbonate as eluant. More
recently, Hobart et al.[16] were able to oxidize Pr(III) to Pr(IV)
in 5.5M K_2CO_3 solution. This is possibly the first evidence of the
presence of Pr(IV) in an aqueous system stabilized by complex formation. In sea water, carbonate plays an important role in complexation of cerium and uranium possibly in their highest oxidation
state of four and six respectively[17], beside being a complexing
agent for other metal ions in sea water.

Although we have normally used a large excess of carbonate
in our spectrometric work, it is found that a ratio of 1:100 for
the lanthanides to carbonate is sufficient to dissolve the carbonate precipitates to form the soluble anionic complex. Such a ratio
(or even a higher factor) is not uncommon for trace elements (lanthanides) to carbonate in geological and biological fluids.

It is interesting that a simple ligand like carbonate is capable of producing such a remarkable hypersensitivity for
$^5D_0 \rightarrow {}^7F_2$ transition of Eu(III) (Fig.4). This criterion was exploi-
ted by Sinha[18] to detect minute amount ($\sim 10^{-7}$M) of Eu(III) as
anionic carbonato $[Eu(CO_3)_4^{5-}]$ complex. One word of caution here,
such commendable low limit of detection is reached using a double
monochromator and a photon counting system employing a cooled de-
tector (Spex Fluorolog) and the detection limit is system (spectrometer) dependent. However, it is hoped that the $[Eu(CO_3)_4^{5-}]$ system
will find its analytical use. The extreme hypersensitivity of the
carbonate and other systems is possibly one of the reasons why
many minerals lend themselves[19] to spectrofluorometric detection
of several lanthanides when these are present only in the ppm range.

In order to compare the effect of carbonate ($O-CO_2^{2-}$) ligand,
we have systematically studied the closely related ligands like
formate ($H-COO^-$) and acetate (H_3C-COO^-) and several inorganic ligands complexes of Eu(III) and Gd(III). Table 5 and 6 summarize
the results. It would be seen that there is hardly any change in
spectrum profile and in intensity (Fig.7) with the addition of
inorganic ligands like F^-(5×10^{-4}M), Cl^-(3.08M), Br^-(3.02M) to a
solution of Eu(III) aquo ion (1.12×10^{-2}M).

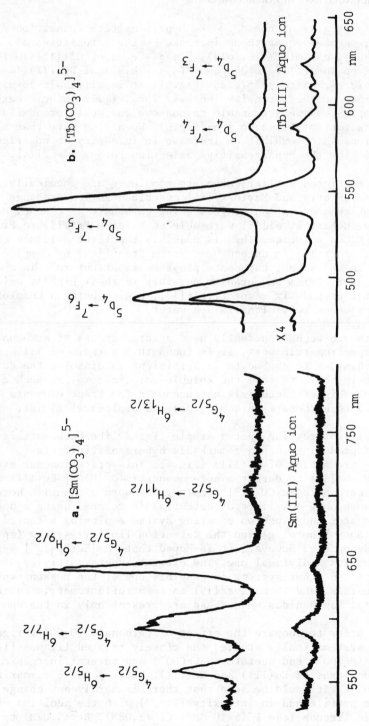

a. $[Sm(CO_3)_4]^{5-}$

Sm(III) Aquo ion

$^4G_{5/2} \leftarrow {}^6H_{13/2}$

$^4G_{5/2} \leftarrow {}^6H_{11/2}$

$^4G_{5/2} \leftarrow {}^6H_{9/2}$

$^4G_{5/2} \leftarrow {}^6H_{7/2}$

$^4G_{5/2} \leftarrow {}^6H_{5/2}$

b. $[Tb(CO_3)_4]^{5-}$

Tb(III) Aquo ion

$^5D_4 \leftarrow {}^7F_3$

$^5D_4 \leftarrow {}^7F_4$

$^5D_4 \leftarrow {}^7F_5$

$^5D_4 \leftarrow {}^7F_6$

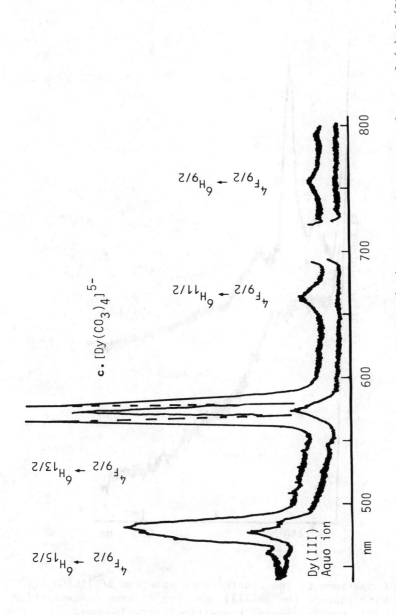

Fig. 5 Comparison of the fluorescence spectra of the anionic tetracarbonato complexes of (a) Sm(III) (Exct. 406 nm) (b) Tb(III) (Exct. 380 nm) and (c) Dy(III) (Exct. 354 nm) with their respective aquo ions at the same concentration. Note that the intensity scale of Tb(III) aquo ion is multiplied by a factor of 4.

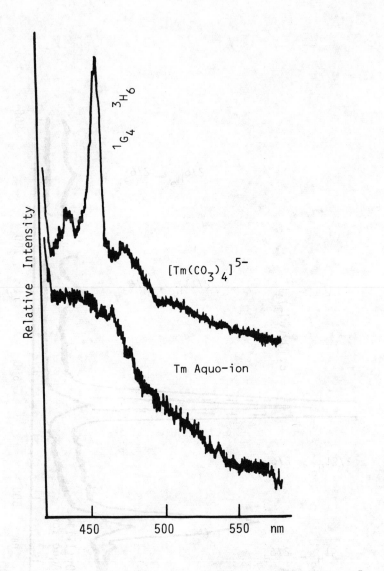

Fig. 6 Comparison of the fluorescence spectrum of $[Tm(CO_3)_4]^{5-}$ complex with that of the Tm(III) aquo ion of same concentration (1.15×10^{-3}M) showing $^1G_4 \rightarrow {}^3H_6$ transition(Exct. 360 nm).

Table 5. Change in the Fluorescence Intensity of Eu(III) due to
Complex Formation at Room Temperature

System	pH	$^5D_o \rightarrow {}^7F_1$	$^5D_o \rightarrow {}^7F_2$
Aquo ion	3.57	41.1	9.4
Fluoride	4.02	42.1	9.7
Chloride	3.25	42.7	11.4
Bromide	3.16	43.2	11.7
Iodide*	–	13.0	4.0
Nitrate	3.64	46.1	35.6
Thiocyanate	3.64	27.0	37.0
Tetra-Carbonate	11.83	45.0	264.0
Formate	7.68	46.9	78.5
Acetate	9.08	46.8	153.6

* Freshly prepared solution at 8^oC
[Eu] = $1.12 \times 10^{-2}M$, all ligand concentration 3M except for fluoride,
[F] = $10^{-4}M$.
Intensities are expressed as peak heights.

It is well known from potentiometric and other physico-chemical
measurements[20] that Cl^- and Br^- ligands form very weak complexes
with the lanthanides (Eu-Cl, $\log K_1 = 0.80$; Eu-Br, $\log K_1 = 0.58$[20]),
whereas the complex formation with F^- ion is rather strong (Eu-F,
$\log K_1 = 3.19$)[21]. It was difficult to increase the F^- ion concentration to any comparable degree as that of Cl^- or Br^- in our
system without precipitating the metal.

However, addition of NO_3^-(3.07M), SO_4^{2-}(satd), SCN^-(3.08M), and
CO_3^{2-}($\sim$3M) (Fig.4,8) drastically altered the profile and intensities
of the fluorescent transitions, especially the ratio of $^5D_o \rightarrow {}^7F_1$
and $^5D_o \rightarrow {}^7F_2$ transitions.

Contrary to the halides, ligands like NO_3, SO_4, and SCN form

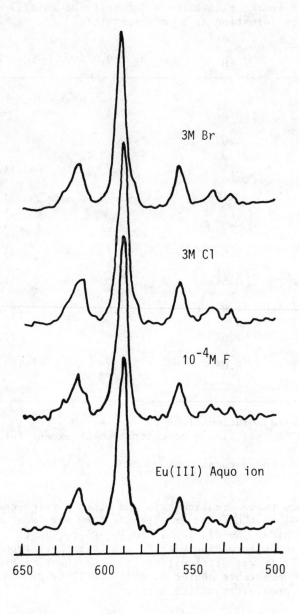

Fig. 7 Effect of the addition of halides on the fluorescence
spectrum of Eu(III) aquo ion(0.011M)(Exct. 392 nm).

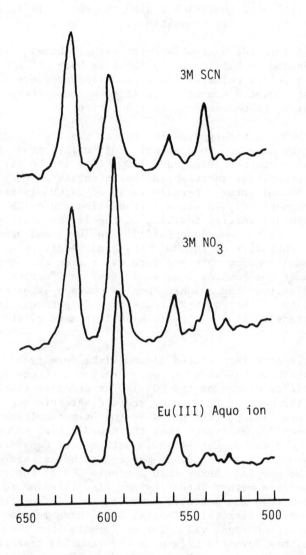

Fig. 8 Change in the spectrum profile of Eu(III) aquo ion(0.011M) on addition of nitrate and thiocyanate(Exct. 392 nm).

stronger complexes with the lanthanides (Eu-NO$_3$, logK$_1$=2.0; Eu-SO$_4$, logK$_1$=1.54; Eu-NCS, logK$_1$=5.05[20]). The effect of strong complexation by these ligands is evident from Table 5, which shows not only an increase in the intensity of the $^5D_0 \rightarrow {}^7F_2$ transition of Eu(III) but also the change in intensity ratio of $^5D_0 \rightarrow {}^7F_2/^5D_0 \rightarrow {}^7F_1$ transitions.

The Eu(III)-iodide system is interesting. The mixture (1.12x10^{-2}M Eu and 3M KI) is stable about an hour at $\sim$ 8°C but slowly becomes yellow coloured. This oxidation proceeds at a much faster rate at elevated temperature. At room temperature ($\sim$20°C) it was impossible to obtain any clear solution.

During the preliminary experiments, it was found that gradual addition of carbonate ion to a solution of Eu(III) produced decrease in the intensity of the $\sim$ 592nm band ($^5D_0 \rightarrow {}^7F_1$) in the first instance with concomitant increase in the intensity of the $^5D_0 \rightarrow {}^7F_2$ transition ($\sim$620nm) after a certain ratio of Eu/CO$_3$ is reached. The $^5D_0 \rightarrow {}^7F_1$ is magnetic dipole allowed transition and quite sensitive to the environment. Initial addition of the ligand destroys the centro-symmetry of the eneaaquo-Eu(III) ion (D$_{3h}$) and hence the decrease in the intensity of the $^5D_0 \rightarrow {}^7F_1$ band. Addition of more ligand (carbonate) removes more and more water from the aquo ion with the formation of the complex ion. Thus after the "carbonato" precipitate is dissolved the anionic complex shows a stronger increase in the intensity of the $^5D_0 \rightarrow {}^7F_2$ transition (hypersensitivity) although the intensity of $^5D_0 \rightarrow {}^7F_1$ is also increased relative to the aquo ion.

A comparison of the related ligands like formate and acetate with carbonate show that at a 3M concentration of these ligands, the hypersensitive effect of the CO$_3$ is far stronger (Table 5) than that in the case of formate or acetate. At this high concentration of these ligands most probably the tris-complexes are present in solution[1]. We expect that the coordinating chromophores are the bidentate carboxylate ions in both cases. Considering that the methyl group (-CH$_3$) in acetate may act as better sink for drainage of the pumping energy than a single C-H of the formate, one would predict a higher intensity and also lifetime for the formate complex than the acetate. However, this is not the case (Table 5). In my earlier talk (Chapter 3), I have mentioned that the deviation of the logK$_1$ value for Eu-formate from the Inclined W Systematics, may have its origin in the reducing character of the ligand. The low intensity of Eu-formate $^5D_0 \rightarrow {}^7F_2$ transition is most probably due to partial reduction of Eu(III) and/or due to charge transfer reaction in this system. The anionic carbonato complex has a probable composition [Eu(CO$_3$)$_4$]$^{5-}$ and it possibly has an octacoordinated structure. The high hypersensitivity of the $^5D_0 \rightarrow {}^7F_2$ band may be taken as an indication that no water (or hydroxide ion) is bound to this tetracarbonato-Eu(III) complex.

A more convincing proof came from the study of the lifetime measurements and the energy transfer studies on Eu-Tb-tetracarbonate system (see Section V).

In the Gd-series, however, the change in fluorescence intensity follows the normal expected trend i.e., $CO_3 >> HCOO > H_3CCOO$ (Table 6). In the case of Gd-formate system, reduction of the cation and/or the charge transfer of the type mentioned for Eu-formate would not usually take place. Of the inorganic ions, we have investigated Cl^-, Br^-, NO_3^-, SCN^- ions. Chloride and bromide ions did not produce any significant change in fluorescence intensity of Gd(III). In contrast to these, both nitrate and thiocyanate ions quenched the fluorescence of Gd ion to such an extent that no fluorescence spectrum could be recorded (see Section IVb).

IV. LIFETIMES OF AQUO IONS AND COMPLEXES

The change in fluorescence intensity due to complex formation does not provide enough information on the nature of the species present in solution. However, a second parameter, the lifetime of the system (eq.1), often provides additional information on the

Table 6. Change in Fluorescence Intensity of Gd(III) due to
    Complex Formation at Room Temperature

System	pH	$^6P_{7/2} \rightarrow {}^8S_{7/2}'$
Aquo ion	5.93	36.9
Chloride	5.53	34.8
Bromide	5.60	31.7
Tetra-Carbonate	12.58	121.3
Formate	7.66	69.4
Acetate	8.85	64.4

$[Gd] = 1.08 \times 10^{-2}M$, all ligand concentration 3M.
Nitrate and thiocyanate totally quenched Gd(III) fluorescence.
Intensities are expressed as peak heights.

number of species present in solution.

$$\ln I = \ln I_0 - (1/\tau)t \qquad (1)$$

The lifetime(τ) is measured from the decay curve by plotting log I vs time (t). If only one species is present in solution the decay curve shows an unique slope. When more than one fluorescing species is present in solution, the decay curve may be resolved into a number of slopes characteristics of the species.

Lifetimes of several lanthanide aquo ions have been measured by various workers[2,21-23]. We have recently been engaged in measuring the time resolved fluorescence spectra and lifetimes of the lanthanide aquo ions and their complexes at different tempe- rature and pH range using the Perkin Elmer model LS 5 emission spectrophotometer. Our measured values for the aquo ions are com- pared with the best value from the literature in Table 7. We found large discrepancy between our value for Gd(III) aquo ion and that quoated in the literature (2 msec) and usually attributed to Kon- drateva[24].

Table 7. Comparison of Observed Lifetime(τ) of the Aquo Ions with Literature Data

Aquo ion	pH	Lifetime (μsec)	
		This work	Literature
Sm (0.1M)	2.22	8	2.3 [2]
Eu (0.1M)	2.0	110	110 [21]
Gd (0.01M)	1.58	203	2000 [2,24]
Tb (0.01M)	6.3	425	400 [2]
Dy (0.01M)	5.70	7	2.4 [2]

We have repeated our measurements on Gd(III) aquo ion and checked the instrument by rerunning the Eu(III) aquo ion sample, only to confirm our value of $\tau = 200\mu$s for Gd(III) aquo ion at pH 1.58 and 22°C. This τ value almost doubled on increasing the pH to 5.93 (23°C) (Table 7). At the present time we do not have

any ready explanation why our values are lower than those of
Kondrateva. However, she commented on the temperature dependence of the lifetime for Gd(III) aquo ion and an exponential
decrease of τ between 2 to 95°C[24].

Our measured values for Sm(III) and Dy(III) aquo ions are
also somewhat higher (Table 7) than those quoted by Stein and Würzberg[2]. I feel that our values are much closer to reality, because
of the fact that one is able to obtain reasonable good fluorescence
spectra for these two ions in aqueous solution. If the lifetime
of these ions are as short as 2μsec as claimed by Stein and Würzberg[2], I would not expect these ions to show any fluorescence
in aqueous solution. Fig. 9 shows the time resolved spectrum of
Sm(III) aquo ion (pH 2.22) on excitation with 398nm radiation.
The lifetimes of $^4G_{5/2}$ level of Sm(III) and $^4F_{9/2}$ level of Dy(III)
are about the same as that of the energy transferring level $^5D_1$
of Eu(III) aquo ion (τ=9μsec).

Nature of the Eu(III) Aquo Ion

The decay curves of the aquo ions follow the classical first
order exponential decay law (eq. 1). It has been pointed out earlier that the presence of multiple species in solution is indicated
by the change in slope of the decay curve. The fluorescence decay
of the aquo ions has been measured over a time span of several
times τ, and we have observed first order exponential decay characterizing a single species for each lanthanide ion at our experimental conditions (22°C, pH 2.0). It is usually assumed that
most lanthanide ions in aqueous solution are nonacoordinated,
$[M(OH_2)_9]$. This is especially true for the isolated aquo ions in
crystalline matrices like ethylsulphate and bromate[25].

As the lifetime is characteristic of a species, we felt that
it would be interesting to compare the lifetimes of isolated
$[Eu(OH_2)_9]$ ion in different matrices with the aquo Eu(III) ion
in order to gain further information on the nature of the Eu(III)
species in aqueous solution.

Both in ethylsulphate and bromate, the lanthanide aquo ions
are nonacoordinated. However, these structures contain extensive
hydrogen bond system[25]. This has an effect of lowering the symmetry from hexagonal space group. But this effect is so small that
such lowering of symmetry cannot be detected at room temperature
by X-ray diffraction measurements. However, Hellwege and coworkers[26-29] observed the effect of lowering of the hexagonal
symmetry for $Eu(OH_2)_9(BrO_3)_3$ from the optical spectral measurements even at room temperature. We have endeavoured to obtain
isolated $[Eu(OH_2)_9]$ ion in another crystalline matrix, and with
the help of Prof. Léa Zimmer of the University of São Paulo synthesized $[Eu(OH_2)_9](CF_3SO_3)_3$ complex. Recent crystal structure

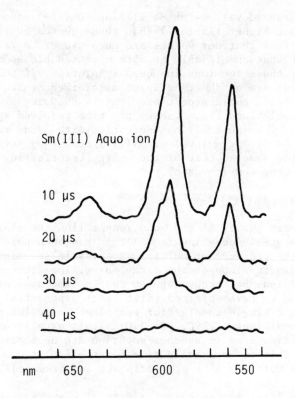

Fig. 9 Time resolved fluorescence spectra of 0.1M Sm(III) aquo ion at pH 2.22 excited with 398 nm radiation. The spectra were recorded after 10, 20, 30 and 40 μsec delay from the begining of the pulse to the begining of observations. The lifetime was calculated from the decay of the 596 nm transition of Sm(III) ion.

determination of Nd and Ho analogues[30] showed the crystals to be hexagonal (P63) and contain $[M(OH_2)_9]$ units with little hydrogen bonding. The Eu(III) complex is supposed to be isostructural.

A comparison of the lifetime (τ) of the ^5D_o level of $[Eu(OH_2)_9]$ ion in trifluoromethanesulphonate ($\tau=102\mu sec$) and bromate ($\tau=119\mu sec$)[31] crystals with the aquo ion ($\tau=110\mu sec$) may be made (Table 8). Allowing for the hydrogen bonding and crystalline matrix effect, these values are remarkably close enough to justify the presence of $[Eu(OH_2)_9]$ moiety in solution at a low pH. If we now compare these values with $Eu_2(SO_4)_3.8H_2O$ crystals ($\tau=187\mu sec$)[32], which at one time were thought to contain nona-coordinated Eu(III) ion, we find the value of τ for hydrated sulphate is much higher. This high τ value is indicative of the presence of a species other than the nonacoordinated aquo ion and indeed recent X-ray structural analyses of several lanthanide sulphate octahydrates (M=Pr, Sm, Yb) including $Am_2(SO_4)_3.8H_2O$ showed the presence of octacoordinated $[MO_4(OH_2)_4]$ species[33-36], where the anion (SO_4^{2-}) participate in forming the coordination polyhedron which is intermediate between an antiprism and a dodecahedron. However, a solution of $Eu_2(SO_4)_3$ in water ($\sim 0.1M$) has a τ value of $140\mu sec$[21]. Kropp and Windsor[21] did not mention the pH of their europium sulphate solution. Dissociation of coordinated sulphate anions and partial replacement by water must have taken place in their solution of europium sulphate. One can readily appreciate the sensitiveness of the τ parameter with change in environment (complexation). This sensitivity of τ becomes more apparent by simply changing ordinary water for heavy water (D_2O) (Table 9). The τ_D/τ_H ratio is proportional to the percentage of D_2O present in the mixed solvent. In H_2O solution the main deactivation process is through coupling of OH vibrations. Heller[37] pointed out that the change in vibrational quantum necessary to bridge the energy gap of the fluorescence transition has a large effect on the lifetime. Taking the ν_3 of free water to be $\sim 3450 cm^{-1}$ (the stretching frequency of M-O bond in aquo ion has a value, which is at least ten times lower than ν_3 of free water), we have the following factors for the H_2O solutions of some of the lanthanides:

Transition Energy (cm^{-1})	Factor	Transition Energy (cm^{-1})	Factor
Sm 7500	2.2	Tb 14700	4.3
Eu 12300	3.6	Dy 7800	2.3
Gd 32100	9.3	Tm 6200	1.8

symmetric stretching (ν_1) 3219 cm^{-1}, bending (ν_2) 1627 cm^{-1}, and antisymmetric stretching (ν_3) 3445 cm^{-1} for liquid water, J. H. Hibben, J. Chem. Phys. 5, 166 (1937). Transition energy corresponds to the gap between the fluorescing level and the highest multiplet of the ground state.

Complexation has the same general effect as D_2O, i.e., on replacing the coordinated water by other ligands an increase in
lifetime value is usually observed.

Table 8. Comparison of Eu(III) Lifetimes (τ) in Solid Matrices
with that in Aquo Ion

Complex	Lifetime(μsec)	Ref.
Aquo ion	110	§
$Eu(OH_2)_9(CF_3SO_3)_3$	102	§
$Eu(OH_2)_9(BrO_3)_3$	119	[31]
$[Eu(OH_2)_6Cl_2]Cl$	120	§
$Eu_2(SO_4)_3 \cdot 8H_2O$	187	[32]
$Eu_2(CO_3)_3 \cdot 0.6H_2O$	266	§
EuF_3 (orthorhombic)	540,1600	§
EuOF (rhombohydral)*	32	§

* Sample obtained after heating EuF_3 at 850°C in air.
§ This work

Table 9. Change in Lifetime (τ) of the Aquo Ions on Deuterium
Substitution[2]

M(III)	τ_{H_2O}	τ_{D_2O}	M(III)	τ_{H_2O}	τ_{D_2O}
Sm	2.27	53.6	Eu	110	4000
Dy	2.35	38.	Tb	400	3880

Before leaving this section, we would like to mention that
the lifetime of Tb(III)-ethylsulphate ($Tb(C_2H_5SO_4)_3.9H_2O$)
(410μsec[31]) is almost the same as that of Tb(III) aquo ion in
solution (425μsec, Table 7), indicating the presence of $[Tb(OH_2)_9]$
moiety in aqueous solution. As expected, the lifetime of octacoordinated Tb(III) in $Tb_2(SO_4)_3.8H_2O$ crystals is much higher
(τ=713μsec)[38] agreeing with the general concept presented for
Eu(III) above.

V.   CHANGE IN LIFETIME DUE TO COMPLEX FORMATION

a)   Eu(III) Complexes

Beside the argument on the relation of the vibrational quanta
and the energy gap presented above, the present author is of the
opinion that kinetic effect could be an important factor. By kinetic effect, we mean here the exchange of "free" bulk water with the
coordinated ones. In the immediate neighbourhood of the lanthanide
ion the water molecules are "organized" with the oxygen dipoles
pointing towards the metal ion. These organized water molecules
probably from a sort of distorted tricapped prism around Eu and
Tb (where we are sure of the presence of eneaquoion, and may be
true for Sm, Gd, and Dy ions as well) describing the primary coordination spere. Such organization falls off rapidly with distance
until the structure of bulk water takes over. This is depicted
pictorially in Fig.10. The water exchange rate is very fast indeed,
although the rate decreases from La to Lu in the lanthanide series.
This exchange mechanism together with the rotation and of course
the vibration of the ligands (water) constitute an effective "sink"
for the energy and deactivation of the excited states causing shorter
lifetime.

Interaction between a ligand (hydrated) and the aquo cation
with the formation of a complex (Fig.10) requires removal of bonded
water molecule from the first coordination spere of the cation.
With increasing ligand concentration, the aquo cation will tend to
loose more and more bonded water (in a sense will be "dehydrated")
until the highest order complex $[M(OH_2)_{9-X}(L)_X]$ (X = maximum number
of attached ligand) predominates. Other factors being the same,
the complexed species will be more or less shielded by the ligands
and will have less interaction with the surrounding (bulk water)
with concomitant increase in lifetime compared to the aquo cation.

We have earlier seen that $\sim$3MCl^- and Br^- ions have virtually
no effect (Table 5) on the intensity and the spectrum of Eu(III)
ion. The same conclusion may be drawn by comparing the τ values
of Cl (110μsec) and Br (111μsec) (Table 10) with the aquo ion.
We do not believe that at such a high halide concentration (3M)
no complexation with Eu(III) has taken place. But the complex
formed may be a solvent-separated -ligand -complex, otherwise

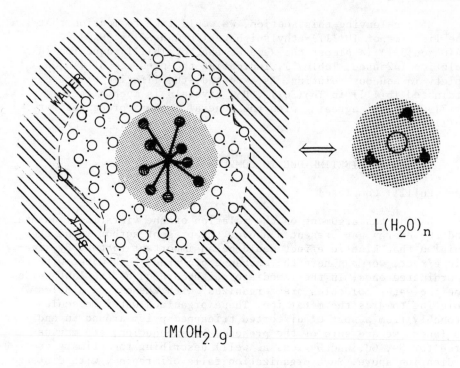

$[M(OH_2)_9]$

Fig. 10 Pictorial depiction of a hydrated lanthanide(III) aquo ion and its interaction with a ligand (hydrated).

an increase in Eu(III) lifetime is expected. When the Cl^-ions are directly bound to the Eu(III) ion as in the case of $[EuCl_2(OH_2)_6]^+$ species, which has a distorted square antiprismatic structure (Eu–Cl = 2.76, Eu–OH_2 = 2.37–2.52Å), an increase of lifetime to 120μsec is observed (Table 8). This τ value is in good agreement with that observed by Dieke and Hall[31] and Bonrath et al[38]. In our model the Eu(III) chloro or bromo complexes in aqueous solution may be visualized as having one or two layers of water molecule between Eu(III) and the halides. In contrast to the Eu(III) system addition of 3M Cl^- to Gd(III) aquo ion increased the lifetime from 203μsec to 660μsec (Table 11) indicating considerable metal–ligand interaction. Considering lifetime criteria, we can safely conclude that the time average symmetry of the chloro complex (in 3M halide) is not the same for Eu(III) and Gd(III).

In the case of fluoro complex, we have observed a decay curve (Fig.11) having two slopes, although the inflection is very weak. From the slopes we have calculated two different lifetimes τ_1=110μsec, $\tau\tau_2$=114μsec, corresponding to aquo ion and the fluoro complex respectively. The polarizability of F^- ion is probably strong enough to replace a part of the coordinated water molecule

Table 10. Change in Lifetime of Eu(III)(^5D_o) on Complexation

System	Lifetime(μsec)	τ_c/τ_{aq}
Aquo ion	110	1.00
Fluoride	111,114	1.01,1.04
Chloride	110	1.00
Bromide	111	1.01
Iodide*	52*	0.47
Nitrate	127	1.56
Thiocyanate	62	0.56
Sulphate	72+	0.66
	140§	1.27
Tetra-Carbonate	465	4.23
Formate	235	2.14
Acetate	337	3.06
	170§	1.55

* Freshly prepared solution at ∿8°C. The low value of τ indicates decomposition.
+ For a mixture of 0.01MEu(III) in saturated K_2SO_4 solution.
§ For ∿0.1M solution[21].
  pH and ligand concentration identical with Table 5.

from the first coordination sphere. It is a pity that a wide range F^- ion concentration dependence study was not possible without precipitating a part of the Eu(III) aquo ion. We take the observed τ value of 114μsec as a reasonable one for a monofluoro complex of Eu(III)(Eu=1.12x10^{-2}M, F^-=5x10^{-4}M, pH=4.02).

In this connection it was interesting to study the fluorescence decay of solid EuF$_3$. A sample of EuF$_3$ (99.5%), purchased from Alfa Product, Massachusetts, USA was used. The decay curve exhibits

Table 11. Change in Lifetime(τ) of Gd(III)($^6P_{7/2}$) on Complexation

System	Lifetime(μsec)	τ_c/τ_{aq}
Aquo ion		
pH 1.58	203	1.00
pH 5.93	419	2.06
Chloride	660	3.25
Bromide	205	1.01
Tetra-Carbonate	1460	7.19
Formate	1764	8.69
Acetate	1380	6.80
[Gd(OH$_2$)$_6$Cl$_2$]Cl solid		
313nm peak	7200	35.96
350nm peak	7700	37.44

pH and ligand concentration identical with Table 6.

two different slopes (Fig.12) yielding lifetime values (^5D_o) of 540 and 1600μsec, showing the presence of two different Eu(III) sites in the solid. Fluorides of Sm and Eu could be obtained both as hexagonal (P$\bar{3}$c1) and as orthorhombic (Pnma) forms [39,40]. Powder X-ray pattern of our sample agreed closely with the ortho-rhombic form of EuF$_3$[39]. We expected to convert this orthorhombic EuF$_3$ to hexagonal form by heat treatment[41] and accordingly heated the sample to 850°C and then quenched it to 18°C. The lifetime of the ^5D_o level drastically reduced to a value of 32μsec (Fig.12) after this heat treatment. The X-ray powder diagram exhibited fewer lines, which corresponded to the rhombohedral (R$\bar{3}$m) modification of EuOF[42]. It is rather unfortunate that our EuF$_3$ sample got oxidized during heating but it again demonstrates the extreme sen-sitiveness of the lifetime parameter with composition and site symmetry.

We have earlier seen that both nitrate and thiocyanate ligands produce an increase of ~ 4 times the intensity of the hypersensitive $^5D_0 \to {}^7F_2$ transition (Table 5). Both of these ligands act as rather strong complexing agents[20], and hence we would generally expect an increase in lifetime of the $^5D_0$ level of Eu(III). However, nitrate has an extra advantage of being able to behave as a bidentate ligand capable of removing more than one water molecule from the first coordination sphere. In a 3M KNO_3 solution (Eu:NO_3=1:300), we measured a τ value of 127µsec, comparable to the τ value of the solid state hydrated trichloride (Table 8). Upto a concentration of 0.1M NO_3^- the lifetime value ($\sim$116µsec)[21,43] is not much higher than the aquo ion value and comparable to the monofluoro complex mentioned earlier. However, Haas and Stein[43] reported a τ value of 127µsec for solutions containing 0.5-1M Eu(NO_3)$_3$. The τ value

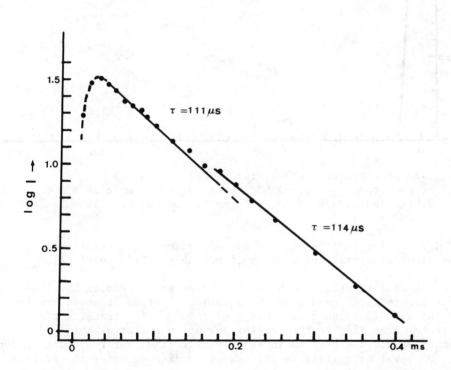

Fig. 11 Fluorescence decay curve of 0.0112M Eu(III) in the presence of 5×10^{-4}M[F^-] at pH 4.02. The excitation frequency was 392nm and the decay was measured at 593nm.

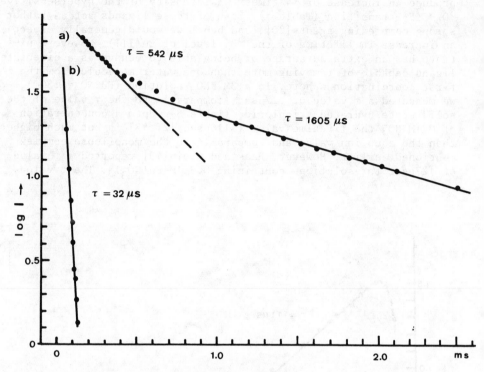

Fig. 12   Fluorescence decay curve of solid EuF$_3$
(a) at room temperature (b) after heating the sample at 850°C
for 24hrs. and quenching to 18°C (Exct.   396nm, Em.   594nm).

of 127μsec for Eu(III)-nitrate system indicate the presence of
higher order nitrate complex (most probably a tris complex).

It is interesting to note that thiocyanate complex has lower
lifetime (62μsec) than even the aquo ion, whereas it shows rough-
ly the same increase in intensity of the $^5D_o \rightarrow\, ^7F_2$ transition as the
nitrate. According to the complex formation criteria mentioned
above, we would expect an increase in lifetime. For the deactivation
of ^5D_o level of Eu(III) in thiocyanate medium we offer the following mechanism:

(1) coupling of the excited state of the complex (Eu-NCS) produced
    due to charge transfer between thiocyanate and Eu(III)
(2) presence of a deactivating heavy atom (sulphur)
(3) vibrational coupling

Such deactivation mechanism may have played a part in deactivating

(τ=72μsec) a solution (0.0112M) of Eu(III) present in a saturated solution of K_2SO_4 (Table 10), whereas a $\sim$0.1M $Eu_2(SO_4)_3$ solution has a τ value of 140μsec[21]. Both thiocyanate and sulphate complexes of Eu(III) possess charge transfer bands in the UV region (NCS 292 and SO_4 240nm)[44]. We would like to emphasize that in the case of the thiocyanate complex the difference in energy between the CT and ^5D_o emitting level ($\sim$17kK) almost coincides with the energy gap between ^5D_o and $^7F_o(^7F_1)$ levels (Table 1), thus a reabsorption of the fluorescence energy is also probable.

b) Gd(III) Complexes

We shall directly proceed by observing that no Gd(III) fluorescence could be measured in aqueous solutions containing either SCN^- or NO_3^- ions (3M)(SectionIII). Unlike the case of Eu(III), Gd(III) with $4f^7$ ground configuration does not produce charge transfer complex with thiocyanate ion, the first absorption maxima for SCN^- and NO_3^- ions occur at 213 and 204nm respectively. Thus, the deactivation process in Gd(III) thiocyanate and nitrate systems is probably due to factor (2) and (3) mentioned above. In the case of nitrate ion, however, $n \rightarrow \pi^*$ absorption takes place at 313nm, exactly at the wavelength where $^6P_{7/2} \rightarrow ^8S_{7/2}$ fluorescent transition of Gd(III) ion occur. Hence, reabsorption of the fluorescence may be a deciding factor in the quenching process. The $n \rightarrow \pi^*$ transition in carbonate occurs at a much higher energy ($\sim$46kK) and does not interfere with the lanthanide fluorescence.

In contrast to the Eu(III)-chloride system, addition of 3M Cl^- to 0.0108M Gd(III) in aqueous solution (pH 5.53) produced a drastic change in the lifetime (Table 11) although exhibiting no change in fluorescence intensity. The observed τ value of 660μsec is about 1.5 times higher than that of the aquo ion at roughly the same pH and about 3 times as high as that at a pH 1.58. We believe that this increase in lifetime is not due to a strong complexation effect rather due to abstraction of water molecules from the first coordination shere of Gd(III) aquo ion.

It is of some interest to compare these results with solid complex $[Gd(OH_2)_6Cl_2]$ in gadolinium trichloride hexahydrate, which has the same structure as the Eu(III) analogue. Fig.13 shows the decay curve for the $^6P_{7/2}$ level (313nm) of Gd(III) in this salt ($GdCl_3.6H_2O$ purchased from Aldrich Chemical Co. Wisconsin and of 99.999% purity). The measured lifetime is very high 7.3msec (Table 11) which compares well with the calculated value for the radiative lifetime of this level (10.9msec)[45]. However, Kropp and Windsor[21] reported a value of 5.4msec based on integrated absorption measurements of the chloride salt in aqueous solution. From their studies on the deuterium enhancement ratio, they concluded that for gadolinium non-radiative (quenching) process does not involve OH modes.

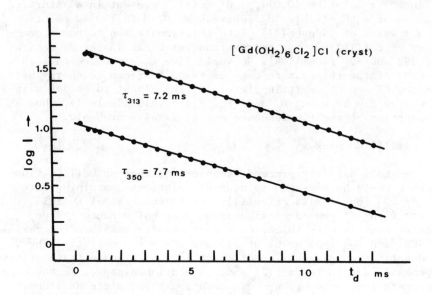

Fig. 13 Fluorescence decay curve of (a) 313nm and (b) 350nm bands in crystalline $GdCl_3.6H_2O$ at room temperature (Exct. 271nm).

We have, however, made an interesting observation during our studies on the time resolved fluorescence spectra of solid $GdCl_3.6H_2O$. Normal fluorescence and time resolved spectra of Gd(III) aquo ion and $GdCl_3.6H_2O$ in the solid state contain three peaks at about 304($32890cm^{-1}$), 314($31850cm^{-1}$), and 348($28740cm^{-1}$)nm corresponding to transitions $^6P_{5/2} \rightarrow {}^8S_{7/2}$, $^6P_{7/2} \rightarrow {}^8S_{7/2}$ and water (OH) vibration mode when excited with near monochromatic radiation of 272nm (Fig.14). The broad and weak band around 350nm has been observed by others[46] and identified as the vibronic level provided by OH stretching of water molecule.

The question is then why a Raman like peak of water appears around 350nm when Gd(III) aquo ion is excited with 272nm($36760cm^{-1}$) radiation. One would expect the water Raman peak to appear around 302nm for this radiation, and indeed the intensity of the observed 304nm fluorescence ($^6P_{5/2} \rightarrow {}^8S_{7/2}$) peak of Gd(III) contains contribution from the water Raman peak (time resolved spectra confirmed this). Could it be that the ∿350nm peak is an overtone or combination peak? Again the energy difference between 314nm and 350nm, e.g. $3280cm^{-1}$ is close to symmetric stretching (ν_1) mode of liquid water ($3219cm^{-1}$).

In order to obtain further information we did the following experiment. We excited a sample of water and Gd(III) aquo ion

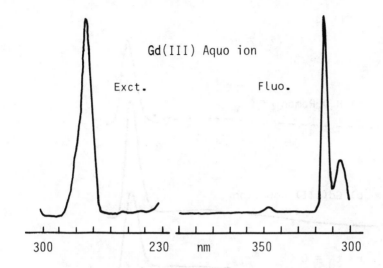

Fig. 14 Excitation (Em. monochromator at 313nm) and fluorescence (Exct. 272nm) spectra of 0.011M Gd(III) aquo ion at pH 5.93.

$(1.08 \times 10^{-2}$, pH 5.93) with near monochromatic radiation of 314nm. Both water and aqueous Gd(III) solution showed a broad peak at $355nm(28170cm^{-1})$ with a half width of $712cm^{-1}$ when the instrument (PELS5) was run on the fluorescence mode (Fig.15). Time resolved spectra of the Gd(III) aquo ion was then run giving a lifetime value $\tau \sim 7\mu sec$ for 355nm band (Fig.15) on 314nm excitation. The half width of the time resolved $(t_d = 20\mu sec)$ peak at 355nm remained the same as that of the water peak and Gd(III) peak mentioned above.

We were pleasantly surprised to see the 350nm peak of solid sample of $GdCl_3.6H_2O$ persisting much beyond the normal lifetime of a vibronic level. Fig.16 shows time resolved spectra of solid $GdCl_3.6H_2O$ at 1 and 5msec after excitation with 271nm radiation. The decay curve (Fig.13) of 350nm fluorescence peak gave a lifetime value of $\tau=7.7msec$, virtually identical with the 313nm $(^6P_{7/2} \rightarrow {}^8S_{7/2})$ transition ($\tau=7.2msec$). This allowed us to identify the excited level of 350nm fluorescence to be the $^6P_{7/2}$ of Gd(III) decaying to a quasi ground state $\sim 3280cm^{-1}$ above the actual electronic ground state $(^8S_{7/2})$ (Fig.17), provided by the strong coupling of the bonded water molecules in $[Gd(OH_2)_6Cl_2]^+$ moiety. The strict selection rule is apparently not obeyed in Gd system. Coupling of vibronic transitions to the electronic transition for the d and f group complexes is not uncommon[47].

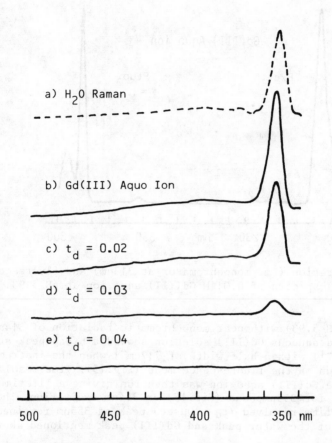

Fig. 15 (a) Raman spectrum of water (b) normal fluorescence
spectrum of 0.011MGd(III) aquo ion and time resolved fluorescence
spectra of the same Gd(III) aquo ion after (c) 0.02ms (d) 0.03ms
(e) 0.04ms delay from the beginning of the pulse to the beginning
of observation. 314nm excitation was used all through.

c)   Carbonate and Other Ligands

We have earlier mentioned that we were pleasantly surprised
with strong hypersensitive transitions in Sm, Tb, Dy, Tm, and
especially of Eu in tetracarbonato complex (Table 4). The lifetimes of the tetracarbonato complexes of these lanthanides are
compared with those of the aquo ion in Table 12. An increase in
lifetime by a factor of 2-4 is evident. Beside being bidentate
ligand, carbonate forms strong complex and the anionic complex
$[M(CO_3)_4]^{5-}$ is probably octacoordinated with no water in the first

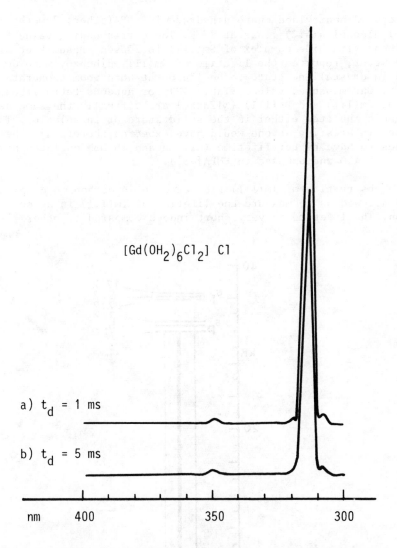

Fig. 16   Time resolved fluorescence spectra of crystalline
GdCl$_3$.6H$_2$O at room temperature (Exct.   271nm)   (a) delay time 1ms
(b) delay time 5ms.

coordination sphere. The bonded carbonate molecules then act as a
kind of insulating sheath round the lanthanide ions preventing
non-radiative energy migration.

Only limited lifetime data are available on Sm(III) and Dy(III)
systems especially in aqueous solution. Freeman and Crosby[48]
reported τ values of 14 and 12μsec for Sm(III) and Dy(III) complexes

of the type dibenzoylacetonate dihydrate in EMPA(ether: 3-methyl-
pentane: alcohol 4:4:2) glass at 77°K. The corresponding value for
tris-dibenzoylmethide complex of Sm(III) is 20μsec. Bhaumik et al
[49], however, reported the lifetime of Sm(III)-dibenzoylmethide
complex in crystalline state to be 15μsec at both room temperature
and 90°K. Our measured values (Table 12) for aqueous tetracarbonate
system of Sm(III) and Dy(III) (∿12μsec) are virtually the same as
the organic chelates either in the solid state or in solution. This
is rather interesting as one would have expected the organic chelate
complexes to have higher lifetime (cf. Eu and Tb DBM chelates have
τ values of 430 and 630μsec in EMPA[48]).

The observation of Tm(III) fluorescence in carbonato complex
(Fig.6) allowed us to measure the lifetime of Tm(III) in aqueous
solution. The lifetime is very short indeed compared to other

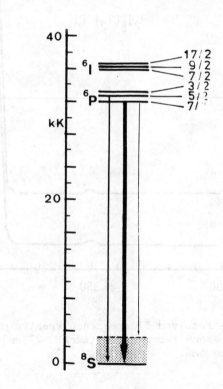

Fig. 17 Energy levels involved in the fluorescence of Gd(III)
aquo ion and of hydrated chloride at room temperature. The transition to the quasi ground state (dotted line) provided by the vibrational level of the bonded water is also shown.

Table 12. Change in Lifetime (τ) of the Anionic Tetracarbonato
Complexes of the Lanthanide Ions

M(III)	τ_{aq} (µsec)	τ_{CO_3} (µsec)	τ_{CO_3}/τ_{aq}
Sm	8	12	1.5
Eu	110	465	4.23
Gd	203	1460	7.19
Tb	425	1100	2.59
Dy	7	11	1.57
Tm	–	7	–

lanthanide ions. A τ value of 7µsec is calculated for the lifetime of the $^1G_4$ level of Tm(III) in $[Tm(CO_3)_4]^{5-}$ complex. This
to my knowledge is the first measurement of the lifetime of a
Tm(III) complex in aqueous solution.

We have also investigated the formate and acetate complexes
of Eu(III) and Gd(III)(Tables 10,11). The non-reducible Gd(III)
ion shows the expected trend of τ value formate>acetate. The lower
τ value for the acetate complex may be explained as due to energy
tunnelling via the CH_3 vibration of the acetate group (in the case
of formate only one C–H bond is present). In the case of Eu-formate,
the reducing character of formate ion in producing charge transfer transition is mainly responsible for the reverse trend, for-
mate<acetate, in τ value. The earlier reported[21] τ value of 170µsec
for a ∿0.1M Eu(acetate)$_3$ solution should be considered as due to
partial dissociation of the tris-complex. Our measured value of
337µsec in 3M acetate is most probably due to the tris-complex.

d) $[Eu(CO_3)_4]^{5-}$ Complex *par excellence*, Eu:Tb:CO$_3$ System

It often happens that the presence of a second lanthanide ion
in the system produces profound effect in the intensity and lifetime of the original lanthanide complex.

We were interested to see if any energy transfer from Tb(III)
to Eu(III) in anionic carbonato complex takes place, as we observed
[18] no change in Eu(III) fluorescence intensity in presence of

large excess of La and Tb. Accordingly we measured the lifetime
of 1:1 mixture of $[Eu(CO_3)_4]^{5-}$ and $[Tb(CO_3)_4]^{5-}$ in aqueous solution (0.0046M Eu, 0.0050M Tb and 3M carbonate) by exciting the
mixture with the excitation frequencies of both Tb(380nm) and
Eu(393nm) and monitoring the Tb(547nm) and Eu(618nm) emissions.
The relevant fluorescence lifetime data are summarized in Table 13.
We observed (Fig.18) initially no change in fluorescence lifetime
of $^5D_0$ level of Eu(III)(τ=479μsec compared to 465μsec for
$[Eu(CO_3)_4]^{5-}$ on exciting the system at Eu(III) excitation band $^5L_6$
(393nm). The lifetime of the $^5D_4$ level of Tb(III) at this excitation
is $\sim$8μsec. This clearly suggests virtually no loss of energy from
the $^5L_6$ level via any of the ^5D_J levels ($^5D_2$of Eu(III) is closer
to $^5D_4$ of Tb(III); Fig.3) of Eu(III) to Tb $^5D_4$ level. On the other
hand, excitation of the mixture with 380nm ($^5D_3$ of Tb(III)) radiation produced drastic drop in Tb(III) lifetime (610μsec) compared to $[Tb(CO_3)_4]^{5-}$ (Table 12) and only a small increase ($\sim$16%)
of Eu(III) $^5D_0$ lifetime. Allowing that Eu(III) complex may also
be excited (rather inefficiently) with 380nm light causing fluorescence, we see that the presence of Eu in this mixed system tun-
nels away a great part of the energy from the $^5D_4$ level of Tb(III).

The important feature of this observation lies in the fact
that the experiment demonstrates the migration of energy from one
lanthanide to another (Tb$\rightarrow$Eu) in hydroxylic (aqueous) solvent
without the need for an organized lattice. Obviously, in the present case the energy migration is occuring over a rather long
distance quite efficiently. Energy transfer from various donor
cations to the emitting lanthanide ions in doped crystal and condensed phase is, however, well documented[50]. Many years ago
Sinha et al[51] observed unusually strong red fluorescence characteristic of Eu(III) in crystalline Gd(dip)$_2$Cl$_3$.H$_2$O doped with $\sim 10^{-3}$
mole fraction of Eu(III) and Tb(III) [dip=2,2'-dipyridyl], and
commented on the transfer of energy from chelated Gd(III) to Eu
and Tb. Such transfer of energy did not take place in ethanolic
solution of the same doped-Gd(III)-chelate. Soon after Kleinerman
and Choi[52] were able to demonstrate from temperature dependent
lifetime measurements of similar phenanthroline complexes that a
ligand triplet exciton migration mechanism was operating in Sinha's
complexes.

In our mixed carbonate system, the energy transfer from Tb to
Eu is expected to be a slow process and a finite time is required
for energy to leak from Tb($^5D_4$) to Eu($^5D_0$). We noticed an initial
rise in red fluorescence intensity of Eu(III)(618nm) when Tb(III)
is pumped with 380nm radiation. From the risetime curve we have
estimated the exchange time (time to reach the maximum intensity,
t_{max}) to be $\sim$25μsec. Such risetime is usually characteristic of
doubly doped (Eu:Tb) inorganic system in the solid state (Na$_{0.5}$
Tb$_{0.45}$ Eu$_{0.05}$ WO$_4$ [53]), whereas for the phenanthroline systems
mentioned above [52] t_{max} is of the order of hundreds of μsec.

We believe that in our anionic Eu-Tb-CO$_3$ system direct energy transfer from $^5D_4$(Tb) $\rightarrow ^5D_0$(Eu) is operative.

Table 13. Observed Lifetime(τ) of a 1:1 Mixture of the Tetra-
Carbonato Complexes of Eu(III) and Tb(III)

Excitation(nm)	Emission(nm)	Lifetime(μsec)	τ/τ_{CO_3}
380(Tb)	547(Tb)	610	0.55
	618(Eu)	543	1.16
393(Eu)	547(Tb)	<8	-
	618(Eu)	479	1.03

VI.   ON THE TRAIL OF BASTNASITE

Bastnasite (MFCO$_3$) and parisite (CaM$_2$F$_2$(CO$_3$)$_3$)(M=lanthanide) are new types of ores providing raw materials for the production of the lanthanides. Whereas in most other ores the lanthanides are present as trace or rare amounts (ppm range, see chapter 12,13), a lanthanide concentration of 10-12 percent is by no means uncommon in bastnasite. It is also interesting that in certain parts of the world large bastnasite deposits have been found. One such massive deposit containing probably several million tons of bastnasite is found at Mountain Pass, California and processed by Molycorp, now supplies the bulk of world's lanthanide requirements.

Our knowledge in elementary lanthanide chemistry tells us that ligands like F$^-$ and CO$_3^{2-}$ ions would have precipitated the lanthanides when acting separately long before the concentration of free lanthanide ions reached even a few percent. Yet we find large bastnasite deposits. Some kind of synergistic action must have taken place in nature. While carbonate in low concentration precipitates hydrous lanthanide carbonates, a lanthanide: carbonate ratio of 1:50-60 is sufficient to keep the metal ions in solution. In this connection we thought that our observation of the formation of soluble anionic complexes of the lanthanides in carbonate solution may have played an important role in nature. Accordingly, we investigated the ternary system lanthanide-fluoride-carbonate using Eu(III) as probe. In the earlier section we have established the

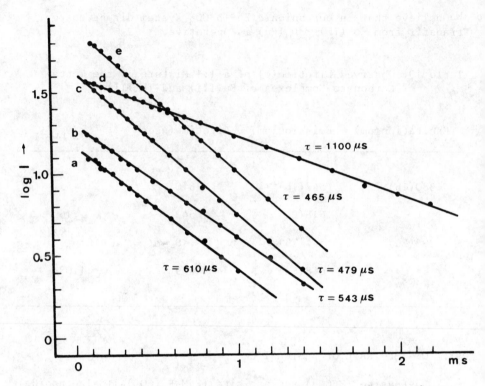

Fig. 18 Fluorescence decay curves of a 1:1 mixture (pH 12.6) of
$[Eu(CO_3)_4]^{5-}$ and $[Tb(CO_3)_4]^{5-}$ (a) Exct. 380nm(Tb); Em. 547(Tb)
(b) Exct. 380nm(Tb); Em 618nm(Eu) (c) Exct. 393nm(Eu);
Em. 618nm(Eu) compared with (d) $[Tb(CO_3)_4]^{5-}$ Exct. 380nm;
Em 547nm and (e) $[Eu(CO_3)_4]^{5-}$ Exct. 393nm; Em. 618nm.

spectroscopically rather robust nature of $[Eu(CO_3)_4]^{5-}$. Thus a
change, especially decrease, in the lifetime value of $^5D_0$ level
would be a strong indication of the formation of a mixed (ternary)
complex. A general reaction of the following type would be expected (ignoring the charges)

$$[Eu(CO_3)_4] + nF \rightleftarrows [EuF_x(CO_3)4-y] + (n-x)F + yCO_3 \quad (2)$$

and finally $[EuF_x(CO_3)4-y] \rightarrow EuFCO_3$ $\qquad\qquad\qquad$ (3)

It is even more curious that a five fold negative moiety of the
anionic carbonato complex would react with a negative fluoride
ion!

We have studied the change in fluorescence intensity with the
variation of ligand concentrations keeping one of them constant.

We have also measured the more sensitive parameter τ. Fig.19 and
Table 14 summarize the results. Addition of small amount of F^- ion
to anionic carbonato complex until a Eu:F ratio of 2:1 did not
produce any change in lifetime of $[Eu(CO_3)_4]^{5-}$ complex (465µsec).
At a Eu:F ratio of ∿1 we observed noticeable change in τ value
(456µsec) as well as two slopes in the decay curve. On increasing
the Eu:F to 1:100 the two slopes of the decay curve almost dis-
appeared (Table 14) and a further increase to 1:150 gave a single
slope decay curve (Fig.19) with τ=608µsec indicating the presence
of a single species. Plot of the lifetimes vs log$[F^-]$ produced a
discontinuity at 1:1   Eu:F ratio (Fig.20) indicating the formation
of a mono-fluoro-Eu(III)-carbonato complex. However, the steep
rise of the curve at higher $[F^-]$ concentration (Fig.20) is indicative

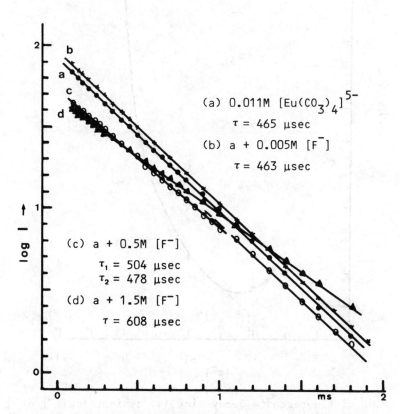

Fig. 19 Change in the fluorescence decay time of $[Eu(CO_3)_4]^{5-}$
(0.011MEu, 3MCO_3) with gradual addition of $[F^-]$ (a) no fluoride
(b) $5x10^{-3}MF^-$ (c) $5x10^{-1}MF^-$ (d) 1.5MF$^-$ (Exct. 393nm; Em. 618nm;
pH 12.8).

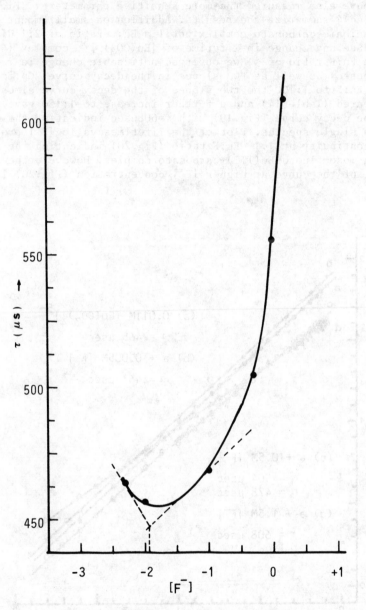

Fig. 20 Plot of fluorescence decay time (τ) against $\log[F^-]$ of
the ternary $Eu(III)(0.011M)-CO_3(3M)-F$ system showing break at
$[F^-]\sim0.01M$ giving a $[Eu]:[F^-]$ ratio of 1.

Table 14. Change in Lifetime(τ) of the Anionic Tetra-Carbonato
Complex of Eu(III) with the Addition of Fluoride Ion

Fluoride(M)	Lifetime(μsec)
0	465
5×10^{-3}	463
1×10^{-2}	456,422
1×10^{-1}	467,459
5×10^{-1}	504,478
1.0*	556,547
1.5	608

* Experimental data may be fitted to a single slope giving
$\tau=550\mu$sec which is statistically indistinguishable from the two
τ values derived by fitting two straight lines to the observed
data points. This fluoride concentration may be the turn over
point from the two slopes to a single slope characterizing an
unique species.
[Eu]$=1.12 \times 10^{-2}$M, [CO$_3$]$=$3M

of the replacement of more CO$_3$ ions from the coordination sphere.
The final reaction leading to MFCO$_3$ (bastnasite) in eq. 3 may be
initiated in many ways: by the change of pH, decomposition of the
[MF$_x$(CO$_3$)4-y] complex with evolution of CO$_2$ at higher temperature,
disproportionation or reaction with other calcium salts leading to
parisite formation. We are investigating more fully these ternary
systems and we shall report the results elsewhere. In conclusion
we would like to add that systems containing large amount of carbonate and fluoride ions are capable of keeping lanthanides from
precipitating for a long period of time.

EXPERIMENTAL

All high resolution fluorescence spectra and the integrated
intensities were measured with a Spex Fluorolog equipped with a
photon counting device. The recording of the time resolved spectra

and fluorescence lifetime measurements in solution and in the solid state were performed with a Perkin-Elmer LS 5 Luminescence Spectrometer provided with a pulsed source and quantum corrected photomultipliers for recording corrected excitation spectra. The spectra of the solid samples were obtained by front surface illumination using a special holder developed by Sinha, which fits in the liquid sample holder of the LS 5.

All lanthanide salts used were better than 99.9% purity unless otherwise mentioned. All ligands used in this study were of puriss p.a. quality. K_2CO_3 was used for making stock carbonate solution.

ACKNOWLEDGEMENT

I wish to thank Perkin-Elmer Co., Bodenseewerk in Überlingen and Dr. Ringhardtz in particular for providing me with a LS 5 Luminescence Spectrometer. Mr. Massenbach of Perkin-Elmer, Berlin, is thanked for his personal interest.

REFERENCES

1. S. P. Sinha, Complexes of the Rare Earths, Pergamon Press, Oxford, 1966.

2. G. Stein and E. Würzberg, J. Chem. Phys. 62, 208 (1975).

3. W. T. Carnall, P. R. Fields, and K. Rajnak, J. Chem. Phys. 49, 4412, 4424, 4443, 4447, 4450 (1968).

4. U. V. Kumar, D. R. Rao, and P. Venkateswarlu, J. Chem. Phys. 66, 2019 (1977).

5. L. G. De Shazer and G. H. Dieke, J. Chem. Phys. 38, 2190 (1963).

6. S. P. Sinha, Spectrochin Acta 22, 57 (1966).

7. S. P. Sinha and H. H. Schmidtke, Mol. Phys. 10, 7 (1965).

8. S. P. Sinha, Section Lecture XIIIth Int. Conf. Coord. Chem. (Cracow-Zakopane) 1970; published in Novels in Coordination Chemistry (Ed. B. Trzebiatowska), PWN Publishers, Wrocław, 1974, pp. 101-110.

9. S. P. Sinha, J. Inorg. Nucl. Chem. $\underline{33}$, 2205 (1971).

10. S. P. Sinha, P. C. Mehta, and S. S. L. Surana, Mol. Phys. $\underline{23}$, 807 (1972).

11. P. Möller, G. Morteani, and F. Schley, Lithos $\underline{13}$, 171 (1980).

12. I. O. Nyambok, Origin and Distribution of the Elements, Pergamon Press, Oxford, 1979, pp. 533-539.

13. S. P. Sinha, Plenary Lecture, Academy of Science São Paulo, Anais VI Simposio Anual ACIESP, $\underline{36}$, 1 (1982).

14. C. K. Jørgensen and B. R. Judd, Hol Phys. $\underline{8}$, 281 (1964).

15. H. S. Sherry and J. A. Marinsky, Inorg. Chem. $\underline{3}$, 330 (1964).

16. D. E. Hobart, K. Samhoun, J. P. Young, V. E. Norvell, G. Mamantov, and J. R. Paterson, Inorg. Nucl. Chem. Letters $\underline{16}$, 321 (1980).

17. K. Schwochau, Nachr. Chem. Tech. Lab. $\underline{27}$, 563 (1979).

18. S. P. Sinha, Fresenius Z. Anal. Chem. $\underline{313}$, 238 (1982).

19. P. Möller and S. P. Sinha, Fortschr. Miner. $\underline{60}$(1), 144 (1982).

20. S. P. Sinha, Europium, Springer-Verlag, Berlin, 1967, Chapter 5, Part 1.

21. J. L. Kropp and M. W. Windsor, J. Chem. Phys. $\underline{42}$, 1599 (1965).

22. P. K. Gallagher, J. Chem. Phys. $\underline{43}$, 1742 (1965).

23. Y. Haas, G. Stein, and E. Würzberg, J. Chem. Phys. $\underline{58}$, 2777 (1973).

24. E. V. Kondrateva and G. S. Lazeeva, Optics Spectrosc. (Engl. Trans.) $\underline{8}$, 67 (1960).

25. J. Albertsson and I. Elding, Acta cryst. $\underline{B33}$, 1460 (1977).

26. K. H. Hellwege and H. G. Kahle, Z. Physik $\underline{129}$, 85 (1951).

27. K. H. Hellwege, Angew. Chem. $\underline{65}$, 113 (1953).

28. H. G. Kahle, Z. Physik $\underline{155}$, 129 (1959).

29. H. G. Kahle, Z. Physik $\underline{155}$, 145 (1959).

30. C. O. P. Santos, E. E. Castellano, G. Vicentini, and L. C. Machado, Abstract VI Annual Symposium Acad. Sci. Sao Paulo, Brasil, Nov 9-11, 1981, pp. 41-42.

31. G. H. Dieke and L. A. Hall, J. Chem. Phys. 27, 465 (1957).

32. B. Rinck, Z. Naturforschg. 3a, 406 (1948).

33. E. G. Sherry, J. Solid State Chem. 19, 271 (1976).

34. N. V. Podberezskaja and S. V. Borisov, Russ., J. Struct. Chem. (Engl Ed.) 17, 164 (1976).

35. L. Hiltunen and L. Niinistö, Cryst. Struct. Comm. 5, 561 (1976).

36. J. H. Burns and R. D. Baybarz, Inorg. Chem. 11, 2233 (1972).

37. A. Heller, J. Amer. Chem. Soc. 88, 2058 (1966).

38. H. Bonrath, H. Heber, K. H. Hellwege, S. Hüfner, and H. Lämmermann, Naturwissenschaft. 48, 713 (1961).

39. A. Zalkin and D. H. Templeton, J. Amer. Chem. Soc. 75, 2453 (1953).

40. R. E. Thoma and G. D. Brunton, Inorg. Chem. 5, 1937 (1966).

41. S. P. Sinha, Struct. Bonding 25, 69 (1976); see for a general review on structure and bonding in lanthanide complexes.

42. B. Tanguy, M. Vlasse, and J. Portier, Rev. Chim. Min. 10, 63 (1973).

43. Y. Haas and G. Stein, J. Phys. Chem. 75, 3668 (1971).

44. J. C. Barnes, J. Chem. Soc. (London) 3880 (1964).

45. W. T. Carnall, Handbook on the Physics and Chemistry of Rare Earths (Ed. K. A. Gschneidner and L. R. Eyring) 3, 171 (1979).

46. Y. Haas and G. Stein, Chem. Phys. Letters 11, 143 (1971).

47. M. Cieślak-Golonka, A. Bartecki, and S. P. Sinha, Coord. Chem. Rev. 31, 251 (1980).

48. J. J. Freeman and G. A. Crosby, J. Phys. Chem. 67, 2717 (1963).

49. M. L. Bhaumik, H. Lyons, and P. C. Fletcher, J. Chem. Phys. 38, 568 (1963).

50. R. Reisfeld, Structure Bonding 30, 65 (1976).

51. S. P. Sinha, C. K. Jørgensen, and R. Pappalardo, Z. Naturforschg. 19a, 434 (1964).

52. M. Kleinerman and S. Choi, J. Chem. Phys. 49, 3901 (1968).

53. G. E. Peterson and P. M. Bridenbaugh, J. Opt. Soc. Amer. 53,
1129 (1963).

DISCUSSION

ASCENSO (question)

Do you have any explanation for not observing fluorescence from
Tm(III) aquo ion?

SINHA (answer)

At present I do not have a convincing answer, but the reason may
lie in the fact that the lifetime of the excited state of Tm(III)
aquo ion is even shorter than that for Sm(III) or Dy(III).

CARNALL (question)

Your lifetime data on Gd(III) aquo ion is in disagreement with the
results of Haas and Stein.

SINHA (answer)

Yes, our measured lifetime for the $^6P_{7/2}$ excited state of Gd(III)
aquo ion is apparently shorter than that reported by Stein and coworkers [23]. I do not understand this. We have, however, used
99.999% pure $GdCl_3 \cdot 6H_2O$ (Aldrich Chemical Co.) and adjusted the
pH of our solutions with Puriss p.a. HCl. The instrument used was
Perkin-Elmer LS 5, a pulsed source spectrofluorometer. Lifetime
measurements of the solids $GdCl_3 \cdot 6H_2O$, $EuCl_3 \cdot 6H_2O$ and Eu(III) and
Tb(III) aquo ions using this instrument gave values agreeing excellently with the literature data. Thus the problem of impurity
and insturmental mis-performance could be excluded. Unfortunately,
Haas and Stein never specified the purity, pH and temperature of
their Gd-system and they have used a kind of modified flash photo-
lysis apparatus for measuring the lifetime [23]. It is well known
that both Gd(III) and Tb(III) aquo ions show temperature dependence of the fluorescence intensity and lifetime [24]. My feeling
is that the Gd samples of Haas and Stein may have been contaminated
with Tb(III). If I may further comment, the excited state lifetime
of ∿3 μs for Sm(III) and Dy(III) aquo ions seems to be too short
for observing a reasonable fluorescence from these ions (compare
with Tm(III) as discussed in the text). But as we all know, both
Sm(III) and Dy(III) aquo ions exhibit reasonable fluorescence to
be recorded under either constant excitation or pulsed excitation
as shown in the figures. I believe that our measured values
(∿ 8 μsec) seem to be more realistic.

AMBERGER (question)

If a complex has a center of symmetry, would this affect the life-time?

SINHA (answer)

In general, yes. But one has to be extremely careful when talking about symmetry in solutions. Even if the complex in the crystal-line state possesses an approximate center of symmetry, this does not guarantee that it will retain the same symmetry in solution (especially in aqueous solution). In non-aqueous solvent (such as acetonitrile) often the complex is "shielded" from the neighbou-ring interaction, unless the solvent itself coordinates with the complex. The nitriles are very good solvent for preparing the hexachloro, hexabromo type complexes. If the complex has a center of symmetry, it will affect the magnetic and electric dipole allowed transitions differently.

SABBATINI (question)

What is the lifetime of the $^5D_1$ excited state of Eu(III) and which transition did you use to measure it?

SINHA (answer)

The lifetime of $^5D_1$ level is about 10 μsec for the aquo ion. I have used the strongest transition in this series e.g., $^5D_1 \rightarrow {}^7F_2$ occuring around 555 nm for the aquo ion. It is actually immaterial which of the $^5D_1 \rightarrow {}^7F_J$ transitions one uses.

SABBATINI (question)

I found in some Eu(III) cryptates an absorption band at about 300 nm with molar extinction coefficient of the order of 100-200. Other Eu complexes ($EuCl_6^{3-}$ in acetonitrile) exhibit bands with the intensity of the same order of magnitude and attributed to LMCT transition. Can the bands observed in Eu(III) cryptates be attri-buted to the same type of transitions? Why is their intensity so low in comparison to the CT bands observed in d-transition metal complexes?

SINHA (answer)

The bands observed around 300 nm for complexes of Eu(III) with crown ethers and cryptates are indeed the ligand → metal charge transfer band. These genuine CT bands in the lanthanide complexes are generally weaker than those in the d-transition metal complexes because of poor overlap between the ligand orbitals and the 4f orbitals. At one time it was also argued that the high symmetry of the hexachlorides is probably responsible for weaker intensity of the absorption bands in general.

SABBATINI   (question)

If the UV bands observed in the cryptates are the LMCT bands, would the energy transfer from these CT states to the emitting f→f state be expected to have high efficiency?

SINHA   (answer)

We would expect the fluorescence intensity to be higher if the energy transfer from the CT donor state to the emitting 4f level (say $^5D_0$ or $^5D_1$ of Eu(III)) proceed via radiationless process. This is somewhat similar to the intramolecular energy transfer process from the bound ligand to the central lanthanide ion so well investigated during the sixties [see Sinha, Complexes of the Rare Earths, Pergamon Press, Chapter 8]. Pumping at these LMCT bands is expected to produce higher fluorescence intensity than the corresponding aquo ions.
Competing with the radiationless energy transfer (RET) process are the radiative process and the absorption of the excitation energy to a higher state. The efficiency of the RET process will depend on the mechanism, temperature, solvent and of course on the nature of the central lanthanide ion.

THE USE OF LANTHANIDE IONS AS NMR STRUCTURAL PROBES

J.R. Ascenso   and   A.V. Xavier

Centro de Química Estrutural, I.S.T.
Av. Rovisco Pais, 1000 Lisbon, Portugal

The scope and limitations of the use of lanthanide ions as NMR structural probes is discussed. The application of the method is reviewed and examples of determination of structures in solution are presented. A procedure for the obtention of quantitative conformation of flexible molecules is described.

1. INTRODUCTION

Since an understanding of chemical and biochemical processes requires an extensive knowledge of molecular structures, the development of physical techniques which can provide structural information is of great importance.

For many years the most used NMR methods for obtaining conformational information of small molecules in solution relied on the analysis of vicinal coupling constants (1) and on the measurements of nuclear Overhauser effect (2). The conformation information available from these two methods is limited to fragments of the whole molecule due to the short range of the interactions involved.

The use of lanthanide ions as NMR shift reagents was first reported by Hinckley in 1969 (3). During the thirteen years after the initial contributions, numerous studies have been done in the field. The experimental results have been interpreted and structural analyses using the lanthanide probe method (4-6) are now becoming a routine NMR technique.

These studies have been summarized in some excellent reviews:

501

S. P. Sinha (ed.), Systematics and the Properties of the Lanthanides, 501–540.
Copyright © 1983 by D. Reidel Publishing Company.

Reuben (4) surveyed the literature up to 1972 on lanthanide
shift reagents, Dobson and Levine (5) reviewed the applications
of lanthanides to the quantitative analyses of molecular conformations and more recently Inagaki and Miyazawa (6) reviewed recent
developments in this field.

We will describe the basic principles of this technique,
and show how lanthanides can be used for quantitative analyses
of the whole molecular 'structure', within the range ~ 10A from
the lanthanide. Some examples of the determination of the
structure in solution will be presented. Limitations of the method
and complementary techniques will be discussed.

2. BASIC PRINCIPLES

2.1 Lanthanide Ions as Structural Probes

Since the 4f electrons are well shielded by the 5s and 5p
electrons, the chemical properties of the lanthanide ions are
very similar, contrary to the case of the transition metal ions.
On the other hand, the magnetic properties of lanthanide ions
vary appreciably along the series. The ions La^{3+} and Lu^{3+} are
diamagnetic, and among the paramagnetic ions, Pr^{3+}, Eu^{3+} and Yb^{3+}
have short electron spin relaxation time ($\tau_S < 10^{-12}$ sec), and can
be used as shift probes. The ions Eu^{2+} and Gd^{3+} have long relaxation times ($\tau_S > 10^{-10}$ sec), and induce large broadenings with
negligible shifts which make them good candidates as broadening
probes. Between these two extreme groups, Dy^{3+} and Ho^{3+}, still
with short electron spin relaxation times, give rise to large
paramagnetic shifts as well as appreciable broadening effects.

2.2 Contact Shifts

Contact shifts are induced by unpaired electrons of a
paramagnetic ion through a spin polarization mechanism of the
atomic nuclei of complexed molecules. This interaction is
isotropic due to much faster electronic spin relaxation rates
than the hyperfine coupling constants $A/\hbar$. For lanthanide ions
the expression for these shifts is given by the following
equation (7)

$$\delta_c = \frac{\delta\nu}{\nu_o} = \frac{\beta J(J+1)g(g-1)}{\gamma_N \hbar 3kT} A \tag{1}$$

where k is the Boltzmann constant, T is the absolute temperature,
β is the Bohr magneton, J is the quantum number for the total
angular momentum, g is the Landé factor and γ is the magnetogiric
ratio of the nuclei. For Eu^{3+} and Sm^{3+} with appreciably populated
low lying excited states, a correction must be made on the values

of the average spin polarization for the ground state
$<S_z>=-\beta J(J+1)g(g-1)H_o/3kT$, to take into account the contribution
of these excited states (8).

It has been shown by Reuben and Fiat (9) that A is
approximately constant along the series, and has a lower value
than that of the transition metal ions, for the same type of
nuclei. As shown in Table I the relative theoretical values of
the contact shift agree well with the experimental ones, and ions
like Gd^{3+}, Tb^{3+} and Eu^{3+} exhibit the largest contact shift in the
series (8,9).

Table I. Relative contact shifts[a]

Ln^{3+}	Theoretical values[b]	Experimental values[c]
Ce	-3.4	
Pr	-10.4	-14.6
Nd	-15.7	-17.2
Pm	-14.0	
Sm	0.2	
Eu	37.4	
Gd	110.3	121.1
Tb	111.4	128.1
Dy	100.0	100.0
Ho	79.3	84.2
Er	53.8	55.8
Tm	28.7	29.2
Yb	9.1	7.7

[a]Scaled to 100 for Gd^{3+}; [b]Golding et Halton (8); [c]Contact
shifts of the ^{17}O resonance of $H_2^{17}O$ as corrected for pseudocontact shifts. Reuben and Fiat (9).

2.3 Pseudocontact Shifts

These shifts result from local fields induced through space
by the magnetic moment of the ion (which is derived from the
magnetic susceptibility). These local fields are not averaged to
zero over all orientations due to the large magnetic anisotropy
of the magnetic susceptibility tensor. The origin of these shifts
has been treated theoretically by several authors (10,11). A

magnetic nucleus at a site (r,θ,ϕ), senses a pseudocontact shift given by

$$\delta_{pc} = \frac{\delta \nu}{\nu_o} = D(3 \cos^2\theta-1)/r^3+D'\sin^2\theta\cos2\phi/r^3 \qquad (2)$$

where D and D' are constants related to the principal components χ_x, χ_y, χ_z of the susceptibility tensor, $D \propto \chi_z -\bar{\chi}$ and $D' \propto \chi_x-\chi_y$. If the symmetry is axial, D'=0 and the expression for pseudocontact reduces to

$$\delta_{pc}=D(3\cos^2\theta-1)/r^3 \qquad (3)$$

having a form similar to the well-known McConnell-Robertson equation for the pseudocontact shift induced by the transition metal ions (12). The relative values of D are given in Table II.

Table II. Relative pseudocontact shifts[a]

Ln^{3+}	Theoretical values[b]	Experimental values[c]
Ce	-6.3	
Pr	-11.0	-7.1
Nd	-4.2	-2.6
Pm	2.0	
Sm	-0.7	-0.2
Eu	4.0	4.0
Tb	-86	-50
Dy	-100	-100
Ho	-39	-39
Er	33	12[d]
Tm	53	8.2[d]
Yb	22	10[d]

[a]Scaled to -100 for Dy^{3+}; [b]Bleaney et al.(13), at room temperature; [c]C_{H_5}' in 5'-Cmp, Williams et al.(14), at room temperature; [d]The low values of the shifts in these particular complexes will be discussed later.

The lanthanide complexes are normally not axially symmetric in crystals. X-ray structure determinations on the complexes $Ho(Dpm)_3(4-picoline)_2$ (15) and $Eu(Dpm)_3(pyridine)_2$(16) have shown these 8-coordinate complexes to have only a C_2 symmetry axis. Even more asymmetric situations were observed in crystals of several LnEdta.$_3$H$_2$O complexes (17). The crystal structure of these molecules is devoided of any real symmetry element.

The marked magnetic nonaxility of various complexes is also evident from the anisotropy in the magnetic susceptibility tensors, that were determined by Horrocks (18) for all the Ln(Dpm)$_3$(4-picoline)$_2$ series at 298°K.

Situations in which axial symmetry model applies have been observed in most complexes in solution, through the constancy of the shift ratios of any two different nuclei along the series (5). Consequently, it is important to inquire why the magnetic susceptibility tensor of a lanthanide complex acquires effective axial symmetry in solution, if in the crystal this tensor is highly asymmetric.

Briggs *et al.*(19) have correlated this apparent axiality to internal rotation of the ligand (free or with a three-fold or higher barrier) about the lanthanide donor bond. These rotations effectively average out the non-axial part D' of equation (2) for the pseudocontact shifts. However for more rigid complexes as are LnEdta or Ln(Dpm)$_3$(4-picoline)$_2$ complexes, it is hard to accept this model for effective axial symmetry. In these cases the suggestion made by Horrocks (20) that the axial symmetry may result of an averaging due to a fast equilibrium between different geometric isomers of the lanthanide complex is more acceptable.

The effect of chemical exchange on observed shifts will be discussed in section 3.2.

2.4 Selective Broadening

Paramagnetic lanthanide ions enhance selectively the relaxation rates of the nuclei of the substrate. The ground state of Gd^{3+} and Eu^{2+} is a spin only state, there is no zero field splitting, and the electronic spin relaxation times (τ_s) are quite long, $\tau_s < 10^{-10}$sec. Consequently for these two ions the dipolar relaxation rates are enhanced according to the Solomon-Bloembergen equations (21)

$$\frac{1}{T_{1M}} = \frac{2}{15} \frac{\gamma_I^2 S(S+1) g^2 \beta^2}{r^6} \left(\frac{3\tau_c}{1+\omega_I^2 \tau_c^2} + \frac{7\tau_c}{1+\omega_S^2 \tau_c^2} \right) \qquad (4)$$

$$\frac{1}{T_{2M}} = \frac{1}{15} \frac{\gamma_I^2 S(S+1) g^2 \beta^2}{r^6} \left(4\tau_c + \frac{3\tau_c}{1+\omega_I^2 \tau_c^2} + \frac{13\tau_c}{1+\omega_S^2 \tau_c^2} \right)$$

where γ_I is the magnetogiric ratio of the magnetic nuclei(I), r
is the distance between the nuclei and the lanthanide ion, ω_I and
ω_S are the Larmor procession frequencies of the magnetic nuclei(I),
and the electron spin (S), respectively. The correlation time τ_c
for dipole-dipole interaction is given by

$$\frac{1}{\tau_c} = \frac{1}{\tau_R} + \frac{1}{\tau_M} + \frac{1}{\tau_S}$$

where τ_M and τ_R are the life time and rotation correlation time
of the lanthanide ligand complex. Scalar contributions to relaxation rates of protons induced by Gd^{3+} and Eu^{2+} have been found to
be experimentally negligible (22), possibly due to the weak
interaction of the 4f orbitals of these ions with those of the
ligand nuclei. However Merbach *et al.*(23) have shown that for
nuclei directly bound to the paramagnetic ion, as are the ^{17}O
nuclei of water molecules coordinated to the lanthanide ion, the
situation may be much more complicated and the scalar term may
have an important contribution.

Equations (4) have also been found to be experimentally
applicable to paramagnetic lanthanide ions other than Gd^{3+} and
Eu^{2+}, in spite of the anisotropy of the susceptibility tensors
(24,25). The isotropy of the relaxation for the meta and para
proton nuclei of dipicolinate (Dpa) in presence of different
lanthanide ions was observed by Alsaadi *et al.*(24). The ratios
of T_1 relaxation rates for the rigid complexes monodipicolinate
given in Table III are reasonably constant, within the experimental
error,for the different lanthanides. Its average value, 2.3+0.4
agrees well with the value of $(r_p/r_m)^6$, assuming a Ln-N bound
length in the range 2.4-2.7A. These measures indicate that relaxation rates of these nuclei are isotropic, since they do not depend
upon angular factors.

Table III. Relaxation rates ratios for the
protons of LnDpa complexes at
room temperature and 90MHz (24)

Ln^{+3}	$\dfrac{1/T_{1m}}{1/T_{1p}}$
Pr	2.5+0.5
Nd	2.2+0.3
Tm	2.1+0.3
Yb	2.3+0.3

We have obtained a similar conclusion from the observation
that the ratios of the relaxation rates of LnEdta rigid complexes
are constant for the first part of the series (see Table IV). No
results are presented for the last members of the series because
isostructurality is not preserved along the series, as a
consequence of the loss of a coordinated water molecule in the
region of Eu to Tb complexes (25).

Table IV. Relaxation rates ratios for the protons of LnEdta
complexes at 79°C and 100 MHz (25)

Ln^{3+}	$\dfrac{1/T_{1acA}}{1/T_{1et}}$	$\dfrac{1/T_{1acB}}{1/T_{1et}}$
Ce	0.88+0.09	1.10+0.11
Pr	1.08+0.11	1.00+0.10
Nd	1.05+0.10	1.16+0.12
Sm	0.97+0.09	1.01+0.09
Eu	1.25+0.19	1.14+0.13

These results indicate that the contribution of anisotropy
in susceptibility, derived from second order effects of crystal
field to the relaxation rates, is negligible in comparison to the
contribution of the isotropic first order Curie term. If the
crystal field splittings are lower than kT, the Brownian motion
energy, as found for many lanthanide complexes (26) then, this
conclusion is perfectly justifiable.

We have applied equations (4) to the determination of the
values of electronic relaxation times τ_s of lanthanide aquoions
and of small complexes, like LnEdta and LnCdta (25). In these
complexes water molecules are weakly bound to the lanthanide
ions, $\tau_M \sim 10^{-8}$sec, and so they exchange rapidly with the free
water molecules. In these circunstances the observed relaxation
rates in the bulk site follow equation (5).

$$\frac{1}{T_{iobs}} = \frac{P_M q}{T_{iM}} - \frac{1}{T_{iA}} \qquad i=1,2 \tag{5}$$

As a large excess of free ligand molecules exists

$$P_M q = \frac{q[M]}{[S]}$$

where P_M is the mole fraction of bound ligand to the metal ion,
q is the number of molecules coordinated to the paramagnetic ion,
$[M]$ and $[S]$ are the metal and ligand concentrations and $1/T_{iA}$
represents the diamagnetic contribution to the relaxation rates.
In most conformation analysis problems these conditions are met.

The values of the paramagnetic relaxation times T_{iM} at a
given frequency and temperatures may be obtained from the slope
$q/[S]T_{iM}$ of the linear plot of $1/T_{iobs}$ versus the metal concentration, if q and $[S]$ are known from other sources.

The values of τ_S were obtained from the observation of the
longitudinal relaxation rates of the protons of water molecules
at 23 ± 1^oC and 18MHz. The values of r used in equation (4) were
derived from X-ray data of Shanoon and Preitt (27), and a value
of q=9 for the aquoions was chosen, as being the value generally
accepted (28).

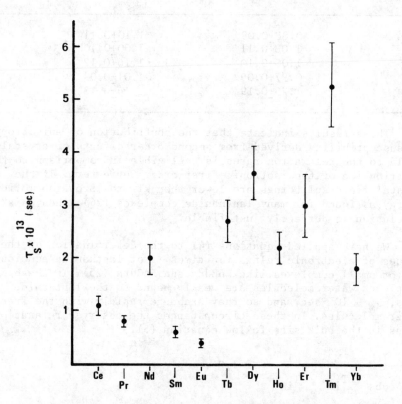

Figure 1. Values of τ_S for the lanthanide aquoions at
          23^oC and 18 MHz.

The relaxation times τ_S derived from aquoions (Figure 1) are all very low, and the trend is similar to that given by Alsaadi *et al*. (29). The error bars in the Figure correspond to an average error of 15% estimated from several determinations of τ_S for the same aquoion. Recently LaMar *et al*. (30) measured the protons T_1 relaxation rates of the t-butyl group of Dpm in a series of non labil quelates, Ln(Dpm)_3 (3,5-lutidine)$_2$. These relaxation rates were found to be predominantly dipolar in origin, and since the structure of the quelates is known it was possible to calculate the values of τ_S.

The linear dependence of $1/T_1$ with μ_{eff}^2 (Figure 2) shows that τ_S values for the second half of the Ln(Dpm)_3 (3,5-lutidine)$_2$ series are nearly constant. However, a bad fit is observed for the two ions of the first half (Pr,Eu). An enlarged representation of the same type using the ethylenic protons of the first half of the LnEdta series (25) (Figure 3) shows that no such linear correlation is observed.

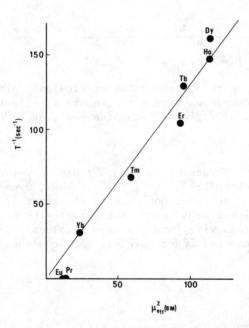

Figure 2. Plot of T_{1M} relaxation rates of t-butyl (Dpm) resonance versus the square of the effective magnetic moment of various lanthanides, at 25°C and 100 MHz (30).

On a theoretical ground the results obtained for these aquo-ions are supported by the fact that usually non-Kramers ions have lower relaxation times than the Kramers ions, and that for ions

with low lying excited states, as are Eu^{3+} and Sm^{3+}, the values
of τ_S should be even more reduced. These observations are evident
from our results in Figure 1.

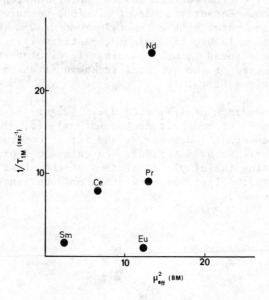

Figure 3. Plot of T_{1M} relaxation rates of ethylenic protons of
light Ln Edta complexes versus the square of effective
magnetic moment of various lanthanide ions, at 79°C and
100 MHz.

The values of τ_S presented in Figure 1 are very low for
lanthanide shift probes (Pr^{3+}, Eu^{3+}, Yb^{3+}). Thus, even in small
complex molecules the correlation time τ_c is primarily governed
by τ_S and no significant broadening is observed with these ions.
On the other hand, ions with intermediate τ_S values (Dy^{3+}, Ho^{3+})
and high magnetic moments, induce significant broadening as well
as large shifts.

3. APPLICATIONS

3.1. Types of Lanthanide Shift Reagents (LSR)

Complexes of dipivaloyl methanates (Dpm) with lanthanides
are widely used as shift reagents ($Pr(Dpm)_3$,$Eu(Dpm)_3$,$Yb(Dpm)_3$)or
as broadening reagents ($Gd(Dpm)_3$)in organic solvents. Substitution of tertiary butyl groups of Dpm by perfluoro alkyl groups
gives a new class of strongly acidic shift reagents ($Ln(Fod)_3$),
more soluble in organic solvents, and which bind relatively well

to weakly basic substrates. This important group of LIS has been applied to a wide range of problems. Shift probes have been extremely useful in the resolution of ambiguities of complex spectra (31), in spectral assignments (32) and in resolution of racemic mixtures (33). Bulkness of these ligands attenuates the interaction with substrate molecules, and consequently the shifts induced are mainly pseudo contact in origin, and these probes are suitable for molecular conformation analysis (34).

In aqueous solutions inorganic salts of lanthanides have been successfully used. In acid solutions, nitrates, chlorides and perchlorides are used due to their weak tendency to form ion pairs. However, at pH > 6, lanthanide ions precipitate as hydroxides. Consequently, other types of aqueous solution probes, e.g., Edta complexes of lanthanides, have been used in conformational studies. This group of probes binds weakly to carboxylate groups and to double ionized phosphate groups, and fast exchange conditions have been observed in studies with nucleotides(35) and α-amino acids (36), and other monocarboxylates (37).

Reilley et $al.$ (38) have compared the utility of Ln Edta complexes as shift probes with that of the complexes formed with the macrocycle ligands 1,4,7,10-tetraazacyclododecane-N,N',N'',N'''- -tetraacetate (Dota) and 1,4,7-triazacyclononane-N,N',N'',N'''-tria- acetate (Nota) (Figure 4) with respect to the sifts induced on charged(^{23}Na$^+$),neutral (^{2}H and ^{17}O of deuterated water) and negatively charged (^{35}Cl$^-$) substrates.

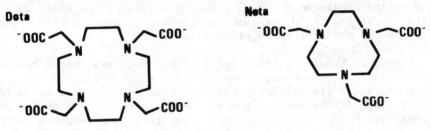

Figure 4. Structures of Dota and Nota ligands.

Both Ln Dota and Ln Nota complexes possess axial symmetry but Dota complexes proved to be better since the contact shifts induced by these last complexes in ^{23}Na$^+$, ^{35}Cl$^-$ and ^{2}H nuclei are much smaller. Even for ^{17}O nuclei of water molecules, the contact contribution was shown to be negligible. The explanation for these facts lies on a more crowded carboxylate region in Ln Dota complexes than in Ln Nota or Ln Edta complexes, which prevents the approach of the ligand and makes the contact shifts induced by Ln Dota negligible. Work is in progress in our laboratory with other tetraaza macrocycles (39) to find their ability as shift and relaxation probes.

3.2. Requirements for Quantitative Conformation Analysis

In order to use the lanthanide shift reagents (LSR) to get quantitative answers to conformation problems, several preliminary points must be taken into account. Firstly, the number and stoichiometry of the complex species present in solution must be studied. Secondly, the pseudo contact shifts for the 1:1 complex species must be obtained. Finally, effective axial symmetry must be proved.

The principles of the available methods for determination of stoichiometry and stability constants are well known (40). Generally the concentration (or relative concentrations) of free and complex species in solution are measured as the total concentration of LSR and substrate are varied. Formation of complexes with lanthanide ions has been investigated by a variety of techniques.

High resolution NMR methods based on lanthanide induced shifts are of great use for the studies of the lanthanide binding sites and for determination of the stoichiometry and stability of the complexes formed. Other methods than NMR include ultraviolet and visible spectroscopy (41), that are very insensitive to the nature and number of substrate molecules coordinated due to the low shift of the weak bands observed for the f-f transitions in these complexes. Enhancements on the luminescence of Eu(III) and Tb(III) ions on binding to a ligand are also capable of providing a detailed knowledge of the metal ions binding sites (28), and finally electron spin resonance and proton relaxation enhancements may also be used for these studies (22).

Most lanthanide shift reagents (LSR) form weak mixed complexes with substrate molecules, generally with a 1:1 stoichiometry. However, a 1:2 stoichiometry has also been demonstrated for a few cases (42). As the shifts induced for the same nucleus of the substrate may be different for the two species a quantitative analysis needs an evaluation of the stoichiometry. When more than one species is present in solution, the chemical equilibrium of the complex formation of a lanthanide shift reagent (L) and a substrate (S) is expressed by the following schemes, under the assumption that the formation of LS_3 and higher order species may be neglected,

$$L + S \underset{\leftarrow}{\overset{K_1}{\rightarrow}} LS$$

$$LS + S \underset{\leftarrow}{\overset{K_2}{\rightarrow}} LS_2$$

where K_1 and K_2 are the binding constants for the 1:1 and 1:2 complexes. If fast exchange conditions prevail, the observed shift δ_{obs} is expressed in terms of the fully bound shifts δ_{LS} and δ_{LS_2} of the two species as

$$\delta_{obs} = \frac{[LS]}{[S_o]} \delta_{LS} + 2 \frac{[LS_2]}{[S_o]} \delta_{LS_2} \qquad (6)$$

where $[S_o]$ refers to the total concentration of substrate.

Various methods have been proposed for the determination of the binding constants and the fully bound shifts of LS and LS_2 complexes (6). The most widely used is the incremental dilution method, in which $[S_o]$ is kept constant and the shift reagent is added stepwise. Reliable results are obtained by this method in those cases where a one-step equilibrium is involved, as is the case of the titration of 3',5'-cAmp with Dy^{3+} (Figure 5). In this case we have only contribution of 1:1 species for the total shift and

$$\frac{1}{\delta_{obs}} = \frac{1}{\delta_{LS}} + \frac{1}{K_1[L]}$$

Thus a plot of $1/\delta_{obs}$ versus $1/[L]$ is linear, with slope $1/K_1$ and intercept $1/\delta_{LS}$. Other graphical methods or computer programs may be used to fit the titration curves. For ions that produce large broadenings only small $[LS]/[L_o]$ ratios can be used.

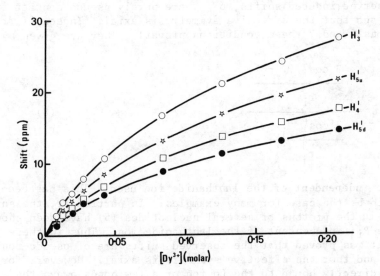

Figure 5. Titration curves for 3',5'-cAmp (25 mM) at pH 2.0 and 24°C with Dy^{3+} (72).

When a two-step mechanism is involved, it appears that two sets of titrations should be done, in order to get reliable results (42). In a first titration, successive dilutions are made by addition of solvent to the original solution. The molar ratio $\rho = [L_o]/[S_o']$ is kept constant, whereas $[S_o]$ is lowered through the experiment. In a second titration, successive additions of substrate solution to the original solution are made, and ρ is gradually decreased to zero. Since these two methods have quite different patterns with respect to the complex formation (43), simultaneous least square fit of the two types of titration curves to equation (6) enhances the reliability of the parameters δ_{LS}, K_1 and δ_{LS_2}, K_2 thus obtained. The fully bound shifts obtained for the 1:1 species may now be used in the calculation of pseudo contact shift ratios, necessary for the conformation analysis (Section 3.3).

The shift ratios R_{ij} are given by

$$R_{ij} = \frac{\delta^i_{obs}}{\delta^j_{obs}} \tag{7}$$

If the stoichiometry is known to be 1:1, the R_{ij} values can be obtained for any degree of complex formation. Thus, knowledge of the fully bound shifts is not necessary.

The shift ratio method is also important to prove both that the observed induced shifts, δ_{obs}, are purely pseudo contact in origin and that the effective symmetry is axial. Indeed, if from equations 3 and 7 these conditions prevail. They are given by

$$R_{ij} = \frac{\dfrac{3\cos^2\theta_i - 1}{r_i^3}}{\dfrac{3\cos^2\theta_j - 1}{r_j^3}} \tag{8}$$

and are independent of the lanthanide ion used. This has been shown to be the case for many examples. In particular, the shifts induced in the protons of several nucleotides (5) have been shown to give R_{ij} independent of the lanthanide ion. Thus in these cases it was proved that the observed shifts are of pseudo-contact origin and that the effective symmetry is axial. However, for nuclei directly bound to the Ln ion or a few bonds away, the observed shifts may have a strong contact contribution (9,44,45,46). In this last case the shift ratios will change markedly with the lanthanide ion used.

In order to show that the dependence of R_{ij} with the lanthanide ion used is due to contact contributions a close

analysis of the expected relative contributions to the observed
shift of pseudo contact and contact must be carried out. Indeed,
the patterns for contact and pseudo contact contributions along
the lanthanide series are quite different (Figure 6).

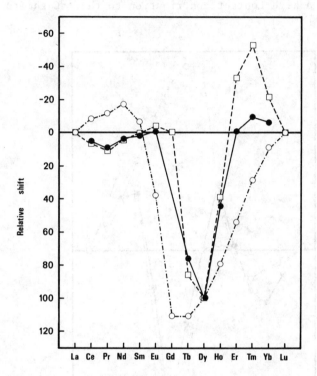

Figure 6. Relative shifts of ethylenic protons of Ln Edta
     complexes at 95°C (——). Calculated contact shifts (-·-)
     and calculated pseudo contact shifts (---). All the shifts
     are normalized to 100 for dysprosium.

     Thus, using two lanthanide ions with contact contributions
of the same sign and pseudo contact shifts of opposite signs
(e.g., Eu^{3+} and Ho^{3+}), a first indication of the importance of
the contact term can be obtained. However, a full analysis of
these contributions usually needs the study of the whole series.

     In Figure 6 we compare the relative shifts of the ethylenic
protons of Ln Edta complexes at 95°C with the theoretical values
of contact and pseudo contact for the whole series. These values
are given in Tables I and II. With the exception of the last
three members of the series, which we shall discuss later, the
agreement with the calculated pseudocontact curve is good. This

agreement shows that the contact contribution to the observed
shifts of ethylenic protons of Ln Edta complexes is negligible.
The same comparative process was applied to the ^{13}C shifts of
these complexes, but as is shown in Figure 7 for ethylenic carbons,
a poor agreement between calculated and experimental curves exists
as a result of a high contact contribution to the ^{13}C shifts of
these nuclei.

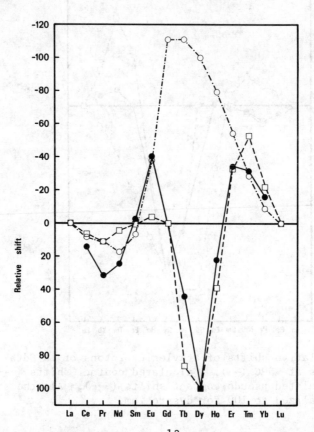

Figure 7.   Relative shifts of ^{13}C nuclei of ethylenic backbone of
Ln Edta complexes at 78°C (——). Calculated contact shifts
(–·–) and calculated pseudo contact shifts (---). All the
shifts are normalized to 100 for dysprosium.

Since contact shifts do not contain any geometric information
(see equation 1), it is desirable to design experiments with shift
reagents which possess low contact to pseudo contact shift ratios
or to design methods for separation of contact from pseudo contact
shifts.

The choice of a LSR for structural purposes is easily done since it is possible to evaluate the contact to pseudo contact ratios for the different lanthanides. These theoretical ratios are given in Table V. As is shown in this Table, Yb^{3+} ion possesses the lowest contact to pseudo contact shift ratio, and for this reason it is the best shift probe for structural searches. On the contrary, Eu^{3+} ion possesses the highest shift ratio and its use is not advisable in cases were contact shifts are expected.

Table V.   Ratios of contact and pseudo contact shifts[a]

Ln^{3+}	Theory[b]	Obs[c]	Obs[d]
Pr	0.10	0.34	0.31
Nd	0.40	0.91	0.61
Sm	−0.03		−0.26
Eu	1	1	1
Tb	−0.14	0.17	−0.26
Dy	−0.11	0.13	−0.13
Ho	−0.22	0.09	−0.25
Er	0.17	0.30	0.27
Tm	0.06		0.06
Yb	0.04	0.09	0.06

(a) Scaled to unity for Eu^{3+};   (b) Golding and Halton (8) and Bleaney (10);   (c) Gansow *et al.* (45) absolute values of ratios; (d) Ajisaka and Kainosho (48).

On the other hand several methods were developed in order to separate contact and pseudo contact shifts.

Dobson *et al.* (44) have described a process of obtaining the pseudo contact   ratios R_{ij} for a nucleus i with a total observed shift δ^i_{obs} that contains a variable contact contribution from lanthanide to lanthanide, and whose pseudo contact shift is δ^i_{pc} . If there is another nucleus j with negligible contact contribution and total shift δ^j_{pc}, and if the hyperfine constant A for i nucleus does not vary along the series, then a plot of $\delta^i_{obs}/\delta^j_{pc}$ against $J(J+1)g(g-1)/\delta^j_{pc}$ is linear, and its intercept gives the pseudo contact ratio $R_{ij} = \delta^i_{pc}/\delta^j_{pc}$. This method has been used by Dobson *et al.* (44) for separating contact contributions in ^{31}P resonance of 5'-cytosine monophosphate (5'-Cmp). The shift ratio (H'_5 as reference nucleus) found 6.0±1.0 was subsequently used in conformation analysis of 5'-Cmp.

As another example,we describe the application of this method
for separating the contact contribution in induced shifts of
ethylenic carbons of Ln Edta complexes (25). A pseudo contact
ratio (ethylenic protons as reference nuclei) of 1.90±0.1 was
obtained for the nuclei of the ethylenic backbone of Edta, which
agrees well with the value of this ratio derived from X-Ray, as
shown in Table VI.

Whenever gadolinium shifted resonances are not too broadened
a different approach to separate contact from pseudo contact can
be used. Ajisaka and Kainosho (48) have observed large contact
shifts induced by Gd(Fod)$_3$ in the ^{13}C resonances of various amines
nitriles and alcohols. The contact shifts induced by any lanthanide
Fod quelate could be evaluated by multiplying the contact shift
induced by Gd(Fod)$_3$ by an appropriate factor α. The appropriate
value of α was found by making the pseudo contact shift values to
satisfy the geometric factors of each nucleus in question. The
validity of this process was checked by comparing the separated
pseudo contact shitfs with those calculated by Bleaney (10)(Table
II). The correlation is good, as is illustrated in Figure 8 for
the α carbon of octylamine. Furthermore, the ratios of contact
to pseudo contact for this nucleus agree closely with the
theoretical ones and with the ratios obtained by Gansow et $al.$(45)
from the ^{1}H and ^{13}C shifts induced by Ln(Dpm)$_3$ on isoquinoline
ligand, as is shown in Table V.

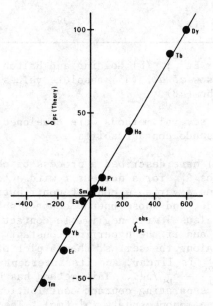

Figure 8. Comparison of theoretical and experimental shifts
          for α carbon of octylamine. The solid line is a least
          square fit (48).

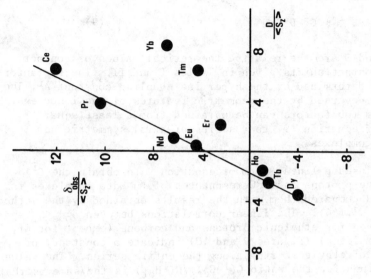

Figure 10. Contact and pseudo contact separation plot for ethylenic carbons of LnEdta complexes. Solid line is a least square fit.

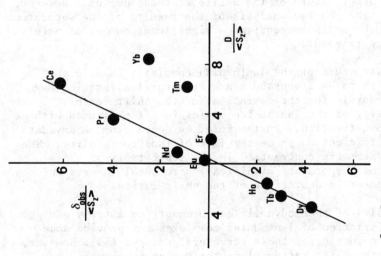

Figure 9. Contact and pseudo contact separation plot for ethylenic protons of LnEdta complexes. Solid line is a least square fit.

In another method developed by Reiley *et al.* (49), the observed total shift δ^i_{obs} of a nucleus i of a substrate bound to a LSR is expressed as

$$\delta^i_{obs} = F^i <S_z> + G^i D \qquad (9)$$

where $<S_z>$ and D are the relative theoretical values of contact and pseudo contact shifts given in Tables I and II. The F^i factor is primarily determined by the hyperfine coupling constant A^i. The G^i factor is governed by the geometric features of the i nucleus. Both of these two factors can be obtained from a least square adjustment of equation 9 to a series of axial symmetric and isomorphous complexes.

We have used a linear form of equation 9 to obtain the G^i values for the protons and ^{13}C resonances of Ln Edta complexes in solution, and compared them with the results obtained by the method of Dobson *et al.*(44). The linear correlations between $\delta^i_{obs}<S_z>$ and $D/<S_z>$ observed for ethylenic protons and carbons (except for Er, Tm and Yb complexes) (Figures 9 and 10) indicate a constancy of the geometry of ethylenic ring along the entire series. The value of 1.90 obtained for the ratio $G(^{13}C_{et})/G(^1H_{et})$ is the same as that obtained from the separation method of Dobson *et al.* (44) and agrees with that given by X-Ray structural determinations of these complexes (Table VI). The low value of the ratio F/G for the ethylenic protons (Figure 9) confirms our previous assumptions on the pseudo contact nature of the shifts of these nuclei. However, care must be taken on the analysis of the results of the separation plots, since the analysis requires a significant number of points in order to be relevant.

The isostructure of the lanthanide complexes is also an important point to be discussed since structural alteration may also be responsible for the deviations of the shift ratios. The isostructurality of the lanthanide series is often assumed without a previous investigation. Furthermore, evidence from X-Ray data may not be sufficient to probe the lack of isostructurality. The slight but systematic contraction in the ionic radii of the lanthanide ions may produce significant structural changes which will be reflected in deviations of the shift ratios.

The results of thermodynamic measurements on entropy and enthalpy of formation of lanthanide complexes may provide some experimental evidence for these structural changes (50). However, constancy of the shift ratios along the series is a sufficient condition for probing isostructurality. This constancy has been observed for the protons of mononucleotides (5), for pyridine and for picoline (52), for tetraaza mycrocycle ligands (32) and for uramildiacetic acid and its derivative series (52). For any

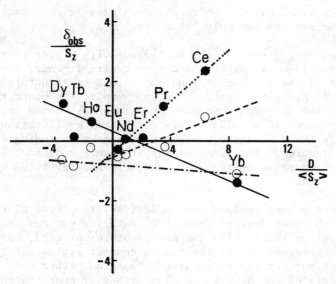

Figure 11. Contact and pseudo contact separation plots for
methylenic protons of acetate groups of LnEdta complexes.
The lines are least square fits for the following halves of
the series: $\cdots\bullet\cdots$ ac_B 1st half; $-\bullet-$ ac_B 2nd half;
$--o--$ ac_A 1st half; $-\cdot o\cdot-$ ac_A 2nd half.

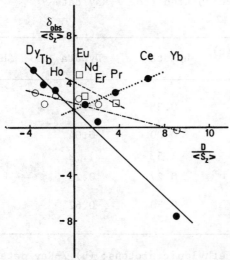

Figure 12. Contact and pseudo contact separation plots for
methylenic carbons and carbonyl carbons of LnEdta complexes.
The lines are least square fits for the following halves of
the series: $\cdots\bullet\cdots$ ^{13}ac 1st half; $-\bullet-$ ^{13}ac 2nd half;
$-\square-$ $^{13}CO_2$ 1st half; $-\cdot o\cdot-$ $^{13}CO_2$ 2nd half.

other nuclei with contact contributions a previous separation is necessary,before one can solve this problem.

On considering results due to ligands with higher sterio-chemical requirements, as Edta and some cyclic ligands, the contraction of the lanthanide ions may result in structural modifications along the series. As an example, the contact separation plot for Edta complexes shows a break in the region of Eu to Tb (Figures 11 and 12). This structural modification affects particularly the carbon and proton atoms near the binding region of water molecules. The atoms of the ethylenic ring in the opposed region of the complex molecule are not very much affected by this structural break.

When the pseudo contact ratios derived from these plots are compared with those derived from X-Ray data for various LnEdta complex species (Table VI), the agreement is not good, but for most of the observed nuclei,the pseudo contact ratios determined in the first part of the series,tends to agree better with those ratios calculated for HLnEdta species with four coordinated water molecules. However, in the second part of the series the values compare better to the calculated ratios for complex series containing only three coordinated water molecules.

Table VI. Comparison of experimental pseudo contact ratios R_{ij}[a] with those calculated from X-Ray data[b]

Atom	Experimental[c]	LaEdta	DyEdta	HLaEdta
ac_B	0.40 -0.17	0.38	0.37	0.48
ac_A	0.19 -0.03	0.00	-0.04	-0.04
^{13}et	1.90	1.53	1.73	1.65
^{13}ac	0.38 -0.22	0.27	0.23	0.35
$^{13}CO_2^-$	-0.50 -0.90	-0.50	-0.62	-0.35

(a) Ratios relative to ethylenic protons; (b) X-Ray data from ref.(17); (c) for each atom,the first value refers to the first half of the series and the second,to the last members of the series.

Another example where the variation of the shift ratios has been initially atributed to structural alterations is that of the complexes of lanthanide ions with α-amino acids.

Levine *et al*. (54) used the indole-3-ylacetate as a model
for studying the conformation of α-amino acids and small peptides
in aqueous solution at pH=6. The fact that the shift ratios for
members of the lanthanide series before Gd(III) remain constant
demonstrates effective axial symmetry for these complexes. For
the later members of the lanthanide series the shift ratios change
from ion to ion, with a strong deviation for Tm^{3+} ion. This was
interpreted as a change in the coordination pattern of the carboxylate group, being bidentate in the first part of the series and
monodentate in the Tm^{3+} ion, with an equilibrium of the two types
of complexes for the middle ions. This explanation was subsequently
extended to the coordination of α-amino acids and small peptides
(55). From measurements of ^1H and ^{13}C shifts of alanine amino
acid in the presence of various aquoions,Sherry and Pascual (56)
reached a different conclusion. They have proposed a monodentate
coordination mode for the light lanthanides and a bidentate mode
for the heavier ones.

Table VII. Dissociation constants for alanine complexes (47) [a]

	Pr^{3+}	Yb^{3+}
$K_1(M)$	0.235±0.015	0.310±0.019
$K_2(M)$	0.94 ±0.15	1.24 ±0.17

(a)   At 39°C and pH = 4.5±0.3

A correct interpretation of the amino acids data was done by
Elgavish and Reuben (47). They have determined the stoichiometry
and the stability of Pr^{3+} and Yb^{3+} complexes of L-alanine, by
performing a group of titrations at constant concentration of
amino acid and another group of titrations at constant lanthanide
concentrations. The results of Table VII show the formation of a
1:1 species and also of a 2:1 species much less stable than the
1:1 species. The constancy of the dissociation constants along
the lanthanide series was taken as an indication of isostructurality
within the series. The isostructurality of α-amino acid complex
series was also suggested by the contact and pseudo contact
separation plots of ^1H and ^{13}C shifts of sarcosine (CH_3 NH CH_2COOH)
as induced by lanthanide ions (47). These plots give good linear
correlations between $\delta^1_{obs}/<S_z>$ and $D/<S_z>$ (except for Er, Tm and Yb
complexes). Thus, the F^1 and G^1 parameters of equation 9 remain

constant along the series and of course no structural changes involving the amino acid ligand should occur. These results are also confirmed by the ^{13}C and ^{1}H longitudinal relaxation rates induced by lanthanide ions in this amino acid (57). In fact, all the distance ratios obtained from the sixth root of the ratios of induced relation rates (Table VIII) exhibit a remarkable constancy along the series. The experimental distance ratios listed in Table VIII agree closely with those ratios calculated from a bidentate model for the coordination of the carboxylic group (proton: CH_3/CH_2, 1.23; carbon: CH_2/CO_2^-, 1.41; CH_3/CO_2^-, 1.48). The same conclusion was obtained by Kieboom *et al.* (58) by measuring the [13]C T_1 relaxation rates of acetate ion in the presence of low concentrations of Gd^{3+}. The experimental distance ratio $CH_3/CO_2^- = 1.52$ determined, was also consistent with a bidentate mode of coordination for the acetate group.

Table VIII.   Experimental ratios of distances between lanthanide ions and magnetic nuclei of sarcosine (a)

Ln^{3+}	ρ	Proton CH_3/CH_2	Carbon-13 CH_2/CO_2^-	CH_3/CO_2^-
Pr	0.093	1.21	1.34	1.58
Nd	0.095	1.20	1.42	1.76
Gd	0.002	1.26	1.34	1.69
Tb	0.045	1.23	1.37	1.71
Dy	0.047	1.21	1.38	1.74
Ho	0.061	1.22	1.33	1.68
Er	0.047	1.20	1.32	1.67
Tm	0.057	1.17	1.38	1.67
Yb	0.083	1.19	1.30	1.64
Mean		1.21 ±0.03	1.35 ±0.04	1.68 ±0.07

(a) Obtained from the sixth root of the ratios of longitudinal relaxation rates (57).

These results do not rule out the possibility of a monodentate coordination mode for the acetate group in these monocarboxylate compounds as was shown by Elgavish and Reuben (47) for sarcosine amino acid. The relative populations for the monodentate complex, 0.026 and for the bidentate one, 0.974, were obtained by a method previously described by those authors (59). The lanthanide oxygen distance obtained in this way, 2.85 A, agrees with crystallographic

data on carboxylate complexes (60), but gives an average distance
for methyl carbon to lanthanide, 5.37 A, which does not agree with
the value, 4.14 A obtained from shift and relaxation data, assuming
a unique position for the lanthanide ion. It thus appears that a
very small contribution, only 2.6% of monodentate coordination, is
sufficient to render the determination of a unique lanthanide
position uncertain.

As we pointed out earlier the finding of a variation of the
R_{ij} along the lanthanide ion series, can in principle be due to:
1) variation of the complex structure along the series;
2) contribution from the contact shift mechanism; and 3) lack of
axial symmetry. Analysis of the α-amino acid complexes data (47)
shows that isostructurality prevails along the series. Furthermore
analysis of the induced shifts of the ethylenic protons of LnEdta
have shown that the variation of R_{ij} observed for the last three
lanthanide ions (Er,Tm and Yb) can not be caused by a contribution
from a contact mechanism. Thus, these deviations should be due to
a lack of axial symmetry. When the symmetry is not axial the second
term of equation 2 must also be considered. Since D and D'
parameters of this equation are not proportional to each other and
change appreaciably along the series (18), the shift ratios of the
same nuclei for different lanthanide ions do not remain constant.

This is supported by the determination of the contribution of
the axial and non axial terms of equation 2, for the proton shifts
of the methyl group of 4-picoline in $Ln(Dpm)_3$ (4-picoline)$_2$ complex
series. In this isomorphous series, the contribution of the non
axial term was found to be less than 15%. However, for thulium
this contribution was found to be much higher. Although this is not
a general result, it shows that even for a case where the geometric
axial factor is about 26 times greater than the non axial geometric
factor, the high value of D' when compare to D as for thulium,can
be responsible for the variation of the shift ratios.

Another point we should discuss concerns the possibility of
conformation changes of the ligand molecule on binding to a LSR.
This is a crucial point since if there is such a change the lantha
nide probe method can not give us information about the ligand
conformation but only about the structure of the complex.

Changes on the conformation of a substrate ligand by addition
of a lanthanide may be monitored by measuring the chemical shifts
of the ligand nuclei, the vicinal coupling constants, and by
performing NOE experiments in the absence and in the presence of
a diamagnetic lanthanide ion. As an example, Dobson et al. (68)
have shown that on addition of La^{3+} to 5'-Amp nucleotide (Figure
13) the proton chemical shifts and the proton vicinal coupling
constants remain essentially constant. This indicates that no

conformation changes takes place on binding a lanthanide ion to
the phosphate group of this molecule.

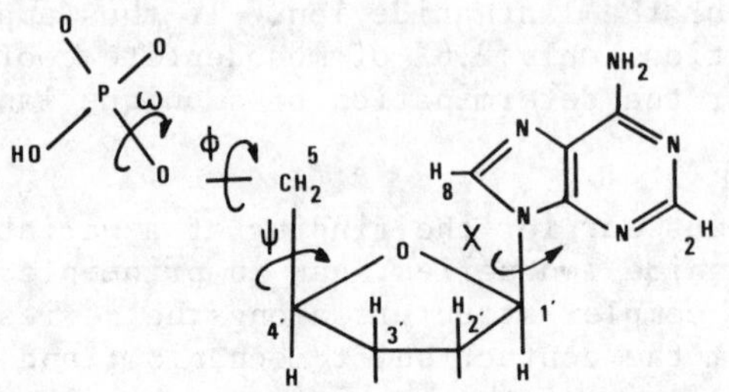

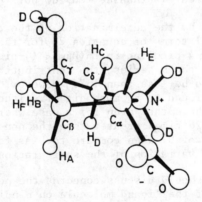

Figure 13.   Molecular structure of 5'-Amp, showing the bonds
     around which rotations may occur.

The same type of conclusion were obtained by Inagaki *et al.*
(61) for hydroxy-L-proline amino acid in aqueous solution (Figure
14). The spin-spin splittings of the spectra of this molecule
are not altered upon complexation with La^{3+}. Even the long range
coupling constant (2.22 Hz) for the pair H_B and H_C protons in a
W-type arrangement does not change upon complex formation.

Figure 14.   Molecular conformation of hydroxy-L-proline derived
     by the lanthanide probe method (61).

3.3.   Conformation Analysis by the Lanthanide Probe Method

Once we have discussed some important points related to the
applicability of the lanthanide probe method, the next task is
concerned with the search for a molecular conformation which best
fits simultaneously the shift and relaxation ratios. We will
discuss here some examples of computation methods applied to
rigid and flexible molecules.

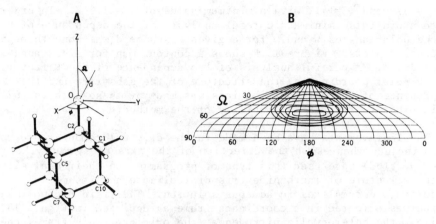

Figure 15.   (A) Coordinated system for orientation of 2-adamantanol
(B) Agreement factor plot for the oxygen lanthanide distance
of 2.9 Å, using Yb(Dpm)₃ data. The contours show regions of
constant agreement factor. The outer contour is at 15% and
the inner contour is at 3%; the contour intervals are 3% (62).

     Essentially, two types of computation methods have been
discussed in the literature for finding the location of the
lanthanide ion in rigid molecules. Willcott *et al.*(62) have
developed a process in which the structure of a rigid molecule
like 2-adamantanol is described via a cartesian frame, with origin
at the oxygen directly coordinated to the lanthanide ion.   The
location of the lanthanide ion is defined by the three classic
polar coordinates, the distance d and the angles Ω and ϕ of Figure
15 A, and the direction of the axial axis is taken as coincident
with that of the lanthanide oxygen bond. The location of the
lanthanide is changed through a computer programme, which for a
fixed distance d increases the two polar angles, and for each
location of the lanthanide ion calculates the geometric factors
$(3\cos^2 \theta_i - 1)/ r_i^3$ for all the observed nuclei of the substrate.
These features are then scaled against the observed shifts δ_i^{obs}
and a set of calculated shifts $\delta_i^{calc}$is obtained. For assessing
the goodness of the fit, the normalized variance R is used,

$$R = \left[\frac{\sum\limits_{i} (\delta_i^{calc} - \delta_i^{obs})^2}{\sum\limits_{i} (\delta_i^{obs})^2} \right]^{1/2}$$

If the observed shifts are known to different degrees of reliability
a weighting factor may be introduced in the definition of R.   For
the shifts induced by Yb(Dpm)₃ in the nine proton resonance of
2-adamantanol,the best location was achieved for d = 3.33 Å,

$\Omega = 57^{\circ}$ and $\phi = 141\%$ with a minimum value of R = 1.3 %. In order
to avoid false minima a careful analysis of the dependence of R on
the polar angles Ω and ϕ for a given distance d,was done through
contour plots. Figure 15 B shows a contour map for a distance
of 2.9 A. A careful analysis of the dependence of the R factor on
the polar coordinates, the structure of the substrate and the
presence of contact contributions was made by these authors (62),
in order to test the reliability of the method.

Improvements have been made in the original computer programmes
to include changes in the direction of the axis of symmetry.Reilley
et al. (63) have used this type of programmes on their analysis of
the structure of lanthanide tris dipicolinate and its methyl
derivative complexes in aqueous solution. The rigid dipicolinate
ion was oriented in a coordinate frame as depicted in Figure 16,
i.e., the heterocyclic nitrogen at the origin of this frame, one
ligand atom ($\underline{C}OO^-$) along the $-z$ axis and a second ligand atom (C_α)
in the x z plane. In a second step the location of the lanthanide
ion was found through a successive modification of the three polar
coordinates (d,Ω, ϕ) of the lanthanide ion and the two angles,α
and β, which define the direction of the axial axis, until a good
agreement exists between calculated and experimental G^1 parameters
($G^1 = (3 \cos^2 \theta_i - 1)/ r_i^3$),for all the observed nuclei of the
ligand. The best location, with R = 3,5 %, gives a tricapped
trigonal prismatic structure for the complex, in which the three
pyridinic rings were found to be tilted at an angle of 40°-41°
with respect to an equatorial plane that contains the 3 nitrogens
and is prependicular to the C_3 axis. This structure is very close
to that obtained by X-ray (64).

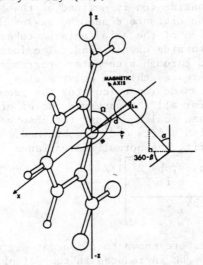

Figure 16. Orientation of Dpa ion in a coordinate frame, and
     illustration of d,ϕ,Ω,α and β parameters.

The results for tris complexes have been confirmed by
Alsaadi *et al.* (24) on a detailed NMR study of the structures of
mono, bis and tris dipicolinate lanthanide complexes. The ratios
for the shifts of meta to para protons for the three types of
complexes were found to remain constant along the lanthanide series,
on a clear evidence for isostructurality, dipolar nature of the
shifts, and existence of axial symmetry in all of them. The axial
symmetry in mono and bis complexes seems to arise from internal
rotations of the pyridinic rings around N-Ln bond, which averages
the shifts of the two meta protons at high temperatures. The values
of canted angles derived from the shift ratios δ_m/δ_p for tris and bis
complexes are 46 ± 1° and 47 ± 1°, very close to the value of 40°-41°
obtained by Reilley *et al.* (63) for the tris complexes. Like the
tris complexes the bis complexes adopt a tricapped trigonal
prismatic stucture (completed with three water molecules) in which
the angle N-Ln-N is 135 ± 1°, a value slightly greater than that of
120° for a perfect trigonal prism.

Similar structure search processes have been applied to the
determination of the location of the lanthanide ion in complexes
with the cyclic ligand Dota (52). The lanthanide was found to be
inside a cavity with a C_4 symmetry axis at a distance of $\simeq2.85$ Å
from the nitrogen atoms.

Barry *et al.* (65) described a programme, MSEARCH, to be used
for molecules containing rigid units. The programme uses as input
parameters the molecular atomic coordinates of the rigid units of
the molecule as well as the ratios of experimental dipolar shifts
and relaxation rates to those of a reference nucleus. The search
is done by changing the position of the lanthanide and the direction
of the symmetry axis, by changing the parameters d, Ω, ϕ, α and β of
Figure 16 in given steps. For each position of the lanthanide the
programme calculates the values of $(3 \cos^2 \theta_i - 1)/ r_i^3$ and of $1/r_i^3$
and takes their ratios to a given reference nucleus. These ratios
are then compared to those derived from experimental shift and
relaxation ratios. The overall goodness of the fit relies on the
tolerances that are imposed to the experimental ratios, according
to their experimental errors. If the calculated ratios for all
observed nuclei are within the set of tolerances a solution is
obtained.

This approach has been applied to the molecule of cholesterol
in chloroform using $Ln(Dpm)_3$ as probes (65). After carrying out
several searches with successively lower values of tolerances, a
restricted number of solutions with very similar conformations
were printed out. A member of this group is taken as representative
of the conformation of the molecule. For the cholesterol molecule,
the metal was found to be coordinated to the oxygen of the OH group
and the axis of symmetry was directed along the M-O bond.

The first approach to the conformational problem of flexible
molecules in solution by the lanthanide probe method was tried by
Barry *et al.* (35) for the case of 5'-Amp mononucleotide at pH=2.0.
In this molecule there are four bonds about which rotation is
possible (Figure 13). These bonds are then rotated in chosen
steps (usually 4°) by a computer programme BURLESK, so that a
large number of new molecular conformations is generated. The
X-ray conformation is used as a starting point in this generation
process. Each of these conformations can then be tested against
shift and relaxation data in the manner described for the MSEARCH
programme, when the location of the lanthanide and the direction
of the axis of symmetry are known. A bidentate coordination
fashion was found for the double ionized phosphate group, with the
lanthanide located in the bissector of the PO_2^- group at a distance
of 2.9 Å from the phosphorus atom, and the axis of symmetry
located along this direction.

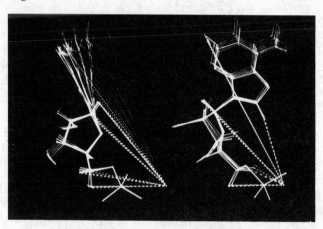

Figure 17.   Ortogonal views of the acceptable conformations
        of 5'-Amp (35).

In the 5'-Amp several groups (families) of closely related
conformations were consistent with shift data, but only one of
these families was consistent with Gd^{3+} relaxation data (Figure 17)
as a consequence of the tolerance limits imposed to the NMR shift
and relaxation ratios. This family of solutions gives a distorted
2' endo conformation for the ribose ring, a conformation about $C_{5'}$-
-$C_{4'}$ bond close to gauche trans (ψ = 139°), a conformation about
$C_{5'}$-$O_{5'}$ bond close to gauche-gauche (ϕ = 167°), a lanthanide
position defined by an angle ω = 45° about the $O_{5'}$-P bond and an
anti conformation for the adenine base around the glycosidic bond.
Figure 18 gives further details about the most stable local
conformations of a 5'-ribonucleotide. The structure obtained is
very similar to the X-ray one and it was shown to be essentially
independent of pH and temperature (66). Very similar conformations

were also obtained for other 5'-ribonucleotides in aqueous solution
(5). However, when the conformations derived by the lanthanide
probe method are compared with those derived from independent
methods, namely from coupling constants analysis, relaxation
measurements and NOE observations, the agreement is not always good.
This is the case observed when there is a conformation equilibrium.
In effect, it is known from vicinal coupling constant analysis (67)
that some 5'-ribonucleotides exist in solution as a conformation
equilibrium. The ribose ring exists predominantly in a 3' endo(N)$\rightleftarrows$
$\rightleftarrows$ 2' endo(S) equilibrium and the rotomers gg and g'g' for rotation
about the $C_5{'}-C_4{'}$ and $O_5{'}-C_5{'}$ bonds being preferred (Figure 18).
The base exists also in a syn$\leftrightarrow$anti equilibrium about the
glycosidic bond, the anti position being in general the most
populated (68).

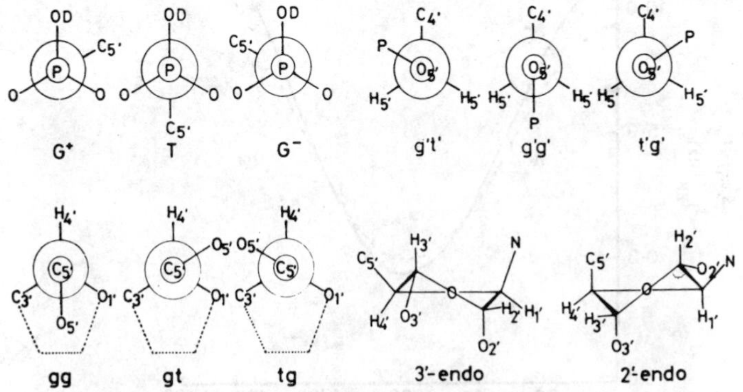

Figure 18. Most stable local conformations of a 5'-ribonucleotide

It is apparent from these results that the lanthanide method
gives a description of a flexible molecule in terms of an average
conformation with loss of information compared to an identification
of individual conformers and their lifetimes. This is the reason
why the overall features of the molecular conformation agree with
those derived from other methods, but for the local features the
agreement is not always good.

In a recent paper (69) Jardetzky discussed the meaning of the
conformations derived from NMR parameters of flexible molecules.
There are only two situations in which a real interpretation of
the conformation problem of flexible molecules can be done from
average NMR data: (1) the case of rigid molecules, with only one
conformer, and (c) those cases of flexible molecules, with several
conformers with finite lifetimes.

The first situation is typified by sufficiently rigid molecules.
as are hydroxy-L-proline and 3',5'-cyclic nucleotides for which
the conformation derived from lanthanide probe method matches the
results of coupling constant analysis.

The molecular conformation of hydroxy-L-proline in solution
(Figure 14) was derived by Inagaki *et al.* (61) from the lanthanide
induced shifts and relaxation rates. The plot of the six vicinal
constants of protons of proline ring against the corresponding
dihedral angles determined from previous conformation (Figure 19)
shows a reasonable agreement with curves calculated by the Karplus
equation using the coefficient of Gerig and McLeod (solid line)(70)
and those of Altona and Sundaralingam (dashed line) (67).

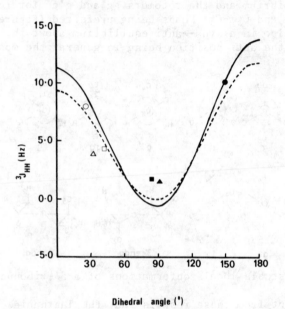

Figure 19. Plot of vicinal coupling constants of hydroxy-L-proline
   against the dihedral angles: ● J_{AE}; ○ J_{BE}; □ J_{AF}; ■ J_{BF};
   ▲ J_{CF}; △ J_{DF} (61).

The cyclic mononucleotides which also bind a lanthanide ion
at the phosphate group are fairly rigid structures with internal
rotation about the glycosidic bond. The conformation of 3',5'-
-cTmp nucleotide in D_2O at pH = 2.02, was derived by Inagaki *et al.*
(71) using the lanthanide probe method. As for the case of the
proline ring of hydroxy-L-proline a reasonable agreement exists
between seven vicinal coupling constants measured in the two inter-
locking rings of this conformation and the empirical curve
calculated by the Karplus equation.

Coupling constant analysis does not help very much in the
elucidation of the syn←→anti equilibrium of the base in nucleotides
and only the result of NOE experiments may confirm the orientation
given by the lanthanide search. In 3',5'-cAmp (72) the lanthanide

search gives an orientation of adenine base dependent on the pH.
At pH=2.0 the orientation is essentially anti, but at pH=5.5 the
adenine base becomes deprotonated (pK=3.8) and stays preferentially
in a syn position. The anti orientation of adenine base at pH=2.0
was checked by NOE observations. Saturation of H_2 resonance
produces changes in the intensity of H_8 but not in H_2, as was
expected from an anti orientation of the base. At pH=5.5 saturation
of H_2 produces simultaneously a change in the intensity of H_2 and
H_8, and no definitive conclusion can be drawn. The consistence of
these two methods of analysis in this type of molecules is very
important for the elucidation of molecular conformation equilibrium
in more flexible molecules.

The situation in which a flexible molecule exists in solution
as a mixture of conformers is no doubt the most common one. In
this case the average conformation derived from NMR data has no
real significance if the individual conformers are not known.
However, if individual conformers are known by independent methods
one can relate average NMR parameters with the geometry of
individual conformers. Birdsall *et al*. (73) used this last process
to estimate the relative populations of the dominant conformers of
the flexible nicotinamide mononucleotide (NMN) by fitting the
lanthanide induced shifts to a mixture of conformations thought to
be present in solution from the results of coupling constants
analysis, NOE observations and theoretical potential energy
calculations.

This procedure was adopted by Geraldes and Williams (74) in
order to get a quantitative picture of the conformation state of
5'-Amp in solution. In the first step of this analysis the ribose
ring was taken as existing in a mixture of 50% N and 50% S
conformers and the phosphate backbone as existing in a single
conformation close to that derived from coupling constants ($\phi=180^{\circ}$,
g'g'; $\psi=60^{\circ}$, gg). The geometries of these conformers were then
generated from atom coordinates of the crystal structure. In a
second step the position of the lanthanide was changed by rotation
about the P-O$_5$' bond (ω) by means of the BURLESK programme. The
pseudo contact shifts and distances were then calculated for each
lanthanide position and for each conformer and were compared with
the observed shift and relaxation ratios. For 100% of gg rotamer
it was not possible to fit the observed ratios, but by increasing
the weight of gt and tg conformers to 30%, in agreement with
coupling constant measurements, an excellent fit was achieved for
both shift and relaxation ratios.

The final step of this search process is the investigation of
the syn and anti ranges of the adenine base conformation, using
contour diagrams for the R factors of shift and relaxation parameters of the base protons. The mixture of conformations for the
ribose ring and exocyclic group also gives a much better minimum

for R factors of shifts and relaxation parameters (Figure 20). The best solution is then defined by a family of solutions in the anti region centered in the value $(\chi_N, \chi_S) = (30^\circ, 30^\circ)$, showing that the adenine base in 5'-Amp is quite rigid. This type of analysis has been extended by these authors to other mononucleotides and to dinucleotides (75).

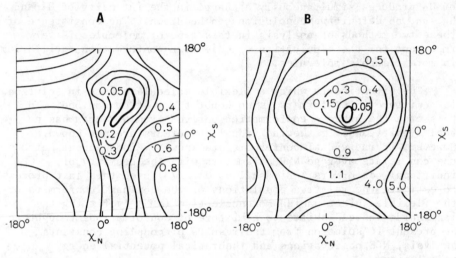

Figure 20.   Contour diagrams of the R factor for base protons) in
        5'-Amp.   (A) shift data;   (B) relaxation data;   (A,B) 70% gg+
        + 30% (gt+tg) (74).

Inagaki *et al.* (76) have performed a similar analysis for the conformation of uridine-5'-monophosphate (5'-Ump) in solution. These authors have written a programme named PCS2R that uses simultaneously the shift, relaxation and coupling constant data of a flexible molecule to determine the mixture of conformers that best fit this group of experimental parameters. For 5'-Ump the best fit found is the mixture of four conformers shown in Table IX. It is clear from this table that the total population of gg conformers amounts to 75%, that pucker of ribose ring is 43% 3' endo and 2' endo and that the base is essentially anti, which is a situation very similar to that depicted for 5'-Amp. It is also evident that there is a clear correlation between the local conformation of the exocyclic group and the pucker of the ribose ring, the 3' endo conformer being associated only with the gg conformer of exocyclic group, while the 2' endo conformer does not show any specific choice. This correlation was previously suggested by analysis of spin coupling constants (77) and by conformation energy calculations (78).

Table IX. Populations of the four major conformers of
5'-Ump (76)

$C_{4'}-C_{5'}$	Ribose ring	Base	Population[a]
gg	3' endo	anti	46%(12%)
gg	2' endo	anti	29%(11%)
gt	2' endo	anti	14%(5%)
tg	2' endo	anti	11%(8%)

[a] Standard deviations in populations in parentheses.

REFERENCES

1. M.Karplus, J.Am.Chem.Soc., 85, 2870 (1963).
2. J.H.Noggle, R.E.Schirmer, The Nuclear Overhauser Effect, Academic Press, N.York (1971).
3. C.C.Hinckley, J.Am.Chem.Soc., 91, 5160 (1960).
4. J.Reuben, Prog.NMR Spectrosc., 9, 1 (1979).
5. C.M.Dobson, B.A.Levine, in "New Techniques in Biophysics and Cell Biology" (R.H.Pain, B.E.Smith, eds.) Vol.3, Wiley, N.York, 1976, pp.19-91.
6. F.Inagaki, T.Miyazawa, Prog.NMR Spectrosc., 47, 67 (1981).
7. W.B.Lewis, J.A.Jackson, J.E.Lemons, H.Taube, J.Chem.Phys., 36, 694 (1962).
8. R.M.Golding, M.P.Halton, Aust.J.Chem., 25, 2577 (1972).
9. J.Reuben, D.Fiat, J.Chem.Phys., 51, 4909 (1969).
10. B.Bleaney, J.Mag.Res., 8, 91 (1972).
11. R.M.Golding, P.Pyykkö, Mol.Phys., 26, 1389 (1973).
12. H.M.McConnell, R.E.Robertson, J.Chem.Phys., 29, 1361 (1958).
13. B.Bleaney, C.M.Dobson, B.A.Levine, R.B.Martin, R.J.P.Williams, A.V.Xavier, J.Chem.Soc.(Chem.Commun.), 791 (1972).
14. C.D.Barry, C.M.Dobson, R.J.P.Williams, A.V.Xavier, J.Chem. Soc.(Dalton Trans.) 1765 (1974).
15. W.Dew.Harrocks,Jr., J.P.Sipe III, J.R.Luber J.Am.Chem.Soc., 93, 5258 (1971).
16. R.E.Cramer, K.Seff, J.Chem.Soc.(Chem.Commun.) 400 (1972).
17. J.L.Hoard, B.Lee, M.D.Lind, J.Am.Chem.Soc., 87, 1612 (1965); B.Lee, Ph.D.Thesis, Cornell University, N.York (1967); L.R. Nassimbeni, M.Robert, W.Wright, J.C.van Niekerk, D.A.McCallum, Acta Cryst., B 35, 1341 (1979).

18. W.DeW.Horrocks,Jr., J.D.Sipe III, Science, 117, 994 (1972).
19. J.M.Briggs, G.P.Moss, E.W.Randall, K.D.Sales, J.Chem.Soc. (Chem.Commun.) 1180 (1972).
20. W.DeW.Horrocks, J.Am.Chem.Soc., 96, 3022 (1974).
21. N.Bloembergen, L.O.Morgan, J.Chem.Phys., 34, 842 (1961).
22. R.A.Dwek, R.E.Richards, K.G.Morallee, E.Niebor, R.J.P.Williams, A.V.Xavier, Eur.J.Biochem., 21, 204 (1971); S.H.Koenig, M. Epstein, J.Chem.Phys., 63, 2279 (1975).
23. R.V.Southwood-Jones, W.L.Earl, E.Newman, A.E.Merbach, J.Chem. Phys., 73, 5409 (1980).
24. B.M.Alsaadi,F.J.C.Rossoti, R.J.P.Williams, J.Chem.Soc. (Chem. Commun.) 527 (1977); B.M.Alsaadi, F.J.C.Rossoti, R.J.P. Williams, J.Chem.Soc.(Dalton Trans.) 597 (1980).
25. J.Ascenso, Ph.D.Thesis, I.S.T., Lisbon (1981).
26. G.H.Dieke, Spectra and Energy Levels of Rare Earth Ions in Crystals (H.M.Crosswhite and H.Crosswhite, eds.) Wiley, N.York, 1968, Chapter 13.
27. R.D.Shanoon, C.J.Preitt, Acta Cryst., B 25, 925 (1969).
28. W.DeW.Horrocks,Jr., D.R.Sudnick, Acc.Chem.Res., 14, 384 (1981).
29. B.M.Alsaadi, F.J.C.Rossoti, R.J.P.Williams, J.Chem.Soc. (Dalton Trans.), 2151 (1980).
30. P.D.Burns, G.N.LaMar, J.Mag.Res., 46, 61 (1982).
31. S.P.Sinha, J.Mol.Struct., 19, 387 (1973).
32. R.B.Lewis, E.Wenkert, in "Nuclear Magnetic Shift Reagents" (R.E.Sievers, ed.), Academic Press, N.York, 1973, pp.99-127.
33. C.Kutal, in "Nuclear Magnetic Shift Reagents", (R.E.Sievers, ed.) Academic Press, N.York, 1973, pp.87-98.
34. M.R.Willcott, R.E.Lenkinski, R.E.Davis, J.Am.Chem.Soc., 94, 1742 (1972).
35. C.D.Barry, A.C.T.North, J.A.Glasel, R.J.P.Williams, A.V.Xavier, Nature, 232, 236 (1971).
36. G.A.Elgavish, J.Reuben, J.Mag.Res., 42, 242 (1981).
37. C.M.Dobson, L.O.Ford, S.E.Summers, R.J.P.Williams, J.Chem.Soc. (Faraday II) 71, 1145 (1975).
38. C.Bryden, C.N.Reilley, J.C.Desreux, Anal. Chem., 53, 1418(1981).
39. R.Delgado, J.J.R.Frausto da Silva, Talanta, 29, 815 (1982).
40. F.J.C.Rossoti, H.Rossotti, Determination of Stability Constants, McGraw Hill, N.York (1961).
41. K.B.Yatsimirskii, N.K.Davidenko, Coord.Chem.Rev., 27, 223 (1979); G.Geier, C.K.Jørgensen, Chem.Phys.Letters, 9, 263 (1971).
42. F.Inagaki, S.Takahashi, M.Tasumi, T.Miyazawa, Bull.Chem.Soc. Japan, 48, 853 (1975).
43. F.Inagaki, M.Tasumi, T.Miyazawa, Bull.Chem.Soc. Japan, 48, 1427 (1975).
44. C.M.Dobson, R.J.P.Williams, A.V.Xavier, J.Chem.Soc. (Dalton Trans.) 23, 2662 (1973).
45. O.A.Gansow, P.A.Loeffler, R.E.Davis, R.E.Lenkinski, M.R. Willcott, J.Am.Chem.Soc., 98, 4250 (1976).
46. O.A.Gansow, P.A.Loeffler, R.E.Davis, M.R.Willcott, R.E.

Lenkinski, J.Am.Chem.Soc., 95, 3390 (1973).

47. G.Elgavish, J.Reuben, J.Magn.Reson., 39, 421 (1980).

48. K.Ajisaka, M.Kainosho, J.Am.Chem.Soc., 97, 330 (1975).

49. C.N.Reilley, B.W.Good, R.D.Allendoerfer, Anal.Chem., 48, 1446 (1976).

50. G.Geier, U.Karlen, Helv.Chim.Acta, 54, 135 (1971); G.Anderegg F.Wenk, Helv.Chim.Acta, 54, 216 (1971);

51. J.N.LaMar, J.W.Faller, J.Am.Chem.Soc., 95, 3817 (1973).

52. J.F.Desreux, C.N.Reilley, J.Am.Chem.Soc., 98, 2105 (1976); J.F.Desreux, Inorg.Chem., 19, 1319 (1980).

53. J.Ascenso, M.Cândida T.A.Vaz, J.J.R.Frausto da Silva, J.Inorg. Nucl.Chem., 43, 1255 (1981).

54. B.A.Levine, J.M.Thornton, R.J.P.Williams, J.Chem.Soc.(Chem. Commun.) 16, 669 (1974).

55. B.A.Levine, R.J.P.Williams, Proc.Royal Soc. (London) A 345, 3 (1975).

56. A.D.Sherry, E.Pascual, J.Am.Chem.Soc., 99, 5871 (1977).

57. G.A.Elgavish, J.Reuben, J.Am.Chem.Soc., 100, 3617 (1978).

58. A.P.G.Kieboom, C.A.M.Vijverberg, J.A.Peters, H.van Bekkum, Recl.Trav.Chem.,96, 315 (1977).

59. R.E.Lenkinski, J.Reuben, J.Am.Chem.Soc., 98, 4065 (1976).

60. S.P.Sinha, Struct.Bonding (Berlin), 25, 69 (1976).

61. F.Inagaki, M.Tasumi, T.Miyazawa, J.Chem.Soc. (Perkin II), 167 (1976).

62. M.R.Willcott, R.E.Davis, Science, 190, 850 (1975).

63. C.N.Reilley, B.M.Good, J.F.Desreux, Anal.Chem., 47, 2111 (1975).

64. J.Albertsson, Acta.Chem.Scand., 24, 1213 (1970); J.Albertsson, ibid, 26, 985 (1972).

65. C.D.Barry, H.A.O.Hill, P.Sadler, R.J.P.Williams, in "Nuclear Magnetic Resonance Shift Reagents, (R.E.Sievers, ed.) Academic Press, N.York, 1973, pp. 173-195.

66. C.D.Barry, J.A.Glasel, R.J.P.Williams, A.V.Xavier, J.Mol.Biol. 84, 471 (1974).

67. C.Altona, M.Sundaralingam, J.Am.Chem.Soc., 95, 2333 (1973); D.B.Davis, S.S.Danyluk, Biochemistry, 13, 4417 (1974).

68. M.Dobson, C.F.G.C.Geraldes, G.Ratcliffe, R.J.P.Williams, Eur. J.Biochem., 88, 259 (1978).

69. O.Jardetzky, Biochim.Biophys.Acta, 621, 227 (1980).

70. J.T.Gerig, R.S.McLeod, J.Am.Chem.Soc., 95, 5725 (1973).

71. F.Inagaki, S.Takahashi, M.Tasumi, T.Miyazawa, Bull.Chem.Soc. Japan, 49, 611 (1976).

72. C.D.Barry, D.R.Martin, R.J.P.Williams, A.V.Xavier, J.Mol.Biol. 84, 491 (1974).

73. B.Birdsall, N.J.M.Birdsall, J.Feeney, J.Thornton, J.Am.Chem. Soc., 97, 2845 (1975).

74. C.F.G.C.Geraldes, R.J.P.Williams, Eur.J.Biochem., 85, 463 (1978).

75. C.F.G.C.Geraldes, R.J.P.Williams, Eur.J.Biochem., 97, 93(1979).

76. F.Inagaki, M.Tasumi, T.Miyazawa, Biopolymers, 17, 267 (1978).

77. F.E.Hruska, in "Conformation of Biological Molecules and

Polymers", The Jerusalem Symposia on Quantum Chemistry and
Biochemistry" (E.D.Bergmann and B.Pullman, eds.) Vol.3,
Academic Press, N.York, 1973, p.345.
78. N.Yathindra, M.Sundaralingam, Biopolymers, 12, 297 (1973).

DISCUSSION

GROUP II (question)

Could you describe one or two examples of the results on the use of lanthanide shift reagents and the advantages obtained by using them?

ASCENSO (answer)

There are many examples in the literature showing particular advantage of using a specific lanthanide shift reagent. I have here two examples that show how the NMR spectra are perturbed by the addition of a lanthanide shift reagent. In the first case, the addition of successive amounts of $Eu(dpm)_3$ to a solution of cyclooctatetraene epoxide enhances tremendously the resolution of the NMR spectra of this compound. By doing spin-spin decoupling it was also possible to assign the seventeen peaks of the shifted spectra of this molecule. In the second case the assign unit of ten resonances in the spectrum of cis-4-tert-butyl-cyclohexanol in $CDCl_3$ by addition of $Eu(dpm)_3$ proved to be very useful in confirming the presence of this ring in androstan-2ß-ol steroid.

MÖLLER (question)

Would it be possible to measure the individual rare earths in frozen melts? Would it be possible to find out if the lanthanides are present in a complexed form in a melt, i.e. surrounded by silica or alumina in a complex?

ASCENSO (answer)

You will have experimental difficulties. You need homogeneous magnetic field inside the probe. Special probe is needed. There are measurements on solid samples. Here you would expect ion-ion dipolar interactions. With heteronuclear magnetic resonance you would be able to do this, but the natural abundance of your ions must be high for good sensitivity.

GROUP II (question)

When you use a lanthanide complex as a shift reagent do you assume that the complex is non-labile or do you assume that the original complex is replaced by bonding of the lanthanide ion to the organic molecule of interest?

ASCENSO (answer)

Of course I am assuming that the lanthanide shift reagents are non-labile complexes so that the complex entity is present in solution. Only in these circumstances special properties like so-lubility in organic solvents, relative high stability constants of 1:1 complexes compared to 1:2 or higher complexes, and low con-tact contribution to the shifts are obtained. The binding to the substrate molecules should be sufficiently weak in order that no conformation change occurs in the substrate molecule. Strong bon-ding will produce an increase in non-axial symmetry which is not a case for easy solution.

NETZ   (question)

Could you give an estimate of the amount of error in relaxation rates?

ASCENSO   (answer)

Errors are generally within $\sim$ 20 %, a value which is normally ac-cepted.

URLAND   (question)

It concerns the theory by Bleaney for pseudocontact shifts. As I recall it was derived in 1972 [J. Mag. Res. 8, 91 (1972); see also J.C.S. Chem. Comm. 791 (1972): Editor]. I am surprised to see that this theory is still in use.

ASCENSO   (answer)

There are some corrections introduced by Golding [Aust. J. Chem. 25, 2577 (1972), Mol. Phys. 26, 1389 (1973): Editor] which include other crystal field terms. The correction amounts to $\sim$ 10 %, and that is within our error limit.

SINHA   (question)

What kind of shift you got for the dipicolinates?

ASCENSO   (answer)

Two meta protons shifted in opposite direction. The para proton has much lower shift. Shifts are also temperature dependent.

SINHA   (question)

Say you have a pure hydrocarbon as substrate. Would you get enough interaction to see a significant shifting of the spectrum?

ASCENSO   (answer)

As you know, you have a small binding constant. The perturbation depends also on the amount of lanthanide you have in solution, so that you can choose the best concentration to see the best effects.

Lanthanide Geochemistry

THE LANTHANIDES AS GEOCHEMICAL TRACERS OF IGNEOUS PROCESSES:
AN INTRODUCTION

Jean-Clair DUCHESNE

Géologie, Pétrologie, Géochimie, Université de Liège,
B-4000 SART TILMAN, Belgium

ABSTRACT

Igneous processes – crystallization or partial fusion – which
imply mineral/melt relations strongly control the REE distribution
in the various materials of the Earth's crust. The concept of mine-
ral/melt distribution coefficient, which can be determined experimentally, permits the use of mathematical models to describe in a
more quantitative way partial fusion and crystallization. This
leads to a better understanding of the petrogenetic mechanisms and
also of the nature of rocks and minerals in the source region of
magma formation. Nd isotope studies can be used as a geochronolo-
gical tool as well as to determine the value of Sm/Nd, i.e. the
REE distribution, in the past.

1. INTRODUCTION: GENERAL CHARACTERS OF THE RARE EARTH ELEMENTS
   IN NATURE

The rare earth elements (REE) form a coherent group in nature.
Their chemical properties vary progressively with the atomic number. They display regular distributions which need major geological
processes to be modified.

Minerals in which REE are major constituents are exceptional.
In normal conditions, REE occur as trace elements in common rock-
forming minerals. They can thus be studied in a large variety of
rocks and provide a better understanding of the formation and evolution processes of these rocks.

S. P. Sinha (ed.), Systematics and the Properties of the Lanthanides, 543–560.
Copyright © 1983 by D. Reidel Publishing Company.

Since REE are present at a very low level of concentration (ppm to ppb), they satisfy Henry's law: concentrations can be used in place of activities. This fundamental property permits the use of REE in quantitative modeling of petrogenetic phenomena. They can be considered as "tracers" of physico-chemical processes.

Among the REE, Sm possesses a radioatcive isotope ^{147}Sm which decays in ^{143}Nd with a half-life sufficiently long to be used as a geological clock. Moreover, measurement of ^{143}Nd/^{144}Nd ratios in rocks of different ages permits to reconstitute the evolution in the past of the Sm/Nd ratio in the source region of the rocks. This makes available a "chemical fossil" to measure the intensity of a fractionation process in the past.

These "tracer" and "fossil" properties have found many applications in geology. It is the purpose of this paper to present the basic principles of the methodology of REE geochemistry in magmatic processes. For more detailed information, the reader is referred to the excellent review papers of Hanson [1] on petrogenetic studies, O'Nions et al. [2] on Nd isotope applications, Irving [3] on experimentally determined distribution coefficients, Taylor and McLennan [4] on the continental crust and Allegre [5] on the integration of trace elements and isotopes studies in plate tectonic. Extensive geochemical data can be found in Felsche and Herrmann [6] and in Haskin and Paster [7]. A review of the REE geochemistry in non-magmatic processes is presented elsewhere in this book by Möller.

2. THE IMPORTANCE OF IGNEOUS PROCESSES IN THE GEOCHEMICAL EVOLUTION OF THE EARTH.

The composition of chondritic meteorites, particularly the carbonaceous chondrites, is generally considered to represent the non-volatile element composition of the solar system (cosmic abundance) and of the bulk Earth.

The primitive mantle results from a differebciation process at high temperature, not well known but certainly early, from which the zonal structure of the Earth was generated with a core, a mantle and a crust. It is admitted that this process did not fractionate the REE. Thus the chemical composition of the primitive mantle of the Earth is the chondritic composition multiplied by a constant factor. Normalization of the REE contents of a sample relative to the chondritic composition - Coryell-Masuda's classical diagram - cancels the Oddo-Harkins effect and, at the same time, gives a quick appreciation of the degree of fractionation relative to the primeval composition.

TABLE 1

Three sets of REE abundances (ppm) in chondritic meteorites

At.No	Element	[1]	[2]	[3]
57	La	0.330	0.378	0.367
58	Ce	0.88	0.976	0.957
59	Pr	0.112	---	0.137(a)
60	Nd	0.60	0.716	0.711
62	Sm	0.181	0.230	0.231
63	Eu	0.069	0.0866	0.087
64	Gd	0.249	0.311	0.306
65	Tb	0.047	---	0.058(a)
66	Dy	---	0.390	0.381
67	Ho	0.070	---	0.085
68	Er	0.200	0.255	0.249
69	Tm	0.030	---	0.036(a)
70	Yb	0.200	0.249	0.248
71	Lu	0.034	0.0387	0.038

[1] from Haskin et al. [8]: values for a composite sample of nine chondrites
[2] from Masuda et al. [9]: values for the Leedey chondrite
[3] from Taylor and McLennan [4], on the basis of carbonaceous chondrite (C-1) contents (multiplied by 1.5 to allow for volatile loss)
(a) interpolated value from neighbouring elements in Coryell-Masuda diagram

The REE distribution in the crust - oceanic or continental - can be studied in terms of equilibrium relations between a magma and minerals. For the oceanic crust is generated in the oceanic ridges by the cooling down of a tholeiitic basalt, produced at depth by partial fusion of the mantle. The continental crust grows laterally through addition of andesitic volcanics arcs produced by subduction of oceanic crust as well as through addition of various magmatic products (basaltic or andesitic) coming from the mantle. According to the uniformitarian principles, the first continental crust must also have been andesitic in composition. In its evolution it tends to differentiate into an upper part, granitic in average composition, and a lower more basic part, made of metamorphic rocks of the granulite facies and of basic to intermediate intrusions. The differentiation takes place during the orogenic periods, through transfer towards the upper levels of eutectical granitic melts formed by partial fusion of sedimentary rocks (anatexis).

Evolution of the mantle also implies mineral-melt relations. The mantle permanently loses various products which rise to form the crust and, particularly, increase the volume of the continents. It thus progressively acquires a more residual character.

Mineral-melt relations can also explain the great variety of rocks which are met in the crust. They are generated by differentiation, that is fractional crystallization, of the various types of magma: basalts (alkali or tholeiitic), andesite and granite.

Moreover, since REE are relatively insensitive to supergenic processes (weathering, erosion, transportation, deposition and diagenesis) - REE have low solubility, low mobility and very short residence time in the ocean -, their distribution in sedimentary rocks reflects the average distribution in the Earth's crust. Similarly, they also have low mobility in metamorphic processes which can be considered isochemical with regards to the REE among many other elements.

It can be concluded that in the geochemical cycle of the REE, sedimentary and metamorphic factors play a negligible role in modifying the REE distribution, the latter being basically controlled by igneous processes. A better understanding of the REE in nature therefore requires the knowledge of their behaviour in igneous processes, that is in mineral-melt systems.

3.   THE DISTRIBUTION COEFFICIENT

The concept of distribution coefficient of an element between a mineral and a melt permits to describe the igneous processes in terms of simple mathematical equations.

The distribution coefficient of an element between a mineral and a melt in equilibrium is $K_d = C_m/C_L$ where C_m is the concentration of the element in the mineral; C_L the concentration in the liquid.

Since the concentration of REE in rock-forming minerals and magmas is very low, Henry's law is satisfied and K_d depends only on temperature, pressure and composition of the mineral and melt. The dependence on pressure is much lower than on temperature and can be ignored.

The nature of the mineral, as will be seen below, strongly influences the K_d values for a given REE. Most minerals in nature have variable compositions - they are solid solutions. Magmas also display a range of chemical compositions: basaltic, andesitic or granitic. The dependence of these factors on the K_d values is however either predictable or sufficiently low to be ignored.

When several minerals, in fractions X, are in equilibrium with a liquid, partition of the element between solid phases and melt is expressed by the bulk distribution coefficient: $D = \Sigma \, X \, K_d = C_S/C_L$, where C_S is the concentration of the element in the solid.

4. THE MODELS

Several mathematical expressions have been proposed to describe the two major igneous physico-chemical processes, viz. formation of a magma by partial fusion and evolution by crystallization. The most "popular" models among trace element petrologists [10] are:

(i) the Nernst-Berthelot equation to describe partial fusion and
(ii) the Rayleigh equation for fractional crystallization.

(i) Partial fusion or batch melting

There is much evidence that partial fusion in nature maintains equilibrium between solid and liquid. Simple mass balance gives:

$$C_L = C_o/D + F(1-D) \tag{1}$$

where C_o is the concentration of the element in the initial solid; F is the fractionation of melt; D is the bulk distribution factor between the solids and the melt at the time of removal of the melt.

More sophisticated processes have also been proposed: fractional fusion, with or without collect of the successive drops of liquid, modal or not modal [11], with exhaustion of a phase, with incongruent melting [12], etc... They require the knowledge of much more parameters than for simple batch melting and thus are more difficult to use. Since generation of magmas takes place in deep seated conditions, the mantle or the lower crust, pressure-temperature conditions and chemical compositions of the starting material are only known with uncertainty, and so are the successive steps of the evolution of the system. In batch melting, the only prerequisite is the knowledge of the nature and proportions of the solid phases which were last in equilibrium with the magma. The simplicity of the batch melting model and our inability to control all the parameters which enter into the other models obviously account for its success.

(ii) Fractional crystallization

The Rayleigh law which only implies instantaneous equilibrium between the solids and melt gives:

$$C_L = C_o \ F^{\ D-1} \tag{2}$$

where C_o is the concentration in the initial liquid, and the other symbols, the same as in the previous equation. It must be emphasized that the equation is only valid if D is kept constant.

Equilibrium crystallization is also a possible mechanism. But fractional crystallization is more realistic. In natural systems, continuous re-equilibration during crystallization of the solids with the melt is very often hampered either by rapid cooling or by subtraction of the crystallizing minerals, due to density differences with the magma.

5. DISTRIBUTION COEFFICIENTS AND QUANTITATIVE MODELING

The K_d values for the REE except Eu - i.e. the Ln^{3+} - are very low in feldspars, olivine, ortho- and clinopyroxenes and Fe-Ti oxides which form a great variety of common igneous rocks (fig. 1). Actually, the Ln^{3+} have a strong magmatophile character. In the igneous processes where these minerals play a role, the bulk partition coefficient D is always very low and negligible relatively to 1.

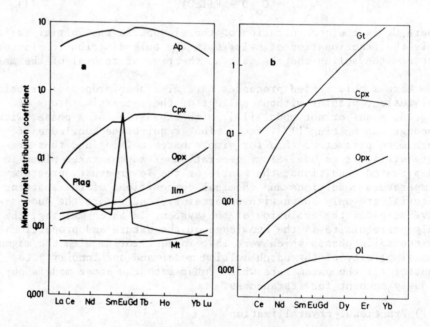

Fig. 1 (a) Set of mineral/melt distribution coefficients for plagioclase (Plag), clinopyroxene (Cpx), ilmenite (Ilm), magnetite (Mt) and apatite (Ap), calculated from a noritic cumulate of the

Bjerkrem-Sogndal lopolith (Duchesne and Roelandts [13]) and used in modeling the REE behaviour in the igneous differentiation (Roelandts and Duchesne [14]).
(b) Set of mineral/melt distribution coefficients for garnet (Gt), clinopyroxene (Cpx), orthopyroxene (Opx) and olivine (Ol), proposed by Hanson [1] for mantle rocks.

It follows that the two models are greatly simplified and equations (1) and (2) become:

$$C_L/C_o = 1/F$$

The Ln^{3+} concentrations thus provide a straightforward index of crystallization or of fusion [15].

Since minerals with very low K_d values are stable in the P,T conditions of the crust and equilibrate with a large variety of melts of various compositions, the igneous processes - fusion or fractional crystallization - will not significantly modify the Ln^{3+} distribution. The absolute amount of Ln^{3+} will be changed but the different elements will not be fractionated (e.g. the La/Yb ratio remains constant).

The overall behaviour can however be modified if some particular minerals, very enriched in heavy or light REE or in all REE, play a role in the process. For instance, fig. 1b gives the variation of K_d between garnet and melt. It can be seen that heavy REE have $K_d > 1$ and light REE have $K_d < 1$, the difference between La and Yb spreading over 2 orders of magnitude. The presence in the mineral assemblage even of minute amounts of garnet will thus lead to quite different values of the bulk distribution factor for La and Yb.

A classical use of this property is made in the interpretation of the REE distribution in alkali basalts and in tholeiitic basalts. The La/Yb ratio is high in the former and slightly less than unity in the latter. This is consistent with the hypothesis of the presence of garnet in the solid residue in equilibrium with alkali basalt in the source region of the mantle where partial fusion takes place. On the other hand, garnet must be absent from the assemblage which last equilibrates with tholeiitic basalt. Since garnet in the mantle is stable at higher pressures than the other Al-minerals, it can be concluded that the source region of the alkali segregation is deeper than that of the tholeiitic basalt. Similarly the light REE depletion in oceanic tholeiitic basalts cannot be explained if the source rock has a chondritic distribution. The suboceanic mantle must also show a light REE depletion similar to the tholeiites.

Another example is the role of apatite which can have very high K_d values. Although values as high as 100, reported in the literature, seem overestimated °14§, one has to take into account sets of K_d varying from say 5 to 50 depending on the composition of the silicate melt. Here the difference between La and Yb is not important. Apatite will thus not fractionate the REE but its major role will be to completely change the course of the evolution in magmatic crystallization. Instead of the enrichment that one can expect due to the magmatophile character of the REE, a decrease can be observed in the successive liquids of the fractional crystallization process when apatite is present in the minerals which crystallize in an amount sufficient to give a bulk partition coefficient higher than 1.

5.1 The europium anomaly

In the oxydo-reduction conditions which prevail in magmas, europium can exist as bivalent or trivalent ions. Eu^{2+} with an ionic radius of 1.12 Å follows Sr^{2+} (1.13 Å) while Eu^{3+} has a behaviour similar to the other Ln^{3+}. The K_d values for Eu^{2+} and Eu^{3+} are thus different for the same mineral and depend on the crystallochemical aptitude of the mineral to accomodate the two ions differently.

Discrimination between Eu^{2+} and Eu^{3+} is particularly important in plagioclase (fig. 1) and in K-feldspar. They always present a higher total Eu content than for the other Ln^{3+}. Liké Sr^{2+}, Eu^{2+} substitutes for Ca^{2+} in the feldspar lattice more easily than do the Eu^{3+} and the other Ln^{3+}.

This interesting property has a major application in igneous petrology. Plagioclase is not stable in the upper mantle at depths where partial fusion takes place. Since the upper mantle presents a chondritic distribution of the REE, with no Eu depletion or enrichment compared to the other Ln^{3+}, the silicate melts produced by partial melting of such material will also have a normal Eu content. Indeed, alkali or tholeiitic basalts or andesitic magmas, to speak only of the most common ones, do not present any Eu anomaly (fig. 2). Consequently, when the magma crystallises at lower pressure conditions, that is at a higher level in the crust, where plagioclase is stable, the crystallization of the plagioclase will deplete the magma in Eu to an extent depending on the K_d values for Eu^{2+} and Eu^{3+} between plagioclase and melt and on the Eu^{2+}/Eu^{3+} ratio in the melt, i.e. on the oxygen fugacity (fO_2) of the magma [16].

Thus, provided the fO_2 can be estimated (e.g. on the basis of the Buddington and Lindsley [17] fO_2-barometer, corrected for possible deuteric readjustment [18], the exact amount of plagioclase which has previously crystallized can be determined.

It must however be recalled that the other minerals which precipitated with the plagioclase have also to be considered. The
bulk partition coefficient has to be taken into consideration.
Besides dilution effect of the other minerals, some minerals can
also balance the effect of the plagioclase if they present a lower
K_d for Eu than for the other REE. It is the case of minerals like

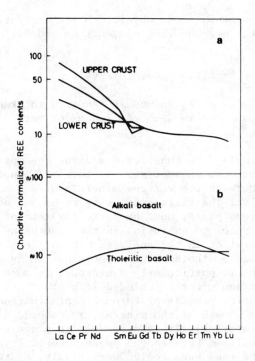

Fig. 2 (a) Chondrite-normalized REE contents of the upper and lower
continental crusts after Taylor and McLennan [4]; the total crust
REE distribution is represented by the unlabelled curve. It parallels the REE distribution in andesitic rocks which is also similar to the distribution in the total crust in the Archaean period
of the Earth's evolution.

(b) Schematic REE distribution in two important types of
upper-mantle generated magmas of basaltic composition: oceanic
tholeiitic basalt, which is depleted in light REE relatively to
the chondritic abundance and thus shows a higher Sm/Nd than in
chondrites; alkali basalt, which has equilibrated at depth with
a garnet-bearing mineral assemblage.

apatite (Roelandts and Duchesne [14], clinopyroxene (fig. 1),
garnet in basaltic melts and amphibole which "prefer" Eu^{3+} than
Eu^{2+}. Actually the exact K_d values for Eu^{2+} and Eu^{3+} in these
minerals should be known accurately in order to control their influence on the Eu anomaly of magmas. This requires the determination of Eu^{2+} and Eu^{3+} in minerals and melt down to the ppm level.
Such determinations have not yet been achieved and, in the present stage of knowledge, it is assumed in the calculations, following Philpotts [19], that:

$$K_d^{Eu^{2+}} = K_d^{Sr^{2+}} \quad \text{and} \quad K_d^{Eu^{3+}} \quad \text{is obtained by interpolation between}$$

the K_d values for the neighbouring elements Sm and Gd.

5.2 Determination of partition coefficients

The exact knowledge of K_d values is required in order to use
the various equations described above. Two major methods are employed.

a. – Natural rock samples are considered and the concentrations are
determined either between phenocrysts and matrix (K_d between minerals and melt) or between the various minerals of a cumulate rock
(K_d between minerals). The first difficulty to be solved is a technical one: the various phases must be perfectly separated – by
physical methods – prior to analysis. In the case of an element
in very low concentration in the analysed mineral (e.g. magmatophile or incompatible), the smallest contamination by a material
rich in this element is particularly dangerous. It must also be
decided whether a trace mineral included in another one has been
formed by an exsolution process or through contamination by foreign
material during the growth of the mineral. The second difficulty
to be solved is theoretical. It is difficult to prove that equilibrium has been realized between the phases. Zoned crystals must
be rejected. Because they have cooled very slowly, plutonic rocks
should be preferred to volcanic rocks as far as true adcumulus
rocks can be selected. Modeling of layered sequence of rocks
(Duchesne and Demaiffe [20], Duchesne [21]) is a complementary
approach to the phenocrysts-matrix method on porphyritic volcanic
rocks.

b. – The K_d values are determined in experimental systems (see Irving [3]). Equilibrium can be closely achieved by reversal or re-
equilibration experiments. The homogeneity of the trace element
distribution can be checked by means of microprobe analysis. However, the low detection limits of the method requires working at
relatively high level of concentrations (higher than say 0.1 %)
which for many elements do not follow Henry's law. Particular
methods such as the ion probe, the differential dissolution technique or the "track mapping" (autoradiography) have also been used

for REE. Cathodo- and thermoluminescence are promising methods to measure the Eu^{2+} and the Eu^{3+} in various REE-rich minerals, especially apatite (Mariano and Ring [22]; Baumer et al. [23]).

6. GEOCHEMICAL APPLICATIONS OF Nd ISOTOPES [2]

^{147}Sm decays to ^{143}Nd with a half-life of about 10E11 years ($\lambda = 6.54 \times 10E-12/y$). According to the law of radioactive decay, the total number of ^{143}Nd atoms at time t in a given system, is:

$$^{143}Nd = (^{143}Nd)o + {}^{147}Sm\ (e^{\lambda t} - 1)$$

with ^{143}Nd the number of ^{143}Nd atoms at t = 0, ^{147}Sm the number of ^{147}Sm at the time t and λ the decay constant. Since ^{144}Nd is a non-radiogenic isotope, the number of ^{144}Nd has not changed with time and, after division, the equation becomes:

$$^{143}Nd/^{144}Nd = (^{143}Nd/^{144}Nd)o + (^{147}Sm/^{144}Nd)\ e^{\lambda t} - 1 \tag{3}$$

In a diagram with coordinates $^{143}Nd/^{144}Nd$ and $^{147}Sm/^{144}Nd$, equation (1) is a straight line of intercept $(^{143}Nd/^{144}Nd)o$ and slope $(e E\lambda t - 1)$. It is called an isochron (fig. 3). All systems

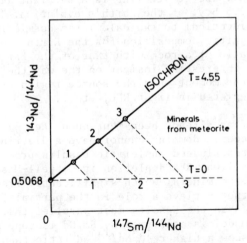

Fig. 3 Isochron diagram (see equation (3)): Sm/Nd evolution diagram for seperated minerals of a meteorite.

of different Sm/Nd ratios will plot on the same isochron if they were isotopically homogeneous - same $(^{143}Nd/^{144}Nd)o$ - t years ago, and if they since remained closed to any migration of Sm and/or Nd. This first condition is fulfilled when the rocks result from the fractional crystallization of the same magma or when they were

metamorphosed at the same time. Since Nd is chemically very similar to Sm, the tendency of Nd to migrate out of the system is very
low. Moreover, since the REE have very low mobility the closed
system condition can also be fulfilled. Thus the Sm/Nd can be used
to determine the age of a rock. It is particularly powerful in
dating ancient basic rocks (meteorites, Archaean rocks, lunar
rocks), for which the other methods, such as Rb-Sr or U-Pb, are
less efficient.

Equation (3) can be simplified, because t is always very
small:

$$^{143}Nd/^{144}Nd = (^{143}Nd/^{144}Nd)o + (^{147}Sm/^{144}Nd) \, t \qquad\qquad (4)$$

In $^{143}Nd/^{144}Nd$, and t coordinates, equation (4) is a straight
line of slope proportional to the Sm/Nd ratio of the system (fig.
4). The initial ratio $(^{143}Nd/^{144}Nd)o$ is the intercept in the isochron diagram.

The evolution of the $^{143}Nd/^{144}Nd$ in the mantle has been determined by measuring the initial ratio of rocks of various ages
coming from the mantle and assuming an initial ratio of the system,
4.55 billion years ago, equal to that of the solar system measured
on meteorites (fig. 4). The growth line is a straight line at
least for the first 2.5 b.y. of the Earth's evolution. Its slope
gives a Sm/Nd ratio identical to the calculated cosmic ratio and
thus confirms a chondritic composition for the Earth. Comparison
of the $^{143}Nd/^{144}Nd$ ratio of a rock with that of the Earth at the
same moment provides interesting insight on the way the continents
evolve and grow, on the moment when the source region of the suboceanic mantle was depleted in light REE, etc...

For instance, it has already been mentioned (see 5) that the
upper mantle under oceanic domains cannot have a distribution similar to the chondrites. Indeed tholeiitic basalts produced by
partial fusion show a light REE depletion in Coryell-Masuda diagram (fig. 2b). Since no mineral which significantly fractionates
the REE - such as garnet - plays a role in the partial fusion process, the distribution in the basalts must parallel that of the
source region. This model is confirmed by Sm/Nd isotope studies.
The oceanic basalts show a higher $^{143}Nd/^{144}Nd$ ratio than the value
predicted by the chondritic model of fig. 4. It can be inferred
that sometime in the past after the end of the Archaean period
2.5 b.y. ago, a differentiation process acted to increase the
initial Sm/Nd ratio of the suboceanic upper mantle. A complementary process - generating material with lower Sm/Nd - can also be
traced in the formation of granitic bodies over that period.

It is also fascinating to observe a similar "revolution" in
the continental crust in post-Archaean times. According to Taylor

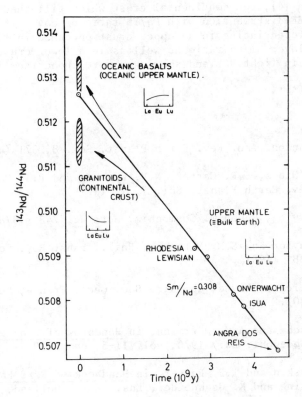

Fig. 4 Evolution of the $^{143}Nd/^{144}Nd$ ratio with time for a constant Sm/Nd ratio (see equation (4)): the values of the $^{143}Nd/$
^{144}Nd initial ratio and the age of mantle-derived basic Archaean
rocks of different ages: Isua (Greenland), Onverwacht (S. Africa),
Rhodesia (Zimbabwe) and Lewisian (Scotland) plot on/near a
straight line of slope corresponding to the Sm/Nd ratio of chondrites and which starts from the point Angra Dos Reis representing the most primitive value of the $^{143}Nd/^{144}Nd$ ratio (0.5068
and T = 4.55 b.y.) (see fig. 3). This diagram indicates that
during the Archaean period (4.5 to 2.5 b.y.), the mantle REE
distribution was and remained chondritic. The post-Archaean evolution is more complex. Present-day oceanic basalts plot above
the chondritic upper mantle curve and granitoids in the continental crust plot under the curve. This indicates that the Sm/Nd
ratio in the source rocks of these materials is not chondritic
and that they derived from already differentiated rocks.

and McLennan [4], the continental crust which till then had an andesitic REE distribution with La/Yb ratio of ca. 5 (fig. 2a) started differentiating in an upper crust enriched in light REE and with a lower Sm/Nd ratio as well as in a lower crust relatively depleted in light REE and showing a more chondrite-like Sm/Nd ratio.

REFERENCES

[1] G.N. Hanson, Ann. Rev. Earth Planet. Sci. 8, 371 (1980).

[2] R.K. O'Nions, S.R. Carter, N.V. Evenson and P.J. Hamilton, Ann. Rev. Earth Planet. Sci. 7, 11 (1979).

[3] A.J. Irving, Geochim. Cosmochim. Acta 42, 743 (1978).

[4] S.R. Taylor and S.M. McLennan, Phil. Trans. Roy. Soc. London A301, 381 (1981).

[5] C.J. Allègre, Livre jubilaire Soc. Géol. France, Mem.h-s. 10, 87 (1980).

[6] J. Felsche and A.C. Herrmann, in Handbook of Geochemistry (K.H. Wedepohl, Ed.) 1970, vol. II-5, chap. 39.

[7] L.A. Haskin and T.P. Paster, in Handbook of Rare Earths (L. Eyring and K. Gschneider, Eds.) North Holland, 1978, no 2.

[8] L.A. Haskin, M.A. Haskin and F.A. Frey, in Origin and distribution of the elements (L.H. Ahrens, Ed.) Pergamon, Oxford, 1968, p. 819.

[9] A. Masuda, N. Nakamura and T. Tanaka, Geochim. Cosmochim. Acta 37, 238 (1973).

[10] S.R. Hart and C.J. Allègre, in Physics of magmatic processes (R.B. Hargraves, Ed.) Princeton University Press, 1980.

[11] D.M. Shaw, Geochim. Cosmochim. Acta 34, 237 (1970).

[12] J. Hertogen and R. Gijbels, Geochim. Cosmochim. Acta 40, 313 (1976).

[13] J.-C. Duchesne and I. Roelandts, unpublished.

[14] I. Roelandts and J.-C. Duchesne, in Origin and distribution of the elements (L.H. Ahrens, Ed.) Pergamon, Oxford, 1979, p. 199.

[15] M. Treuil and J.L. Joron, Soc. Ital. Miner. Petrol. 31, 125 (1975).

[16] M. Drake, Geochim. Cosmochim. Acta 39, 55 (1975).

[17] A.F. Buddington and R.F. Lindsley, J. Petrol. 5, 310 (1964).

[18] J.-C. Duchesne, J. Petrol. 13, 57 (1972).

[19] J.A. Philpotts, Earth Planet. Sci. Lett. 9, 257 (1970).

[20] J.-C. Duchesne and D. Demaiffe, Earth Planet. Sci. Lett. 39, 249 (1978).

[21] J.-C. Duchesne, Contr. Miner. Petrol. 66, 175 (1978).

[22] A.N. Mariano and P.J. Ring, Geochim. Cosmochim. Acta 39, 649 (1975).

[23] A. Baumer, D. Lapraz, J.-C. Duchesne and W.E. Klee, Abst. I.M.A. meeting Varna, in press (1982).

DISCUSSION

CARNALL   (question)

It seems to me that according to your model you had chondritic
material at low temperature and then temperature builds up. Now
we have molten metallic materials mainly iron at the center. Certainly there is a chance for Fe to extract some rare earths or
mixture of metals and I have no idea whether there is a solubility
problem of rare earths in metal mixtures. What evidence there is
that rare earths cannot be distributed down into the core.

DUCHESNE   (answer)

The analysis of metallic meteorite shows the presence of very
little rare earths. The formation of the core of the earth has
not changed much the distribution of the rare earths. We cannot
go to the core, of course. We have some meteorites which can be
equivalent to the core material and can fit the right composition
of the earth's core.

EDITOR   (note)

The lanthanide-iron binary systems, including Sc and Y have been
investigated in some details. Solubility of Ce in Fe is about
0.4 wt. percent between 815 and 1015° C. For a review on the binary phase diagrams of the lanthanides with Fe, Co, Cu, Cr, Ir,
Rh, Ru, V, U, and others see C.E. Lundin in The Rare Earths (F.H.
Spedding & A.H. Daane, Editors), John Wiley, New York, 1961, Chapter 16.

CARNALL   (comment)

I guess my point is that according to the model you are using I do
see the relevance of the meteorite materials that we find at the
surface now and what is in the core. As I understood your model
that was meteoritic material alright but at rather low temperature coming together and then the process of heating began and at
the end we have the material at the center of the earth. So I
think that the process by which that Fe and other materials ended
up at the center of the earth is not a process that one could
model by looking at the chondritic materials that are present on
the surface of the earth.

DUCHESNE   (answer)

But it is possible to correlate all kinds of chondrites. Chondrites
are parts and fragments of a broken planet. So if you analyze the
metallic chondrites, you may take it that you analyze the core of
a big planet. As you do not find rare earths in these metallic
materials, you may conclude that the core of the earth is depleted
of rare earths.

BREWER   (comment)

The problem is that of the equilibrium between slag and metal.
Several years ago there was a dispute amongst geologists whether
the lanthanides and the actinides will be concentrated at the
crust or at the core. The agreement was that all heavy elements
will be concentrated in the core because of the gravitational
effect. What actually happens that the major elements do differentiate according to density: oxygen concentrating upwards and
Fe and heavy elements concentrating downwards. So you have effectively a very strong oxygen gradient going upwards and then if
you represent the thermodynamics of the multicomponent systems,
you find that the density effect is overwhelmed by the fact that
the activity of the actinides and lanthanides is greatly reduced
if oxygen concentration increases. Here you have the oxygen gradient going upwards to the crust essentially pulls the lanthanides (actinides), if you have equilibrium, upto the crust. This
probably is the major factor in concentrating the lanthanides and
actinides to the crust and not to the core.

CERNY   (comment)

All these relates to the very classic distribution of chemical
elements proposed by Goldschmidt into 4 classes. Goldschmidt when
he was formulating the basic rules of geochemistry, he was deri-
ving his evidences from different classes of meteorites as well
as from his own profession in the steel foundries and from the
behavior of the elements in molten metal vs. the slag, which is
as close as you can get to the modelling of the behaviors of the
elements in the original differentiation and layering inside the
earth. So what Professor Brewer told us is perfectly correct; the
strictly gravity consideration is overwhelmed by the tendency of
the elements to combine with different anions or to stay in the
native states. In the case of rare earth elements, they are very
efficiently withdrawn from the metallic meteorites or metallic
core of the earth by their strong affinity towards oxygen and
rising to the upper crust.

CERNY (question)

What lines of evidences are available either analyzing natural
plagioclase or synthesizing them under controlled conditions that
it is really the Eu^{2+} that goes into the plagioclase and not the
3+.

DUCHESNE (answer)

I think it is a matter of ionic radii. It is difficult experimentally to determine the exact amount of Eu^{2+} and Eu^{3+} in these
minerals. I had some data on the thermoluminescence studies on
apatite and it was possible to obtain well defined signals from
Eu^{2+} but not from Eu^{3+}. We do not know as yet if Eu^{3+} is giving
any thermoluminescence signal. But it is quite curious as according to our model apatite will be enriched in Eu^{3+} - whereas it
is only Eu^{2+} that we see.

SINHA (comment)

The origin of the thermoluminescence peaks is somewhat different
than the normal photoluminescence. The mechanism of thermoluminescence is rather complicated. There are the electron trapping
centers which liberate electrons on heating and causing peaks on
the thermoluminescence curves. However, these electrons can be
used to excite the luminophore ions. Merz (Phys. Rev. $\underline{162}$, 217,
235 (1967)) has used this technique quite effectively for investigating luminescence of rare earths in fluorite. The heating, however, does not directly induce luminescence. If the emission peak
is a broad structureless peak, it would be difficult to assign it
without making comparison with the synthetic samples containing
various activator ions. The lanthanide ions usually give sharp
peaks.

13

LANTHANOIDS AS A GEOCHEMICAL PROBE AND PROBLEMS IN LANTHANOID GEOCHEMISTRY
DISTRIBUTION AND BEHAVIOUR OF LANTHANOIDS IN NON-MAGMATIC-PHASES

Peter Möller

Geochemie, Hahn-Meitner-Institut für Kernforschung
Berlin GmbH, D-1000 Berlin 39, Fed. Rep. Germany

ABSTRACT

Ca minerals exhibit a large variety of Ln distribution patterns. These reflect the pTx conditions in the source area, along
the pathlines of the fluids as well as in the geochemical traps
where the Ca minerals crystallized. A very general model for Ln
fractionation is discussed which essentially considers the distribution coefficients to be dependent on the Ln-complexing ligand
concentration in the fluids. Its validity is demonstrated by results
on Ln coprecipitation with Ca-oxalate.

From results of a model calculation of Ln fractionation the
construction of a variation diagram is proposed in which fields
typical for pegmatitic/magmatic, hydrothermal and sedimentary formations of Ca minerals are outlined. These diagrams look different
for each mineral. Primary and secondary crystallization as well as
solution-rock interaction lead to different trends in Ln fractionation. Various reasons which might cause anomalous Eu and Ce values are discussed.

Among the applications of Ln as a geochemical probe (i) the
evaluation of different modes of formations of fluorite and magnesite deposits are outlined; (ii) an example of applying Ln fractionation in calcite is given assisting to decipher the mode of
formation of a particular type of base metal deposits; (iii) tracing of a hydrothermal activity in a cassiterite deposit and generation of ore bearing fluids are outlined; (iv) alteration of magmatites are discussed; (v) and finally the variations of Ln patterns
in oceanic Fe/Mn concreations with geological time are considered.

S. P. Sinha (ed.), Systematics and the Properties of the Lanthanides, 561–616.
Copyright © 1983 by D. Reidel Publishing Company.

1.   INTRODUCTION

Among the various trace elements in Ca-minerals Ln are always present. Their concentration levels are

(i)   well below 1000 ppm in most minerals
(ii) the ionic radii of the tervalent Ln correspond with that of Ca^{2+} (fig.1) and
(iii) many of their chemical and physico-chemical properties change aperiodically within their series.

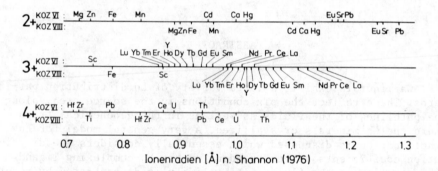

Fig.1 Graphical display of geochemically relevant ionic radii for the coordination numbers VI and VIII in the range of $(0,7-1,2)10^{-8}$ cm [1]

In nature Ln mostly occur substitutionally in Ca-minerals like apatite, calcite, fluorite, and rock forming silicates like biotite and feldspars. There are only few exceptions where they form Ln-minerals like monazite or bastnaesite in considerable amounts. For graphical display of Ln concentrations Coryell et al [2] suggested a normalization procedure by which the Ln concentration in the sample is devided by the respective concentration in chondrites. This has now become convenient. Values for normalization might be taken from Herrmann [3] or Haskin [4]. Such normalized Ln distribution patterns allow a quick grouping of the analyzed samples and assist their interpretation with respect to various genetical aspects.

When dealing with the Ln fractionation in non-magmatic

systems like in postmagmatic, metamorphic, diagenetic, and sedimentary processes following topics have to be discussed:

- complex formation of Ln in solutions;
- the behaviour of Ln in coprecipitation processes;
- probable mechanisms of ion substitution in host minerals by Ln;
- the Ce and Eu anomalies;
- development of variation diagrams as a tool for the genetical
  interpretation of Ln fractionation during mineral formation;
- various geochemical applications of Ln as a geochemical probe.

   Before starting these subjects I would like to make you fa-
miliar with what I am dealing with. Different from magmatic rocks
(see references by Duchesne, this volume) Ln distribution patterns
in fluorite and calcite show large variations depending on the
mode of mineral formation. In fig.2 and 3 some typical Ln distribution patterns are given to illustrate their variations. Some
systematic changes can already be seen:

(i) changes of the light Ln (LLn) are much more pronounced
    than those of the heavy Ln (HLn) in hydrothermal minerals.
    This leads to the fractionation of HLn vs LLn; the ratio
    of concentration of HLn and LLn vary considerably within
    a given mineral deposit.

(ii) Temperature and chemical composition of the liquids de-
     termine the concentration of Ln in the liquid and thereby
     that of the solids crystallizing from them. Highest
     amounts of LLn are found in magmatic carbonates and peg-
     matitic fluorite, followed by hydrothermal formations. In
     sedimentary (including diagenetic and evaporitic) minera-
     lizations lowest concentrations are found.

(iii) In hydrothermally formed fluorite and calcite typical va-
      riations are observed. Within a given mineral occurence
      the LLn sometimes decrease whereas the HLn relatively in-
      crease in concentrations. In case of fluorite often a
      conspicuous increase particularly of Tb is observed among
      the Ln usually analysed [5-8].

(iv) Mineral samples from occurences for which there is clear
     petrological evidence that the minerals have been remo-
     bilized or that they crystallized from metamorphic solu-
     tions in open fissures their Ln pattern can increase from
     LLn to HLn dramatically.

This short discussion of observation on Ln fractionation in two
typical and very abundant minerals indicate that we will find rather large effects on Ln fractionation in non-magmatic systems.

   The Ln fractionation in non-magmatic processes is much more
pronounced than in magmatic systems. This goes with the general

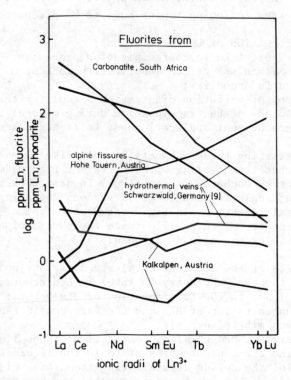

Fig.2 Some typical averaged Ln distribution patterns in fluorites
of different geneses illustrating their wide range of variations.
Those patterns without reference have not been published yet.

trend according to which fractionation coefficients are expected
to increase with decreasing temperature of the system.

2.   MODEL OF LN FRACTIONATION

2.1. Complexation of Ln in Solutions

In geochemical relevant systems the Ln might become com-
plexed by ligands like Cl^-, F^-, SO_4^{2-}, OH^-, HCO_3^-, CO_3^{2-} etc. Many
of the relevant formation constants have been studied at room
temperature and normal pressure. It was found that the formation
constants of fluoro-[13-17], hydroxo-[18-20], carbonato-[21],
and fluoro-hydroxo-[19] complexes increase, those of the sulfato-
complexes [22,23] decrease a little bit from La to Lu.

Many authors [14-17,19] studied fluoro-, hydroxo- and fluoro-
hydroxo-complexes of La, Ce, Nd, Eu, Tb, Er, Yb, Lu in solution
up to 3M NaCl at temperatures up to 65°C. They found that in

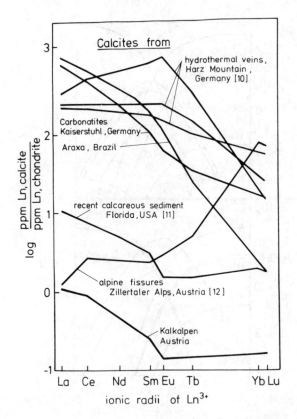

Fig.3 Some typical averaged Ln distribution patterns in calcites
of different geneses illustrating their wide range of variations.
Those patterns without reference have not been published yet.

case of the fluoro-Ln complexes the stability constant passes
through a maximum at Tb for those elements studied. This was observed in solutions with pH values low enough to avoid significant complex formation with OH^- ligands. At pH values >7 hydroxo-
complexes become dominant. Using his formation constants Koß [19]
estimated the amounts of the various Ln species in F-OH-bearing
solutions as a function of pH and pF at room temperature (fig.4+5).
Taking Ce and Tb as representatives of the LLn and HLn, respectively, these diagrams show the dominance of uncomplexed LLn at
pH<7 and pF>3 and the relative amount of various complexed species
under all concentrations of F^- and OH^-. The aquo-complexes will
not be considered in this context. The tendency of Cl^- ions to
form complexes with Ln seem to be rather small [24-26]. Recently
Sinha (see this volume) could prove the existence of aquo-,
fluoro-carbonato-complexes by fluorescence-spectroscopy. Complex
stability constants in NaCl media (1-4M) at temperatures in the

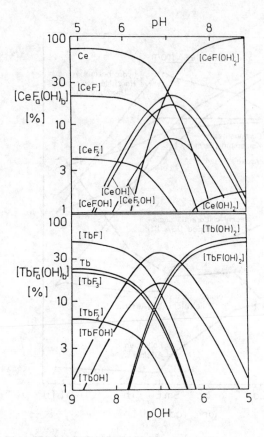

Fig.4 Estimated amounts of various ionic species as a function of pH at $pF = -\log(^aF) = 3$. The electric charges of the species have been omitted. (Reproduction from a thesis [19])

range of 200–400°C are lacking. Thus geochemists are left with extrapolation over wide ranges of temperatures in order to discuss the general trends of complex formation of Ln in hydrothermal systems. Since there is not much experience the extrapolated results must be considered very cautiously. Only recently Becker (unpublished) determined the stability constant of the monofluoro-Ln-complexes at 250°C and 1000 bar and in 1M NaCl. Although the absolute value increases, the trend is similar to that at 60°C.

2.2. Coprecipitation of Ln

Experimental studies of coprecipitation of Ln have been performed by Feibush, Rowly, and Gordon [27] in the Ce-Ln-oxalate system, by Purkayasta and Bhattacharyya [28] and Matsui [29] in the Ca-Ln-oxalate system. It was found that coprecipitation can

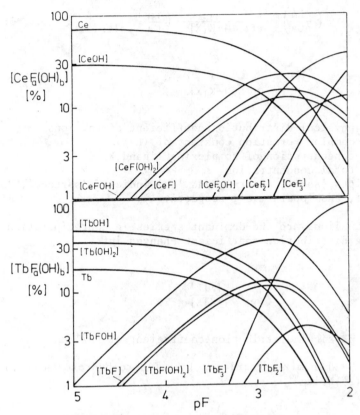

Fig.5 Estimated amounts of various ionic species as a function of
pF = -log(^{a}F) at pH7. The indication of the electric charges of
the species has been omitted. (Reproduction from a thesis [19])

be satisfactorily described by Doerner-Hoskins law [30]. The operating distribution coefficient λ_{op}, was found to be strongly
dependent on the electrolyte concentration and pH of the system.
The order of distribution coefficient under comparable condition
in solution was as follows: Lu>Eu>Ce. This is the reverse order
of the basicity of the Ln or of the solubility of their oxalate
salts [29]. Marchand [31] and Becker [17] tried to determine Ln
distribution coefficients in the Ca-Ln-fluoride system at very low
pH values. Due to experimental difficulties they failed to find
reliable distribution coefficients.

In hydrothermal systems complexes of Ln with X=F⁻, OH⁻, and
CO_3^{2-} seem to be the important species in natural environments.
For the Ln^{3+} to be coprecipitated with a Ca-mineral the Doerner-
Hoskins law writes

$$\left(\frac{\Delta Ln}{\Delta Ca}\right)_{surf} = \frac{\lambda}{1+K[X]} \cdot \left(\frac{Ln}{Ca}\right)_{soltn} \tag{1}$$

$$\lambda_{op} = \frac{\lambda}{1+K[X]} \tag{2}$$

with = physical distribution coefficient of Ln^{3+} ions
K   = complex formation constant of (LnX)
[X]   = concentration of complexing ligand X
Ln,Ca= total concentration
subscripts 'surf', 'soltn', and 'op' indicate 'surface', 'solution', and 'operative', respectively.

If LnX^+ ions are the dominant species in coprecipitation the operating distribution coefficient changes into

$$\lambda'_{op} = \frac{K[X]\lambda'}{1+K[X]} \tag{3}$$

λ' = physical distribution coefficient of LnX^+ ion

When mixed complexes like fluoro-hydroxo-Ln complexes form eq.(2) writes

$$\lambda''_{op} = \frac{\lambda''}{1+K''[X][Y]} \tag{4}$$

K'' = complex formation constant of $LnXY^+$.

If the pH, however, is constant λ''_{op} in eq.(4) reduces to an equation similar to eq.(2).

Whether Ln^{3+} or LnX^{2+} is the dominant species in the distribution process depends on their individual values. This will be influenced by the sign of the surface charge of the growing crystals. A negative surface charge could be built up by adsorption of anions in excess to the cations. Since carbonate and sulfate is very common in hydrothermal solutions a negative surface charge is most probable [32]. Such a surface will reject anions, but attract highly positive cations. Thus the availability of highly charged Ln species will be one of the dominant factors that controls the distribution of Ln between solids and liquids. In fig.6

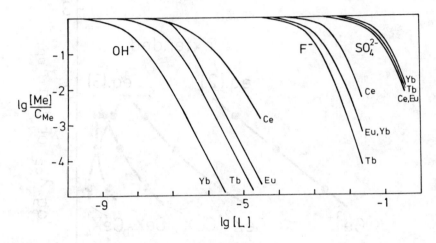

Fig.6 The dependence of the ratio of individual free Ln ions
and their concentration C_{Me} in solution as a function of ligand
concentration (L = OH$^-$, F$^-$, SO$_4^{2-}$). This figure has been reproduced from a thesis [19]

the ratio of uncomplexed and total Ln concentration is given as a
function of the ligand concentration showing different ordering
of Ln in solutions containing OH$^-$, F$^-$, and SO$_4^{2-}$ ions [19].

    Using the reported [29] operative distribution coefficients
in the Ca-Ln-oxalate system which are given as a function of the
acetate concentration in solution it can be shown that λ_{op} seems
to follow eq.(2) extended for further complex species i

$$\lambda_{op} = \frac{1}{1+\Sigma K_i [X]^i}$$ (5)

Here it is expected that

$$\log \lambda_{op} \quad vs \quad 1/(1+\Sigma K_i [X]^i)$$

yields a straight line for the experimentally determined λ_{op}
values from [29]. This is found in fig.7. If, however, one
of the complex species i is dominating in the coprecipitation,
then it follows from the extended eq.(3) that for this species i

$$\log \lambda_{op,i} \quad vs \quad K_i [X]^i/(1+\Sigma K_i [X]^i)$$

must be linear. It can be seen from fig.7 that linearity decreases

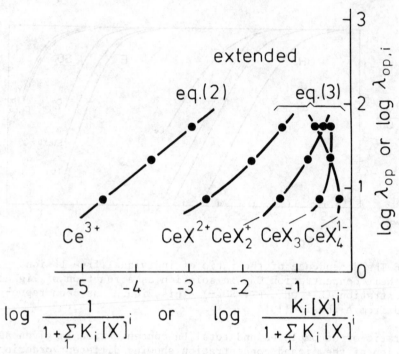

Fig.7 Evaluation of the influence of complex formation on coprecipitation of Ce species with Ca-oxalate in (0.5-2M) acetate solutions. Using the complexity constants for μ = 2M and μ = 0.1M
[33] the relevant constants for μ = 0.5M have been linearly in-
terpolated. For μ = 0.5, 1.0 and 2.0M acetate the terms $K_i[X]i/(1 + K_i[X]^1)$ for i = CeX^{2+}, CeX_2^+, CeX_3, and CeX_4^- and $1/(1 + K_i[X]^1)$
for Ce^{3+} have been calculated. The best linearity is found for
Ce^{3+} ions.

in the series

$$Ce^{3+} > CeX^{2+} > CeX_2^+ > CeX_3$$

For CeX_4^- an inversion of the slope (fig.7) is found. In the Ca-
Ln-oxalate system it is thereby demonstrated that the process of
coprecipitation is probably best described by the Doerner-Hoskins
law using uncomplexed species only. Taking this system for typical
it will be assumed that in Ln coprecipitation with Ca-minerals
only Ln^{3+} have to be considered.

2.3. Ln Fractionation in a Closed System

For the Ca-Ln-oxalate system Matsui [29] found λ_{op} being in the range of 5-160 in closed system depending on the Ln, pH, and ligand concentrations. Extrapolation to [X] equal zero suggests that in systems with negligible Ln complexation λ >1000 have to be expected. λ_{op} increased from La through Lu under constant conditions although the complex stabilities of acetate- and formiate-complexes do not change much within the Ln series.* Different from the experimental conditions in [29] a model will be suggested in which the complex forming ligand is part of the crystals. Then [X] in eq.(2) is not constant throughout the crystallization. Such a condition is realistic for all carbonate and fluorite mineralizations in nature. In order to estimate the behaviour of the Ln during coprecipitation with fluorite in a closed system following data are used in the model calculation:

$$\lambda_{LLn} = 10; \quad K_{LLn} = 500 \ M^{-1}$$

$$\lambda_{HLn} = 10; \quad K_{HLn} = 5000 \ M^{-1}$$

The ratio K_{LLn}/K_{HLn} is realistic since it is the same as found by Koß [19] and Becker [17]. λ_{LLn} and λ_{HLn} are unknown. The value assumed is based on the fact that Ln are becoming enriched in fluorite. The value of the solubility product $[Ca][F]^2 = 10^{-9}$ controlling the end of crystallization is taken from [34]. The calculation starts from a supersaturated solution with concentrations yielding ten times the solubility product. The initial concentrations are therefore

$$[Ca^{2+}] = 1.5 \ 10^{-3}M$$

$$[F^-] \quad = 2.6 \ 10^{-3}M$$

The degree of precipitation is defined by $p = 1-[F^-]/[F_0]$. For stepwise calculations eq.(5) is used. Fig.8 shows the trends of the two ratios LLn/Ca and HLn/Ca as a function of the degree of fluoride precipitation. Since in geochemical systems the degree of precipitation is unknown it has been suggested earlier [5,35, 36] to construct a variation diagram like that in fig.9. All necessary data are taken from fig.8. The advantage of this diagram (fig.9) is that its two variables can be found by chemical analysis. The increase of the HLn/LLn ratio is related to the degree of precipitation and describes the Ln fractionation. This ratio
* see chapter 3.

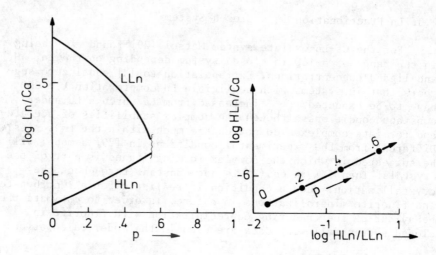

Fig.8 Model calculation on Ln fractionation during Ln coprecipitation with fluorite. Results of calculation for the ratios of
HLn/Ca and LLn/Ca in the fluorite as a function of degree of fluorite precipitation. Assumed data for this calculation are given in
the text.
p = degree of precipitation

Fig.9 variation diagram. This is derived from the estimated data
in fig.8. It illustrates that in a closed system the degree of
precipitation is related to the ratio HLn/LLn. Both ratios in this
diagram can be obtained by chemical analyses.

might be termed fractionation index. The HLn/Ca ratio is chosen as
a second parameter in order to have a reference to the less variable HLn concentration levels in the Ca-minerals. This ratio is
somewhat related to the liquid phase by eq.(1). From fig.9 it
might be seen that low HLn/LLn ratios indicate early, high ratios
later stages of crystallization in closed systems. The trend development during crystallization will be very similar for all Ca-
minerals due to the similarity in the fundamental processes controlling Ln distribution. However, the HLn/Ca ratio might be different for different Ca minerals of similar genesis.

2.4. Ln Fractionation Dynamic Systems

In the closed system discussed so far, the 'later stages of
crystallization' are identical with 'later in time'. Transferring
these results into dynamic (open) systems like hydrothermal vein
mineralizations the 'later stages' are not any more identical with
'later in time'. This is illustrated in fig.10 [37]. Fig.10a shows
one half of a cross section through a fluorite vein. From the high-

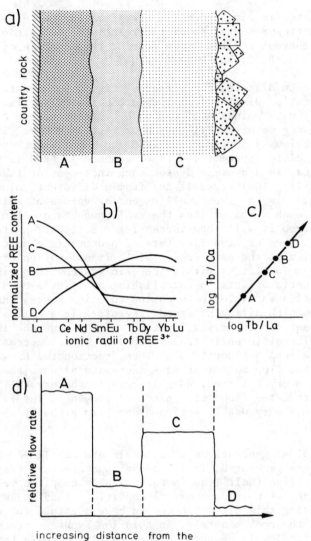

Fig.10
a) Schematic cross section through one half of a hydrothermal fluorite vein showing four coloured layers: A, B, C, and D (simplified redrawing of a vein profile from Marienschacht/ Wölsendorf/Germany described by Jacob [6].
b) Schematic distribution pattern of fluorite samples from the different coloured layers A, B, C, and D.
c) The Tb/Ca - Tb/La variation diagram (schematically). Like in fig.9 a diagonal line is obtained, but the sequence of data points on this line cannot be interpreted in terms

of 'later in time'. Details are given in the text.
d) Schematic presentation of the relative flow rate of the
   hydrothermal solution when forming the fluorite of the
   layers A, B, C, and D in fig.10a.

ly variable Ln distribution patterns (fig.10b) the data points
in the variation diagram (fig.10c) are derived. In fig.10b the
sequence of data points ACBD cannot be interpreted in terms of
increasing degree of fractionation or 'later in time' in general.
This was obviously possible in fig.8 and 9 which represented results in a closed system. The time sequence is ABCD as to be seen
from fig.10a. In a dynamic system each increment of fluorite in
the sample (fig.10a) crystallized from a different volume of solution. Following a given small ascending volume of solution on
its way through a vein system the small amounts of fluorite crystallizing from it will show increasing fractionation from bottom
to top. If, however, the flow rate of hydrotherm changes with
time, fluorite of the same degree of Ln fractionation crystallizes
at different localities within the vein system. This might be due
e.g. to kinetic effects in establishing equilibria of all kinds
between solids and solution when the solution passes through the
vein system with different speed. Therefore in dynamic systems
the time sequence in crystallization and the sequence in mineralization with differently fractionated Ln is not necessarily the
same. Since in A a fluorite with less fractionated Ln was found,
then in B the flow rate must have decreased between the times
when A and B crystallized. When it came to the crystallization of
the fluorite C the flow rate increased somewhat. The fluorite D
formed from a very weakly active hydrotherm probably when it came
to its end.

When older generations of minerals are mobilized saturated
solutions are generated. These move along pores and fissures until
they crystallize their load in tectonically supplied veins in
which changes of physico-chemical conditions lead to supersatu-
ration. During this migration the Ln have a chance to re-equili-
brate with the rock minerals. In this ion exchange process the
concentration levels of the LLn more easily decrease than those
of the HLn. This is explained by their less stable complexes in
the fluids. The new generations of minerals do not show much
variation in HLn. This is schematically shown in fig.11 [37].

Interaction of hydrothermal solutions with Ca-rich rocks
such as limestones or dolostones result in an exchange of Ca-
and Ln-ions between solids and fluids. If Ln-rich hydrothermal
fluids react in this way with limestones which are always rather
low in Ln because of their biogenic-marine origin, the resulting
hydrothermal solution will be lowered in its Ln/Ca ratio. Thus
any Ca mineral crystallizing from such a solution will exhibit

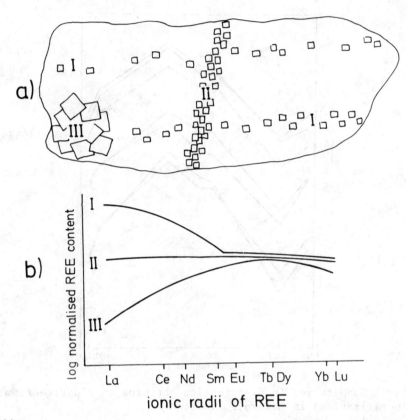

Fig.11
  a) Schematic drawing of a hand specimen according to results on
     sedimentary fluorite in the Nördliche Kalkalpen [5].
     Roman figures indicate three different generations.
  b) Trends of Ln distribution patterns in the fluorite of all
     three generations from the above sample.
  1: Finely dispersed fluorite in a limestone showing sedimentary
     setting,
 11: slightly remobilized fluorite on fissures,
111: bulky fluorite crystals.
     Redrawn after results in [5]

lower Ln/Ca ratios and subparallel Ln patterns (fig.12) both depending on the degree of reaction of the hydrotherm with the carbonate host rocks.

   A very similar but less pronounced trend will be observed
when fluids contact clay-rich layers. The clay minerals act as
ion exchangers collecting preferentially the highly charged Ln.

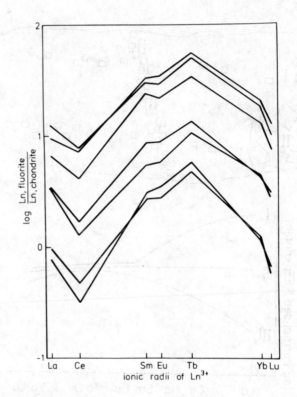

Fig.12 Subparallel Ln distribution patterns in fluorite from
mineralizations in Mesozoic limestones.
Sicily [7]

A very low level of Ln concentrations in the crystallizing Ca-
mineral will result.

2.5. The Variation Diagram

The foregoing discussion suggests that three different trends
should be observable. They are graphically displayed in the variation diagram (fig.13) by arrows. The arrows point towards increase of fractionation or of fluid-wall rock interaction. Making
use of eq.(1) but assuming conditions in solution at which λ_{op}
for each Ln is $\pm$ constant at least over a certain range of crystallization then the integrated Doerner-Hoskins law writes for two
different Ln:

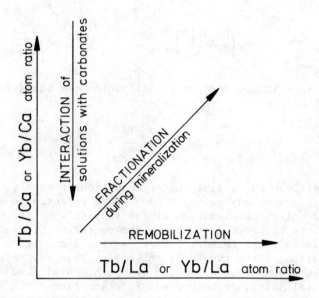

Fig.13 Schematic presentation of three different trends of Ln fractionation; diagonal: primary crystallization; horizontal: remobilization or secondary crystallization; vertical: Ln-ion exchange with wall rocks minerals.

$$\left(\frac{a}{A}\right)_x = a_0 \lambda_a \left(1-p\right)^{\lambda_a - 1} \tag{6}$$

$$\left(\frac{b}{B}\right)_x = b_0 \lambda_b \left(1-p\right)^{\lambda_b - 1} \tag{7}$$

with

$$a \equiv [HLn] \qquad b \equiv [LLn] \qquad A \qquad [Ca^{2+}]$$

$$p = \text{degree of precipitation defined on } [Ca^{2+}]$$

Combining eq.(6) and (7) and rearranging

$$\frac{a}{A} = \beta \cdot \left(\frac{a}{b}\right)^{\frac{\lambda_a - 1}{\lambda_a - \lambda_b}} \tag{8}$$

with

$$\beta = \left(\frac{(b_0\lambda_b)^\alpha}{a_0\cdot\lambda_a}\right)^{\alpha-1} \qquad\qquad \alpha = \frac{\lambda_a-1}{\lambda_a-\lambda_b} \qquad\qquad (9)$$

The logarithmic form of eq.(8) yields the basis for the plots such as fig.12:

$$\log\frac{a}{A} = \log\beta + \frac{\lambda_a-1}{\lambda_a-\lambda_b}\cdot\log\frac{a}{b} \qquad\qquad (10)$$

Zero slope is obtained under conditions where λ_a is unity but λ_b can have any value. Since λ_b is expected to be much greater than unity, complexation of the species with the concentration b must be rather high to bring the operative λ_b value down to zero. Infinite slope results for $\lambda_a = \lambda_b$, a condition for linear correlation of the two Ln in the solid. This may happen in a very rapid precipitation process or if the total load of a solution is crystallizing. In both cases fractionation of Ln is at minimum. Changes in total Ln concentrations with progress of crystallization might occur due to either mixing of two types of solution or to solution-rock interaction both in variable amounts. Any slope m in between leads to the general expression:

$$\lambda_a = \frac{m\lambda_b-1}{m-1} \qquad\qquad (11)$$

from which follows for

$$m \gtrless 1 \qquad\qquad \lambda_b \gtrless \frac{1}{m}$$

λ_a can vary considerably.

Although this discussion is based on the rather weak argument that λ_a and λ_b are constants in eq.(10) and (11) - which is not strictly true for Ln coprecipitation with fluorite or calcite - the foregoing considerations might help in the interpretation of the various trends found in the variation diagram (fig.13).

2.6. Probable Mechanisms of Ca-Substitution by Ln

Following reactions might be considered for Ca substitution in fluorite

$$3Ca_{xx}^{2+} + 2Ln^{3+} = 2Ln_{xx}^{3+} + \square + 3Ca^{2+} \qquad\qquad (12)$$

$$Ca_{xx}^{2+} + 2\overset{.}{Ln}^{3+} + F^- = Ln_{xx}^{3+} + F_{xx}^- + Ca^{2+} \tag{13}$$

$$2Ca_{xx}^{2+} + Ln^{3+} + Na^+ = Ln_{xx}^{3+} + Na_{xx}^+ + 2Ca^{2+} \tag{14}$$

xx indicates ions in the lattice
☐ cation vacancy

Reaction (12) describes substitution of $3Ca^{2+}$ ions by $2Ln^{3+}$ and one cation vacancy in order to balance the charges and lattice sites of a perfectly assumed anion lattice. In reaction eq.(13) Ca^{2+} is substituted by LnF^{2+}. In this case charge neutrality is maintained by putting the extra F^- ion into an interstitial position. The substitution of $2Ca^{2+}$ by $(Ln^{3+} + Na^+)$ ions leaves all lattice positions occupied but with differently charged ions. Na^+ is suggested to be the most favoured ion in reaction (14) since it is extremely abundant in hydrothermal solution and its radius matches with that of Ca^{2+} quite well (fig.1). In a set of fluorite samples we searched for evidence of reaction (14). The sample material was collected from three neighboured hydrothermal fluorite occurences in Sardinia. The ground material was treated with hot HNO_3 (1:1) for 5 min. in order to dissolve accessoric carbonates and other minute mineral components as well as to clean the material from Na^+ that usually is present in liquid inclusions. In fig.14 the molar sums of those Ln per gram fluorite which could be determined by neutron activation analysis were plotted vs. the number of moles of Na^+ per gram fluorite. Fig.14 illustrates a rather good linear relationship between Na_{xx}^+ and Ln_{xx}^{3+} which meets the origin. Although a strong correlation is found, reaction (14) is not really fulfilled since Na^+ is in excess by a factor of about 20. Therefore this correlation is necessary but not sufficient to prove the substitution mechanism of reaction(14). Assuming that reactions (14) and (15)

$$Ca_{xx}^{2+} + F_{xx}^- + Na^+ = Na_{xx}^+ + o + Ca^{2+} + F^- \tag{15}$$

act simultaneously, fig.(13) can be fully explained. Reaction (15) creates anion vacancies (o) in the lattice by substituting (CaF^+) by Na^+. The Ln pattern of each group of samples used in fig.14 were subparallel; i.e. no fractionation of Ln was observed within each group. In a steady state system the Doerner-Hoskins law for Ln and Na distribution can be combined yielding

$$\left(\frac{Ln}{Na}\right)_{xx} = k\left(\frac{Ln^{3+}}{Na^+}\right)_{soltn} \tag{16}$$

Eq.(16) suggests a linear relationship between Ln and Na although two different substitution processes are going on. However, cor-

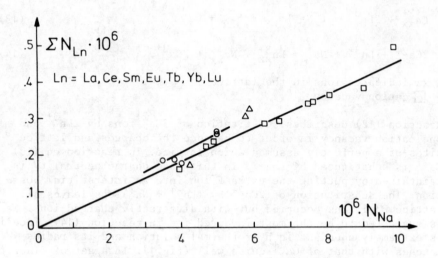

Fig.14 Correlation of the molar sum of determined Ln per gram
fluorite vs. number of moles of Na^+ion per gram fluorite.

related variations of Na and Ln contents in the fluorite at even
constant concentration in solutions might be expected due to kinetic effect in the growth mechanism, too. Such effects have been
studied by Lorens [38] who demonstrated experimentally that the
distribution coefficients of Sr, Cd, Mn, and Cl in calcite depend
very much on the growth rate of crystals in the range of 10^{-2} -
20 $CaCO_3$ layers/min.

3.   THE Ce AND Eu ANOMALIES

Different from the systematic fractionation of the tervalent
Ln ions are those which result from changes in the state of oxidation. Eu^{3+} can be easily reduced and Ce^{3+} is readily oxidized
under favourable conditions. Due to the large changes in their
respective ionic radii (fig.1) a conspicuous anomaly results in
the Ln distribution patterns of Ca minerals. The anomalies are defined by the actually observed normalized values and the ones
which might be found by interpolation from the Ln distribution
patterns. Fig.15 represents a set of pattern with either Eu-, or
Ce-, or combinations of both anomalies.

In general, negative anomalies are expected to build up if the
fluids from which the mineral grow are deficient in terpositive
Ce and Eu. There is some evidence that coprecipitation is controlled crystallo-chemically [39]. It is general experience that
Ca minerals prefer terpositive Ln ions. The efficiency of Eu in
the distribution patterns of Ca minerals does not necessarily

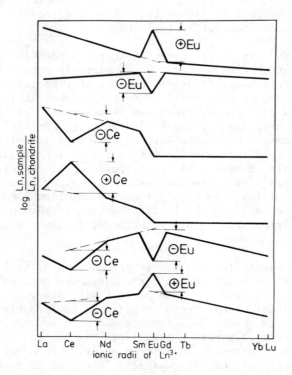

Fig.15 Ln distribution patterns showing (from top to bottom) positive Eu-, negative Eu-, negative Ce-, and positive Ce-anomalies. The last two patterns exhibit a combination of both anomalies.

imply that total Eu has been deficient in solution with respect to Sm and Gd. Eu might have been still present at its normal level or even might have been enriched in Eu^{2+} as e.g. in solutions of low redox potential. Philpotts [40] suggested that Eu^{2+} will behave like Sr^{2+} ions because of the similarity of their ionic radii. This agrees with the observation that Eu^{2+} is preferentially coprecipitated with baryte [39,41] and galena [42].

Deficiencies of Ce in distribution patterns are really due to deficient total Ce in solution. When Ce^{3+} is oxidized it will be easily adsorbed onto any oxyhydrate. Geochemically speaking Ce^{4+} ions will be less mobile than Ce^{3+} ions. However, tracing the Ce^{4+} ions remaining behind in the source areas of such solutions might be rather difficult. This is because the total amount of Ln leached by a hydrothermal solution will generally be low compared to the total amount of Ln left behind in the altered rocks. The oxidation of Ce^{3+} requires a certain oxygen fugacity. Ce deficient solutions

therefore indicate that they are derived from a source region of
oxygen fugacities equal or higher than those required to oxidize
Ce^{3+} under given environmental conditions.

Excess of Eu^{3+} in solutions and henceforth positive Eu anomalies in Ln distribution patterns of Ca minerals result from decomposition of feldspars. Feldspars usually show positive Eu anomalies, the reasons of which will be delt with separately when
Ln fractionation in magmatic processes is discussed (see this
volume). During breakdown of the feldspars (e.g. sericitization)
a considerable amount of the Ln with excess of Eu is released.
Under oxidizing conditions the Eu excess can be inherited by newly formed minerals. However, if these fluids enter reducing environments or turn over to become reducing (e.g. by bacteriologi-
cal activity or mixing with highly reducing solutions) Eu^{3+} might
become reduced to Eu^{2+} and the Ln distribution pattern of Ca minerals crystallizing from such a solution will show a negative
anomaly.

Positive Ce anomalies are not very common in nature. The most
conspicuous ones are found in the ferro-manganese nodules on the
ocean floors which formed in close contact with sea water (hydro-
genetically) [44,46]. These Fe-Mn oxyhydrates preferentially sca-
venge Ce^{4+} from sea water. This process seems to be very effective
(fig.16, curve (a)), since as a result of this the sea water
(fig.16, curve (c)) displays an extreme negative Ce-anomaly. This
negative Ce anomaly of sea water is also inherited to some extent
by autochthonous deep sea sediments [43-47]. The positive Ce anomaly in the hydrogenetic nodules is due to the increased adsorption of Ce^{4+} in comparison with Ln^{3+}ions or their compounds, although all Ln are accumulated in these concretions to a high extent. Since they are in intimate contact with the flowing ground-
water they have a real chance to collect Ln over long time spans
from huge volumes of sea water. Different from the hydrogenetic
Mn-nodules, diagenetic ferromanganese nodules exhibit almost no
or small negative Ce anomalies [49].

Ln distribution pattern of Ca-minerals with negative Ce and
positive Eu anomalies clearly indicate that their solutions were
involved in the breakdown of feldspars under oxidizing conditions.
The new Ca-mineral crystallized in an environment high enough in
oxygen fugacity to prevent reduction of Eu^{3+} to Eu^{2+}ions.

Ln distribution of Ca-minerals pattern with negative Ce and
Eu anomalies can be interpreted as the result of successive processes. Most probably the solutions left Ce behind during first
part of migration under oxidizing conditions. During the second
part of migration these solutions contacted either reducing units
of rocks or sediments or mixed with reducing solutions or became

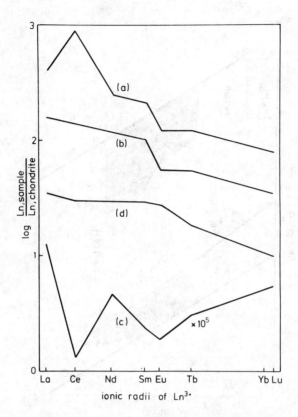

Fig.16 Averaged typical Ln distribution patterns of ferro-
manganese (iron rich) (a) and manganese (iron poor) nodules (b).
For comparison the Ln pattern of sea water [48] (e) and unaltered oceanic basalt (d) is given. The iron rich nodules
preferentially grow by accumulation of oxyhydrates from the
sea water, whereas the iron poor ones are mainly fed from pore
solutions derived from the underlying sediments (unpublished data
from Central Pacific).

reducing due to bacterial $SO_4^{2-} \rightarrow H_2S$ conversion in the environment
in which the mineralization took place.

    In the Eh–pH diagram (fig.17) stability fields for various
Fe and S [50] and Mn [51] species are superimposed. They may
serve as guidelines in the evaluation of appropriate fields in
which Ce^{4+} (probably in a complexed form) and Eu^{2+} are dominant
species in natural environments.

    Calculations of the potential of sea water by using various
redox couples resulted in Eh values between −0.1V and 0.6V [53].
Hartmann und Müller determined 0.4V in bottom sea water from the

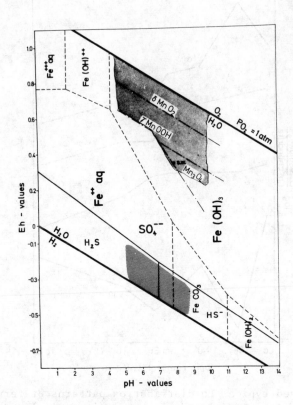

Fig.17 Eh-pH diagram showing stability fields for various Fe, Mn, S species [50,51] in water at 25°C and most probable fields in which either Ce^{3+} is oxidized or Eu^{3+} reduced.
sw = sea water [53]; $^a Fe = 10^{-6} M$

Central Pacific [53]. This value is given in fig.17. In 5.5 M K_2CO_3-solutions the Ce redox potential has been found to be rather low, +0.05V [54]. This low value seems to indicate that Ce^{4+} is stabilized in carbonato-complexes. In comparison with the other Ln Ce is preferentially accumulated in young Fe/Mn concretions (fig.16, curve (a)). If Ce is already present in sea water in its quadrivalent state then this selective enrichment might be ascribed to specific surface adsorption mechanism. On the other hand Hem [51] suggested that the presence of higher Mn-oxides might give rise to some electron transfer reactions. Combining both aspects the stippled area is tentatively proposed to represent the range of Eh and pH values in which Ce^{4+} is dominating.

Neglecting the presence of complexed Eu-species the Eu^{2+}/Eu^{3+}

might be derived from the reactions

$$Eu^{3+} + e = Eu^{2+} \qquad\qquad \varepsilon_{Eu,o} \qquad (17)$$

$$SO_4^{2-} + 8e + 10\ H^+ = H_2S + 4H_2O \qquad \varepsilon_{S,o} \qquad (18)$$

$\varepsilon_{Eu,o}$, $\varepsilon_{S,o}$ are the 'standard' potentials at relevant T,p conditions.

to be

$$\log \frac{Eu^{2+}}{Eu^{3+}} = \frac{BF}{RT} + \frac{1}{8} \log \frac{pH_2S}{aSO_4^{2-}} - 1,25pH \qquad (19)$$

$$B = \varepsilon_H - \varepsilon_{Eu,o} + \varepsilon_{S,o} \qquad (20)$$

From eq.(19) it is evident that changes of a_{H^+} are much more effective on the Eu^{2+}/Eu^{3+} ratio than changes in the ratio of S-species.
It has been often observed that in vein mineralizations with baryte dominating over sulfides [5] the negative Eu anomalies in
accessory Ca-minerals are almost absent. On the other hand negative Eu anomalies in accessory Ca-minerals are very common if
sulfides dominate over baryte [10]. As a rule of thumb it might
be adopted that if Ca-minerals show no negative Eu anomaly they
might have crystallized from solutions with $[SO_4^{2-}] \gg [HS^-]$ and they
will exhibit a negative anomaly when $[SO_4^{2-}] \ll [HS^-]$. The shaded
area in fig.17 indicates the probable Eh - pH range in which Eu^{2+}
dominates over Eu^{3+} ions.

4.   APPLICATIONS TO GEOCHEMISTRY: SOME BRAIN TICKLERS

4.1. General Remarks

Different from lab experiments in the practical work of geochemistry seldom both phases between which the Ln had been distributed are accessible. Since in non-magmatic processes always
solutions take part search for the chemical composition of these
fluids is of particular interest. Due to these fluids gigantic
volumes of rocks undergo chemical reactions because the solids
are not in chemical and/or mineral phase equilibrium with the
contacting solutions. The amount of minerals crystallizing from
solutions after transportation is usually very small on the volume
scale of earth's crust solids. Only the biogenetic formation of
carbonates in the marine environment (about 7% of the sedimentites)

and evaporites are worth mentioning. The fluid phase could give
rise for ± isochemical processes in ± closed system like e.g.
metamorphism with only small changes in bulk composition of the
rocks but considerable changes in the mineralogy of the rocks.
Different from them are allochemical processes in open systems
like e.g. metasomatism where the chemical composition as well as
the mineralogy is changed. Unsolidified sediments undergo diagenesis in which all particles become cemented. In this process
the original sediment partly dissolves and recrystallizes. Very
rarely two different solids are in equilibrium. This is mainly
due to the fact that in most cases the surfaces of the minerals
equilibrate with their common solutions. Diffusion of surface ions
into the bulk crystals is - with the exception of open minerals
(e.g. zeolites, mica, and clay minerals) - negligible. Kinetic
effects of various kinds have to be considered to hinder the estab-
lishment of equilibria, too. With all these restrictions in mind
we will now try to interprete Ln distribution pattern of minerals
and rocks with respect to geochemical processes that caused or
at least influenced the trace element composition of minerals.

4.2. The Use of Variation Diagram

It has already been discussed in which way a variation diagram can be useful. In fig.18 and 19 such diagrams are shown for
fluorite and calcite, respectively.
Fluorite. Since in fluorite samples quite often highest fractionation was found between La and Tb (fig.2) the Tb/La ratio is
chosen as the fractionation index (fig.18). On petrological observations and geochemical considerations three fields could be
marked in the fluorite diagram. Here only the geochemical arguments are given.

Pegmatitic field: Ln become enriched in the residual water-
rich magmatic melts, in particular when they are rich in complex
forming ligands (e.g. F^-, CO_3^{2-}, etc.). Therefore often Ln-minerals
are formed and found in pegmatites (yttrofluorite, bastnaesite).

Hydrothermal field: Minerals crystallizing from hydrothermal
solutions are - by experience - lower in Ln than those in the pegmatites from which stage they may be derived in part. Samples from
nearly all over the world group together in a rather well defined
field. This might be explained by the assumption that most of the
hydrothermal solutions are generated under pT conditions within
rock suites being rather similar in their Ln/Ca ratio. The concentration of complexing ligands will influence the distribution
of Ln between solutions and solids, too. From the work of Koß [19]
it might be suggested that the maximum in the distribution pattern
among the elements usually analysed is near to Tb in acidic solutions where the formation of OH-complexes is negligible. As far as
it is known today, only the monofluoro-complexes exhibit such an

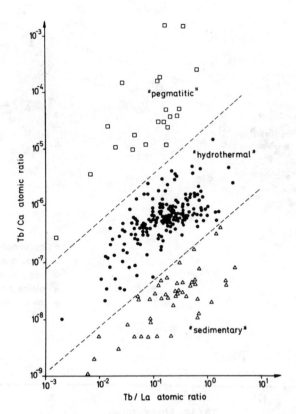

Fig.18 The Tb/Ca - Tb/La variation diagram for petrologically
sorted fluorites from five continents. Only those samples were
excluded which either have undergone exteme remobilization or
the solutions of which re-equilibrated with Ca^{2+}rich rocks prior
to crystallization of the fluorite.

anomalous behaviour. Therefore the presence or absence of a maximum at Tb might become an interesting indicator for the acidity
or basicity of mineralizing solutions, respectively. A direct
prove for the composition of pore waters taking part in regional
metamorphism might be obtained from the analyses of minerals which
crystallized in open fissures in metamorphic rocks. The metamorphic solutions are thought to evaporate after entering the open
fissures. The various crystallizing minerals then ± preserve the
true Ln distribution of the solutions. These solutions are often
drastically enriched in the HLn as might be read from the analytical results on fluorites [55] and calcites [12].

Sedimentary field: Since the Ln concentrations in ocean water

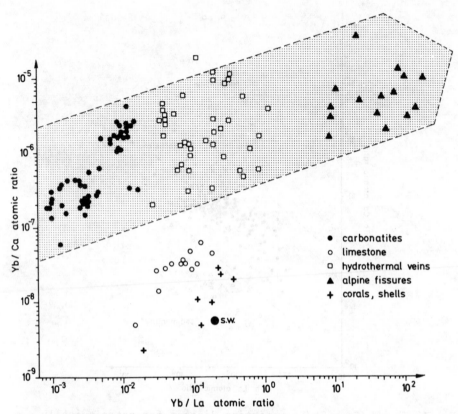

Fig.19 The Yb/Ca - Yb/La variation diagram for petrological sorted
calcites from various localities. Each genetical group covers
distinct fields. The carbonate field includes samples from Europe,
Brazil, Central and South Africa, and India. The hydrothermal
calcite are mainly from the Harz mountain/Germany, those from the
alpine fissures from the Austrian Alps. The corals have been col-
cected from the Bahamas and Florida Keys [57].

are low [48] (fig.16) the same will be expected to be observed in
minerals derived from it. On the other hand F⁻ion concentration
is low in ocean water, too. Therefore direct precipitation will
not be possible. Fluorite formation will occur in evaporating
brines during dolomitization of carbonate sediments. The brines
have lost all the Ln during early precipitation of carbonates and
sulfates. Therefore this process is indicated by extremely low Ln
concentration levels in the resulting fluorite which is mostly
below the detection limit of the most common analytical method of
instrumental neutron activation analysis [56]. Fluorite formed by

metasomatic reactions of F⁻ion rich exhalations with carbonate
sediments in hydrothermally active marine basins show Ln patterns
comparable with those of the carbonates [5]. Although the similarity in Ln patterns pretend cogenetic formation, the fluorite is
epigenetic. In many cases Ln equilibration between carbonates and
fluorites might be established during diagenesis.

Calcite. According to fig.3 fractionation of La and Yb(Lu)
is extreme in calcite. For analytical reasons Yb/La is chosen as
a fractionation index for this mineral. Carbonatites are clearly
recognized and show smallest La/Yb ratios, fig.19; the hydrothermal carbonates from vein mineralization increase in La/Yb ratios
which become highest in the calcite crystallizing from metamorphic
solutions in the Alpine fissures [12]. The biogenic carbonates in
the calcareous sediments are much lower in Yb/Ca [57,58] and show
a Yb/La ratio equal to or higher than in ocean water. Since almost
all primary carbonates in the marine environment are biogenic (i)
the Ca metabolism obscures the direct relationship between Ln in
the hard tissues of organisms and the sea water, and (ii) by far
the greatest amount of carbonates are either magnesian calcites
or aragonite. When these carbonates become low Mg-calcites during
diagenesis this happens under the influence of fresh water. The
often observed increase of Ln contents is mostly due to incorporation of tiny clay particles [57,58].

Ca-phosphates. Primary magmatic phosphates (apatite) can be
clearly distinguished from secondary phosphates (phosphorites)
petrologically and geochemically [36,59]. Marine and terrestrial
phosphorite again can be separated in a Sc/Sr - Yb/La variation
diagram (unpublished).

Data point arrangements. The data points of samples from a
given deposit (mineral occurence) in the variation diagram of
fluorite either line up in one of the three trends which are
indicated in fig.12 or cluster together in one of the three fields
of fig.18.

- The diagonal trend (fig.20) is often found in undisturbed
  gangue mineralizations. Fractionation increases with increasing Tb/La ratio. Lowest and highest values are only
  found by chance and depend therefore on the sample collection
  available from a given deposit.
- The horizontal trend (fig.21) is observed in gangue mineralizations from tectonically active areas [8,61] and sedimentary
  fluorite [5,6]. It represents remobilized fluorite. Depending
  on the migration the Tb/La can increase to very large values
  [8,61].
- The vertical trend (fig.22) is found e.g. in fluorites from
  veins in limestone, dolostone or near to them in silicate
  rocks. Similar trends result when hydrotherms equilibrate with
  clay rich sediments [37]. The highest Tb/Ca ratio represents
  a fluorite that crystallized from the least altered solutions.

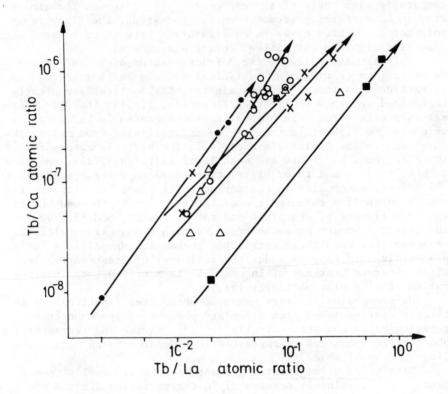

Fig. 20 Alignment of fluorite samples from individual deposits in the variation diagram.

    o Charboniere vein/France [60]
    △ Max mine/Stulln, Germany
    ■ Cäcilie mine/Stulln, Germany
    × Marienschacht mine/Wölsendorf, Germany
    ● Schönfarbiges Bergwerk/Bach, Germany

Arrows indicate increase of Ln fractionation

Plotting the results of a number of samples from a given occurence in the respective variation diagram , their localization in one of the fields helps to characterize the general physico-chemical conditions during mineralization. In case of remobilization or intensive wall rock interaction those samples with e.g. lowest Tb/La or highest Tb/Ca ratio approximately characterize the initial fluids best.

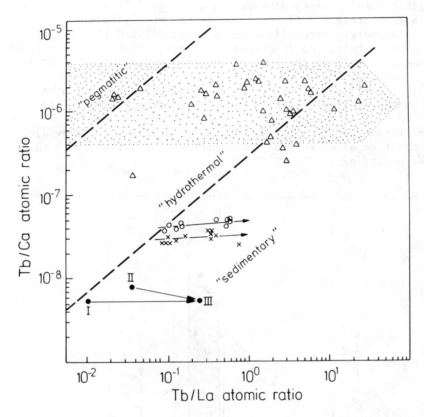

Fig.21 Tb/Ca – Tb/La variation diagram showing four examples of deposits with remobilized fluorite.

- ● sedimentary fluorites from the Nördliche Kalkalpen I, II, III correspond with the generations in fig.11;
- ○ sedimentary fluorite (zebra ore) from the Sierra de Gador/ Spain;
- ✗ sedimentary fluorite (zebra ore) from the Sierra de Baza/ Spain;
- ▲ hydrothermal fluorites from the Chacaltaya, Cordillera Real/ Bolivia [61].

Arrows indicate increase of Ln fractionation

   Probing cross sections through a gangue mineralization varying trends can be obtained depending on the geotectonical and crystallization history. In fig.23 the trends of Tb/Ca and Tb/La are shown

a) in fluorites from a symmetrical, but brecciated fluorite vein
   (Käfersteige mine, Schwarzwald/Germany) [9]; different symbols represent different generations of crystals;
b) in asymetric mineralization with remobilized fluorite in the
   eastern half, and fractionatedly crystallized fluorite in the
   western half of the vein (Gottesehre mine, Schwarzwald/Ger-
   many) [9];
c) in remobilized fluorite in two parallel veins with mainly remobilized fluorite (Clara mine, Schwarzwald/Germany) [9].

The different behaviour of Tb/Ca and Tb/La in fig.23 is evident.
This might be expected from the different possibilities of their
trends according to fig.13.

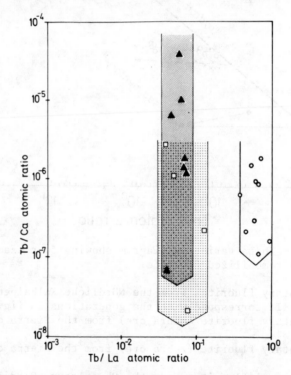

Fig.22 Tb/Ca - Tb/La variation diagram showing three examples of
fluorite crystallizations from solutions which interacted with
carbonate rocks at various degrees prior to mineralization. Arrow
points to increasing extent of interaction.

   pegmatitic fluorite from the cassiterite deposit Zaaiplaats/
   Republic of South Africa [37]
   hydrothermal fluorite from the Pb-Zn deposits at Atacocha,
   Cerro de Pasco/Peru [37]
   hydrothermal fluorite from Sicily [7].

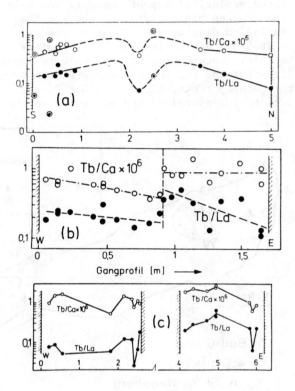

Fig.23 Tb/Ca and Tb/La profiles in cross sections through fluorite
veins of the

        a) Käfersteige mine/Pforzheim
        b) Gottesehre mine/Urberg and
        c) Clara mine, Oberwolfach

All three mines are located in the Schwarzwald/Germany.

    <u>Temperature dependence</u>. The influence of temperature on Ln
distribution patterns of minerals from many occurences within a
large area can probably be derived from the variation diagram, too.
For example the hydrothermal carbonates from the Harz mountain/
Germany show a conspicuous tendency for increasing Tb/Ca and
Yb/La ratios the nearer the hosting veins are to the Brocken-
pluton (fig.24)[10, 62]. Here the plutonism is considered as the
heat source setting up a hydrothermal convection system in the
country rock. Assuming comparable chemical composition of the
hydrothermal solutions in a given area of rather uniform lithology
the discrimination between HLn and LLn in coprecipitation is expected to decrease with increasing temperature. Thus smallest

Fig.24 Simplified geological map of the Harz mountain showing the
location of the Brocken pluton (Br) and the sampling points. In
the calcite variation diagram a systematic trend is to be seen.
This trend is probably due to increasing thermality during generation of the calcite bearing solutions. Calcite samples from
Bad Grund, Clausthal and St. Andreasberg are from vein mineralizations in the country rocks, those from Bad Harzburg and Wurm-
berg from veinlets and miaroles, respectively, in the magmatites.

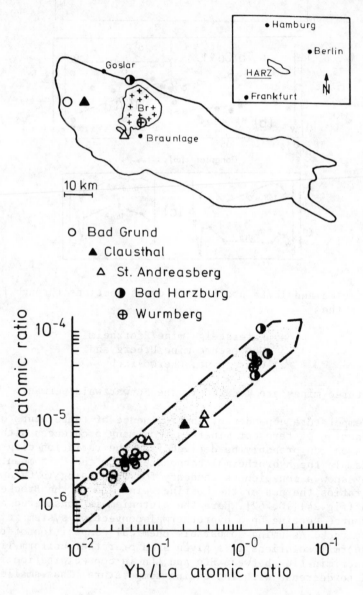

Yb/La ratios will be found farthest away and highest nearest to
the Brocken-pluton/Harz/Germany (fig.24).

4.3. Tracing a Hydrothermal Convection Cell

In the neighbourhood of the outcropping Chacaltaya granitoid/
Bolivia the sectional area of the ascending branch of a hydrothermal convection cell could be traced using some characteristic
features of the Ln distribution patterns in the fluorite [61,63].
The strong remobilization acting upon many of the fluorite samples
can be seen from fig.20. Additionally negative Ce- and highly variable Eu-anomalies have been observed [61]. All representative
points of Chacaltaya fluorites in fig.21 are projected along the
field boundaries on a horizontal line. The range of projection is
intersected starting with '0' for the lowest and ending with '100'
for the highest value. The figures are used as quantitative measures of the degree of remobilization. Furthermore the amount of
the Eu-anomaly was determined. Transferring the remobilization-
and the Eu-anomaly values into a schematic geological map, very
similar structures appear when isolines are drawn (fig.25,26).
The interpretation of these results runs as follows: On the flank
of the dipping granitoid hydrothermal waters ascended and minera-
lized the country rock, a sequence of quartzites and black shales
[61]. The hydrothermal solutions are mainly upheated vadose waters
as indicated in fig.27. With time earlier crystallized fluorite
from the central part of the ascending hydrotherm is mobilized
into parts of the country rock farther away from the center of the
hydrothermal activity. The areas of remobilization '0-40', and of
positive Eu-anomaly - including their deviating inner part - are
considered as the area of ascending hydrothermal solutions. In
the ascending solution Eu is present in excess as indicated by the
positive Eu-anomaly in the fluorite. This indicates that these
solutions were involved in decomposition of feldspars in depth.
The ring-like structure follows from remobilization of fluorite
into the country rock and Eu reduction due to contacts with black
shales. When the hydrothermal activity died down, fluorite -
already mobilized and characterized by a negative Eu-anomaly -
is mobilized again and crystallized in the central part of the
ascending hydrotherm towards the end of the lifetime of the convection cell.

4.4. Behaviour of Ln during Water-Rock Interaction

Ln distribution in solutions. There are only few thorough
studies of the behaviour of the Ln during dissolution of their
host minerals. The final Ln distribution will be controlled by
(i) the complex formation capacity in the solution and (ii) surface
ion exchange phenomena. If the concentration of complexing ligands
is high, HLn will become enriched over the LLn in solution.

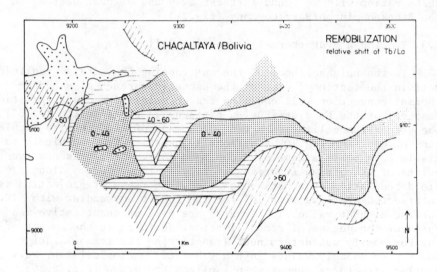

Fig.25 Schematic geological map showing areas of remobilization
of fluorite which are comparable. Details are given in the text.

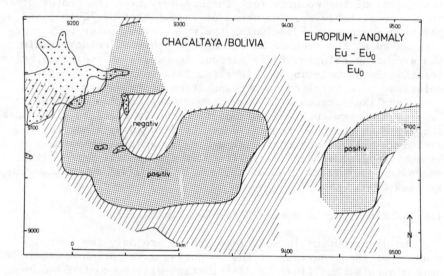

Fig.26 Schematic geological map showing areas of Eu-anomalies in
fluorite which are comparable.

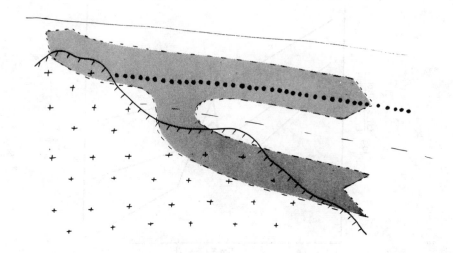

Fig.27 Schematic cross section through the hydrothermal convection
cell on the flank of the Chacaltaya granitoid.
.... = present day surface. The dashed branding arrow indicates
the suggested flow of mainly vadose waters. These enter the granitoid at least in its roof and the flank somewhere at depth.
When the intrusion cools down the convection cell narrows, but
with the centre of the ascending solutions still at the same place.
This naturally involves that in this final stage the earlier remobilized fluorite enters the cycle again. This might explain the
observed remobilized fluorite with its negative Eu-anomaly in the
very centre of the area of the ascending hydrotherm (fig.25,26).

The Ln distribution pattern in minerals formed from such solutions
must reflect this behaviour and it will be done the better the
lower the temperatures. The Ln pattern of the altered rock minerals
or of precipitates from the fluids might serve as a geochemical
probe determining the type and sequence of alteration reactions
within a given system. Graf [42] tried to estimate the Ln patterns
generated in water by reactions of (i) volcanic glass to mica,
(ii) feldspar to mica and (iii) Ln exchange with felsic volcanic
rock. Results for these three different, but very abundant types
of reactions are given in fig.28. Curve (a) and (c) represent the
resulting Ln pattern in solution after mineral reactions took place:
alteration of glass to mica (a) and feldspar to mica (b). If feldspars are involved their Ln with excess of Eu are released together
with other trace elements like Pb and Zn. Such solution may be able
to form Pb-Zn mineralizations in favourable geochemical traps. Then
the galena, PbS, will show a conspicuous positive Eu-anomaly [42],
thereby indicating the origin of Pb from feldspar.

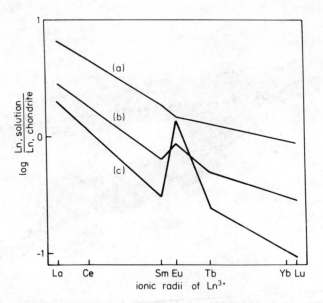

Fig.28 Estimated Ln distribution patterns generated in water by
(a) reaction of volcanic glass to mica, (b) exchange only acting
upon a felsic volcanic rock (30% feldspar, 3% clinopyroxene,
2% orthopyroxene, and 60% volcanic glass), and (c) reaction of
feldspar to mica. The Ln distribution pattern of the felsic rock
is subparallel to (a) but at Ln levels about 20 times higher than
those of curve (a). (Redrawn after data reported by Graf [42].

     Curve (b) in fig.28 represents the estimate of the Ln in
solutions if only ion exchange acted upon a felsic volcanic rock.
It might be mentioned that the solutions equilibrated with felsic
rocks will mostly show higher positive Eu-anomalies than those
equilibrated with mafic rocks. This is due to the rather high
Ln contents of some very mafic minerals (hornblende and biotite)
in comparison with the low Ln contents of feldspars. Whitfield
and Turner [64,65] suggested that the distribution coefficients
of Ln between sea water or river water and crustal rocks is a
function of the ionic interaction of Ln with oxygen. In a crude
way their results seem to confirm that solid state chemistry plays
a major part in controlling the composition of sea water and the
world's average river water. The LLn are preferentially enriched
in most hydrogenous minerals since they accomodate a higher coordination number [65] than others. In general all the Ln are depleted in sea water with respect to river waters indicating that
scavenging of the Ln in the oceans is very efficient [65]. These
arguments explain the Ln distribution pattern of sea water to some
extent (fig.16, curve c).

Ln mobility in alteration reactions. Taylor and Fryer [66] reported on changes of Ln distribution patterns during potassic alteration of a granodiorite-porphyry with subsequent propylitic and phyllitic overprinting (fig.29). During potassic alteration at high temperatures light to medium Ln increase in concentration, the HLn decrease due to their stronger complexation in solutions. Subsequently, with decreasing temperatures, pH and rock/fluid-ratios all Ln are leached during propylitization. In this reaction the existing mineralogy is altered into epidote + calcite + quartz + chlorite. In continuation of leaching at further decreasing temperatures sericitization of feldspars and other minerals (phyllic overprinting) results in still lower Ln concentrations. The over-proportional decrease of Eu was explained to be due to low Eh values as indicated by the formation of pyrite.

Lausch et al [12] reported that the Ln distribution coefficient between gneisses and their metamorphic carbonates in fissures decrease with increase of temperature showing that at lower temperatures light Ln are preferentially distributing into carbonates. Ln distribution patterns of dolomite-calcite pairs in metamorphic marbles show $\pm$ no differences [67]. However, LLn become less enriched in marbles with decreasing temperatures of metamorphism [68]. Magnesite can be formed metasomatically from a limestone precursor. Here the reaction is introduced by a Mg-rich hydrothermal solution. The metasomatic magnesite (fig.30) is very similar in its Ln distribution patterns to those of the limestone [69,70], but differs markedly in the magnesite mobilized into veinlets. The metasomatic magnesite is well distinguishable from sedimentary types (fig.30).

Diagenesis of argillaceous [71] or calcareous sediments [57] does usually not introduce significant changes in Ln contents and their distribution patterns.

Hellmann et al [72] recognized four types of Ln mobility during burial metamorphism of flood basalts:

  - gross and selective light Ln enrichments,
  - scatter of Ln pattern around a primary mean (redistribution)
  - gross Ln depletion resulting in subparallel patterns,
  - selective mobility of individual Ln like Ce, Eu.

These secondary changes of Ln distribution patterns have to be born in mind when drawing petrogenetic conclusions. According to Muecke et al [73] Ln provide a promising tool in the determination of the parentage of metamorphosed and metasomatized rocks up to the upper amphibolite facies of regional metamorphism.

During albitization (spilitization) of oceanic basalts and tholeiites the LLn seem to become enriched relative to the HLn [74,75].

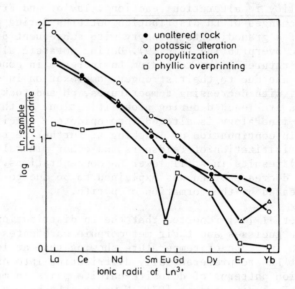

Fig.29 Changes of Ln distribution patterns in a granodiorite-
porphyry due to potassic alteration, propylitization and phyllic
overprinting (redrawn after [66]).

The behaviour of Ln during low temperature submarine weathering
of tholeiitic basalts is not well understood. It is reported that
either (i) the Ln are not affected [76,77] or (ii) the LLn become
considerably enriched in the margin of the pillows with time
whereas the HLn remain unaffected [78] or (iii) the medium to
heavy Ln become relatively depleted in the margin of the pillows [79].
However, since in many cases it is difficult to find out (i) the
difference in the chemical composition between the original margin
and the interior of any pillow and (ii) the extent of the alteration of the fresh product it has been suggested [78] to look into
the chondrite normalized Ln pattern after normalization to constant
Yb for the respective fresh basalt. If so, then the relative increase
of the LLn with increasing time/alteration becomes conspicuous
(fig.31). In the spilites as well as weathered submarine basalts
the interaction with sea water is mostly indicated by the impressed
negative Ce-anomaly.

Nesbitt [80] compared the Ln patterns in a parent granodio-
rite and its moderately altered products. They found the latter
somewhat enriched in HLn, during supracrustal alteration whereas
the extremely altered residual products are depleted in HLn.

Extreme Ln mobility has been observed in some ore formation
processes. In particular CO_3^{2-} and F^- rich solutions are favourable

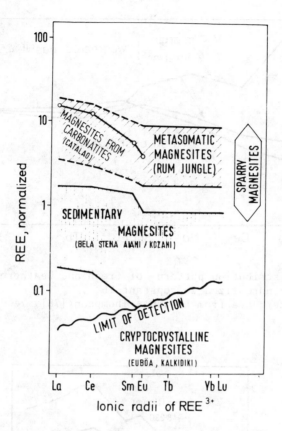

Fig.30 Fields of Ln distribution patterns in magnesite of different
genesis [69,70].
Metasomatic: Sparry magnesite from the Grauwackenzone of the
Northern Kalkalpen/Austria; magnesite from the Rum Jungle Uranium
field/Australia; magnesite from the carbonatite of Catalao/Brazil.
Sedimentary: Magnesite from intramountanous basins (Aiani/Kozani/
Greece and Bela Stena/Yugoslavia).
The crystocrystalline magnesite is from veins in ultra basic rocks.

to mobilize Ln [9,10,61], [66,81,82]. This capability seems to
increase with both increasing temperature and ligand concentration
of the hydrothermal solutions. Mc Lennan and Taylor [81] observed
extreme Ln mobility together with U in metasediments. The Ln pattern (fig.32) show an extreme increase in HLn with increase of U
in the rocks. This was explained by the high mobility of carbonato-complexes of Ln and U in oxidizing alkaline solutions at low
temperatures.

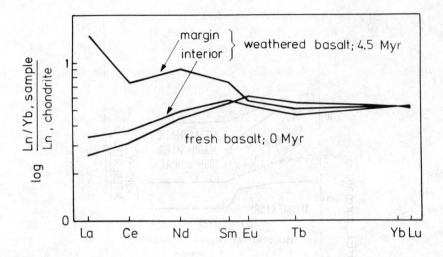

Fig.31 Ln distribution patterns of fresh and weathered tholecitic basalts when normalized to constant Yb
(Redrawn after data from Ludden & Thompson [78])

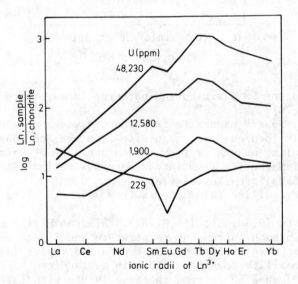

Fig.32 Ln distribution pattern in metasediments showing increasing Tb/La fractionation. The Ln fractionation is somehow related with the U content of the metasediments. Redrawn after [81].

In conclusion: The Ln concentrations in alteration products seem to be strongly dependent on the

- Ln concentration levels in the parent rocks or sediments
- Ln concentration in pore fluids
- concentration of ligands which are able to form Ln-complexes
- volume of solution per volume rock in the reaction
- mineralogy of rocks or sediments

4.5. Ln in Iron Formations and Manganese Nodules

Iron-formations probably allow an insight into changing oxidation states of Ce and Eu with the development of the earth's atmosphere [83]. In fig.33 typical Ln patterns of present day sea water (curve(d)) [48], unaltered glass (curve(c)), Archean (curve(a)) and Proterozoic (curve(b)) iron formations [84] and present day Fe/Mn concreation (curve(e)) are presented showing following features: Ce behaves normal in Archean iron formations whereas it becomes depleted in the younger like in sea water. Eu is mostly en-riched in the old and behaves almost normal in the younger iron formations. This is explained by assuming [83] an Eu^{2+} enriched Archean ocean from which the Fe-oxyhydrates precipitated later when oxygen became available. The anomalous Ce behaviour sets up during early Proterozoic when the redox potential was high enough to oxidize Ce^{3+}ions as it is done today. There seems to have been a time overlap in which Ce became oxidized and Eu still behaved as Eu^{2+}ions [83]. Jakes and Taylor [85] tried to find arguments to explain the excess of Eu in the Archean to Early Proterozoic by predominant weathering of an Eu enriched early crust formed by island arc volcanism. The later depletion of Eu in the younger crust should be due to different processes by which it is formed, e.g. partial melting. Other features of differences are the fact that (i) the old iron precipitates show considerably lower Ln concentrations than present day examples of Fe/Mn concreations and Mn-nodules and furthermore positive Ce-anomalies are rather common today (curve(e),fig.33).

So what has been changed in the marine environment? According to Fryer [83] this indicates "that ferric iron precipitation was not accompanied by trace element scavenging as it is today." It might be suggested that the complex formation capacity of the Precambrian ocean could have changed in the course of the evolu-tion of life. Alternatively, it might reflect in some way changes in the overall formation processes of the earth's crust. The be-haviour of Eu and Ce suggest that about 2000 million years ago sea water was Eu enriched and Ce at normal due to reducing con-ditions in the hydrosphere and the atmosphere. But about this time the change to oxidic conditions took place. At the moment we are far away from understanding completely the development of the

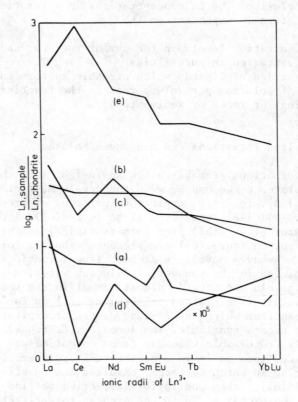

Fig.33 Ln distribution pattern in

    (a) Archean iron formation, Finland [84]
    (b) Proterozoic iron formation, Finland [84]
    (c) basaltic glass, nearly unaltered, Pacific Ocean
        (unpublished data)
    (d) sea water [48]
    (e) Fe/Mn concreations (Central Pacific)
        (unpublished data)

atmosphere and hydrosphere when changing from reducing to oxidic
conditions [86]. But the Eu- and Ce anomalies in old rocks might
serve as a record and help to evaluate this most dramatic and
important chemical change in the earth's development.

5. CONCLUDING REMARKS

The series of the lanthanoids (Ln) represent an interesting chemical probe for studying

(i) fundamental processes of element distribution in phases boundaries under various chemical and p,T conditions;
(ii) some physico-chemical and compositional aspects in fluids from the earth's crust;
(iii) the environmental condition in the source area in which hydrothermal fluids are generated;
(iv) ore formation processes and
(v) the precursors of metamorphic rocks.

For these reasons studies of Ln distribution have become a rather successful tool in various fields of geosciences: In geochemistry the Ln are used to establish new or prove old rules controlling migration of trace elements during the course of geochemical processes; in petrology they are helpful in establishing relationships between different kinds of rocks; in mining geology and ore deposit research they might be used in typifying ore deposit genetically, and in evaluation of concepts of formation of ore deposits. All these aspects taken together will lead to a better understanding of geochemical processes to which the earth's crust is subjected. It is hoped that beside improving our knowledge, strategically new concepts may emerge which might be used in searching for hidden mineral deposits.

A great deal of work has been done in the past, but still more has to follow in order to reach the above mentioned goals.

REFERENCES

1. R. D. Shannon, Acta Cryst., A 32, 751 (1976).

2. Ch. D. Coryell, J. W. Chase, and J. W. Winchester, J. Geophys. Res., 68, 559 (1963).

3. A. G. Herrmann, Handbook of Geochemistry, (K. H. Wedepohl, Ed.), Springer-Verlag, Berlin-Heidelberg, 1969, 39, 57-71.

4. L. A. Haskin, F. A. Frey, and T. R. Wildeman, Origin and Distribution of Elements, (L. H. Ahrens, Ed.), Int. Ser. Earth Sc., 30, 889 (1968).

5. H. J. Schneider, P. Möller, and P. P. Parekh, Mineral. Deposita, 10, 330 (1975).

6. K.-H. Jacob, Deutung der Genese von Fluoritlagerstätten anhand ihrer Spurenelemente - insbesondere an fraktionierten Seltenen Erden, Ph. D. Diss., D 83, Techn. Univ. Berlin, 1974, pp. 99.

7. A. Bellanca, P. Di Salvo, P. Möller, R. Neri, and F. Schley, Chem. Geol., 32, 255 (1981).

8. H. G. Mylius, Lagerstättenkundliche Untersuchungen an Flußspat-Kupfermineralisationen von Cerro Muriano, Cordoba, Spanien, Ph. D. Diss., D 83, Techn. Univ. Berlin, 1982, pp. 132.

9. P. Möller, H. Maus, and H. Gundlach, Die Entwicklung von Flußspatmineralisationen im Bereich des Schwarzwaldes, Jb. Geol. Landesamt Baden-Württemberg, (in press).

10. P. Möller, G. Morteani, J. Hoefs, and P. P. Parekh, Chem. Geol., 26, 197 (1979).

11. L. Haskin and M. A. Gehl, J. of Geophys. Res., 67, 2537 (1962).

12. J. Lausch, P. Möller, and G. Morteani, N. Jb. Miner. Mh., 11, 490 (1974).

13. J. F. Walker and G. R. Choppin, Lanthanide and Actinide Chemistry, Adv. Chem. Ser. (Amer. Chem. Soc. Publ.), 71, 127 (1967).

14. B. A. Bilal, F. Herrmann, and W. Fleischer, J. Inorg. Nucl. Chem., 41, 347 (1979).

15. B. A. Bilal and P. Becker, J. Inorg. Nucl. Chem., 41, 1607 (1979).

16. B. A. Bilal and V. Koß, J. Inorg. Nucl. Chem., 42, 629 (1980).

17. P. Becker, Ionenassoziations- und Verteilungsverhalten von Lanthanoiden bei der Fluoritbildung, Ph. D. Diss., D 83, Techn. Univ. Berlin, 1982, pp. 160.

18. R. Guillaumont, B. Désiré, and M. Galin, Radiochem. Radioanal. Lett., 8, 189 (1971).

19. V. Koß, Untersuchung der Komplexbildung der Lanthanoiden in einem für die hydrothermale Fluorit-Bildung relevanten chemischen System, Ph. D. Diss., D 83, Techn. Univ. Berlin, 1980, pp. 75.

20. B. A. Bilal and V. Koß, J. Inorg. Nucl. Chem., 43, 3393 (1981).

21. J. Dumonceau, S. Bigot, M. Treuil, J. Faucherre, and F. Fro-
    mage, C. R. Acad. Sc. Paris, Sér. C, 287, 325 (1978).

22. R. G. de Carvalleo and G. R. Choppin, J. Inorg. Nucl. Chem.,
    29, 725 (1967).

23. B. A. Bilal and V. Koß, J. Inorg. Nucl. Chem., 42, 1064 (1980).

24. D. F. Peppard, G. W. Mason, and I. Hucher, J. Inorg. Nucl.
    Chem., 24, 881 (1962).

25. T. Sekine, J. Inorg. Nucl. Chem., 26, 1463 (1964).

26. T. Goto and M. Smutz, J. Inorg. Nucl. Chem., 27, 663 (1965).

27. A. M. Feibush, K. Rowley, and L. Gordon, Anal. Chem., 30, 1605
    (1958).

28. B. C. Purkaystha and S. N. Bhattacharyya, J. Inorg. Nucl. Chem.,
    10, 103 (1959).

29. M. Matsui, Bull. Chem. Soc. Japan, 39, 1114 (1966).

30. H. A. Doerner and W. M. Hoskins, J. Amer. Chem. Soc., 47, 662
    (1925).

31. L. Marchand, Contribution à l'étude de la distribution des
    lanthanides dans la fluorine, Ph. D. Diss., Univ. of Orléans,
    1976, pp. 104.

32. P. V. Smallwood, Colloid. Polymer. Sci., 255, 881 (1977).

33. S. P. Sinha, Complexes of the Rare Earth, Pergamon Press,
    Oxford, 1966, pp. 36.

34. G. Strübel, N. Jb. Miner. Mh.,   83 (1965).

35. P. Möller, P. P. Parekh, and H. J. Schneider, Miner. Deposita,
    11, 111 (1976).

36. P. P. Parekh and P. Möller, Nuclear Techniques and Mineral
    Resources, International Atomic Energy Agency, Wien, 1977,
    pp. 353.

37. P. Möller and G. Morteani, On the Geochemical Fractionation
    of Rare Earth Elements during the Formation of Ce-Minerals
    and its Application to Problems of the Genesis of Ore Deposits.

The Significance of Trace Elements in Solving Petrogenetic Problems, (S. S. Augustithis, Ed.), (in press).

38. R. B. Lorens, Geochim. Cosmochim. Acta, 45, 553 (1981).

39. J. W. Morgan and G. A. Wandless, Geochim. Cosmochim. Acta, 44, 973 (1980).

40. J. A. Philpotts, Earth Planet. Sci. Lett., 9, 257 (1970).

41. F. Guichard, Th. M. Church, M. Treuil, and H. Jaffrezic, Geochim. Cosmochim. Acta, 43, 983 (1979).

42. J. L. Graf, Econ. Geol. , 72, 527 (1977).

43. D. Z. Piper and P. Graef, Mar. Geol., 17, 287 (1974).

44. D. Piper, Geochim. Cosmochim. Acta, 38, 1007 (1974).

45. T. J. Barrett, A. J. Fleet, and H. Friedrichsen, (to be published).

46. H. Elderfield, C. J. Hawkesworth, M. J. Greaves, and S. E. Calvert, Geochim. Cosmochim. Acta, 45, 513 (1981).

47. V. Marchig, H. Gundlach, P. Möller, and F. Schley, Mar. Geol., (in press).

48. O. T. Høgdahl, S. Melsom, and T. Bowen, Trace Inorganics in Sea Water, Adv. Chem. Ser. (Amer. Chem. Soc. Publ.), 73, (1968).

49. H. Elderfield and M. J. Greaves, Earth Planet. Sci. Lett., 55, 163 (1982).

50. R. M. Garrels and Ch. L. Christ, Solutions, Minerals, and Equilibria, Harper and Row, New York, 1965, pp. 450.

51. J. D. Hem, Chem. Geol., 21, 199 (1978).

52. H. Ruppert, Chem. Erde, 39, 97 (1980).

53. M. Hartmann and P. J. Müller, Meerestechnik, 5, 201 (1974).

54. D. E. Hobart, K. Samhoun, J. P. Young, V. E. Norvell, G. Mamantov, and J. R. Peterson, Inorg. Nucl. Chem. Lett., 16, 321 (1980).

55. B. Spettel, G. Niedermayr, H. Palme, G. Kurat, and H. Wänke, Fortschr. Miner., 59, Bh. 1, 191 (1981).

56. S. Schulz, Verteilung und Genese von Fluorit im Hauptdolomit Norddeutschlands, Dietrich Reimer, Berlin, 1980, pp. 87.

57. M. Scherer and H. Seitz, Chem. Geol., 28, 279 (1980).

58. P. P. Parekh, P. Möller, P. Dulski, and W. M. Bausch, Earth and Planet. Sci. Lett., 34, 39 (1977).

59. K. Germann, J. M. Pagel, and P. P. Parekh, Z. Dt. Geol. Ges., 132, 305 (1981).

60. C. Grappin, M. Treuil, S. Yaman, and J. C. Touray, Mineral. Deposita, 14, 297 (1979).

61. B. Lehmann, Berl. Geowiss. Abhandl., A 14, 135 (1979).

62. H. Seitz, Aktivierungsanalytische Untersuchungen zur Verteilung von Spurenelementen in Magmatiten und deren Mineralen aus dem Oberharz und Südsardinien, Ph. D. Diss., D 83, Techn. Univ. Berlin, 1980, pp. 142.

63. P. Dulski, B. Lehmann, and P. Möller, Fortschr. Miner., 58, Bh. 1, 21 (1980).

64. M. Whitfield and D. R. Turner, Nature, 278, 132 (1979).

65. D. R. Turner and M. Whitfield, Nature, 281, 468 (1979).

66. R. P. Taylor and B. J. Fryer, Metallization Associated with Acid Magmatism, (A. M. Evan, Ed.), Wiley, New York, 1982, pp. 357.

67. P. Möller, P. P. Parekh, and G. Morteani, Chem. Geol., 13, 81 (1974).

68. J. C. Jarvis, T. R. Wildeman, and N. G. Banks, Chem. Geol., 16, 27 (1975).

69. G. Morteani, P. Möller, and F. Schley, Econ. Geol., 77, 99 (1982).

70. G. Morteani, F. Schley, and P. Möller, Mineral. Deposita Spec., Pub. 3 (in press).

71. S. Chaudhuri and R. L. Cullers, Chem. Geol., 24, 327 (1979).

72. P. L. Hellmann, R. E. Smith, and P. Henderson, Contrib. Mineral. Petrol., 71, 23 (1979).

73. G. K. Muecke, C. Pride, and P. Sarkar, Origin and Distribution

of Elements, (L. H. Ahrens, Ed.), Pergamon, Oxford, 1979, pp. 449.

74. P. L. Hellman and P. Henderson, Nature, 267, 38 (1977).

75. P. A. Floyd, Nature, 269, 134 (1977).

76. F. A. Frey, M. A. Haskin, J. A. Poetz, and L. A. Haskin, J. Geophys. Res., 73, 6085 (1968).

77. J. A. Philpotts, C. C. Schnetzler, and S. R. Hart, Earth Planet. Sci. Lett., 7, 293 (1969).

78. J. N. Ludden and G. Thompson, Nature, 274, 147 (1978).

79. D. J. Terrell, S. Pal, M. M. López, and R. J. Pérez, Chem. Geol., 26, 267 (1979).

80. H. W. Nesbitt, Nature, 279, 206 (1979).

81. S. M. Mc. Lennan and S. R. Taylor, Nature, 282, 247 (1979).

82. S. P. Sinha, Chapter 10, this volume.

83. B. J. Fryer, Geochim. Cosmochim. Acta, 41, 361 (1977).

84. K. Laajoki, Bull. Geol. Soc. Finl., 47, 91 (1975).

85. P. Jakes and S. R. Taylor, Geochim. Cosmochim. Acta, 38, 739 (1974).

86. H. D. Holland and M. Schidlowski, Mineral Deposits and the Evolution of the Biosphere, Springer-Verlag, Berlin, 1982, pp. 332.

DISCUSSION

HÜFNER (question)

You said that it is very difficult to determine Eu^{2+}/Eu^{3+} ratios.
I think it is easy to be done by auger-spectroscopy. It gives you
different signals for the two Eu species and it is a very sensitive method.

MÖLLER (answer)

The concentration level of Eu is in general about 1 ppm in calcite
and fluorite. Since auger-spectroscopy only deals with thin surface layers, the amount of total Eu available for analysis is really very low. I question that under these circumstances the auger-
spectroscopy is the method of choice.

SINHA (question)

What do we know about the Eu^{2+}/Eu^{3+} ratio and potential in silicate
melts?

MÖLLER (answer)

The Eu^{2+}/Eu^{3+} ratios for various melt compositions have been determined by ESR measurements in quenched melts by Morris and co-
authors [Geochim. Cosmochim. Acta 38, 1435 (1974); ibid., 38, 1447
(1974)].

SINHA (question)

I presume, it is a quenched synthetic melt, so it is not the melt
itself of which we know the Eu^{2+}/Eu^{3+} ratio. What is known is that
if you add Eu^{3+} into a silicate melt any impurity might reduce it.
It can be preserved without oxidation for a very long time as
Eu^{2+}-silicate in various forms. On these lines the existence of
Eu^{2+} in nature is not very surprising.

MÖLLER (answer)

My answer will be more a general comment: If we look into the experiments of Morris and co-worker we find the following results
for quenched melts (fig. 1). At an oxygen fugacity of about 10^{-7} bar
and $1600^{0}C$ melts of various given compositions show – after quenching – sytematic changes of the Eu^{2+}/Eu^{3+} ratio (XEu^{2+} in fig. 1
is the relative amount of Eu^{2+}, namely $Eu^{2+}/(Eu^{2+} + Eu^{3+})$. XEu^{2+}
is largest in the alumosilicate melt(An), seems to be negligible
in orthosilicate melts (extrapolation of An-Ln), and can be rather

An : Ca Al$_2$ Si$_2$ O$_8$ Wo : Ca Si O$_3$
Ln : Ca$_2$ Si O$_4$ Fo : Mg$_2$ Si O$_4$
Mo : Ca Mg Si O$_4$ En : Mg Si O$_3$
                                               Di : Ca Mg Si$_2$ O$_6$

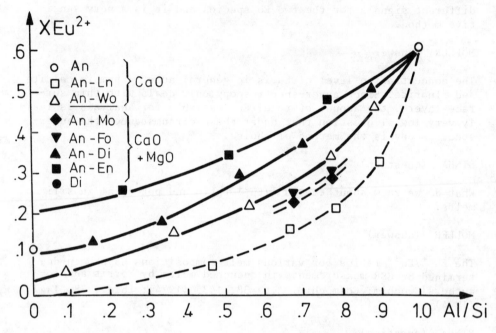

Fig. 1 Relationship between XEu^{2+} and molar ratio of Al and Si in the quenched glasses. XEu^{2+} is the molar ratio of Eu^{2+} related to total Eu in the glasses. All data are taken from Morris and Haskins (Geochim. Cosmochim. Acta 38, 1435 (1974).

low in metasilicate melts (extrapolation of the An-Wo, An-En, and An-Di). Diluting the alumosilicate melts with metasilicates and orthosilicates decreases the XEu^{2+} value. These systematic changes cannot be understood as the effects of Eu^{3+} reducing impurities. Furthermore, one should keep in mind that the oxygen fugacity is always at the same level.

Trying to explain the above results (Möller, in prep.) by considering various chemical equilibria one finds that in the series alumosilicate – metasilicate – orthosilicate, the tendency to stabilize Eu^{2+} in large multidentate complexes decreases. Alumosilicates prove to stabilize Eu^{2+} most effectively. If so, then

the total Eu species not complexed, depends on the concentration
of alumosilicate ligand in the melt. This concentration, however,
is completely independent of the oxygen fugacity of the melt; it
mainly depends on the other cations present in the melt which
compete in complex formation with Eu^{2+} ions. Thus early silicates
crystallizing from alumosilicate bearing melts will show negative
Eu anomalies although the melts might be rather high in oxygen
fugacity. Examples of such early crystallizing minerals with negative Eu anomalies are biotite and apatite in granites. The later
crystallizing feldspars generally show positive Eu anomalies

SINHA   (question)

In your fig. 2 and 3 you were showing Ln distribution patterns in
fluorites and calcites. Have you an idea of the Ln concentrations
in the liquids in equilibrium with these different fluorites and
calcites? Can you be sure that you started from a general pattern
of Ln?

MÖLLER   (answer)

In fig. 2 and 3 I showed Ln distribution patterns of fluorites and
calcites which formed from pegmatitic or carbonatitic melts, hydrothermal solutions and in an sedimentary environment. With the
exception of the marine environment, we know very little about the
Ln concentration in the respective liquids. In sea water the (La/Ca)
ratio is about 8×10^{-9} that of biogenic carbonates 3×10^{-5}. The
comparison of these two ratios indicate tremendous enrichment of
Ln in marine carbonates.

    It has been demonstrated by Marchand[Thesis University of
Orléans (1976) pp. 140], Becker[Thesis Technische Universität
Berlin (1982) pp. 160] and by myself (this paper) that the distribution coefficients of Ln between solid and solutions are rather
high ($\lambda > 1000$ is expected) in systems in which Ln are not complexed
as it is expected according to their results. Since in natural
system Ln are much less fractionated, I therefore would like to
derive that in most hydrothermal systems the Ln are well complexed
and the La/Ca ratio is within one order of magnitude of the ratio
in the crystallizing fluorite or calcite. There is not a general
Ln pattern for solutions or for minerals. But the Ln patterns are
often very characteristic for a mineral formed in a particular environment. This is particularly true for minerals in igneous rocks.
As already indicated in fig. 13 and 28 of my lecture the Ln pattern
will depend on the types of reactions which control Ln distribution.
If you like, igneous and metamorphic rocks do not show such large
variations as it has been observed in minerals which formed form
solutions. It should be borne in mind that the composition of solution is the result of a multistage fractionation of all elements

during migration of carbonate, metamorphic or pore waters in
largely varying environments.

MUECKE (question)

I am a bit worried about your substituting Na^+ into fluorites on
the basis of analysis. As you know, things like fluorite minerals
have quite abundant fluid inclusions. You just told us that these
things have crystallized from very concentrated sodium-chloride
solutions, so when you are measuring Na^+ concentrations in fluorites I was wondering whether you're really not just measuring
the concentration of the Na^+ in the fluid inclusions.

MÖLLER (answer)

I have to apologize for not mentioning that we used finely ground
fluorite ($\sim$30µm), boiled for 5 minutes in 1:1 nitric-acid, to
clean the crystals and get rid of most, probably not all of the
liquid inclusions.

MUECKE (question)

The other thing that worried me a bit is the interpretation of
generations of fluorite or calcite which have grown and the changing of their Ln patterns. You seem to assume that you are dealing
with isothermal systems. As you know, when you look at fluid inclusion temperatures in a vein system you find quite distinct
changes in temperature for various generations. Before you showed
us that there is a thermal dependence on the complex formation and
I just wonder how can one really take a system like that and model
it isothermally.

MÖLLER (answer)

That criticism is alright, but on the other hand you might know
the paper of Richardson and Holland [Geochim. Cosmochim. Acta 43,
1327 (1979)] showing that rather huge amount of fluorite crystallized within a temperature range of about 5°C only. So that is
not much, we are not dealing, let's say, with temperature changes
from 200° to 100°C. Obviously this is only within the 10°C and
that I would proximate as more or less being isothermal with respect to Ln fractionation. Of course, we have to realize that
during mineralization of a vein the temperature might change. But
then the Ln fractionation responds to that. However, at the time
being there are too little data to model the influence of changing
temperature and pressure on Ln fractionation during mineralization.
We observe changes in Ln concentration levels which might or might
not be attributed to temperature changes. But I would like to point
out that not only temperature changes might influence the final
Ln distribution, but we have to consider pH changes in solutions
and kinetic effects during crystal growth as well.

READ   (question)

We were discussing the oxalate precipitation, in case of which it might be difficult to grow large crystals. But in some other cases it is possible. Are there analyses where Ln have been analysed at different distances from the crystal centre by surface analytical techniques like secondary ion mass spectrometry for instance?

MÖLLER   (answer)

I do not know any. However, I would expect Ln concentration gradients in crystal which grow from melts throughout the whole time of solidification.

SINHA   (question)

Can you do exchange experiments with Ca-fluoride and Ln and let the Ln diffuse into the fluorite?

MÖLLER   (answer)

In principle, yes; but one measures surface adsorption only. This process is expected to be different from coprecipitation. Diffusion of adsorbed Ln into Ca minerals like fluorite is negligible even at elevated temperatures in geological time scales.

On well crystallized calcite only one molecular layer takes part in a surface ion exchange process (Sastri and Möller,[Chem. Phys. Letters 26, 116 (1974)] Koss and Möller.[Z. anorg. allg. Chem. 410, 165, (1974)]. If you look up diffusion coefficients of ions in ionic crystals that are reported in literature, you will find that diffusion does not play any important role, except for open crystals like zeolites or micas. Since we are dealing temperatures of about 200^0C, the fluorite crystals do not recognize anything when the solution changes in composition. It is just the surface layer only which might equilibrate easily.

READ   (question)

You mentioned iron-manganese nodules as being the cause of the negative Ce anomaly in sea water and the ocean sediments. These nodules are rather restricted in their occurence. Do you think they are sufficiently abundant to account for the Ce depletion observed?

MÖLLER   (answer)

Not everything that precipitates from sea water shows you that enrichment of Ce. Most of the samples, like pure calcite and authochthonous sediments are deficient in Ce. The relative small

amounts of Mn-Fe concretions in deep-sea sediments show this extreme enrichment of Ce. I do not know if somebody tried a reliable
balance of Ce in hydroxide concretion on the ocean's floor and the
other sediments. This might be really difficult since Ln are supplied to the oceans by rivers from the continents and by alteration
of oceanic basalts. This tremendous alteration (spilitization)
occurs along all ridges at the ocean's floor. How does the Ln
pattern of the alteration solution look like? Unfortunately, I do
not know that. The Ce become concentrated in Mn-Fe concretions as
derived from the atomic ratios of Ce/(Fe+Mn) in sea water and the
hydrogenetic nodules which are about 1×10^{-4} and 6×10^{-4}, respectively. The other Ln are less enriched in this type of nodules.

When these nodules are buried with time, it seems that they
dissolve under chemically reducing conditions. In such cases Ce
re-distributes and equilibrates with minerals deficient in Ce.
This long time process might prevent a precise balancing of Ln in
the sea water-sediment system. I should add that I came to know
of the attempt of Bonnot-Courtois [Chem. Geol. 30, 119 (1980)]
who tried to balance the Ce anomaly by assuming that it is mainly
due to Ce enrichment in palagonized glasses.

SUMMARY AND FUTURE TRENDS IN LANTHANIDES

L. Brewer: GENERAL TRENDS AND SYSTEMATICS

Interest in the lanthanides will surely increase in the future. There are the obvious commercial applications of lanthanide
compounds for hydrogen storage, for magnets and for solid-state
lasers. The lanthanide metals will be used more widely for the
purification of metals by removal of oxygen, sulfur, and other
non-metals. However, their most important application will be in
adding to our fundamental understanding of all materials.

The use of isotopic labels or tracers has been a powerful
tool in all branches of science. The closely related lanthanide
elements can play a role similar to that of isotopes but with
greater sensitivity to changes in chemical environment such as
oxidizing or reducing conditions or the general structural environment.

For non-lanthanide elements, the change of nuclear change
by one unit substantially changes so many factors that affect the
properties of the elements and their compounds that it is often
difficult to separate the roles of each factor such as size, electronic configuration, etc. The study of the properties of lanthanides and their compounds offers the possibility of relatively
small changes and a better fundamental understanding of the role
of each factor. The future studies of the lanthanides will play an
important role in improving our understanding of all elements.

There is one particular area that will attract more attention
in the future. Kenneth Pitzer[Accounts Chem. Res., 12, 27 (1979)]
has pointed out that relativistic terms to bond energies, ionization potentials, and various chemical properties of heavy elements
are required if accurate calculations are to be made. The lantha-

617

S. P. Sinha (ed.), Systematics and the Properties of the Lanthanides, 617–623.
Copyright © 1983 by D. Reidel Publishing Company.

nides and actinides are particularly important for testing the
role of relativistic effects. Various anomalous effects and departures from periodic trends for the lanthanides and actinides
were shown by Pitzer to be due to relativistic effects. Much more
work is needed in this area.

Study of the relativistic contributions for the lanthanides
and actinides where the changes from element to element are
smaller, will help us understand the relativistic contributions
for the 5 d transition metals from Hf to Au and particularly for
the elements from mercury to astatine and the elements following
the actinides.

W.T. Carnall: f-ELEMENT SPECTROSCOPY

EXPERIMENTAL STUDIES - OPTICAL SPECTROSCOPY

Measurements employing lasers should be performed in order
to develop a basis for understanding the intrinsic widths of absorption and emission lines in the crystal-field spectra of the
f-elements.

The photo chemistry of lanthanide compounds has not been discussed at this conference - rather the emphasis has been on photo
physics. With our developing understanding of energy level structure and the availability of high power lasers, there should be
much more attention given to the photochemical consequences of
selective activation in all phases. Actually, photo physics and
photo chemistry should be pursued as parts of the same general
problem of understanding the consequences of the production of
the excited states.

Where opportunities exist, systematic trends in the spectroscopic properties of the lanthanides should be explored with the
actinides where there is some increase in the accessibility of the
f-electrons for bonding and thus the possibility of gaining greater insight into fundamental chemical properties.

THEORY - ABSORPTION AND FLUORESCENCE SPECTROSCOPY

The experimentally determined crystal-field parameters have to be understood in terms of first principle calculations. Promising approaches to this problem such as superposition analysis and angular-overlap calculations should be continued, and molecular orbital calculation of intrinsic parameters should be pursued to supplement work done on $PrCl_3$ by Newman and coworkers.

Energy transfer needs to be explored in much greater depth both in crystals and in solutions in order to understand the dynamics of population and depopulation of f-element energy levels. Much of the present treatment is in terms of generalized formulations which do not actually involve a mechanistic understanding of the problem.

Model calculation of transition intensities between Stark components in the trivalent lanthanides and actinides would be a great asset to the experimentalist and should be pursued along the lines of testing the sensitivity of the parameterization schemes already developed by Morrison and coworkers.

MAGNETIC PROPERTIES

Zeeman splitting in both lanthanide and actinide crystal spectra has been found to be poorly reproduced by model calculation in a number of cases - although in general the correlation is quite satisfactory. Higher order connections need to be examined in a systematic manner. Much greater emphasis needs to be given to testing the correlation between magnetic properties predicted from spectroscopic measurements and those obtained experimentally. Exciting new experiments using MCD spectroscopy are being conducted and can be expected to provide important contribution to structural studies.

NMR

The technique of NMR should become better known in all branches of chemistry. There should be much more attention given to this technique of looking directly at transition metal nuclei instead of looking at the protons and carbons of metallic complexes, since this is a much more direct method of exploring the environment of nuclei. Why not assume that this kind of observation can be extended to lanthanide nuclei. Of course there are problems related to the sensitivity of a particular nucleus and also to the broadening and splitting of lines by the presence of quadrupole moments. However, these kinds of problems are being successfully solved in transition metals ions by special irradiation multipulse techniques.

Other fields in which progress can be expected include the area of large molecules (proteins) where substitution of "silent" metals such as Ca^{2+} by trivalent lanthanides can give considerable information on the bonding sites of these molecules.

After solving some theoretical problems, the lanthanide probe method could be applied to solids in the same way as it is in solutions to provide structural information. Emphasis should be given to the use of Gd^{3+} in NMR imaging or a relaxation agent of water protons in tissues.

P. Möller: LANTHANIDE GEOCHEMISTRY

My dear colleague Jean Claude Duchesne started on the idea that our earth evolved out of solar dusts with Ln distribution very similar to that found in chondritic meteorites. Due to fractionation of all elements in the early earth, chemically different spheres evolved. Today, samples are available only from the outermost parts of the earth. Comparing their Ln distribution patterns with those in the chondrites, which are believed to represent a very primitive matter, indicate a fractionation of Ln. Petrologists and geochemists try to interprete these changes in Ln distribution in terms of partial melting and crystallization, anatexis, metamorphism and/or metasomatism.

In order to understand the Ln distribution in magmatic rocks reliable distribution coefficients are required. The problem with them is that they are dependent on the pTx conditions and that they are different for all minerals. Furthermore, the modelling of crystallization processes is complicated by the fact that the rock forming minerals mostly do neither crystallize at the same time nor one after the other, but mostly with considerable time overlap and in varying sequences and often in repetition of such sequences. Simple models for partial melting and partial crystallization are applied in order to describe the analytical result of Ln fractionation. In a first approach their result seems to be satisfactory. It should be mentioned here that highly sophisticated models have been developed, too. But their increased number of variables does restrict their application.

The fractionation of Ln in magmatic processes has become a powerful tool in geochemistry: it is used
  (i) to characterize the physico-chemical processes which led to the differentiation of magmas into various types of magmatic rocks and
 (ii) to establish the consanguinity of rocks of different petrographical appearance or
(iii) to distinguish between ortho- and para-rocks.

In non-magmatic processes the degree of Ln fractionation is much more pronounced. Here the series of the Ln represent an interesting chemical probe for studying
  (i) fundamental processes of element distribution in phase boundaries under varying p,T,x conditions
 (ii) some physico-chemical aspects in fluid phases in the earth's crust
(iii) chemical processes during ore formation
 (iv) the precursors of metamorphic rocks.
     In order to study the above mentioned topics carefully we need informations on
  (i) behaviour of Ln in fluid phases
 (ii) behaviour of Ln in coprecipitation
(iii) probable mechanisms of ion substitution by Ln in host minerals.

At the moment we are able to model some of the geochemical processes semiquantitatively by using those informations on Ln available. It might be expected that with increase of our knowledge the quality of our models increases and thereby the reliability of our predictions made by them.

A considerable time in our discussion was focused on the interpretation of Ce and Eu anomalies observed so often in Ln distribution patterns. Unfortunately direct measurements of the oxidation states of Ce and Eu are not possible by using methods available at present. However, it seems to be evident that it will not be possible to explain the Eu^{2+}/Eu^{3+} ratio by fO_2 alone. It has to be considered that Eu^{2+} might become stabilized to a considerable extent, especially in silicate and alumosilicate melts. Ce will only be present in its quadrivalent state in surface near solutions when enough oxygen is accessible.

The various examples of water-rock interactions like metamorphism, metasomatism, diagenesis and weathering given in the lectures illustrated that many parameters have to be taken into account when Ln distribution patterns are used in characterization of those alteration processes. The parameters are besides p and T

   (i) Ln concentration levels in parent rocks, sediments and/or
       individual minerals
  (ii) Ln concentration level in the interacting solution
 (iii) concentration of ligands in the fluid phase which forms
       strong complexes with Ln
  (iv) the mineralogical composition of the rocks or sediments
   (v) the volume of fluid phase/per volume of rock or sediment
       in the alteration reaction.

   Finally the application of the Sm decay into Nd should be
mentioned which is coming up very fast at the moment in geochronology. This method has the phantastic advantage that the Sm/Nd
and 140 Nd/144 Nd ratios are much less affected by any alteration process than the respective ones in the more classical method of Rb/Sr decay. The Sm/Nd method seems to open the Precambrian to geochronology.

   Information needed in the future
   (i) data on formation constants of Ln complexes in solutions
       and melts of geochemically relevant composition,
  (ii) thermodynamic approaches for calculations of chemical equi-
       libria in solution and melts at any T p condition,
 (iii) any kind of model which allows to find reliable distribu-
       tion coefficients and complex stability constants,
  (iv) experimental data on physical distribution coefficients be-
       tween liquids and solids,
   (v) direct measurements of the oxidation states of Eu and Ce
       in naturally occurring minerals,   ,
  (vi) a better understanding of what is termed as 'crystallo-
       chemical control' when describing co-crystallization pro-
       cesses and its anomalies,
 (vii) studies on the adsorption of Ln species onto charged sur-
       faces of host minerals,
(viii) studies on the kinetics of Ln incorporation into growing
       crystals (this is not the same as topic vii),
  (ix) informations on the changes of surface lattice site energy
       within the Ln series at the surfaces of host minerals,
   (x) understanding of the often observed Ln-zonation in crystals:
       influence of the kinetics and/or the chemical changes in the
       liquid phase.

   These topics were mentioned during discussions among the geo-
scientists present in this meeting. These do not include the specific problems or finer details in the individual fields of their
research in geoscience. These general points are brought up with
the idea to find assistance from the chemists, physicists and
from other interdisciplinary fields.

L. Niinistö: STRUCTURAL CHEMISTRY

1. More structural data will become available due to
   a) the availability of better instruments and
   b) the growing interests in the chemistry of the rare
      earths in general, and in their organo-metallic com-
      pounds in particular.

2. More emphasis should be directed towards systematic
   studies in order to obtain a better understanding of
   a) different types (classes) of compounds
   b) compounds that are common in nature and are geologi-
      cally important (e.g. silicates, carbonates)
   c) multiligand complexes
   d) complexes with low (4-7) and high (10-12) coordination
      numbers
   e) structures measured at different temperatures (both
      low and high - Ed.)

   These studies will, however, require advanced preparative
   and crystal growth techniques.

3. Results obtained with the X-ray diffraction technique
   should be correlated, when possible, with the chemical
   and physical properties and with data obtained using
   other experimental techniques, the ultimate goal in predicting the properties of the complexes from structural
   data.

4. Further studies should be aimed at establishing a better
   link between the structures in the solid state and those
   in solutions and in melts. Synchrotron radiation technique
   will prove very valuable in structural investigations for
   liquids and melts.

LIST OF POSTERS

1. A MIXED-VALENT TERBIUM IODATE

    John C. Barnes
    Chemistry Department, The University, Dundee DD 14 HN, Scot-
    land

    Dark brown crystals were obtained with the same triclinic
    unit cell as pale pink $Tb(IO_3)_3$ $2H_2O$. How should these be for-
    mulated?

2. THE CIGARETTE LIGHTER FLINT - AN UNDERGRADUATE EXERCISE IN
    X-RAY FLUORESCENCE

    John C. Barnes, John D. Paton and Khalid T. Al-Rasoul
    Chemistry Department, The University, Dundee DD 14 HN, Scot-
    land

    The analysis of mischmetall gives a good introduction to the
    uses and problems of analytical X-ray fluorescence spectroscopy.

3. SOFT X-RAY APPEARANCE POTENTIAL SPECTRA OF RARE EARTHS[*]

    D.R. Chopra
    Physics Department, East Texas State University, Commerce,
    Texas, 75428, U.S.A.

S. P. Sinha (ed.), Systematics and the Properties of the Lanthanides, 625–633.
Copyright © 1983 by D. Reidel Publishing Company.

We have measured the $M_{4,5}$- and $N_{4,5}$-level Soft X-Ray Appearance Potential Spectra (SXAPS) of the lanthanide series rare earths in the 100–1600 eV energy range. The spectra represent the differential excitation probability of the $M_{4,5}$ and $N_{4,5}$ core levels to the unoccupied 4f states. The $M_{4,5}$ spectral region consists of two main peaks superimposed with secondary structures. The $N_{4,5}$ spectra consist of a prominent peak, exhibiting fine structure below the ionization threshold and a broad continuum due to the autoionized levels above. The complexity of the structure is rich for medium rare earths and decreases for light and heavy elements. The measured threshold energies of the SXAPS peaks are determined to be below the corresponding binding energies of the core levels. The M_5/M_4 peak ratios show marked increase with Z towards the end of the series consistent with the j-j coupling transition rates in rare earths.

The spectral features have been explained in terms of the atomic-like transitions involving exchange interaction between the 4f electrons and $M_{4,5}$, $N_{4,5}$ vacancies. The main peaks primarily result from the resonance scattering of both the incident and core electrons into the empty 4f states consistent with the lower binding energies of the core levels observed. Transitions of the type $3d^{10}4f^n + e \rightarrow 3d^94f^{n+2}$ cannot occur in metal Yb and beyond which have a filled 4f shell. The rich resonance structure in the vicinity, and beyond, the $N_{4,5}$ absorption edges arises from the large splitting of the $4d^94f^{n+2}$ configuration. However, this exchange interaction in the case of $M_{4,5}$ levels is relatively reduced because of the decreased overlap between the $M_{4,5}$ and $N_{6,7}$ orbitals. The spin-orbit interval is preserved. Consequently, the line structure is expected to get complex for medium rare earths and to cluster in the near vicinity of the M_4 and M_5 absorption edges as observed.

* Work supported by a grant from the Robert A. Welch Foundation

4.   MEASUREMENTS OF LOCAL 4f MOMENTS OF Ce, Nd AND Eu IONS IN
      DIFFERENT METALS BY TDPAD

G. Netz, H.J. Barth, M. Luszik-Bhadra, D. Riegel
Institut für Atom- und Festkörperphysik, Freie Universität
Berlin, D-1000 Berlin 33

The time differential perturbed angular distribution method (TDPAD) is a suitable tool for the investigation of 4f moments. Local 4f moments of Ce, Nd and Eu ions have been measured by the TDPAD method.

The most important results are:

(i) These ions show 4f instabilities in different metals. The isolated Ce ion shows intermediate valence in all rare earth metals. The Nd ion shows it in Ir and Ta metal.

(ii) Ce ions show a Ce^{4+} behaviour in Al, Ti, Rh, Pd, Te, Ta, W, Os, Ir, and Pt.

(iii) The 4f spin relaxation rates range from 10^{12} to a few $10^{13} s^{-1}$ and may be dominated by hybridization.

5.   RARE EARTH METALS UNDER HIGH PRESSURE

W.A. Grosshans, Y.K. Vohra, W.B. Holzapfel
Universität Paderborn, Experimentalphysik, D-4790 Paderborn,
Federal Republic of Germany

The rare earth metals Sc, Y, La, Pr, and Gd have been studied by Energy Dispersive X-ray diffraction[§]. The regular trivalent RE show the crystal structure sequence hcp → Sm type → dhcp → fcc → dist.fcc with increasing pressure. This sequence can be explained by an increase of the number of d electrons.

There are two remarkable exceptions:

1) About 20 GPa Pr undergoes a transition to the α-Uranium structure. This transition marks the onset of f-bonding.

2) Sc does not enter the RE sequence at all, but transforms to a tetragonal lattice at about 20 GPa.

[§] Part of the work has been done at Hasylab, DESY, Hamburg

6.   MAGNETIZATION AND MÖSSBAUER EFFECT STUDY OF Dy_6Mn_{23} AND Tm_6Mn_{23} AND THEIR TERNARY HYDRIDES

P.C.M. Gubbens, W. Ras, A.M. van der Kraan and K.H.J. Buschow
IRI, Mekelweg 15, 2629 J.B. Delft, The Netherlands

The compounds Dy_6Mn_{23} and Tm_6Mn_{23} and their ternary hydrides were investigated by means of magnetic measurements and [161]Dy and [169]Tm Mössbauer spectroscopy. In both compounds the absorption of hydrogen gas leads to a strong reduction of the bulk magnetization even though the Dy and Tm sublattice remains magnetically ordered at low temperatures, with magnetic moments on the rare earth atoms close to the free ion values.

7. ^{169}Tm MÖSSBAUER SPECTROSCOPY OF TmAl$_2$ AND TmCo$_2$

P.C.M. Gubbens, A.M. van der Kraan and K.H.J. Buschow
IRI, Mekelweg 15, 2629 J.B. Delft, The Netherlands

We have observed spin lattice relaxation of the Orbach type in TmAl$_2$. It is a direct resonant two phonor process between the levels of the $\Gamma_5^{(1)}$ triplet ground state via the real intermediate Γ_4 lying 15 ± 1 K above the ground state. TmCo$_2$ has a first order magnetic phase transition within the magnetic phase ground state $|J_z = 5 >$ and a first excited state $|J_z = 4 >$.

8.   THE Th-REE MINERALIZATION IN EASTERN TURKEY

Taner Saltoglu
Maden Tetkik ve Arama Enstitüsü, Lab. Dairesi, Ankara, Turkey

The Ardıçlı (Sofular-Malatya) Th-REE deposit is located approximately mid-way between the provincial town Sivas and Malatya in eastern Turkey. The mineralization occurs as veins, and is related to alkali syenite intrusion of Uppermost Cretaceous age.

Preliminary studies using microscopic, X-ray diffraction, qualitative electron microprobe methods, and XRF and optical spectrographic analytical techniques have shown that the Th and REE minerals are thorite, britholite and bastnasite with high concentrations of Th and REE (up to 70% Th and 70% REE). More detailed geological, mineralogical and geochemical studies on this mineralization are in progress in the Mineral Research and Exploration Institute of Turkey.

9.   BEHAVIOUR OF THE RARE EARTH ELEMENTS DURING ALTERATION AND
      METAMORPHIC PROCESSES

G.K. Muecke, Department of Geology, Dalhousie University,
Halifax, N.S., Canada

Most Precambrian shield regions and orogénic belts, and to some extent younger mobile belts, consist of rocks which have undergone some degree of alteration or have been modified by metamorphism subsequent to their formation. Their present mineralogy, texture and chemical composition frequently differ substantially from those of their precursor. Any group of elements, such as the rare earths (REE), which has been shown to be a valuable environ-

mental or petrogenetic indicator in unaltered rocks, has the potential of substantially extending our understanding of metamorphic
terrains. However, their application to metamorphic rocks depends
upon whether their distribution and abundance have been substantially affected by alteration and metamorphism. Opinion is presently divided among those investigators who cite evidence indicating
REE mobility [e.g. 1-4] and a majority who consider this group as
essentially immobile during many of these processes [e.g. 5-8].
An inherent problem in the determination of REE mobility is the
very high degree of variability in concentration of these elements
which is possible in rocks of otherwise very similar chemistry.
For example, unmodified basaltic rocks of similar major element
chemistry can show variations in concentration of the light REE
of nearly three orders of magnitude [9].

   The mobility of the REE during alteration and metamorphism
can be studied in a number of ways:
   (i)by direct comparison of the chemical composition of unaltered/
      unmetamorphosed parent material with the products of altera-
      tion and metamorphism;
  (ii)by investigations of metasomatic zones and bodies in altered
      rock bodies;
 (iii)by comparison of the strong linear correlations of some trace
      elements (including REE) in fresh cogenetic igneous suites
      with the correlations of the same elements in altered and meta-
      morphosed suites;
  (iv)by statistical comparisons between unaltered and altered
      rock suites.

   Sample suites which allow an unambiguous application of
approach (i) are relatively rare since the pristine precursors
and the altered/metamorphosed product are seldom found in immediate proximity. An exception are submarine pillow lavas from the
ocean floors which have undergone low-temperature seawater/basalt
interaction. A comparison of their altered margins and relatively
fresh cores shows selective enrichment of the light REE in the
heavily palagonized rims [3]. Enrichment factors range up to 6
for La and marked negative cerium anomalies are characteristic of
most altered patterns, probably reflecting relative REE distributions in seawater. Experimental determinations on seawater/basalt
interactions at elevated temperatures [7] have not been able to
detect any modifications of REE patterns and abundances, despite
substantial changes in the mineralogy of the samples.

   During metamorphism strong compositional gradients in bulk
composition or the fluid phase can give rise to mass transport and
the formation of locally zoned bodies or metasomatic zones which
attain chemical compositions substantially different from their
host or parent material. Where such zones and their hosts can be
recognized, they provide examples of extreme element mobility

during metamorphism and provide a "natural laboratory" to observe
possible REE mobility (approach ii). Muecke et al. [6] have documented examples of metasomatic zones in metabasites which have
retained REE abundances and patterns of their parent rocks despite major changes in such elements as Ca, Na, K, Sr, etc. [Fig.1].

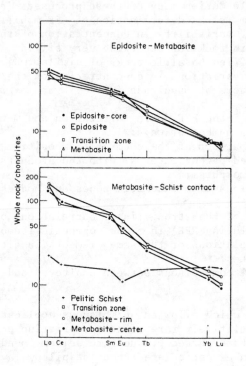

Fig.1 Chondrite-normalized rare earth patterns of rocks associated
      with metasomatic zones. (a) Top: Rare earth patterns of
      metasomatic epidosite lens, transition zone and host meta-
      basite are nearly identical. Metasomatic changes in element
      composition are not reflected by the rare earths. (b)Bottom:
      Rare earth patterns of centre of metabasite sill, metasoma-
      tized rim and transition zone are nearly identical. Adjacent
      pelitic schist is unaffected by the metasomatism and shows
      contrasting composition. From [6].

The introduction of substantial amounts of REE-poor minerals into
porous rocks during alteration/diagenesis will give rise to apparent depletions in REE abundances that have sometimes been cited

as evidence of REE mobility. For example, Hellman et al. [2] report substantially lower REE concentrations in amygdaloidal flow tops (initially porous) than in massive cores of the same metamorphosed flows. Depletion factors for all the REEs are very similar in these cases and are best accounted for by mass dilution by REE-poor mineral phases (ex. quartz, zeolites). Such examples provide further evidence for the general immobility of the REEs, rather than their mobility as proposed by the original authors.

In unaltered cogenetic volcanic suites the light REEs (La to Sm) usually show excellent linear correlations with other large-

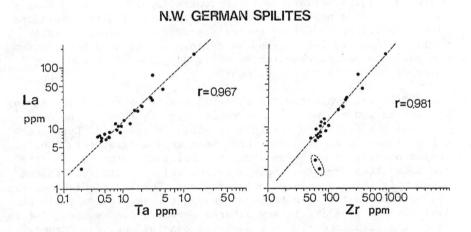

Fig.2 Linear trace element correlations from a metamorphosed suite of volcanic rocks from N.W. German Rhenohercynian Belt. Note the preservation of the excellent linear La-Ta and La-Zr correlations despite major chemical changes. Outlier values probably represent limited mobility of REE and/or Ta and Zr. Data from [12].

ion lithophile (LIL) trace elements [ex. 10]. Investigations of
such linear correlations in altered or metamorphosed volcanic
suites provide an insight into the relative mobility of these elements (approach iii). The excellent correlations of LREE with K,
Rb, Cs and U in unaltered rocks are usually absent in altered/
metamorphosed rocks, whereas the equally good correlations with
Zr, Hf, Y, Nb, Ta and Th are frequently preserved [6,11,12][Fig.2].

The coherent behaviour of REEs and this latter group is best
explained by their relative immobility during alteration and metamorphism. Individual samples within an altered suite which deviate
significantly from such linear correlations may indicate mobility
of one or both elements. However, such deviations can also arise
if, for example, samples from several petrogenetically distinct
suites are mixed. Also, the dilution or addition involving REE-
poor phases would not disturb such original linear correlations,
but can lead to substantial alterations in absolute concentrations
i.e. absence of mobility cannot be equated with the preservation
of pristine concentration values.

Statistical comparisons between unaltered and altered rock
suites [ex. 1] of spatial/temporal/chemical affinity (approach iv)
is fraught with difficulties, particularly when comparisons are
made using measures of dispersion about mean values. The REEs
often show large variations in concentration in unaltered suites
as a result of a number of igneous differentiation processes and
high dispersions do not necessarily provide a measure of later
mobility. Such variations can be further accentuated by dilution
and addition processes which have already been discussed, but
which do not reflect element mobility.

Critical evaluation of the evidence for REE mobility during
alteration and metamorphism of rocks suggests that in the majority
of cases the REEs have remained essentially immobile. Even where
other element concentrations have been seriously affected and no
longer provide information about the original nature of the rocks,
the REEs often preserve their pristine relative distributions and
frequently their absolute concentrations. However, deviations from
this generalization are frequent enough to make it advisable to
test for REE mobility in any altered/metamorphosed suite prior to
using the data to derive petrogenetic conclusions. For cogenetic
igneous suites linear correlations probably provide the easiest
and most reliable test of element mobility.

REFERENCES

[1] D.A. Wood, I.L. Gibson, R.N. Thompson, Contr. Mineral. Petrol.
    55, 241 (1976).
[2] P.L. Hellman, R.E. Smith, P. Henderson, Contr. Mineral. Pe-
    trol., 71, 23 (1979).

[3]   J.N. Ludden, G. Thompson, Earth Planet. Sci. Letters, 43,
      85 (1979).
[4]   P.A. Floyd, Nature, 269, 134 (1977).
[5]   A.G. Herrmann, M.J. Potts, D. Knake, Contr. Mineral. Pe-
      trol., 44, 1 (1974).
[6]   G.K. Muecke, C. Pride, P. Sarkar, Origin and Distribution
      of the Elements, Pergamon Press, Ed. L.H. Ahrens, II, 449
      (1979).
[7]   M. Menzies, W. Seyfried, D. Blanchard, Nature, 282, 398
      (1979).
[8]   T. Koljonen, R.J. Rosenberg, Bull. Geol. Soc. Fin., 47,
      127 (1975).
[9]   R. Kay, P.W. Gast, J. Geol., 81, 653 (1973).
[10] C.L. Allègre, M. Treuil, J.-F. Minster, B. Minster, F.
      Albarède, Contr. Mineral. Petrol., 60, 57 (1977).
[11] G.K. Muecke, I.L. Gibson, P. McGraw, in press.
[12] K.H. Wedepohl, K. Meyer, G.K. Muecke, Intracontinental
      Fold Belts – Case Studies in the Variscan Belt and the
      Damaran Belt, Springer Verlag (1983).

LIST OF PARTICIPANTS

H.D. AMBERGER, Inst. Anorg. Angew. Chemie, Universität Hamburg, Martin Luther King Platz 6, D-2000 Hamburg 13, West Germany

J. ASCENSO, Centro de Quimica Estrutural, Instituto Superior Tecnico, 1096 Lisboa-Cedex, Portugal

J.C. BARNES, Chemistry Department, The University, Dundee DD 14 HN, Scotland

L. BREWER, Department of Chemistry, University of California, Berkeley, California 94720, U.S.A.

A.O. BRUNFELT, Mineralogical Geological Museum, Sars Gate 1, Oslo 5, Norway

W.T. CARNALL, Chemistry Division, Argonne National Laboratory, 9700 South Cass Avenue, Argonne, Il. 60439, U.S.A.

P. CERNY, Dept. of Earth Science, University of Manitoba, Winnipeg, Manitoba R3T2N2, Canada

D.R. CHOPRA, Physics Department, East Texas State University, Commerce, Texas 75428, U.S.A.

L. COUTURE, Lab. (L.O.S.C.) Aimé Cotton, Bât. 505, Campus d'Orsay, F-91405 Orsay-Cedex, France

L.E. DELONG, Department of Physics, University of Kentucky, Lexington, Kentucky 40506, U.S.A.

G. De PAOLI, Ist. Chimica e Technologia, Consiglio Nazionale Ricerche, Corso Stati Uniti, 35100 Padova, Italy

J.C. DUCHESNE, Université de Liège, Géologie, Pétrologie et Géochimie, B-4000 Sart-Tilman par Liège 1, Belgium

A. ENGELEN, Lab. Anal. Anorg. Scheikunde, Celestijnenlaan 200 F, 3030 Heverlee, Belgium

W. GROSSHANS, Experimentalphysik, Universität Paderborn, Warbur-
    gerstr. 100, D-4790 Paderborn, West Germany

P.C.M. GUBBENS, IRI, Mekelweg 15, 2629 J.B. Delft, The Netherlands

M. CETIN GÜLOVALI, Ankara Nuclear Research Center, Besevler,
    Ankara, Turkey

J. HÖLSÄ, Helsinki University of Technology, Dept. of Chemistry,
    SF-02150 Espoo 15, Finland

S. HÜFNER, Fachbereich 11 - Physik, Universität des Saarlandes,
    D-6600 Saarbrücken, West Germany

B. KANELLAKOPULOS, Inst. Heiße Chemie, Kernforschungszentrum
    Karlsruhe, Postfach 3640, D-7500 Karlsruhe, West Germany

C. KHAN-MALEK, Lab. de Radiochimie, I.P.N. Bât. 100, B.P. No. 1,
    F-91405 Orsay-Cedex, France

W. KLEMM, Theresiengrund 22, D-4400 Münster/Westfalen, West Ger-
    many

R. KOKSBANG LARSEN, Chemistry Department, Aarhus University,
    DK-8000 Aarhus C, Denmark

P. MÖLLER, Hahn-Meitner-Institut, Postfach 390128, D-1000 Berlin
    39, West Germany

G. MORTEANI, Inst. Petrologie, Technische Universität Berlin,
    Straße des 17.Juni 135, D-1000 Berlin 12, West Germany

G.K. MUECKE, Dept. of Geology, Dalhousie University, Halifax,
    Nova Scotia B3H3J5, Canada

G. NETZ, Schrockstr. 4, D-1000 Berlin 37, West Germany

N. NEYT, Lab. Anal. Anorg. Scheikunde, Celestijnenlaan 200 F,
    3030 Heverlee, Belgium

L. NIINISTÖ, Department of Chemistry, Helsinki University of Tech-
    nology, Otakaari 1, SF-02150 Espoo 15, Finland

G.J. PALENIK, Department of Chemistry, University of Florida,
    Gainesville, Florida 32611, U.S.A.

K. RAJNAK, Dept. of Physics, Kalamazoo College, Kalamazoo,
    Michigan 49007, U.S.A.

D. READ, Dept. of Geology, University College, Gower Street,
   London WC1, England

J. ROSSAT-MIGNOD, Centre d'Etudes Nucléaires, Grenoble 85X,
   F-38041 Grenoble-Cedex, France

N. SABBATINI, Istituto Chimico "G. Ciamician", Via Selmi 2,
   40126 Bologna, Italy

T. SALTOGLU, Maden Tetkik ve Arama, Enstitüsü, Lab. Dairesi,
   Ankara, Turkey

N. SERIN, Ankara Universitesi, Fak. Fizik Bölümü, Besevler
   Ankara, Turkey

T. SERIN, Ankara Universitesi, Fak. Fizik Bölümü, Besevler
   Ankara, Turkey

H.L. SKRIVER, Risø National Laboratory, Postbox 49, DK-4000
     Roskilde, Denmark

S.P. SINHA, Hahn-Meitner-Institut, Postfach 390128, D-1000 Berlin
     39, West Germany

L. SUCHOW, Chemistry Division, New Jersey Institute of Technology,
   323 High Street, Newark, New Jersey 07102, U.S.A.

S.M. TAHER, Physics Department, Wichita State University, Wichita,
     Kansas 67208, U.S.A.

L. THOMPSON, Dept. of Chemistry, University of Minnesota, Duluth,
   Minnesota 55812, U.S.A.

W. URLAND, Max Planck Institut für Festkörperforschung,
   D-7000 Stuttgart, West Germany

P. VANDEVELDE, Lab. Anal. Anorg. Scheikunde, Celestijnenlaan 200 F,
   3030 Heverlee, Belgium

G. VINCENTINI, Inst. de Quimica, Universidade Sao Paulo, Caixa
   Postal 20780, Sao Paulo, Brasil

L. BARBIERI ZINNER, Inst. de Quimica, Universidade Sao Paulo,
   Caixa Postal 20780, Sao Paulo, Brasil

SUBJECT INDEX